Drought Stress

Usman Khalid Chaudhry
Zahide Neslihan Öztürk • Ali Fuat Gökçe
Editors

Drought Stress

Review and Recommendations

Springer

Editors
Usman Khalid Chaudhry
Department of Agricultural Genetic Engineering
Niğde Ömer Halisdemir University
Niğde, Türkiye

Ali Fuat Gökçe
Department of Agricultural Genetic Engineering
Niğde Ömer Halisdemir University
Niğde, Türkiye

Zahide Neslihan Öztürk
Department of Agricultural Genetic Engineering, Ayhan Sahenk Faculty of Agricultural Sciences and Technologies
Niğde Ömer Halisdemir University
Niğde, Türkiye

ISBN 978-3-031-80609-4 ISBN 978-3-031-80610-0 (eBook)
https://doi.org/10.1007/978-3-031-80610-0

This Springer imprint is published by the registered company Springer Nature Switzerland AG
The registered company address is: Gewerbestrasse 11, 6330 Cham, Switzerland

Contents

About the Editors

Usman Khalid Chaudhry is a Plant Molecular Stress Physiologist with over a decade of research and teaching experience in Agriculture. He holds a Ph.D. in Agricultural Genetic Engineering from Niğde Ömer Halisdemir University and has a rich publication history, particularly in areas concerning climate change and the adaptive responses of various crops to abiotic stresses. Currently, he also serves as the Deputy Director (Technical) at the Pakistan Environmental Protection Agency, Ministry of Climate Change & Environmental Coordination, Islamabad. Dr. Chaudhry is deeply committed to identifying practical solutions for global climate-related challenges, and he contributes his wealth of expertise to this book in his role as an editor.

Zahide Neslihan Öztürk is a Plant Molecular Biologist. She completed her doctoral studies at Ruprecht-Karls-Universitaet Heidelberg, demonstrating her dedication to academic excellence. Dr. Öztürk has a vast and impressive track record in the fields of transcriptomics and functional molecular biology of plants, where her research has contributed significantly to plant genetics. Additionally, her proficiency in bioinformatics enhances her capacity to analyze complex biological data. She also successfully secured and executed numerous projects focused on the screening of vegetables for abiotic stress tolerance, particularly in anticipation of the looming challenges posed by drought conditions. With a rich history of teaching experience, Dr. Öztürk's role as a co-editor for this book enhances the understanding of crop resilience in the face of environmental stresses.

Ali Fuat Gökçe is a distinguished Plant Breeder and Geneticist, holding a Ph.D. from the esteemed University of Wisconsin. As a co-editor of this book, Dr. Gökçe contributes his extensive expertise in plant breeding, with a focus on developing vegetable crops resilient to abiotic stresses. His recent accomplishments include securing and executing multiple high-impact projects on stress tolerance screening. Furthermore, his impressive portfolio of peer-reviewed publications has significantly advanced our understanding of crop resilience under environmental stress.

About the Editors

Reframing Plant Stress Tolerance in the Era of Climate Change

Zahide Neslihan Öztürk, Anı Akpınar, Ahmed Mohamed Algharib, H. Busra Cagirici, Ertuğrul Filiz, Babar Hussain, Amer F. Mahmoud, Tuğdem Muslu, Bahar Soğutmaz Özdemir, Meral Yüce, Naimat Ullah, and Hikmet Budak

Abstract Currently every country is suffering from climate change and despite the ongoing efforts, there are still limited success stories for the improvement of plant abiotic stress resilience. The studies mainly focused on understanding the stress response and tolerance at the molecular level and utilizing it to another genotype with no tolerance. The strong interaction of plants with their environment is the main obstacle here, which prevents the outcomes of small-scale experiments to be repeated in the field. The inclusion of wild genotypes adapted to extreme environments in QTL studies and breeding efforts may stand as a promising approach for the improvement of abiotic stress tolerance. Agricultural practices including biostimulants, micronutrients and nanoparticles, genome editing, and engineering organelles can also lead to reduced yield loss. Understanding how stress memory is initiated and how to make it heritable even under favorable growth conditions is another approach that needs more investigation. The aim of this book chapter is to point out the fact that single efforts by themselves will not be enough for sustain-

Z. N. Öztürk (✉)
Department of Agricultural Genetic Engineering, Ayhan Sahenk Faculty of Agricultural Sciences and Technologies, Niğde Ömer Halisdemir University, Niğde, Türkiye
e-mail: zahideneslihan_ozturk@ohu.edu.tr

A. Akpınar · T. Muslu
Montana BioAgriculture Inc, Missoula, MT, USA

A. M. Algharib
Department of Environment and Bio-agriculture, Faculty of Agriculture, Al-Azhar University, Cairo, Egypt

H. B. Cagirici
Department of Medicine, Stanford University School of Medicine, Standford, CA, USA

E. Filiz
Department of Crop and Animal Production, Cilimli Vocational School, Duzce University, Duzce, Türkiye

B. Hussain
Department of Biotechnology, Faculty of Science and Technology, University of Central Punjab, Lahore, Pakistan

U. K. Chaudhry et al. (eds.), *Drought Stress*,
https://doi.org/10.1007/978-3-031-80610-0_1

ability, and although there is still a gap between breeders, farmers, biodiversity scientists, agricultural engineers, molecular scientists, plant pathologists, and climate scientists, they should combine their knowledge to secure future generations.

Keywords Abiotic stress · Biotic stress · Breeding · Biostimulants · Micronutrients · Epigenetics · Genomics · Chloroplast engineering · Nanotechnology · Genome editing tools

1 Introduction

Agriculture is the most important source of human supply with food, clothing, and medicine, as well as raw materials for industry. It can be affected by many factors, the most important of which is the climate, which significantly affects the productivity. Climate change is described as a long-term shift in climate caused by human and non-human activities (Chandio et al. 2021). Due to increasing human population causing ever increasing environmental pollution and industrial revolution to support life quality, climate change has become the most important challenge facing the world, which is expected to have a significant impact on the agricultural sector by threatening the world's food and nutritional security. Climate change has the potential to modify rain patterns, reduce agricultural water supplies, and increase the frequency of droughts and overall temperature. It can affect agricultural production both in direct (i.e., modification of physical characteristics such as rainfall distribution and temperature levels on specific agricultural production systems), and indirect ways through changes in other species, such as pollinators, pests, disease vectors, and invading species which have indirect effects on agricultural production. So, it is undoubtedly a major driver of change in agricultural sector by having a negative impact on crop productivity.

A. F. Mahmoud
Department of Plant Pathology, Faculty of Agriculture, Assiut University, Assiut, Egypt

B. Soğutmaz Özdemir
Department of Genetics and Bioengineering, Faculty of Engineering, Yeditepe University, Istanbul, Türkiye
e-mail: bahar.sogutmaz@yeditepe.edu.tr

M. Yüce
SUNUM, Sabanci University Nanotechnology Research and Application Centre, Tuzla, Istanbul, Türkiye

N. Ullah
Institute of Biological Sciences (IBS), Gomal University, D. I. Khan, Pakistan

H. Budak
Montana BioAgriculture Inc, Missoula, MT, USA

Department of Agriculture, Arizona Western College, Yuma, AZ, USA

Greenhouse gases (GHGs), which consist of carbon dioxide CO_2 (76%), methane CH_4 (16%), nitrous oxide N_2O (6%), and combination of perfluorocarbons, sulfur hexafluoride and hydrofluorocarbons (2%,) are the main cause of climate change, as excessive emissions of these gases cause global warming, leading to extreme weather events and natural disasters. Currently, it is safe to say that the GHG caused by human activities are the main cause of global warming, as since the pre-industrial era global GHG concentrations in the atmosphere have increased dramatically (Liu et al. 2019). In fact, agriculture is both the victim and the contributor of climate change. Climate change has a significant impact on water and land in agriculture, and can decrease yield by 60% from the perspective of being a victim (Matiu et al. 2017). There are many studies that indicated a decrease in the productivity of major food grain crops such as wheat, barley, sorghum, and maize (Praveen and Sharma 2020). Also, rice crop, which is the main food for many countries, is threatened by a decrease in productivity as the rain-fed rice yield may reduce by 10–14% under deferent scenarios (Boonwichai et al. 2018). Climatic changes may also cause crops to be infested with insect pests and lead to a decrease in global productivity of rice, corn, and wheat by 10%–25% (Deutsch et al. 2018). It also decreases growth in the global availability of nutrients by 19.5% for protein, 13.6% for iron, and 14.6% for zinc (Beach et al. 2019). Pollinators such as bees will also be affected by climate change (Kerr et al. 2015).

On the contrary, agricultural sector is a major contribute to climate change through its production of GHG emissions into the atmosphere, as it contributes 9.3 billion tons of carbon equivalent, which is divided into 5.3 billion from crop and livestock activities within the farm gate, while the remaining 4 billion comes from land use and land use change activities (FAO 2019). At the international level, the agricultural sector produces 13.5% of emissions and is second only to the transport sector in terms of the amount of emissions generated (Mohammed et al. 2019). Therefore, there is an urgent need to reduce emissions from agriculture by adopting many techniques such as added urease enzymes (inhibitors) to urea-based synthetic fertilizers and manure slurry which prevent and reduce anthropogenic ammonia and nitrous oxides emission (Meyer-Aurich et al. 2020). There are many agricultural practices which emissions can be reduced through it, such as, increased carbon sequestration (Locker et al. 2019), reducing tillage intensity (Campbell et al. 2014), planting cover crops (Parkin et al. 2016), adopting extended crop rotations (Shipitalo et al. 2013), and improving nitrogen fertilizer management (Snyder 2017). Plant breeding strategies may be one of the effective ways to reduce emissions, as is the case in the rice plant, whose cultivation produces methane emissions as rice fields are responsible from 11% of global anthropogenic of methane emissions. Recently, the concept of climate-smart agriculture (CSA) has emerged, which has the potential to greatly mitigate agricultural GHG emissions by increasing soil organic carbon stocks and decreasing N_2O emissions (McNunn et al. 2020).

Crop adaptation to a changing climate is a critical component in reducing the impact on food production. As being sessile organisms, plants suffer from changes in their environment which initiates abiotic stress response causing changes in morphology, physiology and cellular metabolism. The main goal of a plant under abiotic stress is to direct all available resources to be able to survive, which eventually leads to severe reductions in yield. Therefore, the aim of crop adaptation to climate change is to increase the tolerance level of plants to abiotic stress conditions to mitigate yield loss. However, the complexity of abiotic stress response in cellular level and the unique responses of plant species to changes in their environment makes both generalizing the response mechanisms for all crops and finding a key element to increase the stress tolerance a very challenging goal. There are quite a large number of studies in literature to understand the changes in morphology and physiology of crops in response to abiotic stress and to understand the abiotic stress response in several omic levels including genomics, transcriptomics (both transcriptome and miRNome), proteomics, metabolomics, ionomics, lipidomics, and phenomics approaches. However, there are limited number of reviews focusing on the possibility of different approaches like chloroplast engineering, epigenomics studies, biostimulants, micronutrients, genome editing and nanoparticles to hamper the yield loss in a changing climate. Therefore, in this review, we aimed to reframe these novel approaches that should be investigated to improve the stress resilience of major crops and wanted to start with emphasizing the neglected thread of increasing number of biotic stress episodes caused by climate change, since occurrence of biotic and abiotic stresses together should also be considered for the goal of adaptation of crops to a changing climate.

2 Effect of Climate Change on the Occurrence of Biotic Stresses

Climate change influences the occurrence, prevalence, and severity of plant diseases. Projected atmospheric and climate change will thus affect the interaction between crops and pathogens in different ways. Climate change affects crop pests and disease susceptibility which in turn affects crop health, and these changes cause deviations in farming practices as to cope with the effects of these changes and to prevent a decline in productivity. In recent years, the importance of biotic stress has increased as a result of climate change enhanced frequency and intensity of weather extremes as well as globalization which has increased the movement of pathogens and pests. Factors associated with both the host pathogen interaction and availability of inoculum may affect spread and incidence of the disease. Warmer, drier, or wetter conditions may all influence the host and/or pathogen, or the interaction in ways that may increase, or reduce the effect of the disease on a plant host. Conversely, in the tropics, longer periods above the maximum temperature for infection and latent development may reduce the risk of disease (Beresford et al. 2020). The classic disease triangle establishes the conditions for disease expansion, i.e., the interaction of a susceptible host, a virulent pathogen and a favorable environment. This relationship is evidenced in the identification of plant disease itself. A plant disease is a dynamic process in which a host and a pathogen intimately related to the environment are mutually influenced, resulting in morphological and physiological changes in the plant. Plant diseases occur only when a virulent pathogen attacks a susceptible host under suitable environmental conditions at an appropriate time.

The close relationship between the environment and diseases suggests that climate change will cause modifications in the current plant health requirements. The impacts can be positive, negative or neutral, since there can be a decrease, an increase or no effect on the different pathogenicity system, in each region. The analysis of the potential impacts of climate change on plant pathology is essential for the adoption of acclimation measures, as well as for the development of resistant cultivars, new control methods or adapted techniques, in order to avoid severe decrease in production (Ghini et al. 2008). Plant pathogens are ubiquitous in natural and managed systems, being among the first to demonstrate the effects of climate change due to the numerous populations, ease of reproduction and dispersal, and short time between generations. Therefore, they constitute a fundamental group of biological indicators that needs to be estimated regarding climate change impacts, besides being responsible for production losses, and potentially threat to agroecosystem sustainability (Garrett et al. 2006). Increase in temperature coupled with high soil moisture creates a warm and humid climate highly favorable to soil-borne pathogens and foliar diseases. For example, increase in ambient temperature has increased potential for rice blast (*Magnaporthe grisea*) incidence and epidemics for both cooler and subtropical rice-growing regions. Similarly, the severity of Septoria leaf spot (*Septoria lycopersici*) and stem rust (*Puccinia graminis*) in oat cultivars has also increased due to temperature sensitivity (Kobayashi et al. 2006). Furthermore, the change in the distribution and range of pathogens and pests due to climate change may increase outbreaks in wider areas (Pureswaran et al. 2018). Thus, despite the contrasting observations, increased disease outbreaks are likely in many areas as climate predictions are equally contrasting in different regions of the world (IPCC 2018).

2.1 Importance of Increasing Biotic Stress Tolerance in Agriculturally Important Crops

Agriculture is an economic activity which is highly reliant on climate and weather in order to produce the food necessary to sustain human life. Agriculture is, however, an activity which is extremely vulnerable to climate change. Biotic stress factors cause major economic losses by reducing yield and quality in agriculture. Climate change aggravates the adverse effects of biotic stress and increases such losses (Savary et al. 2019). It has already, and increasingly, will affect the prevalence and frequency of different plant diseases across a spectrum of important staple and specialty crops, and in natural ecosystems (Burdon and Zhan 2020). With the increasing frequency and intensity of weather extremes due to global climate change, the interaction among plants, pathogens and pests, and the environment has become an important aspect of plant disease and pest outbreaks. Otherwise, extreme weather events may also affect pathogens and pests, further complicating the interaction. Apart from the weather extremes such as droughts and heat waves, mild variations in temperature and precipitation also affect the dynamics of disease and pest outbreaks (Weed et al. 2013). In recent years, changes in disease outbreak due to unfavorable weather conditions have affected wheat and rice production and have caused severe yield losses and reduced grain quality (Tang et al. 2017a). Plant pathogens are spreading globally with latitudinal shifts in their distribution for the last 60 years, largely in line with climate change projections. The damage imposed by these organisms may have serious consequences on economies of the developing world because of greater dependence on agriculture in these countries, as well as limited resources and technological interventions to manage crop diseases (Bebber et al. 2014). Temperature increased infection of *Pinus albicaulis* by *Cronartium ribicola* only at high relative humidity and up to a certain threshold of 11 °C (Thoma et al. 2019). Whereas the incidence of *Phytophtora* spp. on *Corymbia calophylla* did not increase with temperature and only slightly increased with decreasing precipitation (Paap et al. 2017). FAO estimates that annually up to 40 percent of global crop production is lost to pests. Each year, plant diseases cost the global economy over $220 billion, and invasive insects at least $70 billion. Climate change will increase the risk of pests spreading in agricultural and forestry ecosystems, especially in cooler Arctic, boreal, temperate and subtropical regions. For example, a single, unusually warm winter may be enough to assist the establishment of invasive pests. Some pests, like fall armyworm (which feeds on a growing number of crops, including maize, sorghum, millet) and Tephritid fruit flies (which damage fruit and other crops), have already spread due to warmer climate. Others, such as the desert locust (the world's most destructive migratory pest), are expected to change their migratory routes and geographical distribution because of climate change (FAO 2021).

2.2 Progress in Increasing Biotic Stress Tolerance in Plants

Biotic stresses are the damage to plants caused by other living organisms such as bacteria, fungi, nematodes, protists, insects, viruses and viroids. Plants have developed a multitude of defense responses that allow them to adapt, survive and reproduce under stress conditions (Pieterse et al. 2009). The plant immune system consists of successive layers counteracting suppression of defense responses by pathogens through secretion of effector proteins (Hemetsberger et al. 2012). Recognition of the effectors by corresponding resistance-genes (R-genes) belonging to nucleotide-binding leucine-rich repeat (NB-LRR) protein family or the effect of effectors on intracellular host proteins (guarded proteins) results in effector-triggered immunity. This is usually but not always manifested by localized cell death, termed the hypersensitivity response (Coll et al. 2011). Different methodologies have been employed for enhancing climate change-related stress tolerance in plants such as conventional breeding and plant genetic modification (Hu and Xiong

2014). Furthermore, root-associated microorganisms, such as arbuscular mycorrhizal fungi and plant growth promoting bacteria (PGPB), are also important actors in this context and the optimization of their real utilization has a great potential in an innovative and sustainable agriculture context, providing benefits to plant growth and health through an increase in plant nutrition, conferring plant tolerance to abiotic stresses and improving resistance to biotic threats (Busby et al. 2017). On one hand, PGPB can act as "priming stimulus" through diverse mechanisms such as modifications in phytohormonal levels, antioxidant compounds, osmolytes (proline), and polyamines, regulation of stress responsive genes, changes in root morphology, and emission of volatile organic compounds (Alagna et al. 2020). Utilization of chemical compounds as priming agents was developed after it was found to significantly improve plant tolerance against a range of different abiotic stresses (Irani and Todd 2018). Priming refers to a technique in which plants are treated with chemical agents to protect them against environmental effects. It is an adaptive strategy that induces stress resistance and improves plant defense capabilities. The plant then takes defensive measures against the potential attacks and also builds the base for a faster and stronger defense system for future stresses. Priming is not only effective against pathogens, but also biotic stress such as insects and arthropods (Mauch-Mani et al. 2017). Conventional breeding methods still play an important role in the selection of new varieties. However, emerging tools in biotechnology are much needed to maximize the probability of success. One area of biotechnology, molecular marker assisted breeding, has already made significant impact in improving efficiency of conventional breeding. Another newly approach, based on the use of improved recombinant DNA techniques to introduce new traits in early phases of cultivar selection (Budak et al. 2015).

2.3 Current Views on Solving Problems in Agriculture Due to Combined Biotic and Abiotic Stresses

Biotic stresses caused by living organisms, such as fungi, bacteria, nematodes, insects, birds, weeds, and parasitic plants are the most important constraints of agricultural production worldwide. Various ways and means to manage these biotic factors have been developed through intensive research and updated as and when the situation demands. However, no environmentally sustainable solutions exist until now for the control of insect pest and weeds. Understanding the underlying condition of host plants, especially in natural environments is, therefore, a crucial component to understanding pest and disease impacts and stresses the contextual information necessary for effective disease surveillance. The effect of drought and other disturbances on forested ecosystems has been dramatic, with wide scale dieback and decline (Seidl et al. 2020). Disease simulation models frequently used for forecasting can be utilized in simulation studies of spatial and temporal distribution in future climate scenarios. For vector-transmitted pathogens the risk analysis may include the effects of climate change on the vector population (Henry and McNab 2002). Advance prediction of plant diseases through plant disease forecasting systems can be a very helpful solution in minimizing these losses. As the computer programming and literacy is spreading to each and every corner of the world, the development of computer-based plant disease forecasting systems is becoming easier. These systems can not only handle the larger data, but also can generate prediction for large number of diseases via incorporation of huge information of pathogens (Islam 2018). Forecasting is a great tool to know the truth about the emergence of diseases in agriculturally important crops. They predicted with main tomato diseases and showed that climate change will not be favorable for the occurrence of late blight (*Phytophthora infestans*), verticillium wilt (*Verticillium albo-atrum*), and white mold (*Sclerotinia sclerotiorum*), and will not alter the importance of the tomato mosaic caused by the tomato mosaic virus (ToMV) and the septoria leaf spot (*Septoria lycopersici*). In contrast to those diseases, the authors considered that the importance of powdery mildew (*Leveilula taurica*) will increase in all tomato production regions across the country, just as the importance of early blight (*Alternaria solani*), fusarium wilt (*Fusarium oxysporum* f.sp. *lycopersici*), bacterial wilt (*Ralstonia solanacearum*), tomato spotted wilt virus (TSWV), tomato chlorotic spot virus (TCSV), groundnut ring spot virus (GRSV), Chrysanthemum stem necrosis virus (CSNV) and yellow leaf curl virus (Geminivirus). Disease-forecasting models with improved quality are needed to guide farmers. This type of prediction tool may allow farmers to respond timely and efficiently with public-private partnership. Propagation of resistant cultivars is an effective means for the management of several plant diseases and minimizes the damage to these important crops (Mahmoud and Abd El-Fatah 2020). The introduction of new fungicides with a larger effect under rainfall conditions might help to minimize plant disease problems (Hannukkala et al. 2007). The efficacy of fungicides can be increased with proper timing of fungicide application (Bedos et al. 2002). For example, stem rot (*Sclerotium rolfsii*) of peanut were managed by applying fungicide early in the morning improved spray deposition in the lower canopy of the plant (Augusto et al. 2010). Severe weather events like heat waves, erratic rainfall patterns and drought will probably become more common. Simultaneously, climate change will affect diseases. The geographical and temporal distribution of diseases and management methods will have to be modified according to climate change scenarios. New climate-resilient strains of bio-control agents with high competitive ability that can survive in the introduced environment may be needed. Region- and crop-specific strains may have to be incorporated and integrated into pest management approach. Integrated pest management is an ecosystem-based strategy, which emphasizes the long-term prevention of pests or their damage through a combination of methods such as cultural practices, habitat management, use of resistant varieties and biological control (Mahmoud 2016). New chemistries as well as the more effective exploitation of chemical induced resistance agents or resistance elicitors and biological control with the microbiome concept may have the potential to offer more benign and sustainable disease control interventions.

3 Breeding Practices to Improve Abiotic Stress Tolerance

Domestication, originating from the Latin word "*domus*" (house), is the adaptation process of wild plants and animals for human use (Clement et al. 2012). Adaptation of plants to cultivation has a very important place in the birth of modern civilization, as it provided the transition from hunter-gatherer societies to agricultural societies. The domestication of plants stems from Mesopotamia which dates to more than 10,000 years ago and the major cereal crops and vegetables, which human survival is dependent on today, were domesticated from their wild progenitors around 8000–2000 BCE, in various regions of the world (Ross-Ibarra et al. 2007). Through the history of domestication until now, many plants have undergone a series of processes that constructed the genetics and diversity of modern crops (Louwaars 2018). Artificial plant selection, as the most primitive way of breeding, was practiced for thousands of years until the introduction of Mendelian laws in the nineteenth century (Ahmar et al. 2020). During that time, according to archeological studies, it is thought that farmers saved the seeds of best-fitted plants from one generation to another without any knowledge of trait heritage (Doebley et al. 2006). The adaptation of plants to cultivation has started based on selection, where favorable traits such as size, shape, taste, and color were either deliberately or unintentionally selected aiming to increase the suitability of plant species to the human consumption and cultivation (Breseghello and Coelho 2013). Among thousands of plant species, only a small portion is domesticated and in use. It is hypothesized their ease of breeding is the reason that some plant species among thousands were chosen to be domesticated in the first place. For instance, a key trait that affects harvesting and is undesirable in domesticated crops, seed shattering, is controlled by only a few genes, if not one, in many major crop species (Maity et al. 2021).

The limits of plant breeding tremendously enhanced with the understanding of trait inheritance leading to scientific breeding. Plant breeding strategies have improved in thousands of years of agricultural practices and currently, compared to 100 years ago, farmers are producing so much more products using the same amount of agricultural land. Breeding was mostly focused on keeping beneficial traits for a long time, neglecting the interspecific and intraspecific variations. Inbreeding in cross-pollinated crops results in the accumulation of identical alleles which causes loss of fitness of the population over time (Govindaraj et al. 2015). Plants with desired characteristics represent only a small portion of the whole population, therefore domestication reduces the trait variability resulting in less diversity due to narrowing of the genetic base, genetic bottlenecks, compared to their wild progenitors (Tanksley and McCouch 1997). While we managed to domesticate crops over billion years, the selection of favorable traits such as higher yield and nutrition over the defense and stress tolerance genes, caused the loss of diversity resulting in susceptibility against various environmental conditions (Gaillard et al. 2018). Many domesticated plants, such as maize and cauliflower became totally dependent on human intervention as the result of breeding practices so much that they can even no longer propagate themselves in nature (Doebley et al. 2006). Because agricultural production was mostly practiced with short-term concerns, in order not to go beyond the known and take risks, modern agriculture became mostly dependent on five crops accounting for 60% of the total calorie intake globally (Dwivedi et al. 2016). It is estimated that 75% of the agrobiodiversity is lost between 1900 and 2000, due to replacing local varieties and landraces with genetically uniform, high yielding improved, or exotic varieties. In a guideline that the Food and Agriculture Organization of the United Nations (FAO) released in 2019, the genetic erosion of major crops was defined as a severe threat to food security, and farmers were called to grow genetically diverse crops and varieties to provide agriculture sustainability (FAO 2019). On the road to achieving global food security and sustainability, one of the major challenges the world currently facing is developing resilient crop cultivars which can adapt to rapid changes in the environment due to climate change. Resilience and adaptability are found in diversity; therefore, plant breeding strategies are focusing on breeding strategies that enhance the genetic diversity of crops by using donor plants in crossing as rich genetic resources (Khoury et al. 2022). This requires identification of genetic variations among species, and incorporation of them while preserving the traits economically important (Swarup et al. 2021).

Ancestors of modern crops are sources of stress response genes as they came across and bat away various stresses through their evolution. Crop wild relatives (CRWs) are undomesticated ancestors or relatives of modern crops, found in the wild, and possess a high diversity, but relatively fewer desirable traits in terms of yield, size, and shape compared to modern varieties (Fig. 1). They can mostly be found as weeds in undisturbed habitats and are used commonly in breeding studies mostly to improve biotic resistance of modern plants (Dempewolf et al. 2017; FAO 2019). In a recent study, Wang et al. 2020) have identified a gene, SlHAK20, which is responsible for the reduction of salt tolerance in cultivated tomato varieties and have shown the higher activity of the salt tolerance allele SlHAK20 in wild varieties enhanced the salt tolerance whereas the lower activity of SlHAK20 resulted in the loss of salt tolerance in cultivated tomatoes.

The first rounds of domestication formed landraces (locally selected cultivars) by selecting traditional varieties based on their adaptation to local environments. Landraces evolved from wild varieties in various regions of the world under various

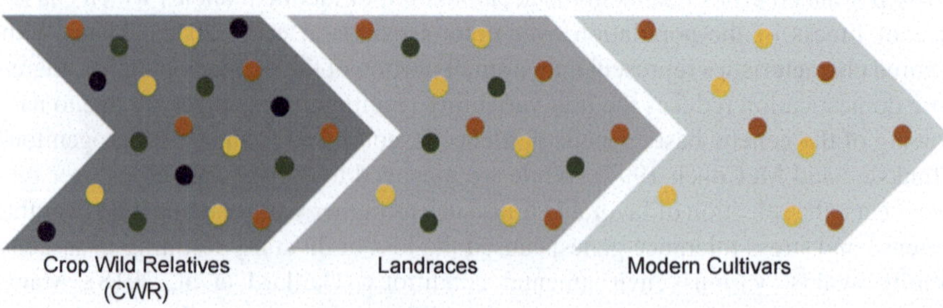

Fig. 1 Genetic bottleneck representation of the loss of diversity through domestication. Each colored dot represents traits contributing the genetic diversity. As domestication is driven from left to right side, the genetic variability decreases

environmental conditions as a result of the migration of our ancestors and are often genetically heterogeneous, more stress-resilient but less yielding compared to modern cultivated gene pools (Adhikari et al. 2022). Landraces provide beneficial genes for adaptation to many stress conditions such as water, salt, and high-temperature (Dwivedi et al. 2016). Identification and mapping of many quantitative trait loci (QTL) from abiotic stress tolerant landraces and CWRs provided many molecular markers to be used in marker-assisted breeding (Dwivedi et al. 2016). In previous studies, landraces of major crops contributed to the identification of key abiotic stress-related genes such as SUB1A gene alleles SUB1A-1 and SUB1A-2 which is associated with submergence tolerant and intolerant lines, respectively (Xu et al. 2006). The identified genes are then introgressed to local rice cultivars using marker-assisted backcrossing and resulted in the creation of a submergence tolerant variety used in Bangladesh, India, Indonesia and Philippines (Dwivedi et al. 2016).

Identification of the genes present in CRWs and landraces that are lost during the domestication and breeding over time can provide the introduction of those variable genes into breeding programs aiming to enhance genetic diversity (Swarup et al. 2021). While the discovery of alleles of interest mostly relies on a high-quality reference genome (Varshney et al. 2021), a single genome may not be adequate to detect structural variations (SVs) that include single nucleotide polymorphisms (SNPs), presence/absence variations (PAVs), or copy number variations (CNVs) as a result of mutation, adaptation, and breeding. Many previous studies identified alterations in gene activity based on SNPs which causes intraspecific phenotype differences against abiotic stresses (Yolcu et al. 2020). Advances in high-throughput sequencing technology provided ease of sequencing many lineages of a species, creating a pangenome, to identify the lineages with beneficial alleles to be used in breeding applications. In previous studies, many structural variations associated with a/biotic stresses were identified as a part of dispensable genomes of many plants using pangenome approaches. Molecular markers are then developed using these SVs which leads to crop improvement (Khan et al. 2020). A recent study has identified various gene variations as a part of the dispensable genome of *Amborella*, which their functions were shown to be mainly associated with salt, cold, and drought responses and may contribute to the adaptation of modern cultivars to various abiotic stress conditions (Hu et al. 2022). With the discovery of molecular markers technology in the 1980s, the way breeding perceived is completely changed. Molecular markers such as simple sequence repeats (SSRs) and single-nucleotide polymorphism (SNP) are immensely used in the introgression studies to transfer genes from PGRs to the germplasm of interest (Nadeem et al. 2018). The use of molecular markers has paved the way for precision breeding, easing QTL mapping, and gene identification and isolation (Salgotra and Stewart Jr. 2020).

Plant genetic diversity can be captured either as in-situ or ex-situ conservation. The screening of plant genetic resources (PGRs) for stress resistance and identification of candidate genes to be included in breeding programs are vital in supporting agricultural sustainability. In-situ conservation is the conservation of plant genetic resources in their own habitats whereas ex-situ conservation is the conservation of plant genetic resources in gene bank facilities by public or private institutions.

Currently, there are over 1750 gene banks with more than 7.4 million accessions, mainly consisting of wild species, breeding lines, and landraces (Khoury et al. 2022). There are still quite a few challenges regarding ex-situ conservation including periodical regeneration of aging accessions, documentation updates, and characterization which is both costly and requires skilled labor. Moreover, identification of the desired trait in millions of accessions is like looking for a needle in a haystack. Therefore, there are some strategies developed to overcome this issue. Firstly, Focused Identification of Germplasm Strategy (FIGS) is an efficient approach in predicting variations in adaptive traits. FIGS uses the environmental data associated with each accession and creates a network of trait-environment relationships providing a quick search (Stenberg and Ortiz 2021). Another tool is the advanced use of machine learning in the analysis of high-throughput data (Varshney et al. 2021). For efficient use of PGRs, both in-situ and ex-situ accessions need to be better-linked, and more accessible by plant breeders and farmers.

Breeding for improving stress tolerance is a challenging process as stress response in plants is a complex mechanism, generally controlled by the interaction of multiple QTLs and plants usually face a combination of stressors simultaneously (Zhang et al. 2018d). Also, a gene successfully functioning in favor of stress tolerance in one species may not be effective in working in another (Yolcu et al. 2020). Enhancing genetic diversity by crossing a local or exotic breed with a wild relative donor may provide resistance against various abiotic stresses by transferring a novel trait, but also in lower yield or unfavorable size and shape of the product. This phenomenon may be caused by either genetic background effects or linkage drag (Swarup et al. 2021). Linkage drag is the introduction of unfavored traits, such as small grain size, together with the gene-of-interest during hybridization. In conventional breeding approaches, backcross breeding with the elite cultivar is applied for a few generations until reaching homozygous alleles for the gene of interest, resulting again in a less diverse genotype (Ross-Ibarra et al. 2007). Other major limitations of conventional plant breeding include the lack of genotypic and phenotypic data on CWRs stored in gene banks, crossability barriers, and poor agronomic performances of generated varieties (Dempewolf et al. 2017). Conventional plant breeding strategies are still time-intensive and require many resources and labor; however, should be introduced to crop improvement in the climate change era, since landraces provides a reliable gene pool for a/biotic stress tolerance and as the main cause of losing these gene pool is the human based breeding studies focusing on the selection for quality and yield (Anwar and Kim 2020).

4 Effects of Biostimulants on Increasing Tolerance

The adverse effects of environmental stresses (e.g., salinity, drought, low and high temperature, acidity, heavy metal pollution, and ultraviolet radiation) are critical factors affecting plant growth and yield in agriculture. Abiotic stresses may be alleviated through improvement of plant growth conditions including water, nutrient, and plant growth regulators (PGRs) supplementations. Besides the classical agricultural practices, the biostimulants have been widely used for this purpose and they increase the yield by optimizing the physiological processes in plants (Yakhin et al. 2017). Proteins, enzymes, amino acids, micronutrients and other molecules could be used as the biostimulants. In addition, natural stimulants are often accepted as biostimulants, such as humic and fulvic acids, protein hydrolases, phenols, and salicylic acid (Chiaiese et al. 2018; du Jardin 2015). A plant biostimulant, any substance or microorganism, promotes plant growth or fitness such as abiotic stress tolerance, crop quality traits, and nutrients content when applied in low quantities. In addition, they can serve as commercial products including mixtures of such substances and/or microorganisms (du Jardin 2015). Kauffman et al. 2007 reported that biostimulants are generally clustered into three main groups based on source and content; amino acid containing products (AACP), humic substances (HS), and hormone containing products (HCP). The bioactive compounds in some biostimulants cause signaling activity or induce signaling pathways in plants. In addition, various amino acids and peptides serve as signaling molecules in the plant metabolism (Yakhin et al. 2017). Biostimulants also enhance plant development and yield by increasing nitrogen, carbon, and sulfur assimilations (Jannin et al. 2012). As a result, it will also continue to be one of the hot topics for agriculture with positive effects on crop quality and performance (Drobek et al. 2019).

4.1 Protein Hydrolysates

Plant-derived protein hydrolysates (PHs) are accepted as one of important biostimulants used in many horticultural and agronomic crops. PHs are defined as mixture of various protein sources that are produced using partial hydrolysis, such as amino acids, oligopeptides, and polypeptides (Schaafsma 2009). PHs are generally obtained through chemical (acid and alkaline hydrolysis), thermal and enzymatic hydrolysis of plant biomass (such as legume seeds and alfalfa hay) and animal wastes (du Jardin 2015). In addition to PHs, some nitrogenous molecules (polyamines, betaines and non-protein amino acids) function as biostimulants (Vranova et al. 2011). du Jardin et al. (du Jardin 2015) stated that PHs positively affects soil fertility indirectly by increasing microbial biomass and activity as well as chelating effects of some amino acids (proline) against heavy metals. In plant nutrition, PHs applications in plant leaves and roots enhance Fe and N assimilations, water and nutrient use efficiencies and nutrient uptake (Cerdán et al. 2009). In addition, PH application has been shown in many studies to reduce reductions in crop loss under abiotic stress conditions, includes drought, salinity, nutrient deficiency and alkalinity (Colla et al. 2015).

4.2 Humic and Fulvic Acids

Humic substances (HS) are organic compounds and are the main components of humus in the soil, which are formed because of the breakdown of plant, animal and microbial residues. HSs are heterogeneous compounds and are categorized according to their molecular weights and solubility of humins, humic acids and fulvic acids. In addition, HSs and its complexes provide interaction between plant roots, microorganisms and organic material in the soil (Rose et al. 2014). HSs are known to have a wide range of roles in plants, such as soil fertility, root nutrition, nutrients uptake, cell enlargement, organ growth, and stress protection (Jindo et al. 2012). Shah et al. 2018 stated that humic substances trigger complex networks at transcriptional and post-transcriptional levels to cope with abiotic stresses, including metal ligation, protein and nitrogen metabolism, sulfur sequestration, photosynthesis and energy metabolism, and membrane trafficking. Many studies have been carried out to show the recovery effects of humic substances, especially in the fight against abiotic stresses in plants, including drought (Fang et al. 2020), salinity (Akladious and Mohamed 2018), heavy metal (Chaab et al. 2016), and heat stresses (Cha et al. 2020).

4.3 Seaweed Extracts

Seaweeds, the members of the marine coastal ecosystems, are macroscopic and multicellular algae. It is estimated that there are approximately 9000 species of macroalgae and the most used group in agricultural activities is brown seaweeds with approximately 2000 species (Khan et al. 2009). Fresh seaweeds are used as organic material and as fertilizer is ancient in agriculture, but recently it is considered to have a biostimulant effect. Seaweed extracts (SEs) contain alginates, carrageenans, polysaccharides laminarin, fucans, phlorotanins, polyphenols, micro- and macronutrients, hormones, sterols, and N-containing compounds (Khan et al. 2009). SEs can be added into soils and hydroponic solutions as well as applying as foliar treatments. In the soil, seaweed polysaccharides contribute to gel formation, creating a suitable ground for cation exchange and fixation, which is beneficial for soil microflora and plant growth-promoting bacteria (du Jardin 2015). Seaweed extracts had both protective properties (antioxidants) and potential to trigger endogenous stress-responsive genes. In *Vigna sinensis* and *Zea mays* plants, seaweed liquid extracts (SLEs) from three Egyptian seaweeds (*Ulva fasciata*, *Cystoseira compressa*, and *Laurencia obtusa*) were reported to have ameliorating effects on seed germination and seedling growth under salinity stress (Hussein et al. 2021). Additionally, it is found that *Kappaphycus alvarezzi* extract treatments in different wheat varieties enhance chlorophyll content and carotenoids, and lead to the increase in root length, and the decrease in electrolyte leakage and lipid peroxidation under salt and drought stresses (Patel et al. 2018). Similarly, seaweed application at different growth and developmental stages in sugarcane has been reported to increase yield and promote thc sucrose accumulation. In particular, the yield increasing effect of the extract has been observed even more striking under drought stress conditions (Chen et al. 2021a, b).

4.4 Beneficial Fungi and Bacteria

From mutualistic symbiosis to parasitism, fungi interact with plant roots in different ways, and the mutualism–parasitism relationship has opened up a path that provides an easy explanation for their evolutionary history (Johnson and Graham 2013). Mycorrhizal fungi contain taxa consisting of heterogeneous groups that can form symbiosis with more than 90% of plant species (Bonfante and Genre 2010). Many scientific studies have been published showing that arbuscular mycorrhizal fungi (AMF) contribute to the growth and development of plants under different stress conditions such as salinity, drought, heat and metals (Birhane et al. 2012). The AMFs enhance the stress tolerance mechanism in plants by upregulating antioxidant system, transcription factors, ABA responsive genes, aquaporin (*AQP*) genes and phytohormone biosynthesis. In addition, it increases nutrient uptakes by modifying root morphology and soil health. In some plant species, AMFs have been shown to reduce the negative effects of drought stress, such as *Capsicum annuum*, sweet potato (Yooyongwech et al. 2016) and strawberry (Moradtalab et al. 2019). Despite exposing to salinity stress, improvements have been noted in some plant species by means of using AMFs, such as date palm (*Phoenix dactylifera* L.) (Ait-El-Mokhtar et al. 2019), and lettuce (Aroca et al. 2013). AMFs are also considered to increase plant growth and development by strengthening the plant defense systems against heavy metal stress (Yousaf et al. 2016). In particular, internal and external origin of fungal hyphae are immobilized, where heavy metals are fixed in the cell walls, stored in the vacuole, or chelated in the cytoplasm (Punamiya et al. 2010). Gavito et al. 2005 reported that AMF-inoculated plants generally grow better than non-AMF inoculated plants under heat stress. Considering combined abiotic stress applications, it has been evident that AMFs applications in tomato and *Ricinus communis* have curative effects on plant development and growth (Zhang et al. 2018a, b, c, d).

Soil bacteria are effective biological agents used in crop production for many years, providing nutrients to crops, promoting plant growth, and improving soil structure. Plant–bacterial interactions in the rhizosphere are important factors affecting soil fertility and plant health (Hayat et al. 2010). In terms of the biostimulating effects, mutualistic endosymbionts of the *Rhizobium* and rhizospheric type PGPRs ('plant growth-promoting rhizobacteria', mutualistic) stand out as bacterial plant biostimulants (BPBs) (du Jardin 2015). PGPRs are also called as plant health promoting rhizobacteria (PHPR) or nodule promoting rhizobacteria (NPR) (Hayat et al. 2010). The PGPR contains *Enterobacter* spp., *Ochrobactrum* spp., *Rhizobium* spp., *Azospirillum* spp., *Bacillus* spp., *Azotobacter* spp., and *Pseudomonas* spp. taxa

(Rouphael and Colla 2020). Particularly, PGPR plant biostimulants, prominent environmentally sustainable alternatives to agrochemicals, are widely used in agricultural and horticultural crops in recent years. These microbial plant stimulants play essential roles in abiotic stress tolerance (upregulation of stress related genes, cytokinin and ABA production, and secondary metabolite production), growth promotion, quality improvement, and disease resistance (Hamid et al. 2021). Moreover, it has been also reported that PGPR biostimulants are effective in the development of abiotic stress tolerance in various crops. In this regard, salt tolerance in rice, wheat, and sunflower (Fatima and Arora 2021), drought stress tolerance in maize and red rice (Filgueiras et al. 2019), cold stress tolerance in bean and wheat (Tiryaki et al. 2019), heavy metal stress tolerance in maize (Rizvi and Khan 2018), and flooding stress tolerance in rice (Etesami et al. 2014) are improved.

5 Importance of Nutrition Availability in Abiotic Stress Tolerance

The nutrition status of the soil determines the status of normal growth and development of the plant. Abiotic stress has dire effect on nutrients availability in the soil, uptake through root hairs in the rhizosphere and translocation in the plants. Extreme temperatures, drought, soil salinity, soil pH and excess of heavy metal accumulation are major adverse environmental conditions that plants often encounter (Schützendübel and Polle 2002). Sufficient amount of nutrients in the soil is mandatory for normal plant growth and development under suitable environmental conditions. The soil would be considered healthier and fertile if requisite amount of nutrients is present in it and only healthy soils result in healthy plants (Pasala et al. 2022). Both deficiency and excess of necessary nutrients in soil can affect plant's normal growth mechanisms even under normal circumstances (Ahanger and Ahmad 2019). Plant nutrition level gets altered under adverse environmental conditions to support tolerance in plants against stress. Abiotic stress is the major among environmental stresses halting normal growth mechanisms by altering the nutrition in plants (Fageria et al. 2008).

Each nutrient required in an appropriate range in plants. An imbalance in this range, could be deficiency or toxicity, would affect plant normal growth. Toxicity occurs when a nutrient is above the sufficiency range than the plant needs, and this result in a decrease in growth of plant or its quality (Sajjad et al. 2021). Deficiency or toxicity of nutrients in plants expressed as characteristic symptoms causes great yield loss especially in crop plants. The availability of nutrients in the soil and mobility to target parts in plants have a major role to protect or avoid the stressful situation created by any abiotic stress such as drought, salinity, extreme temperatures and soil pH (Ahanger and Ahmad 2019). A balanced supply of macro- and micronutrients is essential to decrease the susceptibility of plants to biotic and abiotic stresses. Silicon—an element not belonging to the essential nutrients—plays a key role in the response of plants to biotic stresses (e.g., fungal attack) and a series of abiotic stresses such as drought or heavy metal stress. For example, silicon can decrease the toxic effects of heavy metals or play a protective role against fungal diseases by positively influencing the structure and function of plant cell walls (Noreen et al. 2018).

5.1 Micronutrient Status and Its Role in Tolerance Against Abiotic Stress

Plants experience either poor or excessive availability of micronutrients under abiotic stress, which alters their biochemical composition and minimizes growth and yield. To maintain the yield in crop plants, improvement of the micronutrient status in plants under abiotic stress is very important. Micronutrient availability greatly affects the ability of plants to adapt to unfavorable conditions. Essential micronutrients, such as zinc (Zn), copper (Cu), manganese (Mn), boron (B), iron (Fe), molybdenum (Mo), chloride (Cl), nickel (Ni), and cobalt (Co), have direct roles in plant metabolism (Khan et al. 2018). Most of these nutrients have two or more oxidation states, and therefore participate in oxidation–reduction reactions through electron transport (Karthika et al. 2018). These elements form metalloenzymes, function as catalysts, and are vital in osmoregulation and protection against abiotic stress in plants. Micronutrients protect plants by functioning as constituents and activators of several enzymes in their defense system that are involved in the detoxification of highly reactive oxygen species (ROS) produced during abiotic stress (Apse et al. 1999).

5.2 Effects of Abiotic Stress on Translocation of Nutrition in Plants

The decrease in transpiration rate of water after closing stomata under abiotic stress, specifically drought, can affect the uptake and translocation pull of different micro- and macro-nutrients. Aerial application of these nutrients to foliage can be the best option to avoid from stress. Several environmental factors influence the mineral uptake of plants, including pH, redox potential, and the presence of xenobiotics as well as the temperature and salinity (Zhang et al. 2022a, b). Changes in environmental conditions affect the biological and physiological response of plants. Most important targets of plants are maintained by the ion homeostasis and mineral uptake in stress conditions. Plants may use the different procedures to regulate homeostasis. In general, salt stress decreases the levels of cationic nutrients such as K, Ca, and Mg due to competition for ion transport sites. Drought and salinity stress are physiologically related and the tolerance mechanisms overlap. Metals can interfere with mineral nutrition and change the concentration and composition of plant nutrients. Besides, metals may also alter the conformation of proteins, including transporters, or regulator proteins. Herbicides may disrupt the function and integrity of the cell membrane, and significant ion losses can occur. However, molecular mechanisms and genetic basis of interactions between abiotic stress and mineral uptake is lacking (Apse et al. 1999).

5.3 Micronutrient Application to Increase Tolerance Against Abiotic Stresses

Micronutrients are essential to plant nutrient elements after macronutrients. In addition to soil and foliar application, B, Zn, Mn, Fe, Cu, and Mo have been widely used for seed priming too. The foliar spray of micronutrients either as individual, specific compounds or in combination with other macronutrients and/or micronutrients has been widely employed (Hasanuzzaman et al. 2018). Rather than soil application, a foliar application is more effective at enhancing plant inbuilt mechanisms, plant cell homeostasis, water relations, and productivity during abiotic stresses. However, excess foliar application will lead to abnormalities in crop growth and scorching. Crop- and growth-specific foliar sprays are essential to achieve maximum crop productivity.

In the twenty-first century, nutrient efficient plants will play a major role in increasing crop yields compared to the twentieth century, mainly due to limited land and water resources available for crop production, higher cost of inorganic fertilizer inputs, declining trends in crop yields globally, and increasing environmental concerns. Furthermore, at least 60% of the world's arable lands have mineral deficiencies or elemental toxicity problems, and on such soils fertilizers and lime amendments are essential for achieving improved crop yields. Fertilizer inputs are increasing cost of production of farmers, and there is a major concern for environmental pollution due to excess fertilizer inputs (Apse et al. 1999). Higher demands for food and fiber due to increasing world populations further enhance the importance of nutrient efficient cultivars that are also higher producers. Nutrient efficient plants are defined as those plants, which produce higher yields per unit of nutrient, applied or absorbed than other plants (standards) under similar agroecological conditions. During the last three decades, much research has been conducted to identify and/or breed nutrient efficient plant species or genotypes/cultivars within species and to further understand the mechanisms of nutrient efficiency in crop plants. However, success in releasing nutrient efficient cultivars has been limited (Morton et al. 2019). The main reasons for limited success are that the genetics of plant responses to nutrients and plant interaction. A complete understanding of the physiological, biochemical and molecular mechanisms especially the signaling cascades in response to abiotic stresses in tolerant plants will help to manipulate susceptible crop plants and increase agricultural productivity in the near future. Moreover, advances in genomics strategies including genetic engineering and genome editing have provided new opportunities for crop improvement by employing precise genome engineering for targeted traits in crop plants. However, the selection of the key genes is critical for the success of these approaches.

6 Epigenetics of Abiotic Stress Tolerance

Plants response to stress varies depending on the magnitude and duration of the stress (Srikant and Drost 2021). Short-term stress responses are driven by plant innate adaptation mechanisms at the physiological level to reduce the adverse effects of stress, thereby developing a stress tolerance. Long-term stress responses, on the other hand, include transgenerational alterations in the genome which results in inheritable gene expression changes. Severe or prolonged stress factors can initiate such alterations which regulate the expression of stress responsive genes, that is, epigenetics (Srikant and Drost 2021).

Epigenetic changes cover any heritable genome modifications that do not change the underlying nucleotide sequences in an organism. These changes can be reversible and can result in the activation as well as the inactivation of the genes depending on the stress environment (Tsuji et al. 2006). A well-documented example of epigenetic regulation in plant response to severe abiotic stresses is the key regulatory roles of epigenetic marks in vernalization. During this vernalization process, winter varieties of plants flower in the spring only after a prolonged exposure to cold in the winter (Angel et al. 2011). Epigenetic mechanisms in vernalization has been recently reviewed extensively by (Luo and He 2020).

Abiotic stress factors contribute to the alterations in the epigenetic marks, thus the alterations in the transcription of stress responsive genes leading to the formation of heritable epialleles which subsequently enable plant to withstand stress. Epialleles can be noted by their distinct expression patterns compared to genetically identical genes due to epigenetic changes. The dynamic changes in epigenetic marks on stress-responsive genes alter their chromatin status such that the underlying DNA is either accessible or inaccessible for the transcription factor or RNA polymerase binding (Tsuji et al. 2006). To increase the expression of genes, its promoter must be accessible to regulators to initiate the transcription and the nucleosome complex must be untangled from the DNA itself for an uninterrupted transcription (Zemach et al. 2010). Conversely, epigenetic inhibition of gene expression is established by the methylation of DNA and the condensation of the DNA stretch through the recruitment of histones (Shahbazian and Grunstein 2007). In return, epigenetic marks regulate the expression of stress-responsive genes at the transcriptional or posttranscriptional levels (Lämke and Bäurle 2017). Epigenetic regulation involves DNA methylation, posttranslational modifications of histones, and RNA interference.

6.1 DNA Methylation

DNA methylation is one of the epigenetic marks regulating gene expression in response to environmental stresses in plants. DNA methylation occurs extensively at cytosine residues (Wang et al. 2016). DNA methylation can drive the formation of heterochromatic and euchromatic regions. Heterochromatic regions refer to highly methylated regions which extensively covered with the transposons and other repetitive sequences whereas euchromatic regions being loose condensed through cytosine methylation (Ahmed et al. 2011). The levels of DNA methylation vary among plants depending upon the genome content of the species. For example, up to 50% of the cytosine were shown to be methylated in maize which seem to be due to large numbers of transposons (Palmer et al. 2003). In *Arabidopsis thaliana*, DNA methylation was observed in all the transposons as well as 20% of the genes.

The effects of epigenetic regulation on gene expression depend on not only the type of the epigenic mark but also its position on the genes. Formation of heterochromatic regions in the gene promoter is usually associated with the inhibition of transcription whereas methylation in the gene body does not appear to be associated with the transcription. Rather, gene body methylations are most likely to be moderately expressed (Bewick et al. 2016). Abiotic stresses can induce hypermethylation or hypomethylation of the DNA to regulate gene expression. For example, Hu et al. 2012 showed that cold stress treatment resulted in a genome-wide DNA methylation except that of the knob-associated tandem repeat sequences. These knobs are always methylated; however, under the cold stress, they were subjected to demethylation, and thus activated. Methylation state was recovered after prolonged stress treatment. In another study, Pandey and Pandey-Rai evaluated the effects of UV-B radiation to the DNA methylation dynamics in *Artemisia annua* (Pandey and Pandey-Rai 2015). UV-B treatment enhanced the expression of DOUBLE BOND REDUCTASE 2 (DBR2), a key regulatory gene of artemisinin biosynthesis, through DNA demethylation. In a more recent study, Pandey et al. (Pandey et al. 2019) showed that UV-B-induced stress initiates a genome-wide DNA hypomethylation, which increases the expression of several transcription factors.

Another study reported the role of DNA methylation in drought response (Mao et al. 2015). A genome-wide association study in maize showed that a miniature inverted-repeat transposable element (MITE) inserted in the promoter of NAC gene (*ZmNAC111*) causes the inhibition of gene expression through DNA methylation (Mao et al. 2015). Enhanced expression of *ZmNAC111* conferred drought tolerance in maize seedling. MITE insertion in the promoter of *ZmNAC111* enables cascade of epigenetic regulation, which resulted in the DNA methylation, and thus the inhibition of *ZmNAC111* (Mao et al. 2015). In another study, investigation on drought-induced DNA methylation sites in rice showed that epigenetic marks at the 70% of the sites were reversed after elimination of stress, whereas 29% of the sites retained the drought-induced DNA methylation changes even after recovery (Wang et al. 2011). Another study reported that drought tolerant varieties showed hypermethylation, while the drought-sensitive varieties showed hypomethylation under drought in rice (Gayacharan and Joel 2013). These findings above suggest that the chromatin changes caused by DNA methylation play crucial roles in plants drought tolerance.

6.2 *Small RNA Directed DNA Methylation*

RNA interference (RNAi) is a broad term describing repression of gene expression by the interference of small RNAs (Matzke and Mosher 2014). Small interfering RNAs (siRNAs), a class of small RNAs, are involved in the epigenetic regulation of gene expression. siRNAs can mediate DNA methylation by a mechanism called RNA-directed DNA methylation (RdDM) to enforce epigenetic states (Pontes et al. 2006). RdDM is a unique chromatin modification as it depends on specialized transcriptional machinery including two plant specific polymerases: Pol IV and Pol V (Matzke and Mosher 2014). As is best understood in the model plant *A. thaliana*, DNA methylation can be established de novo through the RdDM pathway (Matzke and Mosher 2014; Pontes et al. 2006). Besides being the first example of small RNA-mediated epigenetic modification, RdDM is among the most complex regulatory mechanisms representing impressive capabilities of plant genomes.

The best-known regulations through RdDM pathway includes response to heat stress. The transcription of heat-responsive genes is dynamically regulated by RdDM (Pontes et al. 2006). Mutations in the Pol VI and Pol V resulted in hypersensitivity under the heat stress and the subsequent recovery after heat stress was halted (Popova et al. 2013). Another study showed that the transcription factor MYB74 is normally silenced through RdDM in Arabidopsis and is desilenced by salt stress treatments (Xu et al. 2015).

6.3 *Histone Modifications*

Covalent modifications to the N-terminal regions of histone proteins constitute another part of the epigenetic mechanisms in plants (Dhar et al. 2014). Histone modification dynamics cause variations in chromatin structure and are important in specification of the chromatin function. Some histone modification, namely acetylation, and phosphorylation, and ubiquitination, has been associated with the activation of transcription, while biotinylation and sumoylation repress the gene expression (Sadakierska-Chudy and Filip 2015). Histone methylations, on the other hand, determine the transcriptional state and expression level depending on the level and the site of methylation. For example, tri-methylation of the 27th lysine residue of histone H3 (H3K27me3) is associated with euchromatin, whereas H3K27me is a heterochromatin marker (Dhar et al. 2014). Similarly, tri-methylation of H3K4 and H3K36 marks activate gene expression, whereas di-methylation of H3K9 and H3K27 represses the transcription (Dhar et al. 2014).

The coordination between histone acetyltransferases (HATs) and histone deacetylases (HDACs) as well as the histone methyltransferases and histone demethylases are important for the adaptation of plants to changing environments. The modifications through histones appear to be dynamic and recovered to the basal levels after the stress factor was eliminated. Histone modifications act fast under stress (Hui Ng et al. 2003) and can influence DNA methylation (Zhu 2008). The dynamic shift in histone modifications and in different epigenetic marks is of great significance to plant response to environmental stress (Zhang et al. 2018a, b, c, d).

Histone modifications are known to be involved in important stress response mechanisms. For example, submergence of rice seedlings causes an increase in H3K4 methylation and H3 acetylation in the submergence-inducible genes *ADH1* and *PDC1* (Tsuji et al. 2006). Similarly, in Arabidopsis, drought stress alters the histone modifications H3K4me3 and K3K9ac on drought-responsive genes, such as *RD19A*, *RD29B*, *RD20*, and At2g20880 (Kim et al. 2008). Furthermore, severe drought stress resulted in the elevated levels of H3K4me3 and H3K9ac in the stress responsive genes compared with the mild treatments of drought (Kim et al. 2012). These results indicated that histone dynamics is dependent on the intensity of drought stress. Elimination of the drought stress resulted in a rapid recovery of H3K9ac to it basal level, while H3K4me3 recovers gradually (Kim et al. 2008). Another study reported that the reduction in H3K27me3 within the gene body of drought-responsive genes is important drought tolerance in Arabidopsis (Ramirez-Prado et al. 2019).

6.4 Stress Memory and Agricultural Applications

Plants have evolved sophisticated mechanisms to adapt to rapidly changing environments. Epigenetic regulation is part of the stress adaptation mechanisms in plants. Epigenetic changes usually return to their pre-stress levels after the elimination of stress exposure. However, in some cases, instead of returning to their pre-stress levels, plants can retain a stress memory even after the withdrawal of stress (Ding et al. 2012). This memory will ensure plant adaptation to similar future stress, which is called stress priming. Priming is one such way for plants to develop a quick and efficient stress response to future recurring stress. The stress memory can be retained for only a short period if the memory depends on the degradation of stress-induced proteins and elements, while the memory can last longer if it involves reprogramming of plants cellular processes.

Recent studies have showed that epigenetic changes are indeed one such stable reprogramming of the plants, and thus can result in prolonged stress memory after the stress is returned to its basal levels. Zheng et al. (Zheng et al. 2017) reported that drought adaptability was improved in 11 successive generations of two rice varieties after induction of drought. They showed that altered DNA methylation states were preserved in the younger generations. The progenies also showed drought-tolerance by limiting water consumption and still maintained the yield. They suggested that developing drought tolerance is the result of alterations in DNA methylation state and is maintained at the progeny even in the absence of the stress (Zheng et al. 2017).

Additional studies showed that plants exposed to a stress could acquire tolerance to a second stress. For example, barley pre-exposed to drought stress developed increased cold tolerance compared to non-primed plants (Li et al. 2016). Similarly, drought priming in wheat at the vegetative state resulted in improved cold tolerance later at the stem elongation stage (Li et al. 2015a, b). In another study, research with spring wheat showed that drought priming during stem elongation induced resistance to high temperatures at the grain filling stage (Wang et al. 2015a, b). These examples confirm that stress-primed plants could possess more active defense systems and improved protection compared to non-primed plants. Due to the rapidly changing environment, these cross-stress tolerance is extremely important for plants to survive and retain yield under unpredictable and diverse stress factors.

Priming and stress memory have been fascinating discoveries for the last decades, providing promising results in plant stress adaptation and tolerance. Accumulated evidence suggest that stress memory alters the chromatin modifications at the stress responsive genes and causes immediate and efficient stress response to future stress. Recent studies also showed that some of these epigenetic marks are even inherited to next generations. Although underlying mechanisms are yet to be discovered fully, stress memory can help to advance epigenome engineering to improve plant tolerance to environmental stress factors and climate changes. An important application of stress memory into crop improvement would be stress priming of the crops to generate an orchestrated outcome to modulate the stress responses and enhance the yield. The advent of CRISPR/Cas9-mediated epigenome editing technology will enable selective modification of epigenetic marks at target loci to enable stress memory to explore the mechanisms underlying epigenetic modifications at the transcriptional and phenotypic levels. The main challenges are to elucidate the plant memory for a specific stress and to retain the epigenetic mark even when unused.

7 Genomics of Abiotic Stress Tolerance

Genomics is a widely used tool to understand the genomic basis of abiotic stress tolerance. In this section, we aimed to summarize the most recent studies with drought, salt and heat stresses, as the knowledge of heritable QTLs defining the tolerance to extreme weather conditions is a requirement for improvement through breeding. It is quite clear that more studies, especially on the combined abiotic stresses, are needed since the QTL studies available in literature as summarized below are mostly based on single stresses only, which does not define the expected effects of climate change on agriculture.

7.1 Genomics Studies for Drought Tolerance

Drought is a major constraint in agricultural production and a serious threat to the food security of upcoming generations. In the past decade, total losses in crop production amounted to almost $30 billion and population projections estimate that 5 billion people, almost half of the projected global population, will have to face living in environments with limited water supplies by 2050 (Gupta et al. 2020). A 40% decrease in available water can cause up to 21% yield loss in the temperate crop wheat, and up to 40% loss in maize (Younis et al. 2020). In rice, about one quarter of yield is lost due to drought each year (Reeger et al. 2021).

Drought stress inhibits photosynthesis, causes stomatal closure and changes in root architecture and arrests growth (Gupta et al. 2020; Younis et al. 2020). At the cellular level, drought activates response pathways for phytohormone signaling, production of antioxidants and protective metabolites and synthesis of peroxidases to prevent oxidative damage to proteins and cell membrane (Gupta et al. 2020). Importantly, the plant hormone abscisic acid (ABA) is a major regulator of biotic and abiotic stress responses including drought stress. ABA-inducible bZIP, MYB, HD-Zip and WRKY families of transcription factors (TFs) are mediators of downstream responses that comprise the expression of several stress-responsive genes. TF families DREB/CBF and NAC also trigger stress responses via an ABA-independent way (Hrmova and Hussain 2021). Additionally, multiple pythohormones can converge or work together in stress responses and, often, growth and development-related pathways share certain upstream regulators (Gupta et al. 2020). At the gene level, drought tolerance is a complex trait and is often affected by several QTLs scattered across genomes. Significant progress have been made in identifying these quantitative loci, mostly on the most-produced and consumed crop species, rice, wheat and maize. Beyond crops QTL studies are rare, even though resilient species, like millet, hold great potential as sources of favorable alleles that can be used in breeding studies (Zenda et al. 2021). While dedicated databases exist to record published QTLs such as Gramene QTL (https://archive.gramene.org/qtl/) and wheatQTLdb (http://wheatqtldb.net/), new QTL studies continue to identify novel loci and alleles.

In a recent rice panel, association mapping identified 14 genomic regions linked to agronomic traits under field conditions with drought episodes, of which 4 on chromosomes 1, 4, 6 and 11 may provide beneficial alleles for breeding for drought tolerance in rice (Beena et al. 2021). In another study, Bulk Segregant Analysis (BSA) using 77 polymorphic markers on a recombinant inbred line (RIL) population, generated from tolerant CR 143–2-2 and susceptible Krishnahamsa genotypes, identified 2 QTLs on chromosome 1 affecting chlorophyll content and 1 QTL on chromosome 3 affecting proline content under terminal drought stress. Importantly, both relative chlorophyll content and proline content correlated positively with grain yield under stress conditions (Barik et al. 2020). Under greenhouse conditions, 8 QTLs were detected that explained 3.3–37.7% of phenotypic variation in yield-associated traits among a mapping population derived from a cross of sensitive and tolerant parents Cocodrie and Nagina 22, respectively, under drought stress. Among these, two QTLs on chromosomes 7 and 8 affecting grain yield under drought stress by 12.6 and 13.3% may be novel (Baisakh et al. 2020). Landraces may exhibit considerable genetic diversity that can be used to mine favorable alleles. Association mapping in 180 Vietnamese landraces, including indica and japonica varieties, identified 17 different QTLs under drought stress in greenhouse conditions, some of which controlling multiple drought-related traits, suggesting pleiotropic effects. While 8 of these QTLs co-localized with previously identified genomic loci, remaining QTLs, which harbor candidate genes including bZIP and AP2 family of TFs, stress-responsive genes and/or photosynthesis-related genes, could be novel (Hoang et al. 2019).

In rice, root architecture, specifically, root length changes upon application of osmotic stress was used to select tolerant Heena and sensitive Kiran varieties from an initial screen of 106 rice plants. Transcriptomic data from these two contrasting varieties highlighted 7 previously identified stress-related QTLs on chromosomes 2, 3, 5, 6 and 12 that contained several stress-responsive genes (Tiwari et al. 2021). In another study, CRISPR-Cas9 knockout of a candidate gene, Os02g0673500, from a previously known drought QTL on chromosome 2 had altered xylem architecture in roots in response to drought, suggesting a negative regulatory function (Reeger et al. 2021). Overexpression of another candidate gene, OsHMGB707, from a previously identified, drought-related QTL on rice chromosome 4 enhanced tolerance against osmotic stress, likely through the activation of stress-responsive genes, including DREB TFs. The drought tolerance of the overexpressor line was also observed under field conditions, making this gene an important candidate for breeding studies (Xu et al. 2021).

As of early February 2022, manually curated WheatQTLdb database (v2.0) lists 1685 and 383 drought tolerance-related QTL in hexaploid bread wheat (*Triticum aestivum*) and tetraploid durum wheat (*Triticum durum*), respectively, identified using diverse mapping populations and methods (http://wheatqtldb.net/drought_new.php). It should be noted that some of these QTL likely represent the same genomic location in different populations, which requires further studies to confirm. Several physiological traits are also known to affect drought tolerance and several QTL for these traits have also been mapped (Tshikunde et al. 2019). As new QTLs are being revealed, ultimately, the identification of causal genes within these QTLs can lead to a better understanding of the underlying mechanisms of drought tolerance and aid in breeding, which is, in most cases, not a trivial task.

Using single chromosome recombinant double haploid lines, which contained segments of chromosome 2A from a drought-sensitive cultivar Yanetzkis Probat in the drought-resistant cultivar S29 background, a recent study further dissected QTLs on bread wheat chromosome 2A, identifying 4 clusters along the chromosome. These QTLs clusters affected several traits under normal watering and drought-stress conditions. Notably, these QTL clusters mostly affected enzyme activity, yield components, chlorophyll/carotenoid contents under drought stress and contained stress-responsive TFs such as the BZIP and WRKY families (Pshenichnikova et al. 2021). Another recent study identified 11 QTLs related to drought tolerance in a RIL population derived from a cross between tolerant Reeder and susceptible Albany varieties, of which 6 were exclusive to drought. Some of these loci had major effects on grain yield, explaining up to 14% of the phenotypic variation (Rabbi et al. 2021). A potentially novel, drought-related QTL, QTKW.ndsu.2D.1, identified in this study affected thousand kernel weight (TKW) by 7.7%; however, this QTL likely co-localizes with another QTL in chromosome 2D identified in an earlier study (Rabbi et al. 2021; Touzy et al. 2019).

In maize (*Zea mays*), 18 QTL for drought response is listed in Gramene QTL database (https://archive.gramene.org/qtl/). A very recent study explored associations between SNP genotypes and metabolomic data in a diverse maize population of 385 lines, which included inbred lines and tropic/subtropic/mixed origin lines, in response to drought. Significantly, 3415 genomic loci were associated with drought-responsive metabolites, 21% of which co-localized with previously published drought QTL that used traditional drought response indices (Zhang et al. 2021). This study demonstrates the utility of the metabolomics data, as a new frontier, in discovering potentially new targets for breeding studies. By combining different methodologies, Zhao et al. (2019a, b) proposed 20 stable drought-related QTLs in maize that can be used in breeding studies. Among these, 4 QTL had pleiotropic effects that can facilitate introduction of multiple favorable traits simultaneously. In addition to maize, few studies on *Setaria* species are undertaken as a model for C4 plants. Phenotyping with thermal and optical imaging in a RIL population of a *Setaria italica* x *Setaria viridis* cross revealed 32 QTLs controlling 7 traits under well-watered and drought stressed conditions. Specifically, 3 QTLs on chromosomes 5, 7 and 9 explained 14–21% of the variance on canopy temperature under drought stress, where the QTL on chromosome 9 likely co-localized with another QTL that affected biomass variation by 13.5% (Prakash et al. 2021).

7.2 *Genomic and Molecular Bases of Salt Tolerance*

More than 6% of the world's land area is salinized (accumulation of excess salts), and cultivated land is under the risk of salinization due to climate change and human activities such as excessive irrigation and land clearing (Ismail and Horie 2017). Salt stress has several detrimental effects on germination, total chlorophyll content, growth and development of plants as it reduces root and shoot lengths, plant biomass, and yield of several plants including major crops such as wheat and rice (Hussein et al. 2021; Tabassum et al. 2021). Salinity inhibited/delayed germination in *A. thaliana* (Lin et al. 2013), caused an average reduction of 62% and 78% in root and shoot weights of bread wheat, and may reduce global rice production by 50% (Tabassum et al. 2021). Salt stress affects the plants in two phases: first, osmotic stress blocks the uptake of water by salt-exposed plants thus causing a physiological drought, and later, direct uptake and accumulation of salts that cause sodium toxicity leading to ionic imbalance, leaf injury and death of plants (Ismail and Horie 2017). The cell injury occurs when Na^+ amount surpasses the cells' ability to extrude/exclude the toxic sodium from the cytoplasm that can be achieved by Na + unloading from the xylem, sequestration of cytosolic Na^+ to vacuoles, and Na + extrusion/exclusion from roots or shoots (maintenance of high K^+/Na^+ ratio in leaves) (Ismail and Horie 2017). The majority of crop plants such as cereals are classified as highly sensitive plants to salt stress (glycophytes). Among these, wheat has low to moderate tolerance to salt stress but rice is highly sensitive to salt stress (Ma et al. 2018; Tabassum et al. 2021).

Sodium exclusion (Nax) or sequestration is proposed as a major salt tolerance mechanisms in in many crop plant species (Hussain et al. 2017; Ismail and Horie 2017). Salt tolerance is a multigenic trait. Therefore, QTL and genome-wide association studies (GWAS)-based marker trait associations (MTAs) for Nax at root and shoot level (tissue tolerance) in crop plants such a durum wheat (Shamaya et al. 2017), bread wheat (Asif et al. 2018; Hussain et al. 2017), rice (Platten et al. 2013), wild rice (*Oryza glaberrima*) (Platten et al. 2013), maize (Zhang et al. 2018a, b, c, d), barley (van Bezouw et al. 2019), soybean (Liu et al. 2016), chickpea (Atieno et al. 2021), cotton (Liu et al. 2015), tomato (Asins et al. 2013), and rapeseed (*Brassica napus* L.) (Yong et al. 2015) have been mapped. In durum wheat, Nax QTL/locus (Nax1) was mapped on chromosome 2A (Lindsay et al. 2004). Later mapping studies have found Nax QTL to be associated with candidate genes such as *Nax2* or *High-affinity* K^+ *Transporter B1* (*HKT1;5-B1*) on chromosome 4B in durum wheat (Shamaya et al. 2017). In bread wheat, Nax QTL were associated with *High-affinity* K^+ *Transporter 1* (*HKT1;4*) or *Nax1* on chromosome 2A (Hussain et al. 2017), *HKT1;1* on chromosome 2B, and 2 *HKT* genes (*HKT2;1* and *HKT2;4*) on chromosome 7A (S. Liu et al. 2015). Other candidate genes associated with Nax include *OsHKT1;5* in *Oryza sativa* and *O. glaberrima* mapped on chromosome 1 (Platten et al. 2013), *ZmHKT1* in maize on chromosome 3 (Zhang et al. 2018a), *HvHKT1;5* in barley on chromosome 4H (van Bezouw et al. 2019), and two *HKTs* (*SlHKT1;1* and *SlHKT1;2*) in tomato on chromosome 7 (Asins et al. 2013).

It is quite evident that *HKT* genes are key regulators of salt tolerance in several plant species belonging to diverse families. The *HKT* genes maintain low sodium content in plant leaves. Indeed, the *OsHKT1;5* gene was cloned in indica rice cultivar Nona Bokra using map-based cloning. The *OsHKT1;5* encodes a Na^+-selective transporter that was localized in the parenchyma cells surrounding the xylem vessels and played key role in regulating K^+/Na^+ homeostasis (Nax) in rice (Ren et al. 2005). Similarly, transfer of a key *Nax* gene (*TmHKT1;5-A* or *Nax2*) from wheat relative *Triticum monococcum* into a commercial durum wheat variety Tamaroi significantly reduced Na^+ in its leaves. This gene encodes a plasma membrane-based Na^+-selective transporter in root cells adjacent to xylem, thus its ideal localization helps to exclude/withdraw Na^+ from the xylem that leads to reduced Na + transport to leaves. In this way, the incorporation of *TmHKT1;5-A* into durum wheat increased the grain yield by 25% when grown in saline field (Munns et al. 2012). Additionally, both *Nax1* and *Nax2* (wheat *HKT1;4* and *HKT1;5*) are documented to reduce the Na^+ accumulation in bread wheat leaves under saline and water logging conditions (James et al. 2011).

However, QTL mapping-based crop improvement is scarce due to minor QTL that control small percentage of phenotypes and lack of diversity in candidate genes due to scarcity of QTL studies for ions other than Na^+ and K^+ under salt stress. However, some studies have reported QTLs and candidate genes for the other cations and anions under salt stress. For example, a major QTL for Cl^- accumulation was identified on chromosome 5A of bread wheat that contributed 27–32% to salt tolerance and was associated with potential candidate genes, chloride channel (CLC) and cation chloride co-transporter (CCC). Additionally, 19 QTLs for Ca^{2+} and Mg^{2+} were mapped on the same locus (Yousaf et al. 2016). In another study, several major QTL for root and shoot Nax, K^+, Zn^{2+}, Ca^{2+}, Mg^{2+}, Mn^{2+}, Cu^{2+}, and Fe^{2+} in bread wheat under salt stress have been mapped on chromosome 2A, 3D, 4B and 6A, and 7A and were associated with candidate genes such as *HKT1;4*, *Chloride channel CLC-e*, *potassium transporter 12* (*KUP12*), *K⁺ outward-rectifying channel* (*SKORs*), *Zn transporter 6*, *Mg transporter NIPA4*, *transparent testa 12*, *auxin transport BIG*, *Peroxidase 12*, *Glutathione S-transferase GSTU6*, *auxin response factor 5-like*, *NAC78*, and *ARF21* (Hussain et al. 2017). In crops other than wheat, QTL and MTAs for salt tolerance and Nax were linked with candidate genes like *salt overly-sensitive 3* (*SOS3*) in barley, *Cation/H⁺ exchanger 1* (*CHX1/GmSALT3*) in soybean (Y. Liu et al. 2016), *CaLG04*, *CaLG05*, and *CaLG06* in chickpea (Atieno et al. 2021), *P-type ATPase 1*, *Na⁺/H⁺ antiporter 1* (*NHX1*), and *Potassium channel AKT2/3* in cotton (Liu et al. 2015), and *RNA binding protein Tudor-SN 1* (*TSN1*) (Yong et al. 2015) in *Brassica napus*.

Additionally, transcriptomics profiling in chickpea and potato (Li et al. 2020a, 2020b) under salt stress revealed the complexity of salt tolerance in plants, and it is conferred by several members of different gene families e.g., *SOS1/Na⁺/H⁺* antiporter, *SOS2/CIPK24*, *SOS3/CBL4*, *SOS4/pyridoxal (PL) kinase*, and *SOS5* genes have key role in Nax under salt stress in chickpea (Li et al. 2020a, b). The above discussion highlights the involvement of multiple gene families and their multiple members in salt tolerance response in plants. Therefore, here we briefly discuss different gene families that are involved in K^+/Na^+ homeostasis in plants and their role in salt tolerance in plants.

There are five major gene families (also classified into seven families) in model plant *A. thaliana* and crop plants that control K^+/Na^+ homeostasis under salt stress (Table 1). These include: (a) High-affinity K+ transporter (HKT) gene family: the members of this gene family are involved in Na^+ unloading in the root xylem and Na^+ loading in the shoot phloem and their overexpression or introduction through cross breeding confers salt tolerance to plants (Munns et al. 2012; Sunarpi et al. 2005). Only 1 *HKT* gene family member has been identified in *A. thaliana* and maize (Sunarpi et al. 2005; Zhang et al. 2018a, b, c, d) while 6 *HKT* members were identified in rice (Ariyarathna et al. 2016) and bread wheat (Ariyarathna et al. 2016; James et al. 2011); (b) Na^+/H^+ exchangers (NHX) gene family: some members of this gene family are situated at tonoplast and control vacuolar K^+ homeostasis while other occur at plasma membrane and exclude sodium from the root (Fukuda et al. 2011; Sharma et al. 2020a, b) (c) Cation/H^+ exchangers (CHX) gene family: different members of this gene family control K^+ accumulation in the root, respond to K+

Table 1 The Arabidopsis, bread wheat, rice and maize K+ channels and transporters with a function in transport of K+ and Na + and K+/Na + homeostasis during salt stress

Family	Abbreviation	Total members[a]	Selectivity	Cellular localization	Functions under salt stress	Reference
High-affinity K^+ transporter	HKT	1, 6, 6, 1	Na^+	Plasma membrane	Na^+ unloading in the root xylem; Na^+ loading in the shoot phloem	Ariyarathna et al. (2016), Chen et al. (2021a), James et al. (2011), Munns et al. (2012) and Sunarpi et al. (2005)
Na^+/H^+ exchangers	NHX	8, 35, 5, –	$K^+/Na^+ \leftrightarrow H^+$	Tonoplast; plasma membrane	Vacuolar K^+ homeostasis; Nax from the root	Barragán et al. (2012), Fukuda et al. (2011) and Sharma et al. (2020a, 2020b)
Cation/H^+ exchangers	CHX	28, 48, 17, 15	$K^+/Na^+ \leftrightarrow H^+$	Plasma membrane	K^+ accumulation in the root; response to K^+ deficiency; leaf Na^+ concentrations	Cellier et al. (2004), Sharma et al. (2020a, b) and van Zelm et al. (2020)
K^+ efflux antiporters	KEA	6, 24, 4, 6	$K^+ \leftrightarrow H^+$	Chloroplast	Affects chloroplast function and integrity; pH regulation, and osmoregulation	Chanroj et al. (2012), Kunz et al. (2014) and Sharma et al. (2020a, b)
K^+ uptake Permease/ high-affinity K^+ Transporter/K^+ Transporter	KUP/HAK/ KT	13, 56, 27, 27	K^+/Na^+	Plasma membrane	K^+ uptake under saline and low K^+ conditions	Ariyarathna et al. (2016), Barragán et al. (2012), Cellier et al. (2004), Chanroj et al. (2012), Cheng et al. (2018), Fukuda et al. (2011), Kunz et al. (2014), Li et al. (2020a, b), Sharma et al. (2020a, b) and Sunarpi et al. (2005)
Two-pore K^+ Channel	TPK	6, 1, 3, –	K^+	Tonoplast	Ca^{2+}-dependent activation and phosphorylation	Borjigin et al. (2021) and Latz et al. (2013)
Shaker-type K^+ Channels	Kv-like/KAT	9, –, 6, 2	K^+ influx	Plasma membrane	Uptake of K^+ in the root; guard cell K^+ uptake	Sato et al. (2009), Su et al. (2005) and Ward et al. (2009)

[a]The number of genes given in table from left to right: *A. thaliana*, bread wheat, rice, and maize, while "–" means the gene members have not been studied in specific plant; where Nax = sodium exclusion

deficiency, and affect leaf Na^+ concentrations under salt stress (Cellier et al. 2004; Sharma et al. 2020a, b); (d) K^+ efflux antiporters (KEA) gene family: members affect the chloroplast function and integrity, pH regulation, and osmoregulation (Chanroj et al. 2012; Kunz et al. 2014; Sharma et al. 2020b); (e) K^+ uptake Permease/High-affinity K^+ Transporter/K^+ Transporter (KUP/HAK/KT) gene family: members are involved in K^+ uptake under saline and low K^+ conditions (Cheng et al. 2018; Zhang et al. 2012); (f) Two-pore K^+ channel (TPK) gene family: it is involved in Ca^{2+}-dependent activation and phosphorylation under soil salinity (Borjigin et al. 2021; Latz et al. 2013); (g) Shaker-type K^+ channels (Kv-like/KAT) gene family: it is involved in uptake of K^+ in the root and guard cells (Sato et al. 2009; Su et al. 2005). The total number of genes in the above-mentioned genes families in *A. thaliana*, bread wheat, rice and maize, their selectivity, and cellular localization are provided in the Table 1. The search of these genes in the mapped QTL regions, their subsequent cloning as reported previously (Munns et al. 2012; Ren et al. 2005) and overexpression can significantly speed up the breeding efforts in crop plants.

7.3 Genomics Studies for Heat Tolerance

Due to their sedentary nature, plants must cope with adverse environmental stressors throughout their lives. While climate change has been one of the hot topics of the last decade, the most recent studies predict that the impact of climate change on crop production will be much more dramatic than anticipated in the previous years (Jägermeyr et al. 2021). Among the severe outcomes of climate change, temperature extremities have been causing probably the most damage to agricultural production. For the continuity of sustainable agriculture, studies focused on the development of heat-tolerant plants are increasing each day as the Earth's temperature is expected to increase 1.5 °C between 2030 to 2052 (IPCC 2018). A study from 2017 has shown the reduction in yield of major crops due to elevated temperatures and estimated the average reduction in global yield for each degree Celsius increase in Earth's temperature as 6% for wheat, 3.2% for rice, 7.4% for maize, and 3.1% for soybean, globally.

Plants are exposed to heat stress at different growth stages. In rice, the grain filling period is when the grain quality is determined but at the same time, it is the most affected period by temperature changes (Liu et al. 2021). Exposure to elevated temperatures has been shown to cause non-favored results such as increased grain chalkiness (Tang et al. 2018), spikelet sterility, and even reduced grain filling (Yan et al. 2021a, b) in higher temperatures. However, the effects of heat in wheat includes limited photosynthesis, increased respiration through stomatal regulation, and reduced kernel weight (Su et al. 2019; Wahid et al. 2007).

Exposure to heat denatures various plant proteins. At normal temperatures, heat shock proteins (HSPs) bind to heat stress transcription factors (HSFs) and inhibit the expression of heat-responsive genes that need to be activated during heat stress. However, these genes are activated upon exposure to high-temperature stress, when misfolded proteins accumulate and bind to heat shock proteins, thus freeing HSFs that activate heat stress responses (Zhang et al. 2022a, b). Heat Shock Proteins (HSPs) are complex and mostly chaperon proteins responsible for the prevention of degradation, denaturation, and aggregation of proteins as a result of stress conditions, and vital for plants for surviving high temperatures (Waters and Vierling 2020). HSPs are classified into five groups: small HSPs (or HSP20), HSP60, HSP70, HSP90, and HSP100, and upregulation of especially sHSPs, HSP60 or HSP70 is associated with improved heat stress tolerance (Xu et al. 2011). Crossing of a heat-tolerant maize line, ZPBL 1304, that synthesizes a 45 kDa HSP with a heat-sensitive maize line, ZPL 389, which lacks the protein provided evidence to this association as in the F_2 generation of this cross, the synthesis of 45 kDa HSP, thus increased ability to recover from heat stress observed. Protein refolding is an important heat stress tolerance mechanism in plants, as well as protein degradation involving various proteases (Huang and Xu 2008).

As a complex trait, heat tolerance is controlled by many QTLs. Many QTL mapping and GWAS studies have successfully uncovered chromosomal loci involved in heat tolerance and reverse genetics approaches identified the causal genes (Wen et al. 2019). Wheat is quite susceptible to heat stress damages as a cool-season crop (Zhai et al. 2021) and damage of heat stress especially when combined with drought may be detrimental to it. Heat tolerance in wheat is controlled by many minor QTL as understood so far (Sun et al. 2022) and there are currently 1812 and 396 heat stress tolerance-related QTL listed in the WheatQTLdb database (v2.0) for hexaploid bread wheat (*Triticum aestivum*) and tetraploid durum wheat (*Triticum durum*), respectively (http://wheatqtldb.net/heat_new.php).

A novel heat stress tolerance locus in wheat that involves both seedlings and adults, TaHST1, is mapped on long arm of chromosome 4A. This region has been shown to possess many deletion polymorphisms in wheat and impairment of TaHST1 locus results in a significant reduction in grain traits such as grain number per spike, grain weight per spike, grain length, and grain width. There are 19 high confidence genes in this locus that are shown to involve in vegetative and reproductive growth of wheat under heat stress conditions (Sun et al. 2022). Various quantification methods are applied when measuring the effect of plant stress. F_v/F_m is one of them which shows the efficiency of light reaction in plants, and any stress that causes photoinhibition lowers the F_v/F_m value (Murchie and Lawson 2013). In wheat, 2 QTLs on 3B chromosome and 1 QTL on the 1D chromosome was identified and shown to increase F_v/F_m of F_2 individuals of crosses between a heat susceptible female parent and heat tolerant male parents. Along with these QTLs, 12 candidate genes that may be directly involved in photosynthesis and/or heat tolerance were proposed (Shamaya et al. 2017). In another study, 144 double haploid lines, derived from Australian wheat varieties Drysdale and Waagan were used to identify two QTL, QHsgw.aww-3B and QHsgw.aww-6B, influencing grain size upon heat stress applied at grain filling period. Both QTL were shown to help preserve flag leaf chlorophyll under heat, thereby increasing plant tolerance (Shirdelmoghanloo et al. 2016).

Exposure to heat stress during flowering times damages the organ viability in rice and wheat, therefore efforts are directed towards shifting flowering time to lower temperatures, either naturally or by genetic interventions (Krishna Jagadish 2020). Early morning flowering (EMF) phenotype may provide some degree of heat stress avoidance (Sheehy et al. 2005) and mitigation of spikelet sterility (Ishimaru et al. 2010). A major QTL on EMF trait in rice, qEMF3, has been identified on chromosome 3 by a marker-assisted breeding approach using wild rice, *O. officinalis*, as the donor parent. qEMF3 has been proposed as a candidate to be included in breeding programs as it's able to provide a shift in flowering time and contribute to heat escape (Hirabayashi et al. 2015). In a 2021 study, a new QTL, qHTT8, for heat tolerance at the flowering period of rice has been defined. Two candidate causal genes LOC_Os08g07010 and LOC_Os08g07440 have been identified on chromosome 8 of $F_{2\text{–}3}$ of crosses between heat-tolerant and heat-sensitive parents (Chen et al. 2021b).

Thermo-tolerance 1(TT1) is a major heat tolerance QTL identified in the rice. *TT1* gene is a highly conserved subunit of 26S proteasome and contributes heat tolerance by eliminating cytotoxic denatured proteins (Li et al. 2015a, b). In a more recent study, an allelic gene to *TT1*, *TT1-2*, has also been identified through map base cloning, using a heat-tolerant donor rice plant. Crossing and genotyping of heat-tolerant N22 and heat-susceptible IR64 rice varieties identified *rlht5.1* QTL for "root length under heat stress" with 20.4% effect and three QTL for "shoot length under heat stress" with an effect of 5.3–10.2%. From a total of 10 QTLs identified for vegetative stage stress tolerance, many putative candidate genes including chaperons, transcription factors, and proteases were annotated (Kilasi et al. 2018).

8 Importance of Organelles in Abiotic Stress Response

Perturbations in organellar function serve as stress sensors that must be accompanied with proper changes in order to maintain cellular homeostasis. Various organelles and cellular compartments can sense and respond to stress conditions; among these, the Unfolded Protein Response (UPR) by ER is a prominent example that targets stress-induced disruption of protein folding, by either promoting chaperons to aid folding or by activating degradation pathways for misfolded proteins (Zhu 2016). Arguably, however, abiotic stresses hit the energy metabolism hardest, through an immediate reduction in overall efficiencies of both photosynthesis and respiration. In response, cells must shift the flow of energy from biosynthetic pathways to stress acclimation until the energy supply is restored (Crawford et al. 2018). The powerhouses of the cell, chloroplasts and mitochondria, are unique in their involvement in stress responses. While the vast majority of proteins that function in these organelles are nucleus-encoded, a minority of proteins crucial to the organelle functioning are encoded by the reduced genomes they still retain from their endosymbiotic origins. A tight coordination between the nuclear and organellar genomes is essential for the survival of the cell. This coordination is ensured by continuous flow of information from these organelles to nucleus, termed as retrograde signaling, and from nucleus to these organelles, known as anterograde signaling.

8.1 Chloroplasts as Generators and Responders of Stress Conditions

Chloroplasts constantly generate oxidative stress as part of the biosynthetic processes taking place within the organelle, including photosynthesis. Since the photosynthetic machinery is highly sensitive to environmental changes, abiotic stresses can readily cause over-excitation of the photosynthetic electron transport chain, leading to an excess of ROS and an oxidative burden that requires further action to overcome. As such, chloroplasts, the major metabolic hubs responsible for the carbon fixation, as well as for the biosynthesis of phytohormones, aromatic amino acids, fatty acids and secondary metabolites, can quickly become a serious threat to the survival of the cell under abiotic stress conditions (Mamaeva et al. 2020; Woodson 2016). Chloroplast responses to abiotic stress conditions include the generation of retrograde signals, suppression of chloroplast import and remodeling the chloroplast proteome.

8.2 Chloroplast Retrograde Signaling

En route to becoming organelles, chloroplasts transferred the majority of their genes to the nuclear genome of the host cell. Consequently, chloroplast genomes of extant plant species typically contain ~100 genes for the thousands of proteins that function inside these organelles. The rest is nuclear-encoded and must be imported into chloroplasts in a highly coordinated manner. Thus, chloroplast to nucleus (retrograde) and nucleus to chloroplast (anterograde) communication are crucial to the proper functioning of the cell. Chloroplast retrograde signaling is an ongoing process that is required for the normal growth and development of a plant, as well as acclimatization to various environmental conditions. Retrograde signaling is generally classified as biogenic, enabling developing chloroplasts to communicate with nucleus for the expression of mostly photosynthesis-related genes, and operational, emanating usually from fully developed chromosomes in response to changes in the environment, including stress factors (Woodson 2016). While biogenic signals are generally related to chloroplast development, genes that are involved often have roles in stress responses as well (Xu et al. 2020; Xu et al. 2019). Notably, the evolution of retrograde signaling might have contributed to the successful colonization of the lands by plants, by enabling them to sense and respond to changes in their terrestrial environment (Zhao et al. 2019a, b).

Under stress conditions, retrograde signals from chloroplast notify the nucleus to primarily downregulate the expression of photosynthesis-related genes to prevent oxidative damage and reduce energy waste (Woodson 2016). Earliest candidates of chloroplast to nucleus retrograde communication were isolated in a forward genetic screen that identified *genomes uncoupled (gun)* mutants that maintain the expression of nuclear-encoded photosynthesis-related genes when chloroplast development is impaired. In Arabidopsis, 6 nuclear *gun* genes code for plastid-localized proteins. While GUN2-GUN6 are involved in tetrapyrrole biosynthesis that is responsible for the production of siroheme, heme, phytochromobilin, chlorophyll a and chlorophyll b (Crawford et al. 2018; Larkin 2016), GUN1 accumulates under stress conditions and is involved in many biological processes, including chloroplast development, differentiation and homeostasis (Shimizu and Masuda 2021; Wu et al. 2018). Perturbations in tetrapyrrole metabolism has been clearly linked to retrograde signaling, though the exact mechanisms are still under some controversy. GUN1, on the other hand, operates through a separate but partially overlapping mechanism with the other GUN proteins and is likely a mediator of multiple biogenic and operational retrograde signals (Jiang and Dehesh 2021; Wu et al. 2018).

Beyond photosynthesis, plastid metabolites can act as operational retrograde signals to induce the expression of stress responsive genes. An intermediate in the methylerythritol phosphate (MEP) pathway for iseprenoid production, methylerythritol cyclodiphosphate (MEcPP) accumulates under certain stress conditions. MEcPP is a known chloroplast-to-nucleus retrograde signal which activates a number of nuclear-encoded general stress response genes, as well as UPR by the ER, both through the action of CALMODULIN-BINDING TRANSCRIPTIONAL ACTIVATOR 3 (CAMTA3) (Crawford et al. 2018; Jiang and Dehesh 2021). MEcPP also affects hormone and light signaling, once again emphasizing the intermingled nature of plant growth and stress responses (Jiang and Dehesh 2021). Another known retrograde signal is 3′-phosphoadenosine 5′-phosphate (PAP), a by-product of sulphur metabolism that accumulates in response to stresses, particularly drought. PAP is normally converted to adenosine monophosphate (AMP) by INOSITOL POLYPHOSPHATE 1-PHOSPHATASE (SAL1) which is deactivated by oxidative stress. Disulphide bridges within the ultrastructure of SAL1 suggest that this enzyme may also act as a redox sensor inside the plastids (Crawford et al. 2018; Watson et al. 2018). Intriguingly, MEcPP and PAP-mediated retrograde signaling may converge over light signaling through the stability and abundance of PHYTOCHROME B (PHYB) (Jiang and Dehesh 2021).

Changes in the redox status and ROS metabolism are also known triggers of chloroplast retrograde signaling (Suzuki et al. 2012). In particular, singlet oxygen ($^1O_2$) is generated at Photosystem II (PSII) reaction center during photosynthesis. Singlet oxygen can also be generated by photoactive chlorophyll precursors upon chloroplast dysfunction. Accumulation of singlet oxygen, as well as non-enzymatic breakdown products of β-carotene, an iseprenoid quencher of singlet oxygen, can specifically activate a number of nuclear genes, known as singlet-oxygen regulated genes (*SORGs*) (Crawford et al. 2018; Suzuki et al. 2012; Watson et al. 2018). Importantly, singlet oxygen can also mark damaged chloroplasts for degradation through ubiquitin-proteasome system or induce programmed cell death, by interacting with chloroplast-localized proteins EXECUTOR 1 and 2 (EX1 and EX2) (Watson et al. 2018; Woodson 2016). Superoxide anion radical ($O_2^{-\cdot}$) and hydrogen peroxide (H_2O_2) also have well-known signaling properties but hydrogen peroxide is the only ROS that can cross membranes and potentially trigger defense responses in neighboring cells. Overall, it is likely that different ROS activates different responses in a specialized or generic manner, by themselves or by crosstalk with other ROS (Crawford et al. 2018).

8.3 *Crosstalk in Organellar Retrograde Signaling*

Mitochondria, the second plant organelle with an endosymbiotic origin, is another source of oxidative stress through the respiratory machinery and contributes significantly to stress responses. Although mitochondrial retrograde signaling is relatively poorly understood, a number of pathways that provide integration of stress signaling from mitochondria and chloroplasts have been described. A hallmark enzyme, nuclear ALTERNATIVE OXIDASE 1 (AOX1) is significantly upregulated in response to stress conditions or mitochondrial electron transport chain inhibitors. Upon mitochondrial dysfunction, AOX1 provides an alternative route in the electron transport chain, thereby protecting against the generation of oxidative stress, at the expense of proton motive force, and hence ATP production (Dourmap et al. 2020). A regulator of AOX1 expression is the transcription factor ABSCISIC ACID INSENSITIVE 4 (ABI4), which is also involved in GUN1-mediated downregulation of photosynthesis-related genes, placing ABI4 at the crossroads of chloroplastic and mitochondrial retrograde signaling (Crawford et al. 2018; Suzuki et al. 2012). Chloroplast and mitochondria retrograde signals may also converge in the nucleus via the Mediator transcriptional co-activator complex. CYCLIN DEPENDENT KINASE E1 (CDKE1) is a component of the Mediator complex, which can sense retrograde signals from both chloroplast and mitochondria and is pivotal to the balance between growth and stress responses (Crawford et al. 2018; Ng et al. 2013).

8.4 *Chloroplast Import and Proteolysis Machineries as Players of Stress Responses*

Post-translational regulation of the import of pre-proteins into chloroplasts has recently been recognized as a means of protection against the effects of abiotic stresses. The SUPPRESSOR OF PPI1 LOCUS1 (SP1), an E3 ubiquitin ligase that resides in the chloroplast outer membrane, is a member of the ubiquitin-proteasome system that regulates Translocon at the Outer envelope membrane of Chloroplast (TOC) components in during development. Under stress conditions, SP1 causes a rapid reduction in the abundance of TOC75, a member of the TOC complex responsible for the import of both photosynthetic and non-photosynthetic proteins into chloroplasts, thereby reducing photosynthesis and the concomitant production of excessive ROS. SP1 overexpression protects Arabidopsis plants against salt and oxidative stresses, but not against temperature or high-light stresses, even though the SP1 expression is responsive to various abiotic stress conditions. This suggests redundant mechanisms that may act on chloroplast import machinery under some abiotic stress conditions (Ling and Jarvis 2015).

Chloroplasts contain a number of proteases located in the inner membrane, stroma, thylakoid membrane and thylakoid lumen, occupying virtually all suborganellar compartments. While some of these proteases have specialized functions such as the STROMAL PROCESSING PEPTIDASE (SPP) and methionine aminopeptidases (MAPs), which are responsible for the removal of N-terminal transit peptides and methionines, respectively, others have roles, including the removal of damaged proteins, that are important for chloroplast development and stress responses (Mamaeva et al. 2020; Watson et al. 2018). For instance, the thylakoid-associated proteases DEGRADATION OF PERIPLASMIC PROTEINS 1 (DEG1) and FILAMENTATION TEMPERATURE-SENSITIVE H (FtsH) are involved in the elimination of photo-damaged photosynthetic proteins under high-light stress conditions. Chloroplast proteases contribute significantly to the remodeling of chloroplast proteome under stress conditions and, notably, they can generate peptide by-products that can act as retrograde signals (Mamaeva et al. 2020).

8.5 The Potential of Chloroplast Engineering in Improvement of Stress Tolerance

In a recent study, an assessment of published microarray datasets from Arabidopsis revealed that genes encoding chloroplast-localized proteins make up 10–20% of all differentially regulated genes in response to abiotic stresses. Due to their extensive involvement in stress responses, manipulation of chloroplasts and chloroplast-related pathways is an attractive tool to engineer plants with improved resilience. Since plant chloroplasts are challenging to manipulate, cyanobacterial or algal systems may provide easier access to identify candidates. In particular, chloroplasts, due to their cyanobacteria-like endosymbiont history, retain considerable similarities in photosynthetic machinery to the cyanobacteria *Synechocystis* sp. PCC6803 and an overall closer phylogenetic relationship to the model green algae *Chlamydomonas reinhardtii*. Albeit a few, these organisms have been used in experimental evolution studies as proxies to the chloroplast function, to identify favorable alleles for traits including stress tolerance (Leister 2017). Importantly, the facultative heterotroph *Chlamydomonas*, with much shorter generation time than plants, can serve as a productive testing ground, for modulating the chloroplast function, including or excluding the photosynthetic machinery (Jackson et al. 2021). Direct manipulation of the chloroplast genome is not trivial, due to the presence of the double membrane enclosing the organelle that is not permeable to several molecules. On the other hand, editing the chloroplast genome may be favorable, mainly due to (i) high efficiency of and a preference for homologous recombination, enabling accurate insertion of the transgene, (ii) high expression of the transgene due to the high copy nature of the organellar genomes, and (iii) the lack of epigenetic regulation that may otherwise silence the transgene. While the manipulation of mitochondrial genome has so far been too challenging, successful application of chloroplast genome engineering have been reported in potato, where integration of *codA* gene from *Arthrobacter globiformis* into the plastid genome enhanced drought tolerance through the accumulation of glycine betaine within plastids (Mahapatra et al. 2021). The transgenic plants had higher chlorophyll content and photosynthetic rate, enhanced activity of antioxidant enzymes and better cell membrane integrity compared to wild-type plants under drought stress. Accumulation of glycine betaine inside the chloroplasts, as opposed to nuclear integration and cytoplasmic expression of the transgene, provided critical protection of the photosynthetic apparatus under drought stress (You et al. 2019). Similarly, the expression of yeast *ArDH* gene from chloroplast genomes in tobacco enhanced tolerance against drought and salt stresses through osmoprotection (Khan et al. 2015). Besides genome editing, modulating the chloroplast metabolism will likely open up future possibilities in crop improvement. An immediate target will be the photosynthetic machinery and its major enzyme Rubisco, catalyzing the initial step of CO_2 fixation, due to its inefficiencies in catalysis and discrimination between CO_2 and O_2. A promising strategy involves increasing CO_2 availability inside chloroplasts by introducing HCO_3^- transporters in the inner-envelope membrane to increase overall efficiency of carbon fixation. Strategies to increase CO_2 availability for the photosynthetic machinery have already been materialized in alien bacterial systems (Mahapatra et al. 2021).

9 Nanoparticles in Abiotic Stress Tolerance Improvement

Nanoparticles (NPs) are various structures of matter having sizes around 1–100 nm at least in one or more dimensions. They can be produced and engineered in different shapes, i.e., core or core-shell, single or multiple sheets, dots, triangles, spheres, conics, singular or multiwalled-tubes, cylindrical hollows, hexagons, using various "bottom-up" (sol-gel, spinning, chemical vapor deposition, pyrolysis, biosynthesis) and "top-down" (nanolithography, mechanical milling, laser ablation, sputtering, thermal decomposition) physical, chemical, and mechanical approaches. They are classified as organic (liposomes, micelles, dendrimers, polymer-based other organic NPs), inorganic (usually metal and metal oxide-based NPs, i.e., gold, copper, iron, silver, aluminum, cobalt, cadmium, iron oxide, aluminum oxide, cerium oxide, magnetite, silicon oxide, and titanium oxide), and carbon-based nanomaterials (carbon dots, graphene and derivatives, carbon nanotubes and derivatives, fullerenes, and nanofibers) based on the primary substrate used in the production (Ealias and Saravanakumar 2017). Unique mechanical, physicochemical, and electrical properties of size-dependent NPs make them superior to their bulk counterparts, for example, high surface area to volume ratio, increased surface reactivity, surface plasmon and localized surface plasmon effect in some metallic particles, extraordinary magnetic properties of nanoscale magnetic materials, and effect of quantum confinement in semiconductors (Khan et al. 2019).

Nanoparticles have been exploited in several fields so far, including electronics and information technology (transistors in nanoscale, random access memory chips, ultra-high-definition displays, flexible and wearable electronics), healthcare and medicine (ultra-small disease diagnosis tools, empowered imaging agents, novel gene sequencing technologies, targeted or encapsulated therapeutics), energy (new catalysts for fuel production, nanostructured solar cells, high-power rechargeable battery systems), environmental remediation (wastewater treatment, hazardous material detection, energy-efficient desalination, air, and soil quality monitoring sensors), transportation (discovery of lightweight and high-strength materials), and everyday materials and processes. Agriculture is another field where nanotechnology has opened new avenues for sustainable crop development and yield improvement. Some NPs were shown to promote plant growth by acting as herbicides, nano-pesticides, and nano-fertilizers (Saxena et al. 2016). Also, they have been shown to enhance plants' tolerance against drought, salinity, and mineral toxicity/deficiencies, known as the significant causes of abiotic stress (Rajput et al. 2021). Among many FeO, ZnO, TiO2, and SiO2 NPs are the ones that have shown promising results in tolerance enhancement in different plant species under a defined stress condition. Here we briefly describe and discuss those NPs recently employed in agricultural studies to combat abiotic stress.

The structure, composition, and characteristic features of FeO-based nanoparticles employed in agricultural studies were reviewed by Rehman and colleagues. The authors further elaborated the application methods of FeO NPs, possible NP uptake mechanisms, corresponding plant responses, and harmful effects at high concentrations. Here we gave only the most recent examples for FeO NPs that successfully improved stress tolerance in tested model plants. FeO (Fe_2O_3/Fe_3O_2) NP treatment in *O. sativa* alleviated the Ar phytotoxicity, increased the synthesis of glutathione (an intracellular thiol molecule acting as a non-enzymatic antioxidant), proline (an amino acid produced more under stress conditions), and phytochelatins (oligomers of glutathione that are important for metal detoxification), and it improved the plant dry mass and length as summarized in Fig. 2 (blue panel) (Bidi et al. 2021). Another

Fig. 2 (continued) Effect of Fe NPs (0, 25, and 50 mg/L) on the activity of catalase (CAT, A), superoxide dismutase (SOD, B), ascorbate peroxidase (APX, C), and glutathione reductase (GR, D) enzymes in rice plants treated with arsenic (As, 0, 25 and 50 μM). Values (means ± SD, n = 5) followed by the same letter are not significantly different ($P < 0.05$; LSD test). Adapted with permission from Bidi et al. (Bidi et al. 2021). **Green panel:** Nanoparticle immobilization on a leaf surface and possible internalization parameters and mechanisms. NPs are shown as black spheres with a red outline. Different leaf structures: the epidermis (ep); mesophyll (me); parenchyma palisade (p.p); spongy parenchyma (s.p); xylem (xy); phloem (ph); sheath cells (sh); vasculature bundle (vb). (Adapted with permission from Avellan et al. 2021)

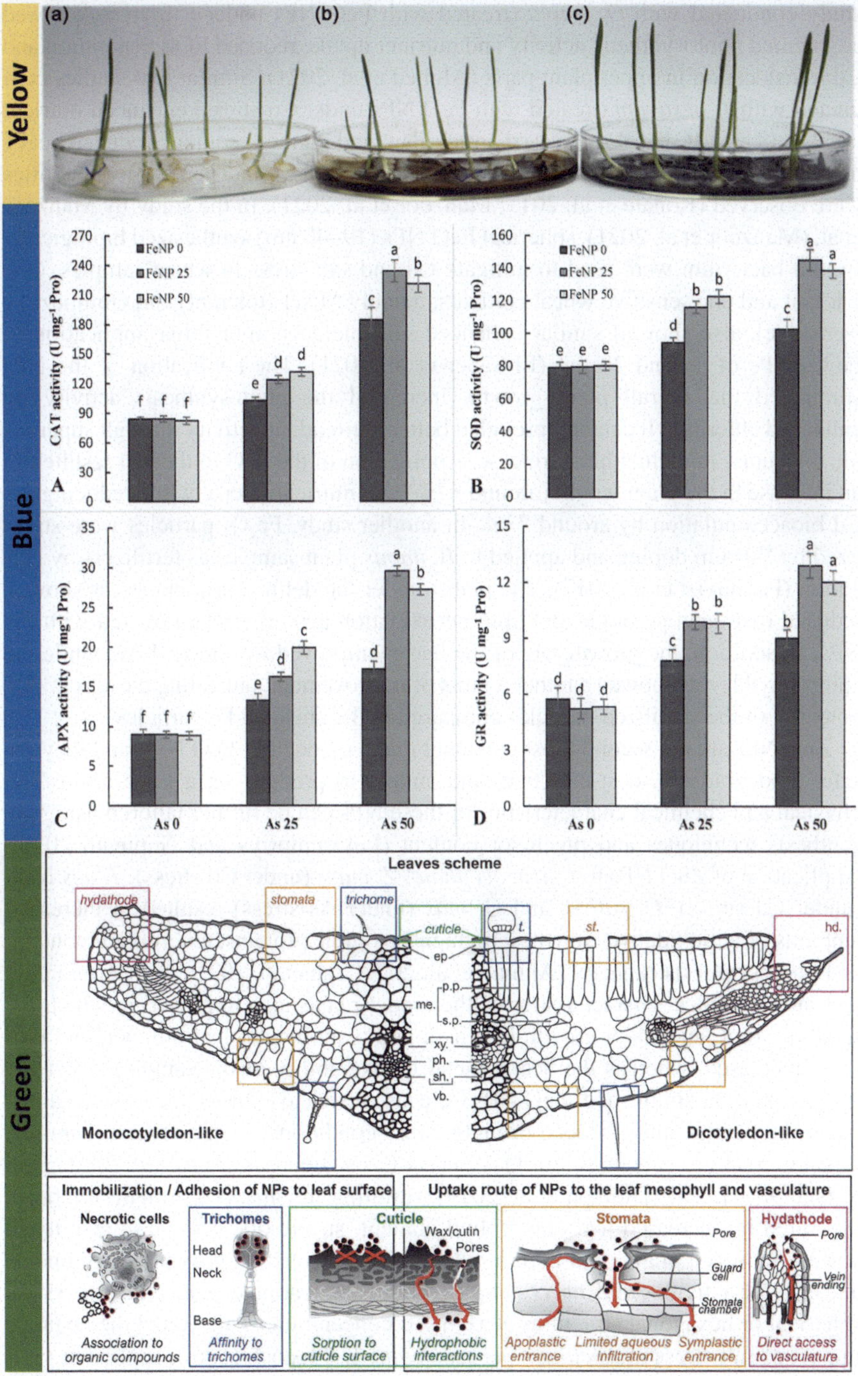

Fig. 2 Examples of NPs employed in agricultural studies for mitigating abiotic stress. **Yellow panel:** Phenotypes of 4-day-old wheat seedlings grown on medium with (**a**) water, 2000 μg/mL of (**b**) GO, and (**c**) G-NH2. Adapted with permission from Chen et al. (Chen et al. 2018). **Blue panel:**

study conducted with *O. sativa* (treated with FeO NPs under Cd stress) showed augmented photosynthetic activity and nutrient uptake reduced ROS generation and Cd translocation in upper plant parts (Ahmed et al. 2021). Similar two studies conducted with *T. aestivum* (treated with FeO NPs under Cd stress) enhanced nutrient uptakes like N, P, and K, increased antioxidant and pigment content, reduced heavy metal transport as well as better superoxide dismutase and peroxidase activities were observed (Konate et al. 2017; Manzoor et al. 2021). In the study by Manzoor et al. (Manzoor et al. 2021), spherical FeO NPs (19–40 nm) synthesized biologically from a bacterium were used to mitigate Cd and salt stress in wheat samples. One tolerant and one sensitive wheat cultivars, namely Misr1 (tolerant) and Gimmeza11 (sensitive), also showed similar enhanced salt tolerance after foliar application of Fe_3O_4 NPs of around 22 nm (El-Saber et al. 2021). The application of the NPs stimulated the overall plant growth, increased the photosynthesis activity by enhanced chlorophyll content, provided better antioxidant activity through superoxide dismutase and glutathione release. Application of the FeO to the soil resulted in an increase in the plant length (around 37%) and nutrient uptake while reducing the Cd bioaccumulation by around 73%. In another study, Fe_2O_3 particles were stabilized by Yttrium doping and applied to *B. napus* plant samples as fertilizers by irrigation (Palmqvist et al. 2017). The plants under the defined drought stress showed reduced hydrogen peroxide and lipid peroxidation activities when treated with the NPs. In addition, the growth rate of the leaves improved by almost 17%, while the chlorophyll levels showed another 5 units of improvement, indicating the agronomic potential of the stabilized particles compared to the chelated Fe particles.

ZnO NPs are the second most prevalent metal oxide after FeO NPs, and they are safe, biodegradable, cost-effective, and simple to produce on a large scale. The physical and chemical characteristics of these NPs can be further tailored using the synthesis techniques and precursor content (Lakshmipriya and Gopinath 2021). Application of ZnO NPs to *L. leucocephala*, *Z. mays* (under Cd stress), *T. aestivum* (under Cd stress), *O. sativa*, and *G. max* (under As stress) resulted in increased amounts of antioxidative enzymes, pigments, soluble proteins, chlorophyll content, and photosynthesis capacity (Ahmad et al. 2020; Yan et al. 2021a, b). Germination and morphological characteristics of the model plants also improved in which the seedling, root, and shoot growth fastened, and plant height, leaf number, and biomass increased, whereas the heavy metal uptake ability of the samples under the stress condition diminished. ZnO NPs were also tested to enhance the growth capacity of the plants under induced drought stress conditions, such as *T. aestivum* var. Lassani-2008 (Adrees et al. 2021), *Z. mays* L. cv. Jidan 27 (Sun et al. 2021), and *S. bicolor* var. 251 (Dimkpa et al. 2019), resulting in enhanced drought tolerance measured by increased leaf chlorophyll content, superoxide dismutase activity in wheat, improved grain yield in *S. bicolor* and augmented photosynthetic pigments and enzyme activities in *Z. mays* samples. Foliar application of ZnO NPs (16–35 nm spherical or hexagonal structures at 10 mg/L concentration) alleviated the negative impact of salt stress in *Abelmoschus esculentus* L. Moench (okra), applied at various concentrations (Alabdallah and Alzahrani 2020). Enhanced superoxide dismutase, catalase activities, and photosynthetic content were observed as the direct beneficial impacts of the applied green synthesized NPs, reducing the total soluble sugar content and proline levels. Treatment of *H. vulgare* L. seedlings directly with ZnO NPs (32 nm, 1, 2, 3, and 4 mg/L concentrations) resulted in enhanced germination and the shoot/root elongation (Plaksenkova et al. 2020). However, it was also noticed that the increased NP concentration caused apparent cell death, especially in the roots, indicating the importance of dose studies and variation of the overall impact in different plant species.

TiO_2 NPs, on the other hand, could be perceived as an inducer for several plants that activate different metabolic pathways involved in stress tolerance. The factors that affect the biological function of these NPs are the same as the previous NPs, like size, shape, and dose (Gohari et al. 2020). Foliar application of TiO_2 NPs to hydroponic *Z. mays* samples reduced Cd bioaccumulation, prompted antioxidant enzyme activities, and upregulated metabolic pathways, such as aspartate, alanine, glycine, threonine, serine, and galactose (Lian et al. 2020). The dry weight of the shoot and root samples under Cd stress also increased following the NP treatment. On the other hand, Ar toxicity was mitigated in *V. radiata* by exogenous TiO_2 NP treatment (Katiyar et al. 2020). In this study, authors also compared chemically synthesized NPs to those synthesized in a greenway in which the green NPs proved to be more effective in upregulating the antioxidant enzyme gene production. Gohari et al. (2020) conducted a greenhouse experiment with *D. moldavica* (an aromatic, medicinal plant) samples treated with different salinity stress conditions. The NPs at different concentrations were introduced three continuous days to Hoagland solution 2 weeks after salinity stress application. Application of TiO_2 NPs (spherical-like particles, 20–30 nm in size, at 100 mg/L) increased antioxidant enzyme activity, reduced H_2O_2 concentration, and yielded the highest essential oil concentration. In the report by Faraji and Sepehri (Faraji and Sepehri 2019), a parametric experiment was conducted with TiO_2 (0–2000 mg/L) and sodium nitroprusside NPs on wheat seedlings under PEG-stimulated drought stress. The application of the selected NPs alone or in combination resulted in a significant enhancement in germination rate, shoot and root lengths, and fresh mass, which overall mitigated the harmful effects of the stimulated drought stress on early seedling growth and germination of the wheat samples. It should also be noted that the growth-enhancing effects of TiO_2 were observed more clearly in the samples under the drought stress condition than in the control samples.

Si and SiO_2 NPs are the basis for a great deal of multidisciplinary research due to their low toxicity (under a specific size and dose condition), stability, and ability to be modified with a range of bio/nano molecules and polymers. Treatment of *P. sativum* (under Cd stress), *O. sativa*, and *G. max* (under Hg stress) with Si and Si and SiO_2 NPs at various sizes and concentrations minimized the metal storage capacity, augmented the plant growth rate, and macro/micronutrient uptake (Cui et al. 2017). Specifically, in the *O. sativa* study where Si NPs were tested at different sizes (19, 48, and 202 nm), the smaller particles performed better in the fight against Cd stress (Cui et al. 2017). The results demonstrated a significant increase in the proportion of the viable cells, calculated as 95.4%, 78.6%, and 66.2% for 19, 48, and 202 nm Si NPs, respectively. Potter et al. (Potter et al. 2021) investigated the drought stress mitigation effect of Si and SiO_2, ZnO, and CuO NPs in a wheat cultivar (*T. aestivum* var. Juniper) where the plants were grown in a nutrient-sufficient solid growth matrix already. The plant samples' seeds were surrounded by *Pseudomonas chlororaphis* O6 bacterium with previous incubation to protect the roots. The samples analyzed after 8-day drought stress showed a decrease in mass (shoot and roots), photosystem II yield, and water content. The NPs did not affect the beneficial microbiome colonization, which was present in 22-day old samples. The rhizosphere solution samples presented an increased metal (nutrient) content released from the NPs in a dose-dependent manner, not affected by the drought stress. This study showed that the drought tolerance of the wheat samples grown in a nutrient-sufficient environment and protected with a microorganism was not improved by the presence of NPs, which may only occur under specific environmental conditions, as previously discussed.

Besides ZnO, TiO_2, FeO, and SiO_2 NPs, which are utilized most for mitigating stress tolerance in plants, there are other types of nanomaterials in metal, metalloid, semiconductor, or carbon structures used for similar purposes in agricultural studies, including but not limited to carbon dots for decreasing heavy metal accumulation capacity (Xiao et al. 2019), graphene to alleviate salinity and alkalinity stresses (Chen et al. 2021a, b), aminated graphene oxide to facilitate seedling growth as seen in Fig. 2 (yellow panel) (Chen et al. 2018), multiwall carbon nanotubes for salt tolerance enhancement (Karami and Sepehri 2018), anionic CeO NPs for scavenging reactive oxygen species in *A. thaliana* (Wu et al. 2017), CuO NPs as micronutrient sources, pesticides, and enhancers of crop stress tolerance (Potter 2020; Van Nguyen et al. 2021), biogenic Se NPs for mitigating both biotic and abiotic stress (Zohra et al. 2021) and chitosan NPs to mitigate drought stress by inducing the antioxidant potential and gene expression of alkaloid biosynthesis (Ali et al. 2021). As seen in the examples, those NPs applied under optimized experimental conditions (at a defined size, shape, surface charge, dose, and application type) could enhance the stress tolerance capacity of plants through restricted HM bioaccumulation, activated molecular and enzymatic pathways, increased antioxidant mechanisms, and several physiological or morphological changes.

Finally, the way NPs are applied to the plants is also critical to fully reveal the beneficial impact of NPs. Various methods have been suggested in the literature to deliver NPs into the plants, including but not limited to seed, root, or pollen incubation, foliar spraying, injection, irrigation, hydroponic treatments, and biolistic guns (Singh et al. 2021). The internalization of the particles takes place either through the root epidermis or the surface interactions via the symplastic route (via cytoplasm continuum) or the apoplastic route (via cell wall continuum) (Sharma et al. 2020a). A recent metadata analysis conducted by Avellan et al. (Avellan et al. 2021) highlights that plant species that were used in the literature to evaluate NP behaviors are mostly dicotyledon (75%), while monocotyledons (such as cereals) are less common. As re-illustrated in Fig. 2 (green panel), NPs interact with different cells and use various mechanisms for translocation (Avellan et al. 2021). Furthermore, the parameters such as NP size, surface charge, hydrophobicity, solubility, exposure time, concentration are correlated to in vivo mobility of the particles within the plant cells, which paves the way for stimuli-driven or targeted delivery of beneficial NPs to the desired plant compartments. Despite the significant number of reports in the field, only in a limited number of studies, the physical and technical characterization of NPs were fully explored and systematically evaluated in physiological, morphological, and molecular levels. On the other hand, NP delivery mechanisms, toxicity, and environmental impact analyses, and structure-function relation on a molecular basis are yet to be explored for further engineering of plant-specific or stress-specific, tunable NP production (Alabdallah et al. 2021).

10 Genome Editing with CRISPR Technology for Abiotic Stress Management

Abiotic stress is a major limiting factor for plant growth and development. The consequences of these environmental stresses severely affect the yield and quality of the crop that results in economic losses and inadequate food supply. In order to solve the forthcoming food insecurity issue, plant scientists have focused on discovering the molecular mechanisms of plant stress response using plant molecular biology and genetics tools besides conventional breeding studies (Hirayama and Shinozaki 2010). Since the pathways involved in plant response against abiotic stress consist of many complex interactions and crosstalks, studies at different omics levels and system biology approach have been utilized to understand this complex mechanism (Mosa et al. 2017). On the other hand, genetic manipulation of plants has both given the scientists an insight for the understanding of plant response mechanisms against a/biotic stress and ability to produce new plant varieties. Classical transgenic technology, for agricultural improvement, has been serving as a tailor to alter the species genome with foreign genes in order to introduce desirable traits while preserving the individual characteristics of the plant. Longer shelf life, higher yield, pest resistance, improved quality, tolerance to a variety of biotic and abiotic stresses are some of the traits that have been applied by plant transgenic technology (Jhansi Rani and Usha 2013). Though many genetically modified plant products have been produced in years, requirement of new technologies for targeted gene manipulation was a need for crop improvement and basic plant research.

The biggest breakthrough of targeted gene manipulation was the invention of double stranded breaks (DSBs) at targeted gene regions in 1983 (Szostak et al. 1983). The common strategy of genome editing depends on introduction of a double-strand break (DSB) in a specifically targeted chromosomal sites by using programmable site-specific nucleases (SSN). After DSB occurs, the DNA repair mechanisms get involved and either non-homologous end joining (NHEJ) or homologous recombination (HDR) repair occurs in the target region of the genome resulting in gene modification (Durai et al. 2005; Gaj et al. 2013). If the break is in the coding sequence of a gene, NHEJ repair, which is the main repair pathway in plants, leads to insertion or deletions (indels) which results in frameshift mutations or knock-outs (Puchta 2005). In cases where a donor sequence with homology to the DSB is present, the cell uses HDR repair pathway which can be used for gene insertion (Bortesi and Fischer 2015).

There are different types of systems used for genome editing such as Zinc Finger Nucleases (ZFNs), Transcription Activator Like Effector Nucleases (TALENs) and Clustered Regularly Interspaced Short Palindromic Repeat/Cas9 (CRISPR/Cas9) which can introduce a DSB in a targeted part of the genome (Jinek et al. 2012). First generation of SSNs was zinc finger nucleases (ZFNs) (Bibikova et al. 2002; Smith et al. 2000). However, it is a costly process and difficult to manipulate ZFN constructs (Ramirez et al. 2008; Sanjana et al. 2012). After ZFNs, as a more promising tool, transcription activator-like effector nucleases (TALENs) system has been adapted from *Xanthomonas* bacteria and it is firstly applied in plants (Boch et al. 2009; Christian et al. 2010; Li et al. 2011). Even if use of TALENs is easier than ZFNs, construction of tandem repeat domains in the TAL proteins is still complicated (Ma et al. 2016). Lastly, from adaptive immunity system of bacterium *Streptococcus pyogenes*, clustered regularly interspaced short palindromic repeat (CRISPR)-associated protein9 (Cas9)-based genome editing tool, CRISPR/Cas9, has been developed (Jinek et al. 2012) and utilized to edit genomes of many organisms including plants (Doudna and Charpentier 2014). CRISPR is advantageous with its efficiency, practical usage, low cost, versatility and methylated DNA cleavage capability (Lozano-Juste and Cutler 2014). Thus, this sophisticated technology has become popular for genome editing.

Among different types of systems, Type II system derived from *Streptococcus pyogenes* which works with one single multifunctional protein (Cas9) for introducing a DSB in the targeted DNA sequence is adapted for use in genome editing. This system requires only two elements to create a targeted break in the genome, the single guide RNA and the Cas9 nuclease (Shan et al. 2020). The single guide RNA (sgRNA), also known as guide RNA (gRNA), is a chimeric molecule that consists of two RNA molecules, crRNA and trans-activating crRNA (tracrRNA), forms a complex with Cas9 protein and recognize the targeted section (spacer) in the genome and leads the way for Cas9 nuclease to bind and cleave the double stranded DNA from that point. The only requirement of Cas9 nuclease is a promoter-adjacent motif (PAM) located in the downstream of the target (Jinek et al. 2012).

Using CRISPR/Cas9, creation of indels, gene or donor DNA sequence insertion, methylation control and activation/repression of specific genes are provided. CRISPR/Cas9 system is not only limited to single gene targeting but also multiple genes in a single organism can be targeted which is referred as multiplexed genome editing. Cas9 can be directed to the corresponding sites in the same cell with multiple engineered sgRNAs (Cong et al. 2013; Iida and Terada 2005). Multiplexed editing is an important and promising application since it provides functional studies for related genes that control complex traits and multiple members of gene families (Xie et al. 2015). The ability of multiplexed targeting has been viewed as a promising approach to develop plant varieties that are resilient to biotic and abiotic stress factors (Mushtaq et al. 2018). It has been shown by Zhang et al. 2016 that six different genes can be targeted at the same time using multiple sgRNAs, a practical solution is open for improving multiplex editing, like an efficient expression system or usage of different Cas enzymes with higher nuclease activity. Besides the native Cas9, different variants like Cas9 nickase (Cas9n) which creates a nick instead of a DSB and dead Cas9 (dCas9) which is catalytically inactive have been developed by point mutations to be employed for different purposes (Cong et al. 2013). Genome editing functions like introduction of a mutation or gene insertion, CRISPR/Cas9 editing tools have become useful for transcriptional gene regulation studies. For transcriptional repression and activation, deactivated Cas9 is used. This system is referred to as CRISPR activator (CRISPRa) and CRISPR interference (CRISPRi). Using both of these systems, transcriptional activators and repressors are bound to Cas9 with a lacking nuclease activity, dCas9. Simultaneously, activation or repressions of multiple genes are also available using this system. In addition to these, fusion of dCas9 with elements like DNA demethylase creates an opportunity for epigenome editing studies (Maeder et al. 2013).

With recent efforts, CRISPR/Cas9 technology has been initiated to transfer this knowledge from lab to field. CRISPR/Cas9 system was initiated for implementation in agriculturally and economically important plants such as *Triticum aestivum* (Y. Wang et al. 2014), *Nicotiana tabacum* (Gao et al. 2014), *Zea mays* (Xing et al. 2014), *Oryza sativa*, *Solanum lycopersium* (Brooks et al. 2014), *Solanum tuberosum* (Wang et al. 2015a, b), *Glycine max* (Jacobs et al. 2015) and *Populus tomentosa* (Fan et al. 2015) with using stable transformation protocols. There are many more studies with different species and traits using CRISPR system, however, several are representing the transient transformation results rather than stable ones.

Improvement of crops for abiotic stress tolerance through genetic engineering or genome editing depends on the elucidation of the functions of the genes and related mechanisms. In order to understand the function of potential genes on abiotic stress tolerance, different studies have been performed using CRISPR technology by targeting genes to verify their role in heat, cold, drought and salinity stress tolerance as well as herbicide resistance and multiple stress tolerance (Bhat et al. 2021). CRISPR system was used in several studies to improve the plants for abiotic stress tolerance. Drought stress tolerance improvement was achieved in *A. thaliana* (Osakabe et al. 2016). Rice was engineered with CRISPR/Cas9 and the mutants showed increased resistance to salt and oxidative stress (Zhang et al. 2020). In rice, also, heat stress tolerance improvement (Miao et al. 2018), development of rice plant with low cesium amount (Nieves-Cordones et al. 2010), and low cadmium accumulation (Tang et al. 2017a, b) in rice plants were achieved. Besides, arsenic tolerance of the rice plant was improved (Wang et al. 2017). Herbicide resistant rice (Sun et al. 2016), maize (Svitashev et al. 2015), and watermelon (Tian et al. 2018) plants were produced. Improved yield was achieved under drought conditions in maize which conferred increased tolerance to drought stress conditions (Shi et al. 2017).

Genome editing via CRISPR/Cas systems is expected to accelerate breeding by its ability to create precise modifications at specific loci. Further studies demonstrated that CRISPR/Cas 9 genome editing in plants has the potential of fighting with biotic and abiotic stress as well as increasing yield and growth.

11 Current Situation and Future Prospects

The improvement of major crops to a changing climate is an ongoing study undertaken by a number of scientists over the last three decades in order to guarantee food security and safety for the future generations. Despite all these efforts, due to the complexity of abiotic stress response and tolerance mechanisms, there is still not a very promising outcome utilized in agriculture. The main problem here is the strong interaction of plant behavior with its environment, which hinders the results observed in small scale experiments to be repeated in the field. It is rather clear that new approaches and agricultural practices should be used for improvement in yield. This review aimed to summarize other possible novel approaches that can be used for this purpose including possibility of utilizing biostimulants, micronutrients and nanoparticles, exciting a heritable epigenetic memory, genome editing and genetically engineering chloroplasts to improve tolerance. Breeding efforts to transfer tolerance to commonly planted crops, although slow and painful, should be carried out based on the QTLs defining such qualities. The efforts to finding major QTLs should include the landraces too, considering the wild genotypes are still the main resource for resilience to both abiotic and biotic stresses. One of the main challenges that needs to be considered is the simultaneous abiotic and biotic stresses, since changing climate increased the occurrence and severity of pathogen attacks. The studies to mitigate yield loss should, therefore, focus on biotic and abiotic resilience together, rather than improving only one resilience. The same problem stands for the abiotic stress tolerance itself, since most, if not, all of the studies focused on response to a single stress at a time, although it is clear that abiotic stresses occur together at the nature.

In literature, there are many studies summarizing the response upon a single abiotic stress of a specific plant genotype or cultivar. Although there is a huge data available, the integration of all these data is still close to impossible due to different stress treatment approaches used by the scientists and the specificity of the plant genotypes in their response mechanisms. One should note that plants adapted to their environment over the years, which led to regionally preferred genotypes/cultivars in agriculture suitable to that climate or microclimate. Therefore, the efforts to improve stress tolerance levels of plants should consider regionally preferred cultivars. The scientific approaches should also consider the climate change scenarios to understand the main requirement for their region. One last point to consider is the fact that although there are enormous number of studies with major crops like wheat, rice and maize, there is a limited effort to improve vegetables, which has to be considered in order to maintain agricultural diversity in the era of climate change.

The Conflict of Interest and Data Availability Statement The authors declare that there is no conflict of interest and data is publicly available.

References

Adhikari S, Kumari J, Jacob SR, Prasad P, Gangwar OP, Lata C, Kumar S (2022) Landraces-potential treasure for sustainable wheat improvement. Genet Resour Crop Evol 69:499–523

Adrees M, Khan ZS, Hafeez M, Rizwan M, Hussain K, Asrar M, Ali S (2021) Foliar exposure of zinc oxide nanoparticles improved the growth of wheat (*Triticum aestivum* L.) and decreased cadmium concentration in grains under simultaneous Cd and water deficient stress. Ecotoxicol Environ Saf 208:111627

Ahanger MA, Ahmad P (2019) Role of mineral nutrients in abiotic stress tolerance: revisiting the associated signaling mechanisms. In: Plant Signaling Molecules: Role and Regulation Under Stressful Environments, pp 269–285

Ahmad P, Alyemeni MN, Al-Huqail AA, Alqahtani MA, Wijaya L, Ashraf M, Bajguz A (2020) Zinc oxide nanoparticles application alleviates arsenic (As) toxicity in soybean plants by restricting the uptake of as and modulating key biochemical attributes, antioxidant enzymes, ascorbate-glutathione cycle and glyoxalase system. Plan Theory 9(7):825

Ahmar S, Gill RA, Jung K-H, Faheem A, Qasim MU, Mubeen M, Zhou W (2020) Conventional and molecular techniques from simple breeding to speed breeding in crop plants: recent advances and future outlook. Int J Mol Sci 21(7):2590

Ahmed I, Sarazin A, Bowler C, Colot V, Quesneville H (2011) Genome-wide evidence for local DNA methylation spreading from small RNA-targeted sequences in Arabidopsis. Nucleic Acids Res 39(16):6919–6931

Ahmed T, Noman M, Manzoor N, Shahid M, Abdullah M, Ali L, Li B (2021) Nanoparticle-based amelioration of drought stress and cadmium toxicity in rice via triggering the stress responsive genetic mechanisms and nutrient acquisition. Ecotoxicol Environ Saf 209:111829

Ait-El-Mokhtar M, Laouane RB, Anli M, Boutasknit A, Wahbi S, Meddich A (2019) Use of mycorrhizal fungi in improving tolerance of the date palm (*Phoenix dactylifera* L.) seedlings to salt stress. Sci Hort 253:429–438

Akladious SA, Mohamed HI (2018) Ameliorative effects of calcium nitrate and humic acid on the growth, yield component and biochemical attribute of pepper (*Capsicum annuum*) plants grown under salt stress. Sci. Hort. 236:244–250

Alabdallah NM, Alzahrani HS (2020) The potential mitigation effect of ZnO nanoparticles on [*Abelmoschus esculentus* L. Moench] metabolism under salt stress conditions. Saudi J Biol Sci 27:3132–3137

Alabdallah NM, Hasan MM, Hammami I, Alghamdi AI, Alshehri D, Alatawi HA (2021) Green synthesized metal oxide nanoparticles mediate growth regulation and physiology of crop plants under drought stress. Plan Theory 10:1730

Alagna F, Balestrini R, Chitarra W, Marsico AD, Nerva L (2020) Getting ready with the priming: innovative weapons against biotic and abiotic crop enemies in a global changing scenario. In: Priming-mediated stress and cross-stress tolerance in crop plants, pp 35–56

Ali EF, El-Shehawi AM, Ibrahim OHM, Abdul-Hafeez EY, Moussa MM, Hassan FAS (2021) A vital role of chitosan nanoparticles in improvisation the drought stress tolerance in *Catharanthus roseus* (L.) through biochemical and gene expression modulation. Plant Physiol Biochem 161:166–175

Angel A, Song J, Dean C, Howard M (2011) A Polycomb-based switch underlying quantitative epigenetic memory. Nature 476:105–109

Anwar A, Kim JK (2020) Transgenic breeding approaches for improving abiotic stress tolerance: recent progress and future perspectives. Int J Mol Sci 21:2695

Apse MP, Aharon GS, Snedden WA, Blumwald E (1999) Salt tolerance conferred by overexpression of a vacuolar Na+/H+ antiport in Arabidopsis. Science 285:1256–1258

Ariyarathna HACK, Oldach KH, Francki MG (2016) A comparative gene analysis with rice identified orthologous group II HKT genes and their association with Na+ concentration in bread wheat. BMC Plant Biol 16(21):21

Aroca R, Ruiz-Lozano JM, Zamarreño AM, Paz JA, García-Mina JM, Pozo MJ, López-Ráez JA (2013) Arbuscular mycorrhizal symbiosis influences strigolactone production under salinity and alleviates salt stress in lettuce plants. J Plant Physiol 170:47–55

Asif MA, Schilling RK, Tilbrook J, Brien C, Dowling K, Rabie H, Pearson AS (2018) Mapping of novel salt tolerance QTL in an Excalibur × Kukri doubled haploid wheat population. Theor Appl Genet 131:2179–2196

Asins MJ, Villalta I, Aly MM, Olías R, Álvarez De Morales P, Huertas R, Belver A (2013) Two closely linked tomato HKT coding genes are positional candidates for the major tomato QTL involved in Na+/K+ homeostasis. Plant Cell Environ 36:1171–1191

Atieno J, Colmer TD, Taylor J, Li Y, Quealy J, Kotula L, Sutton T (2021) Novel salinity tolerance loci in chickpea identified in glasshouse and field environments. Front Plant Sci 12:667910

Augusto J, Brenneman TB, Culbreath AK, Sumner P (2010) Night spraying peanut fungicides II. Application timings and spray deposition in the lower canopy. Plant Dis 94(6):683–689

Avellan A, Yun J, Morais BP, Clement ET, Rodrigues SM, Lowry GV (2021) Critical review: role of inorganic nanoparticle properties on their foliar uptake and in planta translocation. Environ Sci Technol 55(20):13417–13431

Baisakh N, Yabes J, Gutierrez A, Mangu V, Ma P, Famoso A, Pereira A (2020) Genetic mapping identifies consistent quantitative trait loci for yield traits of rice under greenhouse drought conditions. Genes 11:62

Barik SR, Pandit E, Mohanty SP, Nayak DK, Pradhan SK (2020) Genetic mapping of physiological traits associated with terminal stage drought tolerance in rice. BMC Genet 21:76

Barragán V, Leidi EO, Andrés Z, Rubio L, De Luca A, Fernández JA, Cubero B, Pardo JM (2012) Ion exchangers NHX1 and NHX2 mediate active potassium uptake into vacuoles to regulate cell turgor and stomatal function in Arabidopsis. Plant Cell 24(3):1127–1142

Beach RH, Sulser TB, Crimmins A, Cenacchi N, Cole J, Fukagawa NK, Ziska LH (2019) Combining the effects of increased atmospheric carbon dioxide on protein, iron, and zinc availability and projected climate change on global diets: a modelling study. Lancet Planet Health 3(7):307–317

Bebber DP, Holmes T, Gurr SJ (2014) The global spread of crop pests and pathogens. Glob Ecol Biogeogr 23(12):1398–1407

Bedos C, Cellier P, Calvet R, Barriuso E, Gabrielle B (2002) Mass transfer of pesticides into the atmosphere by volatilization from soils and plants: overview. Agronomie 22:21–33

Beena R, Kirubakaran S, Nithya N, Manickavelu A, Sah RP, Abida PS, Siddique KHM (2021) Association mapping of drought tolerance and agronomic traits in rice (*Oryza sativa* L.) landraces. BMC Plant Biol 21 484

Beresford RM, Shuey LS, Pegg GS (2020) Symptom development and latent period of *Austropuccinia psidii* (myrtle rust) in relation to host species, temperature, and ontogenic resistance. Plant Pathol 69:484–494

Bewick AJ, Ji L, Niederhuth CE, Willing EM, Hofmeister BT, Shi X, Schmitza RJ (2016) On the origin and evolutionary consequences of gene body DNA methylation. Proc Natl Acad Sci USA 113(32):9111–9116

Bhat MA, Mir RA, Kumar V, Shah AA, Zargar SM, Rahman S, Jan AT (2021) Mechanistic insights of CRISPR/Cas-mediated genome editing towards enhancing abiotic stress tolerance in plants. Physiol Plant 172:1–14

Bibikova M, Golic M, Golic KG, Carroll D (2002) Targeted chromosomal cleavage and mutagenesis in Drosophila using zinc-finger nucleases. Genetics 161(July):1169–1175

Bidi H, Fallah H, Niknejad Y, Barari Tari D (2021) Iron oxide nanoparticles alleviate arsenic phytotoxicity in rice by improving iron uptake, oxidative stress tolerance and diminishing arsenic accumulation. Plant Physiol Biochem 163:348–357

Birhane E, Sterck FJ, Fetene M, Bongers F, Kuyper TW (2012) Arbuscular mycorrhizal fungi enhance photosynthesis, water use efficiency, and growth of frankincense seedlings under pulsed water availability conditions. Oecologia 169:895–904

Boch J, Scholze H, Schornack S, Landgraf A, Hahn S, Kay S, Bonas U (2009) Breaking the code of DNA binding specificity of TAL-type III effectors. Science 326(5959):1509–1512

Bonfante P, Genre A (2010) Mechanisms underlying beneficial plant – fungus interactions in mycorrhizal symbiosis. Nat Commun 1:48

Boonwichai S, Shrestha S, Babel MS, Weesakul S, Datta A (2018) Climate change impacts on irrigation water requirement, crop water productivity and rice yield in the Songkhram River Basin, Thailand. J Clean Prod 198:1157–1164

Borjigin C, Schilling RK, Jewell N, Brien C, Sanchez-Ferrero JC, Eckermann PJ, Roy SJ (2021) Identifying the genetic control of salinity tolerance in the bread wheat landrace Mocho de Espiga Branca. Funct Plant Biol 48(11):1148–1160

Bortesi L, Fischer R (2015) The CRISPR/Cas9 system for plant genome editing and beyond. Biotechnol Advan 33(1):41–52

Breseghello F, Coelho ASG (2013) Traditional and modern plant breeding methods with examples in rice (*Oryza sativa* L.). J Agric Food Chem 61(35):8277–8286

Brooks C, Nekrasov V, Lipppman ZB, Van Eck J (2014) Efficient gene editing in tomato in the first generation using the clustered regularly interspaced short palindromic repeats/CRISPR-associated9 system. Plant Physiol 166:1292–1297

Budak H, Hussain B, Khan Z, Ozturk NZ, Ullah N (2015) From genetics to functional genomics: improvement in drought signaling and tolerance in wheat. Front Plant Sci 6:1012

Burdon JJ, Zhan J (2020) Climate change and disease in plant communities. PLoS Biol 18(11):e3000949

Busby PE, Soman C, Wagner MR, Friesen ML, Kremer J, Bennett A, Dangl JL (2017) Research priorities for harnessing plant microbiomes in sustainable agriculture. PLoS Biol 15(3):e2001793

Campbell B, Chen L, Dygert C, Dick W (2014) Tillage and crop rotation impacts on greenhouse gas fluxes from soil at two longterm agronomic experimental sites in Ohio. J Soil Water Conserv 69(6):543–552

Cellier F, Conéjéro G, Ricaud L, Doan TL, Lepetit M, Gosti F, Casse F (2004) Characterization of AtCHX17, a member of the cation/H+ exchangers, CHX family, from *Arabidopsis thaliana* suggests a role in K + homeostasis. Plant J 39(6):834–846

Cerdán M, Sánchez-Sánchez A, Oliver M, Juárez M, Sánchez-Andreu JJ (2009) Effect of foliar and root applications of amino acids on iron uptake by tomato plants. Acta Hortic 830:481–488

Cha JY, Kang SH, Ali I, Lee SC, Ji MG, Jeong SY, Kim WY (2020) Humic acid enhances heat stress tolerance via transcriptional activation of heat-shock proteins in Arabidopsis. Sci Rep 10:15042

Chaab A, Moezzi A, Sayyad Go, Chorom M (2016) Alleviation of cadmium toxicity to maize by the application of humic acid and compost. Life Sci J 13(12):56–63

Chandio AA, Jiang Y, Akram W, Adeel S, Irfan M, Jan I (2021) Addressing the effect of climate change in the framework of financial and technological development on cereal production in Pakistan. J Clean Prod 288:125637

Chanroj S, Wang G, Venema K, Zhang MW, Delwiche CF, Sze H (2012) Conserved and diversified gene families of monovalent cation/H+ antiporters from algae to flowering plants. Front Plant Sci 3:25

Chen J, Yang L, Li S, Ding W (2018) Various physiological response to graphene oxide and amine-functionalized graphene oxide in wheat (*Triticum aestivum*). Molecules 23

Chen Z, Niu J, Guo Z, Sui X, Xv N, Kareen HA, Wang Q (2021a) Graphene enhances photosynthesis and the antioxidative defense system and alleviates salinity and alkalinity stresses in alfalfa (*Medicago sativa* L.) by regulating gene expression. Environ Sci Nano 8(9):2731–2748

Chen L, Wang Q, Tang M, Zhang X, Pan Y, Yang X, Liang T (2021b) QTL mapping and identification of candidate genes for heat tolerance at the flowering stage in rice. Front Genet 11:621871

Cheng X, Liu X, Mao W, Zhang X, Chen S, Zhan K, Xu H (2018) Genome-wide identification and analysis of HAK/KUP/KT potassium transporters gene family in wheat (*Triticum aestivum* L.). Int J Mol Sci 19:3969

Chiaiese P, Corrdo G, Colla G, Kyriacou MC, Rouphael Y (2018) Renewable sources of plant biostimulation: microalgae as a sustainable means to improve crop performance. Front Plant Sci 9:1782

Christian M, Cermak T, Doyle EL, Schmidt C, Zhang F, Hummel A, Voytas DF (2010) Targeting DNA double-strand breaks with TAL effector nucleases. Genetics 186(2):756–761

Clement CR, Borém A, Lopes MTG (2012) From plant domestication to breeding. In: Domestication and breeding: Amazonian species

Coll NS, Epple P, Dangl JL (2011) Programmed cell death in the plant immune system. Cell Death Differ 18(8):1247–1256

Colla G, Nardi S, Cardarelli M, Ertani A, Lucini L, Canaguier R, Rouphael Y (2015) Protein hydrolysates as biostimulants in horticulture. Sci Hortic 196:28–38

Cong L, Ran FA, Cox D, Lin S, Barretto R, Habib N, Zhang F (2013) Multiplex genome engineering using CRISPR/Cas systems. Science 339(6121):819–823

Crawford T, Lehotai N, Strand Å (2018) The role of retrograde signals during plant stress responses. J Exp Bot 69(11):2783–2795

Cui J, Liu T, Li F, Yi J, Liu C, Yu H (2017) Silica nanoparticles alleviate cadmium toxicity in rice cells: mechanisms and size effects. Environ Pollut 228:363–369

Dempewolf H, Baute G, Anderson J, Kilian B, Smith C, Guarino L (2017) Past and future use of wild relatives in crop breeding. Crop Sci 57(3):1070–1082

Deutsch CA, Tewksbury JJ, Tigchelaar M, Battisti DS, Merrill SC, Huey RB, Naylor RL (2018) Increase in crop losses to insect pests in a warming climate. Science 361:916–919

Dhar MK, Vishal P, Sharma R, Kaul S (2014) Epigenetic dynamics: role of epimarks and underlying machinery in plants exposed to abiotic stress. Int J Genomics 2014:1–10

Dimkpa CO, Singh U, Bindraban PS, Elmer WH, Gardea-Torresdey JL, White JC (2019) Zinc oxide nanoparticles alleviate drought-induced alterations in sorghum performance, nutrient acquisition, and grain fortification. Sci Total Environ 688:926–934

Ding Y, Fromm M, Avramova Z (2012) Multiple exposures to drought "train" transcriptional responses in Arabidopsis. Nat Commun 3:740

Doebley JF, Gaut BS, Smith BD (2006) The molecular genetics of crop domestication. Cell 127(December):1309–1321

Doudna JA, Charpentier E (2014) The new frontier of genome engineering with CRISPR-Cas9. Science 346(6213):1258096

Dourmap C, Roque S, Morin A, Caubrière D, Kerdiles M, Béguin K, Couée I (2020) Stress signalling dynamics of the mitochondrial electron transport chain and oxidative phosphorylation system in higher plants. Ann Bot 125:721–736

Drobek M, Frąc M, Cybulska J (2019) Plant biostimulants: importance of the quality and yield of horticultural crops and the improvement of plant tolerance to abiotic stress—a review. Agronomy 9(6):335

du Jardin P (2015) Plant biostimulants: definition, concept, main categories and regulation. Sci Hortic 196:3–14

Durai S, Mani M, Kandavelou K, Wu J, Porteus MH, Chandrasegaran S (2005) Zinc finger nucleases: custom-designed molecular scissors for genome engineering of plant and mammalian cells. Nucleic Acids Res 33(18):5978–5990

Dwivedi SL, Ceccarelli S, Blair MW, Upadhyaya HD, Are AK, Ortiz R (2016) Landrace germplasm for improving yield and abiotic stress adaptation. Trends Plant Sci 21(1):31–42

Ealias AM, Saravanakumar MP (2017) A review on the classification, characterisation, synthesis of nanoparticles and their application. IOP Conf Ser: Mater Sci Eng 263:032019

El-Saber MM, Mahdi AA, Hassan AH, Farroh KY, Osman A (2021) Effects of magnetite nanoparticles on physiological processes to alleviate salinity induced oxidative damage in wheat. J. Sci. Food Agric 101:5550–5562

Etesami H, Hosseini MH, Alikhani HA (2014) Bacterial biosynthesis of 1-aminocyclopropane-1-caboxylate (ACC) deaminase, a useful trait to elongation and endophytic colonization of the roots of rice under constant flooded conditions. Physiol Mol Biol Plants 20(4):425–434

Fageria NK, Baligar VC, Li YC (2008) The role of nutrient efficient plants in improving crop yields in the twenty first century. J Plant Nutr 31(6):1121–1157
Fan D, Liu T, Li C, Jiao B, Li S, Hou Y, Luo K (2015) Efficient CRISPR/Cas9-mediated targeted mutagenesis in populus in the first generation. Sci Rep 5:12217
Fang Z, Wang X, Zhang X, Zhao D, Tao J (2020) Effects of fulvic acid on the photosynthetic and physiological characteristics of Paeonia ostii under drought stress. Plant Signal Behav 15(7):1774714
FAO (2019) Voluntary guidelines for the conservation and sustainable use of farmers' varieties/ landraces
FAO (2021) Climate change fans spread of pests and threatens plants and crops, new FAO study. Retrieved from https://www.fao.org/news/story/en/item/1402920/icode/2011
Faraji J, Sepehri A (2019) Ameliorative effects of TiO_2 nanoparticles and sodium nitroprusside on seed germination and seedling growth of wheat under PEG-stimulated drought stress. J Seed Sci 41(3):309–317
Fatima T, Arora NK (2021) *Pseudomonas entomophila* PE3 and its exopolysaccharides as biostimulants for enhancing growth, yield and tolerance responses of sunflower under saline conditions. Microbiol Res 244:126671
Filgueiras L, Silva R, Almeida I, Vidal M, Baldani JI, Meneses CHSG (2019) Gluconacetobacter diazotrophicus mitigates drought stress in *Oryza sativa* L. Plant Soil 451:57–73
Fukuda A, Nakamura A, Hara N, Toki S, Tanaka Y (2011) Molecular and functional analyses of rice NHX-type Na+/H+ antiporter genes. Planta 233(1):175–188
Gaillard MDP, Glauser G, Robert CAM, Turlings TCJ (2018) Fine-tuning the 'plant domestication-reduced defense' hypothesis: specialist vs generalist herbivores. New Phytol 217(1):355–366
Gaj T, Gersbach CA, Barbas CF III (2013) ZFN, TALEN and CRISPR/Cas-based methods for genome engineering. Trends Biotechnol 31(7):397–405
Gao J, Wang G, Ma S, Xie X, Wu X, Zhang X et al (2014) CRISPR/Cas9-mediated targeted mutagenesis in *Nicotiana tabacum*. Plant Mol. Biol 87:99–110
Garrett KA, Dendy SP, Frank EE, Rouse MN, Travers SE (2006) Climate change effects on plant disease: genomes to ecosystems. Annu Rev Phytopathol 44:489–509
Gavito ME, Olsson PA, Rouhier H, Medina-Peñafiel A, Jakobsen I, Bago A, Azcón-Aguilar C (2005) Temperature constraints on the growth and functioning of root organ cultures with arbuscular mycorrhizal fungi. New Phytol 168:179–188
Gayacharan, Joel AJ (2013) Epigenetic responses to drought stress in rice (Oryza sativa L.). Physiol Mol Biol Plants 19(3):379–387
Ghini R, Hamada E, Bettiol W (2008) Climate change and plant diseases. Sci Agric 65:98–107
Gohari G, Mohammadi A, Akbari A, Panahirad S, Dadpour MR, Fotopoulos V, Kimura S (2020) Titanium dioxide nanoparticles (TiO_2-NPs) promote growth and ameliorate salinity stress effects on essential oil profile and biochemical attributes of Dracocephalum moldavica. Sci Rep 10:912
Govindaraj M, Vetriventhan M, Srinivasan M (2015) Importance of genetic diversity assessment in crop plants and its recent advances: an overview of its analytical perspectives. Genet Res Int. 2015(Figure 1)
Gupta A, Rico-Medina A, Caño-Delgado AI (2020) The physiology of plant responses to drought. Science 368:266–269
Hamid B, Zaman M, Farooq S, Fatima S, Sayyed RZ, Baba ZA et al (2021) Bacterial plant biostimulants: a sustainable way towards improving growth, productivity, and health of crops. Sustainability (Switzerland) 13:2856
Hannukkala AO, Kaukoranta T, Lehtinen A, Rahkonen A (2007) Late-blight epidemics on potato in Finland, 1933-2002; increased and earlier occurrence of epidemics associated with climate change and lack of rotation. Plant Pathol 56:167–176
Hasanuzzaman M, Fujita M, Oku H, Nahar K, Hawrylak-Nowak B (2018) Plant nutrients and abiotic stress tolerance. In: Plant Nutrients and Abiotic Stress Tolerance. Springer, p 590

Hayat R, Ali S, Amara U, Khalid R, Ahmed I (2010) Soil beneficial bacteria and their role in plant growth promotion: a review. Ann Microbiol 60(4):579–598

Hemetsberger C, Herrberger C, Zechmann B, Hillmer M, Doehlemann G (2012) The *Ustilago maydis* effector Pep1 suppresses plant immunity by inhibition of host peroxidase activity. PLoS Pathog 8(5):e1002684

Henry M, McNab A (2002) Barley yellow dwarf disease: recent advances and future strategies. In: Proceedings of an International Symposium. Retrieved from http://repository.cimmyt.org/xmlui/handle/10883/1186

Hirabayashi H, Sasaki K, Kambe T, Gannaban RB, Miras MA, Mendioro MS et al (2015) QEMF3, a novel QTL for the early-morning flowering trait from wild rice, *Oryza officinalis*, to mitigate heat stress damage at flowering in rice, *O. sativa*. J Exp Bot 66(5):1227–1236

Hirayama T, Shinozaki K (2010) Research on plant abiotic stress responses in the post-genome era: past, present and future. Plant J 61:1041–1052

Hoang GT, Van Dinh L, Nguyen TT, Ta NK, Gathignol F, Mai CD et al (2019) Genome-wide association study of a panel of Vietnamese rice landraces reveals new QTLs for tolerance to water deficit during the vegetative phase. Rice 12:4

Hrmova M, Hussain SS (2021) Plant transcription factors involved in drought and associated stresses. Int J Mol Sci 22:5662

Hu H, Xiong L (2014) Genetic engineering and breeding of drought-resistant crops. Annu Rev Plant Biol 65:715–741

Hu Y, Zhang L, He S, Huang M, Tan J, Zhao L, Li L (2012) Cold stress selectively unsilences tandem repeats in heterochromatin associated with accumulation of H3K9ac. Plant Cell Environ 35:2130–2142

Hu H, Scheben A, Verpaalen B, Tirnaz S, Bayer PE, Hodel RGJ et al (2022) Amborella gene presence/absence variation is associated with abiotic stress responses that may contribute to environmental adaptation. New Phytol 233(4):1548–1555

Huang B, Xu C (2008) Identification and characterization of proteins associated with plant tolerance to heat stress. J Integr Plant Biol 50(10):1230–1237

Hui Ng H, Robert F, Young RA, Struhl K (2003) Targeted recruitment of Set1 histone methylase by elongating Pol II provides a localized mark and memory of recent transcriptional activity. Mol Cell 11:709–717

Hussain B, Lucas SJ, Ozturk L, Budak H (2017) Mapping QTLs conferring salt tolerance and micronutrient concentrations at seedling stage in wheat. Sci Rep 7(1):1–14

Hussein MH, Eltanahy E, Al Bakry AF, Elsafty N, Elshamy MM (2021) Seaweed extracts as prospective plant growth bio-stimulant and salinity stress alleviator for *Vigna sinensis* and *Zea mays*. J Appl Phycol 33:1273–1291

Iida S, Terada R (2005) Modification of endogenous natural genes by gene targeting in rice and other higher plants. Plant Mol Biol 59(1):205–219

IPCC (2018) Global warming of 1.5°C. In an IPCC special report on the impacts of global warming of 1.5°C above pre-industrial levels and related global greenhouse gas emission pathways, in the context of strengthening the global response to the threat of climate change, sustainable development

Irani S, Todd CD (2018) Exogenous allantoin increases Arabidopsis seedlings tolerance to NaCl stress and regulates expression of oxidative stress response genes. J Plant Physiol 221:43–50

Ishimaru T, Hirabayashi H, Ida M, Takai T, San-Oh YA, Yoshinaga S et al (2010) A genetic resource for early-morning flowering trait of wild rice *Oryza officinalis* to mitigate high temperature-induced spikelet sterility at anthesis. Ann Bot 106:515–520

Islam W (2018) Plant disease epidemiology: disease triangle and forecasting mechanisms in highlights. Host Viruses 5(1):7–11

Ismail AM, Horie T (2017) Genomics, physiology, and molecular breeding approaches for improving salt tolerance. Annu Rev Plant Biol 68(February):405–434

Jackson HO, Taunt HN, Mordaka PM, Smith AG, Purton S (2021) The algal chloroplast as a testbed for synthetic biology designs aimed at radically rewiring plant metabolism. Front Plant Sci 12:708370

Jacobs TB, LaFayette PR, Schmitz RJ, Parrott WA (2015) Targeted genome modifications in soybean with CRISPR/Cas9. BMC Biotechnol 15:16

Jagadish SVK (2020) Heat stress during flowering in cereals–effects and adaptation strategies. New Phytol 226:1567–1572

Jägermeyr J, Müller C, Ruane AC, Elliott J, Balkovic J, Castillo O et al (2021) Climate impacts on global agriculture emerge earlier in new generation of climate and crop models. Nat Food 2(11):873–885

James RA, Blake C, Byrt CS, Munns R (2011) Major genes for Na+ exclusion, Nax1 and Nax2 (wheat HKT1;4 and HKT1;5), decrease Na+ accumulation in bread wheat leaves under saline and waterlogged conditions. J Exp Bot 62(8):2939–2947

Jannin L, Arkoun M, Ourry A, Laîné P, Goux D, Garnica M, Etienne P (2012) Microarray analysis of humic acid effects on *Brassica napus* growth: involvement of N, C and S metabolisms. Plant Soil 359:297–319

Jhansi Rani S, Usha R (2013) Transgenic plants: types, benefits, public concerns and future. J Pharm Res 6(8):879–883

Jiang J, Dehesh K (2021) Plastidial retrograde modulation of light and hormonal signaling: an odyssey. New Phytol 230:931–937

Jindo K, Martim SA, Navarro EC, Pérez-Alfocea F, Hernandez T, Garcia C et al (2012) Root growth promotion by humic acids from composted and non-composted urban organic wastes. Plant Soil 353:209–220

Jinek M, Chylinski K, Fonfara I, Hauer M, Doudna JA, Charpentier E (2012) A programmable dual-RNA-guided DNA endonuclease in adaptive bacterial immunity. Science 337(August):816–822

Johnson NC, Graham JH (2013) The continuum concept remains a useful framework for studying mycorrhizal functioning. Plant Soil 363:411–419

Karami A, Sepehri A (2018) Beneficial role of MWCNTs and SNP on growth, physiological and photosynthesis performance of barley under NaCl stress. J Soil Sci Plant Nutr 18(3):752–771

Karthika KS, Rashmi I, Parvathi MS (2018) Biological functions, uptake and transport of essential nutrients in relation to plant growth. In: Plant nutrients and abiotic stress tolerance, pp 1–590

Katiyar P, Yadu B, Korram J, Satnami ML, Kumar M, Keshavkant S (2020) Titanium nanoparticles attenuates arsenic toxicity by up-regulating expressions of defensive genes in *Vigna radiata* L. J Environ Sci (China) 2020 Jun:92:18–27

Kauffman GL, Kneivel DP, Watschke TL (2007) Effects of a biostimulant on the heat tolerance associated with photosynthetic capacity, membrane thermostability, and polyphenol production of perennial ryegrass. Crop Sci 47:261–267

Kerr JT, Pindar A, Galpern P, Packer L, Potts SG, Roberts SM, Pantoja A (2015) Climate change impacts on bumblebees converge across continents. Science 349(6244):177–180

Khan W, Rayirath UP, Subramanian S, Jithesh MN, Rayorath P, Hodges DM, Prithiviraj B (2009) Seaweed extracts as biostimulants of plant growth and development. J Plant Growth Regul 28:386–399

Khan MS, Kanwal B, Nazir S (2015) Metabolic engineering of the chloroplast genome reveals that the yeast ArDH gene confers enhanced tolerance to salinity and drought in plants. Front Plant Sci 6:725

Khan M, Ahmad R, Khan MD, Rizwan M, Ali S, Khan MJ, Zhu S (2018) Trace elements in abiotic stress tolerance. In: Plant Nutrients and Abiotic Stress Tolerance, pp 137–151

Khan I, Saeed K, Khan I (2019) Nanoparticles: properties, applications and toxicities. Arab J Chem 12(7):908–931

Khan AW, Garg V, Roorkiwal M, Golicz AA, Edwards D, Varshney RK (2020) Super-pangenome by integrating the wild side of a species for accelerated crop improvement. Trends Plant Sci 25(2):148–158

Khoury CK, Brush S, Costich DE, Curry HA, de Haan S, Engels JMM, Thormann I (2022) Crop genetic erosion: understanding and responding to loss of crop diversity. New Phytol 233(1):84–118

Kilasi NL, Singh J, Vallejos CE, Ye C, Jagadish SVK, Kusolwa P, Rathinasabapathi B (2018) Heat stress tolerance in rice (*Oryza sativa* L.): identification of quantitative trait loci and candidate genes for seedling growth under heat stress. Front Plant Sci 9:1578

Kim JM, To TK, Ishida J, Morosawa T, Kawashima M, Matsui A, Seki M (2008) Alterations of lysine modifications on the histone H3 N-tail under drought stress conditions in Arabidopsis thaliana. Plant Cell Physiol 49(10):1580–1588

Kim JM, To TK, Ishida J, Matsui A, Kimura H, Seki M (2012) Transition of chromatin status during the process of recovery from drought stress in *Arabidopsis thaliana*. Plant Cell Physiol 53(5):847–856

Kobayashi T, Ishiguro K, Nakajima T, Kim HY, Okada M, Kobayashi K (2006) Effects of elevated atmospheric CO_2 concentration on the infection of rice blast and sheath blight. Phytopathology 96(4):425–431

Konate A, He X, Zhang Z, Ma Y, Zhang P, Alugongo GM, Rui Y (2017) Magnetic (Fe_3O_4) nanoparticles reduce heavy metals uptake and mitigate their toxicity in wheat seedling. Sustain For 9:790

Kunz HH, Gierth M, Herdean A, Satoh-Cruz M, Kramer DM, Spetea C, Schroeder JI (2014) Plastidial transporters KEA1, −2, and −3 are essential for chloroplast osmoregulation, integrity, and pH regulation in Arabidopsis. Proc Natl Acad Sci USA 111(20):7480–7485

Lakshmipriya T, Gopinath SCB (2021) Introduction to nanoparticles and analytical devices. In: Nanoparticles in analytical and medical devices, pp 1–29

Lämke J, Bäurle I (2017) Epigenetic and chromatin-based mechanisms in environmental stress adaptation and stress memory in plants. Genome Biol 18:124

Larkin RM (2016) Tetrapyrrole signaling in plants. Front Plant Sci 7

Latz A, Mehlmer N, Zapf S, Mueller TD, Wurzinger B, Pfister B, Becker D (2013) Salt stress triggers phosphorylation of the Arabidopsis K+ channel TPK1 by calcium-dependent protein kinases (CDPKs). Mol Plant 6(4):1274–1289

Leister D (2017) Experimental evolution in photoautotrophic microorganisms as a means of enhancing chloroplast functions. Essays Biochem 62:77–84

Li T, Huang S, Jiang WZ, Wright D, Spalding MH, Weeks DP, Yang B (2011) TAL nucleases (TALNs): hybrid proteins composed of TAL effectors and FokI DNA-cleavage domain. Nucleic Acids Res 39(1):359–372

Li XM, Chao DY, Wu Y, Huang X, Chen K, Cui LG, Lin HX (2015a) Natural alleles of a proteasome α2 subunit gene contribute to thermotolerance and adaptation of African rice. Nat. Genet 47:827–833

Li X, Topbjerg HB, Jiang D, Liu F (2015b) Drought priming at vegetative stage improves the antioxidant capacity and photosynthesis performance of wheat exposed to a short-term low temperature stress at jointing stage. Plant Soil 393:307–318

Li X, Tan DX, Jiang D, Liu F (2016) Melatonin enhances cold tolerance in drought-primed wild-type and abscisic acid-deficient mutant barley. J Pineal Res 61:328–339

Li Y, Li Y, Zhu N, Liang X, Bai X, Zheng L, Gao Y (2020a) Silica nanoparticles alleviate mercury toxicity: via immobilization and inactivation of Hg(ii) in soybean (*Glycine max*). Environ Sci Nano 7:1807–1817

Li Q, Qin Y, Hu X, Li G, Ding H, Xiong X, Wang W (2020b) Transcriptome analysis uncovers the gene expression profile of salt-stressed potato (*Solanum tuberosum* L.). Sci Rep 10:5411

Lian J, Zhao L, Wu J, Xiong H, Bao Y, Zeb A, Liu W (2020) Foliar spray of TiO_2 nanoparticles prevails over root application in reducing Cd accumulation and mitigating Cd-induced phytotoxicity in maize (*Zea mays* L.). Chemosphere 239:124794

Lin Y, Yang L, Paul M, Zu Y, Tang Z (2013) Ethylene promotes germination of Arabidopsis seed under salinity by decreasing reactive oxygen species: evidence for the involvement of nitric oxide simulated by sodium nitroprusside. Plant Physiol Biochem 73:211–218

Lindsay MP, Lagudah ES, Hare RA, Munns R (2004) A locus for sodium exclusion (Nax1), a trait for salt tolerance, mapped in durum wheat. Funct Plant Biol 31:1105–1114
Ling Q, Jarvis P (2015) Regulation of chloroplast protein import by the ubiquitin E3 ligase SP1 is important for stress tolerance in plants. Curr Biol 25:2527–2534
Liu S, Lacape JM, Constable GA, Llewellyn DJ (2015) Inheritance and QTL mapping of leaf nutrient concentration in a cotton inter-specific derived RIL population. PLoS One 10(5):e0128100
Liu Y, Yu L, Qu Y, Chen J, Liu X, Hong H, Guan R (2016) GmSALT3, which confers improved soybean salt tolerance in the field, increases leaf Cl– exclusion prior to Na+ exclusion but does not improve early vigor under salinity. Front Plant Sci 7:1485
Liu Y, Tang H, Muhammad A, Huang G (2019) Emission mechanism and reduction countermeasures of agricultural greenhouse gases—a review. Greenhouse Gases Sci Technol 9:160–174
Liu W, Yin T, Zhao Y, Wang X, Wang K, Shen Y, Tang S (2021) Effects of high temperature on rice grain development and quality formation based on proteomics comparative analysis under field warming. Front Plant Sci 12:746180
Locker CR, Torkamani S, Laurenzi IJ, Jin VL, Schmer MR, Karlen DL (2019) Field-to-farm gate greenhouse gas emissions from corn Stover production in the Midwestern U.S. J Clean Prod 226:1116–1127
Louwaars NP (2018) Plant breeding and diversity: a troubled relationship? Euphytica 214:1–9
Lozano-Juste J, Cutler SR (2014) Plant genome engineering in full bloom. Trends Plant Sci 19:284–287
Luo X, He Y (2020) Experiencing winter for spring flowering: a molecular epigenetic perspective on vernalization. J Integr Plant Biol 62:104–117
Ma X, Zhu Q, Chen Y, Liu YG (2016) CRISPR/Cas9 platforms for genome editing in plants: developments and applications. Mol Plant 9(7):961–974
Ma NL, Che Lah WA, Kadir NA, Mustaqim M, Rahmat Z, Ahmad A, Ismail MR (2018) Susceptibility and tolerance of rice crop to salt threat: physiological and metabolic inspections. PLoS One 13(2):e0192732
Maeder ML, Linder SJ, Cascio VM, Fu Y, Ho QH, Joung JK (2013) CRISPR RNA-guided activation of endogenous human genes. Nat Methods 10(10):977–979
Mahapatra K, Banerjee S, De S, Mitra M, Roy P, Roy S (2021) An insight into the mechanism of plant organelle genome maintenance and implications of organelle genome in crop improvement: an update. Front. Cell Dev. Biol. 9:671698
Mahmoud AF (2016) Genetic variation and biological control of Fusarium graminearum isolated from wheat in Assiut-Egypt. Plant Pathol J 32:145–156
Mahmoud AF, Abd El-Fatah BES (2020) Genetic diversity studies and identification of molecular and biochemical markers associated with fusarium wilt resistance in cultivated faba bean (*Vicia faba*). Plant Pathol J 36:11–28
Maity A, Lamichaney A, Joshi DC, Bajwa A, Subramanian N, Walsh M, Bagavathiannan M (2021) Seed shattering: a trait of evolutionary importance in plants. Front Plant Sci 12:657773
Mamaeva A, Taliansky M, Filippova A, Love AJ, Golub N, Fesenko I (2020) The role of chloroplast protein remodeling in stress responses and shaping of the plant peptidome. New Phytol 227:1326–1334
Manzoor N, Ahmed T, Noman M, Shahid M, Nazir MM, Ali L, Wang G (2021) Iron oxide nanoparticles ameliorated the cadmium and salinity stresses in wheat plants, facilitating photosynthetic pigments and restricting cadmium uptake. Sci. Total Environ. 769:145221
Mao H, Wang H, Liu S, Li Z, Yang X, Yan J, Qin F (2015) A transposable element in a NAC gene is associated with drought tolerance in maize seedlings. Nat. Commun. 6:8326
Matiu M, Ankerst DP, Menzel A (2017) Interactions between temperature and drought in global and regional crop yield variability during 1961-2014. PLoS One 12(5):e0178339
Matzke MA, Mosher RA (2014) RNA-directed DNA methylation: an epigenetic pathway of increasing complexity. Nat Rev Genet 15:394–408
Mauch-Mani B, Baccelli I, Luna E, Flors V (2017) Defense priming: an adaptive part of induced resistance. Annu Rev Plant Biol 68:485–512

McNunn G, Karlen DL, Salas W, Rice CW, Mueller S, Muth D, Seale JW (2020) Climate smart agriculture opportunities for mitigating soil greenhouse gas emissions across the U.S. Corn-Belt. J Clean Prod 268:122240
Meyer-Aurich A, Karatay YN, Nausediene A, Kirschke D (2020) Effectivity and cost efficiency of a tax on nitrogen fertilizer to reduce GHG emissions from agriculture. Atmos 11:607
Miao C, Xiao L, Hua K, Zou C, Zhao Y, Bressan RA, Zhu JK (2018) Mutations in a subfamily of abscisic acid receptor genes promote rice growth and productivity. Proc Natl Acad Sci USA 115(23):6058–6063
Mohammed S, Alsafadi K, Takács I, Harsányi E (2019) Contemporary changes of greenhouse gases emission from the agricultural sector in the EU-27. Geol. Ecol. Landsc 4:282–287
Moradtalab N, Hajiboland R, Aliasgharzad N, Hartmann TE, Neumann G (2019) Silicon and the association with an arbuscular-mycorrhizal fungus (*Rhizophagus clarus*) mitigate the adverse effects of drought stress on strawberry. Agronomy 9:41
Morton MJL, Awlia M, Al-Tamimi N, Saade S, Pailles Y, Negrão S, Tester M (2019) Salt stress under the scalpel–dissecting the genetics of salt tolerance. Plant J 97:148–163
Mosa K, Ismail A, Helmy M (2017) Omics and system biology approaches in plant stress research. In: Plant stress tolerance. Springer briefs in systems biology. Springer, Cham
Munns R, James RA, Xu B, Athman A, Conn SJ, Jordans C, Gilliham M (2012) Wheat grain yield on saline soils is improved by an ancestral Na + transporter gene. Nat Biotechnol 30(4):360–364
Murchie EH, Lawson T (2013) Chlorophyll fluorescence analysis: a guide to good practice and understanding some new applications. J Exp Bot 64(13):3983–3998
Mushtaq M, Bhat JA, Mir ZA, Sakina A, Ali S, Singh AK, Bhat R (2018) CRISPR/Cas approach: a new way of looking at plant-abiotic interactions. J Plant Physiol 224–225:156–162
Nadeem MA, Nawaz MA, Shahid MQ, Doğan Y, Comertpay G, Yıldız M et al (2018) DNA molecular markers in plant breeding: current status and recent advancements in genomic selection and genome editing. Biotechnol Biotechnol Equip 32(2):261–285
Ng S, Giraud E, Duncan O, Law SR, Wang Y, Xu L, Ivanova A (2013) Cyclin-dependent kinase E1 (CDKE1) provides a cellular switch in plants between growth and stress responses. J Biol Chem 288(5):3449–3459
Nieves-Cordones M, Alemán F, Martínez V, Rubio F (2010) The *Arabidopsis thaliana* HAK5 K+ transporter is required for plant growth and K+ acquisition from low K+ solutions under saline conditions. Mol Plant 3(2):326–333
Noreen S, Fatima Z, Ahmad S, Athar H-R, Ashraf M (2018) Foliar application of micronutrients in mitigating abiotic stress in crop plants. In: Plant nutrients and abiotic stress tolerance, pp 95–117
Osakabe Y, Watanabe T, Sugano SS, Ueta R, Ishihara R, Shinozaki K, Osakabe K (2016) Optimization of CRISPR/Cas9 genome editing to modify abiotic stress responses in plants. Sci Rep 6:26685
Paap T, Brouwers NC, Burgess TI, Hardy GESJ (2017) Importance of climate, anthropogenic disturbance and pathogens (*Quambalaria coyrecup* and *Phytophthora* spp.) on marri (*Corymbia calophylla*) tree health in Southwest Western Australia. Ann For Sci 74:62
Palmer LE, Rabinowicz PD, O'Shaughnessy AL, Vivekanand SB, Nascimento LU, Dike S, McCombie WR (2003) Maize genome sequencing by methylation filtration. Science 302:2115–2117
Palmqvist NGM, Seisenbaeva GA, Svedlindh P, Kessler VG (2017) Maghemite nanoparticles act as nanozymes, improving growth and abiotic stress tolerance in *Brassica napus*. Nanoscale Res Lett 12:631
Pandey N, Pandey-Rai S (2015) Deciphering UV-B-induced variation in DNA methylation pattern and its influence on regulation of DBR2 expression in *Artemisia annua* L. Planta 242:869–879
Pandey N, Goswami N, Tripathi D, Rai KK, Rai SK, Singh S, Pandey-Rai S (2019) Epigenetic control of UV-B-induced flavonoid accumulation in *Artemisia annua* L. Planta 249:497–514
Parkin TB, Kaspar TC, Jaynes DB, Moorman TB (2016) Rye cover crop effects on direct and indirect nitrous oxide emissions. Soil Sci Soc Am J 80:1551–1559

Pasala R, Kulasekaran R, Pandey BB, Manikanta CHLN, Gopika K, John Daniel PS, Yadav P (2022) Recent advances in micronutrient foliar spray for enhancing crop productivity and managing abiotic stress tolerance. In: Plant nutrition and food security in the era of climate change, pp 377–398

Patel K, Agarwal P, Agarwal PK (2018) Kappaphycus alvarezii sap mitigates abiotic-induced stress in *Triticum durum* by modulating metabolic coordination and improves growth and yield. J Appl Phycol 30:2659–2673

Pieterse CMJ, Leon-Reyes A, Van Der Ent S, Van Wees SCM (2009) Networking by small-molecule hormones in plant immunity. Nat Chem Biol 5(5):308–316

Plaksenkova I, Kokina I, Petrova A, Jermaļonoka M, Gerbreders V, Krasovska M (2020) The impact of zinc oxide nanoparticles on cytotoxicity, genotoxicity, and miRNA expression in barley (*Hordeum vulgare* L.) seedlings. Sci World J. 2020:6649746

Platten JD, Egdane JA, Ismail AM (2013) Salinity tolerance, Na+ exclusion and allele mining of HKT1;5 in *Oryza sativa* and *O. Glaberrima*: many sources, many genes, one mechanism? BMC Plant Biol 13:32

Pontes O, Li CF, Nunes PC, Haag J, Ream T, Vitins A, Pikaard CS (2006) The Arabidopsis chromatin-modifying nuclear siRNA pathway involves a nucleolar RNA processing center. Cell 126:79–92

Popova OV, Dinh HQ, Aufsatz W, Jonak C (2013) The RdDM pathway is required for basal heat tolerance in Arabidopsis. Mol Plant 6(2):396–410

Potter M (2020) Protecting crops from abiotic stress: copper oxide nanoparticle effects on wheat and a beneficial rhizobacterium. Retrieved from https://digitalcommons.usu.edu/etd/7864/

Potter M, Deakin J, Cartwright A, Hortin J, Sparks D, Anderson AJ, Britt DW (2021) Absence of nanoparticle-induced drought tolerance in nutrient sufficient wheat seedlings. Environ Sci Technol 55(20):13541–13550

Prakash PT, Banan D, Paul RE, Feldman MJ, Xie D, Freyfogle L, Leakey ADB (2021) Correlation and co-localization of QTL for stomatal density, canopy temperature, and productivity with and without drought stress in Setaria. J Exp Bot 72(13):5024–5037

Praveen B, Sharma P (2020) Climate change and its impacts on Indian agriculture: an econometric analysis. J Public Affairs 20(1)

Pshenichnikova TA, Osipova SV, Smirnova OG, Leonova IN, Permyakova MD, Permyakov AV, Börner A (2021) Regions of chromosome 2a of bread wheat (*Triticum aestivum* L.) associated with variation in physiological and agronomical traits under contrasting water regimes. Plan Theory 10:1023

Puchta H (2005) The repair of double-strand breaks in plants: mechanisms and consequences for genome evolution. J Exp Bot 56(409):1–14

Punamiya P, Datta R, Sarkar D, Barber S, Patel M, Das P (2010) Symbiotic role of *Glomus mosseae* in phytoextraction of lead in vetiver grass [*Chrysopogon zizanioides* (L.)]. J Hazard Mater 177(1–3):465–474

Pureswaran DS, Roques A, Battisti A (2018) Forest insects and climate change. Curr Forestry Rep 4:35–50

Rabbi SMHA, Kumar A, Mohajeri Naraghi S, Sapkota S, Alamri MS, Elias EM, Kianian S, Seetan R, Missaoui A, Solanki S, Mergoum M (2021) Identification of main-effect and environmental interaction QTL and their candidate genes for drought tolerance in a wheat RIL population between two elite spring cultivars. Front Genet 12:656037

Rajput VD, Minkina T, Kumari A, Harish, Singh VK, Verma KK, Keswani C (2021) Coping with the challenges of abiotic stress in plants: new dimensions in the field application of nanoparticles. Plan Theory 10:1221

Ramirez CL, Foley JE, Wright DA, Müller-Lerch F, Rahman SH, Cornu TI, Joung JK (2008) Unexpected failure rates for modular assembly of engineered zinc fingers. Nat Methods 5(5):374–375

Ramirez-Prado JS, Latrasse D, Rodriguez-Granados NY, Huang Y, Manza-Mianza D, Brik-Chaouche R, Benhamed M (2019) The Polycomb protein LHP1 regulates *Arabidopsis thaliana*

stress responses through the repression of the MYC2-dependent branch of immunity. Plant J 100:1118–1131
Reeger JE, Wheatley M, Yang Y, Brown KM (2021) Targeted mutation of transcription factor genes alters metaxylem vessel size and number in rice roots. Plant Direct 5:e00328
Ren ZH, Gao JP, Li LG, Cai XL, Huang W, Chao DY, Lin HX (2005) A rice quantitative trait locus for salt tolerance encodes a sodium transporter. Nat Genet 37(10):1141–1146
Rizvi A, Khan MS (2018) Heavy metal-induced oxidative damage and root morphology alterations of maize (Zea mays L.) plants and stress mitigation by metal-tolerant nitrogen-fixing *Azotobacter chroococcum*. Ecotoxicol Environ Saf 157:9–20
Rose MT, Patti AF, Little KR, Brown AL, Jackson WR, Cavagnaro TR (2014) A meta-analysis and review of plant-growth response to humic substances: practical implications for agriculture. In: Sparks DL (ed) Advances in Agronomy, vol 124, 1st edn
Ross-Ibarra J, Morrell PL, Gaut BS (2007) Plant domestication, a unique opportunity to identify the genetic basis of adaptation. Proc Natl Acad Sci 104:8641–8648
Rouphael Y, Colla G (2020) Editorial: Biostimulants in agriculture. Front Plant Sci 11:40
Sadakierska-Chudy A, Filip M (2015) A comprehensive view of the epigenetic landscape. Part II: histone post-translational modification, nucleosome level, and chromatin regulation by ncRNAs. Neurotox Res 27:172–197
Sajjad N, Bhat EA, Shah D, Manzoor I, Noor W, Shah S, Ali R (2021) Nitrogen uptake, assimilation, and mobilization in plants under abiotic stress. In: Transporters and plant osmotic stress. Academic, pp 215–233
Salgotra RK, Stewart CN Jr (2020) Functional markers for precision plant breeding. Int J Mol Sci 21:4792
Sanjana NE, Cong L, Zhou Y, Cunniff MM, Feng G, Zhang F (2012) A transcription activator-like effector toolbox for genome engineering. Nat Protoc 7(1):171–192
Sato A, Sato Y, Fukao Y, Fujiwara M, Umezawa T, Shinozaki K et al (2009) Threonine at position 306 of the KAT1 potassium channel is essential for channel activity and is a target site for ABA-activated SnRK2/OST1/SnRK2.6 protein kinase. Biochem J 424:439–448
Savary S, Willocquet L, Pethybridge SJ, Esker P, McRoberts N, Nelson A (2019) The global burden of pathogens and pests on major food crops. Nat Ecol Evol 3(3):430–439
Saxena R, Tomar RS, Kumar M (2016) Exploring nanobiotechnology to mitigate abiotic stress in crop plants. J Pharm Sci Res 8(9):974–980
Schaafsma G (2009) Safety of protein hydrolysates, fractions thereof and bioactive peptides in human nutrition. Eur J Clin Nutr 63:1161–1168
Schützendübel A, Polle A (2002) Plant responses to abiotic stresses: heavy metal-induced oxidative stress and protection by mycorrhization. J Exp Bot 53(372):1351–1365
Seidl R, Honkaniemi J, Aakala T, Aleinikov A, Angelstam P, Bouchard M, Senf C (2020) Globally consistent climate sensitivity of natural disturbances across boreal and temperate forest ecosystems. Ecography 43(7):967–978
Shah ZH, Rehman HM, Akhtar T, Alsamadany H, Hamooh BT, Mujtaba T, Chung G (2018) Humic substances: determining potential molecular regulatory processes in plants. Front Plant Sci 9:263
Shahbazian MD, Grunstein M (2007) Functions of site-specific histone acetylation and deacetylation. Annu Rev Biochem 76:75–100
Shamaya NJ, Shavrukov Y, Langridge P, Roy SJ, Tester M (2017) Genetics of Na+ exclusion and salinity tolerance in Afghani durum wheat landraces. BMC Plant Biol 17:209
Shan S, Soltis PS, Soltis DE, Yang B (2020) Considerations in adapting CRISPR/Cas9 in nongenetic model plant systems. Appl Plant Sci 8(1):1–17
Sharma S, Muddassir M, Muthusamy S, Vaishnav PK, Singh M, Sharma D, Shanmugam V (2020a) A non-classical route of efficient plant uptake verified with fluorescent nanoparticles and root adhesion forces investigated using AFM. Sci Rep 10:19233

Sharma H, Taneja M, Upadhyay SK (2020b) Identification, characterization and expression profiling of cation-proton antiporter superfamily in *Triticum aestivum* L. and functional analysis of TaNHX4-B. Genomics 112(1):356–370

Sheehy J, Elmido A, Centeno G, Pablico P (2005) Searching for new plants for climate change. J Agric Meteorol 60(5):463–468

Shi J, Gao H, Wang H, Lafitte HR, Archibald RL, Yang M, Habben JE (2017) ARGOS8 variants generated by CRISPR-Cas9 improve maize grain yield under field drought stress conditions. Plant Biotechnol J 15:207–216

Shimizu T, Masuda T (2021) The role of tetrapyrrole-and gun1-dependent signaling on chloroplast biogenesis. Plan Theory 10(2):196

Shipitalo MJ, Owens LB, Bonta JV, Edwards WM (2013) Effect of no-till and extended rotation on nutrient losses in surface runoff. Soil Sci Soc Am J 77:1329–1337

Shirdelmoghanloo H, Cozzolino D, Lohraseb I, Collins NC (2016) Truncation of grain filling in wheat (*Triticum aestivum*) triggered by brief heat stress during early grain filling: association with senescence responses and reductions in stem reserves. Funct Plant Biol 43:919–930

Singh A, Tiwari S, Pandey J, Lata C, Singh IK (2021) Role of nanoparticles in crop improvement and abiotic stress management. J Biotechnol 337:57–70

Smith J, Bibikova M, Whitby FG, Reddy AR, Chandrasegaran S, Carroll D (2000) Requirements for double-strand cleavage by chimeric restriction enzymes with zinc finger DNA-recognition domains. Nucleic Acids Res 28(17):3361–3369

Snyder CS (2017) Enhanced nitrogen fertiliser technologies support the "4R" concept to optimise crop production and minimise environmental losses. Soil Res 55:463–472

Srikant T, Drost HG (2021) How stress facilitates phenotypic innovation through epigenetic diversity. Front Plant Sci 11:606800

Stenberg JA, Ortiz R (2021) Focused identification of germplasm strategy (FIGS): polishing a rough diamond. Curr Opin Insect Sci 45:1–6

Su YH, North H, Grignon C, Thibaud JB, Sentenac H, Véry AA (2005) Regulation by external K+ in a maize inward shaker channel targets transport activity in the high concentration range. Plant Cell 17:1532–1548

Su P, Jiang C, Qin H, Hu R, Feng J, Chang J, He G (2019) Identification of potential genes responsible for thermotolerance in wheat under high temperature stress. Genes 10:174

Sun Y, Zhang X, Wu C, He Y, Ma Y, Hou H, Xia L (2016) Engineering herbicide-resistant rice plants through CRISPR/Cas9-mediated homologous recombination of acetolactate synthase. Mol Plant 9(April):628–631

Sun L, Song F, Zhu X, Liu S, Liu F, Wang Y, Li X (2021) Nano-ZnO alleviates drought stress via modulating the plant water use and carbohydrate metabolism in maize. Arch Agron Soil Sci 67:245–259

Sun L, Wen J, Peng H, Yao Y, Hu Z, Ni Z, Xin M (2022) The genetic and molecular basis for improving heat stress tolerance in wheat. aBIOTECH 3:25–39

Sunarpi HT, Motoda J, Kubo M, Yang H, Yoda K, Uozumi N (2005) Enhanced salt tolerance mediated by AtHKT1 transporter-induced Na + unloading from xylem vessels to xylem parenchyma cells. Plant J 44:928–938

Suzuki N, Koussevitzky S, Mittler R, Miller G (2012) ROS and redox signalling in the response of plants to abiotic stress. Plant Cell Environ 35:259–270

Svitashev S, Young JK, Schwartz C, Gao H, Falco SC, Cigan AM (2015) Targeted mutagenesis, precise gene editing, and site-specific gene insertion in maize using Cas9 and guide RNA. Plant Physiol 169:931–945

Swarup S, Cargill EJ, Crosby K, Flagel L, Kniskern J, Glenn KC (2021) Genetic diversity is indispensable for plant breeding to improve crops. Crop Sci 61(2):839–852

Szostak JW, Orr-Weaver TL, Rothstein RJ, Stahl FW (1983) The double-strand-break repair model for recombination. Cell 33:25–35

Tabassum J, Ahmad S, Hussain B, Mawia AM, Zeb A, Ju L (2021) Applications and potential of genome-editing systems in rice improvement: current and future perspectives. Agronomy 11(7):1–24

Tang X, Cao X, Xu X, Jiang Y, Luo Y, Ma Z, Zhou Y (2017a) Effects of climate change on epidemics of powdery mildew in winter wheat in China. Plant Dis 101(10):1753–1760

Tang L, Mao B, Li Y, Lv Q, Zhang L, Chen C, Zhao B (2017b) Knockout of OsNramp5 using the CRISPR/Cas9 system produces low cd-accumulating indica rice without compromising yield. Sci Rep 7:14438

Tang S, Chen W, Liu W, Zhou Q, Zhang H, Wang S, Ding Y (2018) Open-field warming regulates the morphological structure, protein synthesis of grain and affects the appearance quality of rice. J Cereal Sci 84:20–29

Tanksley SD, McCouch SR (1997) Seed banks and molecular maps: unlocking genetic potential from the wild. Science 277(August):1063–1066

Thoma DP, Shanahan EK, Irvine KM (2019) Climatic correlates of white pine blister rust infection in whitebark pine in the greater Yellowstone ecosystem. Forests 10:666

Tian S, Jiang L, Cui X, Zhang J, Guo S, Li M, Xu Y (2018) Engineering herbicide-resistant watermelon variety through CRISPR/Cas9-mediated base-editing. Plant Cell Rep 37:1353–1356

Tiryaki D, Aydın İ, Atıcı Ö (2019) Psychrotolerant bacteria isolated from the leaf apoplast of cold-adapted wild plants improve the cold resistance of bean (*Phaseolus vulgaris* L.) under low temperature. Cryobiology 86:111–119

Tiwari P, Srivastava D, Chauhan AS, Indoliya Y, Singh PK, Tiwari S et al (2021) Root system architecture, physiological analysis and dynamic transcriptomics unravel the drought-responsive traits in rice genotypes. Ecotoxicol Environ Saf 207:111252

Touzy G, Rincent R, Bogard M, Lafarge S, Dubreuil P, Mini A, Praud S (2019) Using environmental clustering to identify specific drought tolerance QTLs in bread wheat (*T. aestivum* L.). Theor Appl Genet 132(10):2859–2880

Tshikunde NM, Mashilo J, Shimelis H, Odindo A (2019) Agronomic and physiological traits, and associated quantitative trait loci (QTL) affecting yield response in wheat (*Triticum aestivum* L.): a review. Front Plant Sci 10:1428

Tsuji H, Saika H, Tsutsumi N, Hirai A, Nakazono M (2006) Dynamic and reversible changes in histone H3-Lys4 methylation and H3 acetylation occurring at submergence-inducible genes in rice. Plant Cell Physiol 47(7):995–1003

van Bezouw RFHM, Janssen EM, Ashrafuzzaman M, Ghahramanzadeh R, Kilian B, Graner A, vader Linden CG (2019) Shoot sodium exclusion in salt stressed barley (*Hordeum vulgare* L.) is determined by allele specific increased expression of HKT1;5. J Plant Physiol 241:153029

Van Nguyen D, Nguyen HM, Le NT, Nguyen KH, Nguyen HT, Le HM, Van Ha C (2021) Copper nanoparticle application enhances plant growth and grain yield in maize under drought stress conditions. J Plant Growth Regul 41:364–375

van Zelm E, Zhang Y, Testerink C (2020) Salt tolerance mechanisms of plants. Annu Rev Plant Biol 71:403–433

Varshney RK, Bohra A, Roorkiwal M, Barmukh R, Cowling WA, Chitikineni A et al (2021) Fast-forward breeding for a food-secure world. Trends Genet 37(12):1124–1136

Vranova V, Rejsek K, Skene KR, Formanek P (2011) Non-protein amino acids: plant, soil and ecosystem interactions. Plant Soil 342:31–48

Wahid A, Gelani S, Ashraf M, Foolad MR (2007) Heat tolerance in plants: an overview. Environ Exp Bot 61:199–223

Wang WS, Pan YJ, Zhao XQ, Dwivedi D, Zhu LH, Ali J, Li ZK (2011) Drought-induced site-specific DNA methylation and its association with drought tolerance in rice (Oryza sativa L.). J Exp Bot 62(6):1951–1960

Wang Y, Cheng X, Shan Q, Zhang Y, Liu J, Gao C, Qiu JL (2014) Simultaneous editing of three homoeoalleles in hexaploid bread wheat confers heritable resistance to powdery mildew. Nat Biotechnol 32:947–951

Wang X, Vignjevic M, Liu F, Jacobsen S, Jiang D, Wollenweber B (2015a) Drought priming at vegetative growth stages improves tolerance to drought and heat stresses occurring during grain filling in spring wheat. Plant Growth Regul 75:677–687
Wang S, Zhang S, Wang W, Xiong X, Meng F, Cui X (2015b) Efficient targeted mutagenesis in potato by the CRISPR/Cas9 system. Plant Cell Rep 34:1473–1476
Wang X, Li Q, Yuan W, Cao Z, Qi B, Kumar S, Qian W (2016) The cytosolic Fe-S cluster assembly component MET18 is required for the full enzymatic activity of ROS1 in active DNA demethylation. Sci Rep 6:26443
Wang F-Z, Chen M-X, Yu L-J, Xie L-J, Yuan L-B, Qi H, Chen QF (2017) OsARM1, an R2R3 MYB transcription factor, is involved in regulation of the response to arsenic stress in rice. Front Plant Sci 8:1868
Wang Z, Hong Y, Zhu G, Li Y, Niu Q, Yao J, Zhu J (2020) Loss of salt tolerance during tomato domestication conferred by variation in a Na+ /K+ transporter. EMBO J 39(10):1–14
Ward JM, Mäser P, Schroeder JI (2009) Plant ion channels: gene families, physiology, and functional genomics analyses. Ann Rev Physiol 71:59–82
Waters ER, Vierling E (2020) Plant small heat shock proteins – evolutionary and functional diversity. New Phytol 227(1):24–37
Watson SJ, Sowden RG, Jarvis P (2018) Abiotic stress-induced chloroplast proteome remodelling: a mechanistic overview. J Exp Bot 69(11):2773–2781
Weed AS, Ayres MP, Hicke JA (2013) Consequences of climate change for biotic disturbances in North American forests. Ecol Monogr 83(4):441–470
Wen J, Jiang F, Weng Y, Sun M, Shi X, Zhou Y, Wu Z (2019) Identification of heat-tolerance QTLs and high-temperature stress-responsive genes through conventional QTL mapping, QTL-seq and RNA-seq in tomato. BMC Plant Biol 19(398):1–17
Woodson JD (2016) Chloroplast quality control – balancing energy production and stress. New Phytol 212:36–41
Wu H, Tito N, Giraldo JP (2017) Anionic cerium oxide nanoparticles protect plant photosynthesis from abiotic stress by scavenging reactive oxygen species. ACS Nano 11:11283–11297
Wu GZ, Chalvin C, Hoelscher M, Meyer EH, Wu XN, Bock R (2018) Control of retrograde signaling by rapid turnover of GENOMES UNCOUPLED1. Plant Physiol 176:2472–2495
Xiao L, Guo H, Wang S, Li J, Wang Y, Xing B (2019) Carbon dots alleviate the toxicity of cadmium ions (Cd2+) toward wheat seedlings. Environ. Sci. Nano 6:1493–1506
Xie K, Minkenberg B, Yang Y (2015) Boosting CRISPR/Cas9 multiplex editing capability with the endogenous tRNA-processing system. Proc Natl Acad Sci USA 112(11):3570–3575
Xing H-L, Dong L, Wang Z-P, Zhang H-Y, Han C-Y, Liu B, Chen QJ (2014) A CRISPR/Cas9 toolkit for multiplex genome editing in plants. BMC Plant Biol 4:327
Xu K, Xu X, Fukao T, Canlas P, Maghirang-Rodriguez R, Heuer S, Mackill DJ (2006) Sub1A is an ethylene-response-factor-like gene that confers submergence tolerance to rice. Nature 442(7103):705–708
Xu Y, Zhan C, Huang B (2011) Heat shock proteins in association with heat tolerance in grasses. Int J Proteomics 2011:1–11
Xu R, Wang Y, Zheng H, Lu W, Wu C, Huang J, Zheng C (2015) Salt-induced transcription factor MYB74 is regulated by the RNA-directed DNA methylation pathway in Arabidopsis. J Exp Bot 66(19):5997–6008
Xu D, Marino G, Klingl A, Enderle B, Monte E, Kurth J, Kleine T (2019) Extrachloroplastic PP7L functions in chloroplast development and abiotic stress tolerance. Plant Physiol 180:323–341
Xu D, Leister D, Kleine T (2020) VenoSA4, a human dNTPase SAMHD1 homolog, contributes to chloroplast development and abiotic stress tolerance. Plant Physiol 182:721–729
Xu K, Chen S, Li T, Yu S, Zhao H, Liu H, Luo L (2021) Overexpression of OsHMGB707, a high mobility group protein, enhances rice drought tolerance by promoting stress-related gene expression. Front Plant Sci 12:711271
Yakhin OI, Lubyanov AA, Yakhin IA, Brown PH (2017) Biostimulants in plant science: a global perspective. Front Plant Sci 7:2049

Yan H, Wang C, Liu K, Tian X (2021a) Detrimental effects of heat stress on grain weight and quality in rice (*Oryza sativa* L.) are aggravated by decreased relative humidity. Peer J 9:e11218

Yan S, Wu F, Zhou S, Yang J, Tang X, Ye W (2021b) Zinc oxide nanoparticles alleviate the arsenic toxicity and decrease the accumulation of arsenic in rice (*Oryza sativa* L.). BMC Plant Biol 21:150

Yolcu S, Alavilli H, Lee B (2020) Natural genetic resources from diverse plants to improve abiotic stress tolerance in plants. Int J Mol Sci 21:8567

Yong H-Y, Wang C, Bancroft I, Li F, Wu X, Kitashiba H, Nishio T (2015) Identification of a gene controlling variation in the salt tolerance of rapeseed (*Brassica napus* L.). Planta 242(1):313–326

Yooyongwech S, Samphumphuang T, Tisarum R, Theerawitaya C, Cha-Um S (2016) Arbuscular mycorrhizal fungi (AMF) improved water deficit tolerance in two different sweet potato genotypes involves osmotic adjustments via soluble sugar and free proline. Sci Hort 198:107–117

You L, Song Q, Wu Y, Li S, Jiang C, Chang L, Zhang J (2019) Accumulation of glycine betaine in transplastomic potato plants expressing choline oxidase confers improved drought tolerance. Planta 249:1963–1975

Younis A, Ramzan F, Ramzan Y, Zulfiqar F, Ahsan M, Lim KB (2020) Molecular markers improve abiotic stress tolerance in crops: a review. Plan Theory 9:1374

Yousaf B, Liu G, Wang R, Imtiaz M, Zia-ur-Rehman M, Munir MAM, Niu Z (2016) Bioavailability evaluation, uptake of heavy metals and potential health risks via dietary exposure in urban-industrial areas. Environ Sci Pollut Res 23:22443–22453

Zemach A, Kim MY, Silva P, Rodrigues JA, Dotson B, Brooks MD, Zilberman D (2010) Local DNA hypomethylation activates genes in rice endosperm. Proc Natl Acad Sci USA 107(43):18729–18734

Zenda T, Liu S, Dong A, Duan H (2021) Advances in cereal crop genomics for resilience under climate change. Life 11:502

Zhai H, Jiang C, Zhao Y, Yang S, Li Y, Yan K, Wang D (2021) Wheat heat tolerance is impaired by heightened deletions in the distal end of 4AL chromosomal arm. Plant Biotechnol J 19:1038–1051

Zhang Z, Zhang J, Chen Y, Li R, Wang H, Wei J (2012) Genome-wide analysis and identification of HAK potassium transporter gene family in maize (Zea mays L.). Mol Biol Rep 39(8):8465–8473

Zhang Y, Liang Z, Zong Y, Wang Y, Liu J, Chen K, Gao C (2016) Efficient and transgene-free genome editing in wheat through transient expression of CRISPR/Cas9 DNA or RNA. Nat Commun 7:1–8

Zhang M, Cao Y, Wang Z, Wang ZQ, Shi J, Liang X, Jiang C (2018a) A retrotransposon in an HKT1 family sodium transporter causes variation of leaf Na+ exclusion and salt tolerance in maize. New Phytol 217:1161–1176

Zhang T, Hu Y, Zhang K, Tian C, Guo J (2018b) Arbuscular mycorrhizal fungi improve plant growth of *Ricinus communis* by altering photosynthetic properties and increasing pigments under drought and salt stress. Ind Crop Prod 117:13–19

Zhang H, Lang Z, Zhu JK (2018c) Dynamics and function of DNA methylation in plants. Nat Rev Mol Cell Biol 19:489–506

Zhang H, Li Y, Zhu JK (2018d) Developing naturally stress-resistant crops for a sustainable agriculture. Nat Plants 4(December):989–996

Zhang ZS, Alfatih A, Wu J, Jan SU, Xia JQ, Xiang CB (2020) Loss of rice PARAQUAT TOLERANCE 3 confers enhanced resistance to abiotic stresses and increases grain yield in field. Plant Cell Environ 43(11):2743–2754

Zhang F, Wu J, Sade N, Wu S, Egbaria A, Fernie AR, Dai M (2021) Genomic basis underlying the metabolome-mediated drought adaptation of maize. Genome Biol 22:260

Zhang G, Zhou J, Peng Y, Tan Z, Li L, Yu L, Yao X (2022a) Genome-wide association studies of salt tolerance at seed germination and seedling stages in *Brassica napus*. Front Plant Sci 12:772708

Zhang H, Zhu J, Gong Z, Zhu JK (2022b) Abiotic stress responses in plants. Nat Rev Genet 23(2):104–119

Zhao C, Wang Y, Chan KX, Marchant DB, Franks PJ, Randall D, Chen ZH (2019a) Evolution of chloroplast retrograde signaling facilitates green plant adaptation to land. Proc Natl Acad Sci USA 116(11):5015–5020

Zhao X, Zhang J, Fang P, Peng Y (2019b) Comparative QTL analysis for yield components and morphological traits in maize (*Zea mays* L.) under water-stressed and well-watered conditions. Breed Sci 69(4):621–632

Zheng X, Chen L, Xia H, Wei H, Lou Q, Li M, Luo L (2017) Transgenerational epimutations induced by multi-generation drought imposition mediate rice plant's adaptation to drought condition. Sci Rep 7:39843

Zhu JK (2008) Epigenome sequencing comes of age. Cell 133:395–397

Zhu J (2016) Abiotic stress signaling and responses in plants. Cell 167(2):313–324

Zohra E, Ikram M, Omar AA, Hussain M, Satti SH, Raja NI, Ehsan M (2021) Potential applications of biogenic selenium nanoparticles in alleviating biotic and abiotic stresses in plants: a comprehensive insight on the mechanistic approach and future perspectives. Green Process Synth 10:456–475

The Impact of Climate Change on Drought Occurrence

Tefide Kizildeniz

Abstract Climate has been fluctuated from the pre-industrial era. Carbon emission has increased the greenhouse gases, which have sparked climate crisis by driving global temperature rises. CO_2 is the principal and most prevalent atmospheric greenhouse gas. Depending on the mitigating measures taken, values might vary from 1.5 °C higher than the present reference period (1986–2005) to more than 4 °C. The greenhouse effect is having an influence on crops. The rising temperatures are causing more frequent Mediterranean agricultural droughts due to reduced water availability for irrigation systems and increased evaporation. Global warming projections anticipate that future circumstances will be significantly drier. Irrigation is a vital resource for growers who cultivate under this climate to enhance soil water resources, may resulting in increased yields without quality losses. For the crops grown under climate change areas to manage the crop growth and quality, controlled deficit irrigation may be applied. In addition to drought, as stress variables function independently and/or combine, its response to climate change-related factors might vary.

Keywords Agriculture · Crops · Drought · Effects of climate change · Elevated temperature · High CO_2 · Vegetative growth · Quality · Yield

1 Introduction

Drought is among the most challenging hydro-climatic risk impacting the planet since its intensity is difficult to evaluate (Wilhite and Pulwarty 2017). It is nearly hard to define a general description of drought since it is an effect-dependent phenomenon with a wide range of impact (Lloyd-Hughes 2014). As a result, distinct

T. Kizildeniz (✉)
Biosystem Engineering Department, Faculty of Agricultural Sciences and Technologies, Niğde Ömer Halisdemir University, Niğde, Türkiye
e-mail: tkizildeniz@ohu.edu.tr

U. K. Chaudhry et al. (eds.), *Drought Stress*,
https://doi.org/10.1007/978-3-031-80610-0_2

drought types must be defined according to the system and sector that are impacted. Wilhite and Glantz (1985) indicated that traditional definitions of drought are broken down into 4 predicaments: meteorological, agricultural, hydrological, and socioeconomic. Considering how drought impacts proliferate across the hydrological cycle, these four drought categories are intimately linked to one another (Changnon and Easterling 1989). Drought impacts are frequently associated with agricultural and water resource industries. They can cause significant economic losses in advanced countries' agriculture sectors due to crop production decreases or crop failure (Sweet et al. 2017; Tian et al. 2018). In severe circumstances, it can also lead to starvation and people migration in poor nations (Grolle 2015). Irrigated agricultural systems may have major repercussions from hydrological droughts as well (Maestro et al. 2016; Vidal-Macua et al. 2018), along with issues with urban water supply, industrial demands, and a decline in hydropower output. Ecosystems can be greatly impacted by drought; sometimes these effects are short-lived, while other times they are structural or permanent (Fig. 1). Drought is influenced by a variety of factors as demonstrated in Fig. 1, including precipitation that only an increase in global precipitation would not always result in fewer droughts.

The fast emission of greenhouse gases was caused by human activities like clearing forests for agricultural and commercial reasons, inappropriate utilization unrefined petroleum products, and enhanced use of air conditioners, all of which have contributed to increase observed in the last couple decades (Morton 2007). Since the industrial revolution, there has been a 25% more CO_2 in the atmosphere as a result of combustion of fossil fuels and extensive destruction of forests (Perera 2018). Certain gas levels, like those of methane, chlorofluorocarbons, and nitrous oxide also have significantly enhanced because of human influence (Roul et al. 2017). It has a significant role in the altering of atmosphere's composition, which cause global warming and alterations in the climatic condition (Lobell et al. 2008). According to analyses of historical climatic shifts, a 1.06 °C increase occurred in main land surface temperature through a period of more than a decade and increase

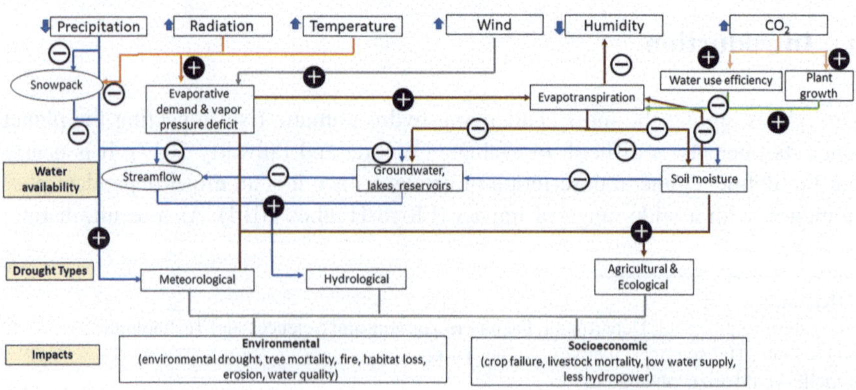

Fig. 1 Climate factors influencing water supply and drought. (Modified from Douville et al. 2021)

0.85 °C has occurred in previous 20 years (IPCC 2014a). Depending on mitigation strategies used, climate forecasts for the end of twenty-first century predict temperature upsurges in a vast variety, from a 1.5 °C rise over to a more than 4 °C rise in world average temperature (IPCC 2014b). Human emissions of greenhouse gases have been the primary cause of the temperature rise. The most significant in terms of volume and impact on the world is carbon dioxide (CO_2; IPCC 2014b) with projections for the turn of the era ranging from 421 to a 936 μL L^{-1}. Its concentration has increased from a pre-industrial level of 280 μL L^{-1} to today more than 400 μL L^{-1} (Meinshausen et al. 2011). The rise in atmospheric CO_2 has effects in the Earth's photosynthetic action in addition to the impacts of global warming (Ziska 2022). Additionally, the stratospheric ozone layer is being destroyed by the buildup of chlorofluorocarbons (CFCs) and other chlorinated and brominated substances, which results in an increase at the Earth's surface in ultraviolet (UV) radiation (Ozcelik 2021).

Thus, global warming, irregularities in precipitation, frequent occurrence of normally unusual severe weather occurrences, melting of arctic ice, an increase in oceanic temperature, sea levels, and weather extremes are all results of the climate change that began in the 1970s (Brown and Funk 2008; Kotschi 2007). Severe weather conditions (droughts and flooding) frequently result in the destruction of crops and a food shortage (Porter et al. 2014; IPCC 2018). As a consequence of this, climate change has an impact on both industrialized and developing nations.

Climate change is seriously endangering humans and biodiversity by impairing agricultural productivity (Siddiqui et al. 2012; Ahmed et al. 2016; FAO 2020; Chandio et al. 2021). It is important in building an economy where agriculture is the primary employment source for the majority of people. The sensitivity of many types of food crops to rising average temperatures is high. Major agricultural crops are being affected by climate changes brought on by rising annual temperatures, shifting rainfall patterns, floods, and dwindling water reservoirs (Appiah et al. 2018; Chandio et al. 2019; Abbas 2022). Almost half of the working force is employed in agriculture, which is a significant output sector in emerging and less developed nations. Aside from the significance of agriculture, jobs, and food security also promotes the expansion of other related industries, alleviates rural poverty, and raises agricultural exports (Tripathi and Prasad 2009; Vyas et al. 2003).

Tramblay et al. (2020) exhibited in Mediterranean region, both the frequency and the spatial breadth of drought events have dramatically increased in recent years. These occurrences are linked to warm temperatures and little precipitation. Given the current climatic warming, the trend could pick in speed in the future (AghaKouchak et al. 2014; Forootan et al. 2019; Raymond et al. 2018; Spinoni et al. 2014). García-Herrera et al. (2019), Schumacher et al. (2019), and Zhao et al. (2019) illustrate how rising temperatures forecasts can boost potential evapotranspiration while decreasing flow. As a result, research on how climate change will affect the water cycles is required to foresee the hazard and effects of hydro-meteorological extremes. A confluence of minimal rainfall (meteorological drought) and higher evapotranspiration results drought in agricultural production. The emergence of an agricultural drought threatens food safety and agricultural production, and it may have long-term socioeconomic effects. Depending on the region, the climate at the time, and the crop species composition, these effects will change (Forzieri et al. 2020). Giorgi et al. (2009) classify the Mediterranean basin as a hotspot because it is thought to be the region most susceptible to climate change, particularly along its southern shore. In the future, it is anticipated that more extreme climate events, like drought, may occur, especially in the basin's semi-arid parts (Achour et al. 2020; Raymond et al. 2018; Tramblay and Hertig 2018; Zeroual et al. 2019, 2020; Zkhiri et al. 2019). Therefore, there is a general need to understand how agricultural drought and water stress affect food and agricultural productivity in internationally.

The extraordinary rise in world population that estimates to cross 9.10 billion by 2050 (FAO 2009) and stagnant agricultural output are contributing to an ongoing rise in world hunger and food insecurity. One of the main goals of the Sustainable Development Goals (SDGs) is to eradicate hunger and provide food security. Water is needed for both human and agricultural purposes have increased due to population growth and socioeconomic activity, endangering the world's current water reserves, especially in dry and semi-arid locations. In order to assure the most effective usage of water in a variety of activities, it is imperative to create plans and scenarios based on realistic predictions.

2 Drought

Drought is defined as a protracted period of little or no rain, which results in a significant hydrologic unbalance and, as a result, scarcity of water, yield loss, reduced water flow and groundwater and soil moisture loss. Although its features differ greatly from one climatic zone to another, drought is a common, recurring occurrence that affects almost all of them (Hulme 1995). The effects of a drought event depend on how the natural occurrence interacts with human demand for water supplies. Recent droughts have highlighted the susceptibility of all communities in response to this inherent danger, but certain factions are more susceptible than individuals. These droughts have occurred in both emerging and industrialized countries. The effects of the drought can be distinguished based on several disciplinary viewpoints such as hydrological, weather-related, and agricultural (Fig. 2).

2.1 Hydrological Drought

The impacts of times of precipitation shortage in underground or surface water holding structures (streams, lakes, reservoirs, and ground water) are referred to as hydrological drought (Wu et al. 2021). Although a lack of precipitation is the cause of all droughts, hydrologists are more interested in how this lack of precipitation affects the hydrologic system as a whole (a watershed or river basin). As rainfall takes longer to occur, deficits begin to manifest in hydrologic network elements.

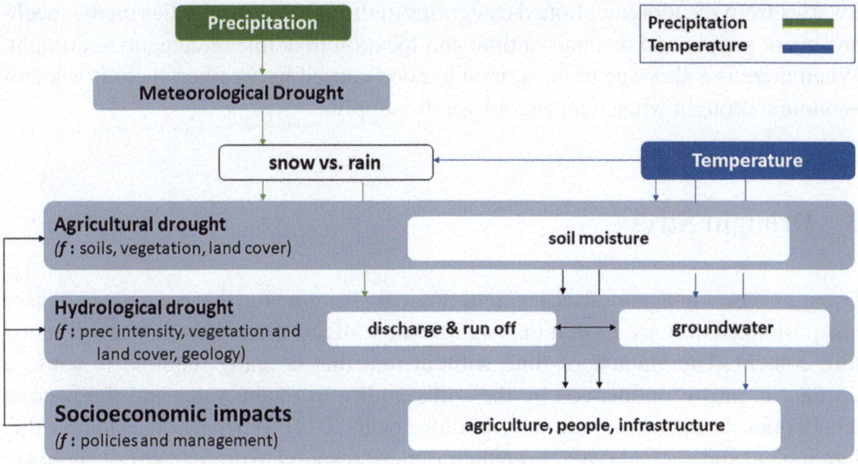

Fig. 2 Drought types, their evolutionary factors and effects are depicted schematically. (Modified from Cook et al. 2018)

2.2 *Weather-Related Drought*

Typically, the degree of dryness and length of the dry period are used to characterize meteorological drought. Thus, according to Hulme (1995), a meteorological drought is a decrease in rainfall supply during a certain length of time relative to a given average condition. Since the atmospheric circumstances that lead to a lack of precipitation vary greatly from place to region, definitions of meteorological drought must take this into account.

2.3 *Agricultural Drought*

Permanent drought is a feature of the driest climates and limited, aridity-adapted plants. Agriculture cannot be practiced in these locations without ongoing irrigation. Seasonal dryness during the 'dry season' is a feature of semi-arid environments. Rain-fed agriculture can only be practiced during the rainy season in places like sub-Saharan West Africa (Engelbrecht et al. 2009). More frequently, the drought is caused by anomalous rainfall failure, which can happen everywhere but is more prevalent in humid and sub-humid climatic zones (Hulme 1995).

2.4 *Socio-economic Drought*

Socio-economic descriptions of drought relate its characteristics to the supply and demand for certain economic items. These characteristics include meteorological, hydrological, and agricultural drought. According to Wilhite and Glantz (1985) that it varies from the aforementioned categories of drought in that it relies on the mechanisms of supply and demand in time and location to define or categorize drought. When there is a shortage of an economic good caused by weather, there is a socioeconomic drought when demand outweighs supply.

3 Drought Stress

Drought is best understood as a meteorological phenomenon that occurs when there is a prolonged absence of rain that significantly affects plant growth and soil moisture content. The amount of time without rain that is really required to cause a drought is mostly influenced by the soil's ability to retain water and the pace at which plants evapotranspiration (Orimoloye et al. 2022). When it comes to how different plant species and types, even their organs, respond to drought stress, there are substantial variations in their physiological and metabolic alterations (Belkhodja et al. 1994). The degree of drought generated by various genotypes relies on the metabolic alterations, or physiological and biochemical responses, that the genotype develops under stress (Kayabaşı 2011).

4 Impact of Drought Stress on Plants

4.1 Mechanical Impact

The instant stress that a plant experiences as its turgor is lost when water is lost from its cells in considerable amounts is mechanical (Time et al. 2018). The liquid-crystalline phase of the plasma membrane results from the cell's aqueous environment. A change in the membrane's structure occurs as water exits the cell. Membranes enter the compact gel phase, where they appear. The membrane in this new structure is less mobile than it was in the liquid-crystalline phase. The volume of the cell shrinks as a result of water loss. Tension-induced plasma membrane rupture is also possible (Božič et al. 2020).

4.2 Metabolic Impact

As water fills the majority of the volume of cells, serves as a transport medium, and serves as a solvent for biological events and processes, and when water is removed from cells, cellular regulation and metabolism are disrupted. Ion buildup that results from cellular water loss can harm cells, disrupt membranes, and result in protein denaturation. The interactions of amino acids with water in the structure of proteins are disrupted as a result of water loss (Božič et al. 2020). The breakdown of nucleic acids like DNA and RNA causes further harm to plant cells under drought stress.

4.3 *Photosynthetic Impact*

Stomatal restrictions, which result from stomatal closure under mild water deficit situations, and other constraints, which often occur under longer and more severe water strains, are the major causes of photosynthetic decline during drought. Stomatal closure, which restricts CO_2 passage towards chloroplasts, is one of the initial reactions to drought (Lima et al. 2002). It has been discovered that plants subjected to a severe water deficit have a reduction in their chloroplasts' photosynthetic electron transport and photophosphorylation capacity (Smirnoff 1993). Drought stress has an impact on photosystems, notably PSII in chloroplasts (Chen et al. 2016).

4.4 *Tolerance Mechanisms Against Drought Stress*

Drought stress causes a variety of physiological, biochemical, and molecular responses in plants, allowing them to adapt to constrained environmental conditions (Arora et al. 2002). The two primary defensive mechanisms elicited in vegetative tissues against drought stress are avoidance and tolerance. The first avoidance technique observed in ephemerals is fleeing. Desert Ephemerals are annual plants that survive the dry season by living only as dormant seeds. Its protoplasm is never exposed to extremely negative water potentials (Kalefetoğlu and Ekmekci 2005). Succulents have yet another avoidance strategy. Such plants can survive for long periods of time without additional hydration thanks to their succulent tissues' ability to store water, which helps them fend off droughts (Salisbury and Ross 1992). Desert evergreens prevent drought by producing osmoprotectants for tissue turgor regulation when there is a water shortage (Mundree et al. 2002).

4.5 *Behavior Towards Drought Stress*

Species and genotypes, duration and severity of water loss, the growth phase, organ and cell type, and cellular compartment (cell wall and cell membrane) all affect how plants react to a water shortage (Bhardwaj et al. 2021). Responses to drought stress can happen quickly for example with a change in a protein's phosphorylation state or slowly for example with a change in gene expression (Jiang et al. 2019). Stress-responsive genes are classified as 'early-response genes' or 'late-response genes'.

Early-response genes are triggered fast (within minutes) and frequently transiently. As all signaling elements are already present, its stimulation does not necessitate the creation of new proteins. Stress activates delayed-response genes more slowly (within hours), and its activation is typically maintained. They account for the vast majority of stress-response genes. While, stress activates delayed-response genes more slowly (within hours), and its activation is typically maintained. They account for the vast majority of stress-response genes (Weinhouse 2021).

5 Climate Change-Related Factors

The total impacts of individual or combination climate change-related factors have been predicted using a number of regional climate models. As a result of increased carbon emissions, certain atmospheric gas levels known as greenhouse gases have increased. These gases are what drive climate change by raising atmospheric temperatures on a global scale. As the main and the most abundant atmospheric greenhouse gas is CO_2, so among the most significant aspects of climate change is considered as atmospheric CO_2. The second-most significant global warming-related factor that is linked with CO_2 is increased air temperature. The rise in atmospheric CO_2 has consequences for the Earth's biological action in addition to the impacts of global warming (Amthor 1995). Aquatic and terrestrial plants may both flourish in certain environments will change due to the effects of escalating CO_2 levels in the atmosphere, elevated atmospheric air temperatures, increased UV radiation, and a number of further modifications (Häder and Barnes 2019).

5.1 *Effects of High Atmospheric CO_2*

Global vegetation is expected directly impacted by rising atmospheric CO_2 levels, including communities of aquatic plants (Greve et al. 2019). Risings in photosynthesis, growth, total biomass, root/shoot ratio (Kizildeniz 2024), and tissue carbon/nitrogen ratio are typical reactions of plants to short-term CO_2 enrichment. Whether these reactions persist depends on how well photosynthetic processes adapt to elevated CO_2 (Collins and Schaum 2021); research has discovered both up- (Drake 1992) and down-regulation of photosynthesis (Amthor 1995) or as known 'acclimation' and related reactions after extended exposure to high CO_2 levels. Evidence points to further possible effects on communities and ecosystems, such as alterations in compositions of plant species, carbon sequestration, deterioration, and nutrient cycling (Clemmensen et al. 2021).

Among the physiological functions most affected by environmental factors and growth circumstances is photosynthesis, which also depends on factors including light type and concentration, temperature, humidity, CO_2, water availability, mineral, and size of the pot. Effects of CO_2 on photosynthesis depend on CO_2

concentration and exposure time. Rubisco is a crucial enzyme for CO_2 fixation in C_3 plants. Given that Rubisco activity is CO_2-limited and that its oxygenation activity is not impeded by the current atmospheric circumstances, photosynthesis rises by more than 50% in the short term at a CO_2 concentration twice that of the ambient level (Salazar-Parra et al. 2012). However, over the long run (days to weeks), this initial stimulus is frequently accompanied by molecular and metabolic changes that lead to a noticeable decline in photosynthetic ability (Morales et al. 2014).

Multiple consequences of rise in atmospheric CO_2 on plants have been noted, involving improved carbon absorption, development, yield, and carbon content (Ainsworth and Long 2005). This is due to a CO_2 fertilization effect that is mediated via photosynthesis. As a result, raised CO_2 might increase the amount of carbohydrates generated from photosynthesis in grains, with starch being the main component (Zhang et al. 2013a, 2013b). Increases in starch concentrations may have a diluting impact on other nutrients, such as proteins, lipids, vitamins, and minerals, as grains are mostly constituted of carbohydrates (mainly in the form of starch). Furthermore, modifications to the photosynthetic system and subsequent transfer from wilting foliage to cereals must be taken into account as the primary processes. Because of the distinct biochemistry of C_3 and C_4 photosynthesis, beneficial impact of increasing CO_2 on photosynthesis is quite visible in C_3 plants like rice and wheat while, less apparent in C_4 crops like maize (Ghannoum et al. 2000). Globally unbalanced plant stoichiometry (compared to pre-industrial periods) is anticipated to result from rising CO_2 levels (Loladze 2002), especially with reference to proteins and minerals like Fe, Zn, and I (or N) (Taub et al. 2008; Medek et al. 2017). Increased CO_2 has been linked to higher non-structural carbohydrates to protein ratios in grains and tubers, resulting to decrease mineral contents in barley, rice, and wheat grains by 6.9, 7.2, and 7.6%, respectively (Loladze 2014). Protein loss owing to high CO_2 was determined to be 15, 10, and 10% for the grain crops barley, rice, and wheat, respectively (Taub et al. 2008). As per their meta-analysis, they looked at how high CO_2 affected wheat grains, the most of minerals (Ca, Cd, Cu, Fe, Mg, Mn, P, S, and Zn) had their concentrations drastically reduced, for example according to Broberg et al. (2017), whereas B and Na had no effect and K had a large rise (<2%). The outcomes of this meta-analysis concur with those of the individual wheat a free-air CO_2 enrichment (FACE) tests (Fernando et al. 2012). Hundreds of millions of people currently consume diets low in two minerals, Fe and Zn, and reports of CO_2-induced decreases in Fe and Zn in important crops' edible sections have been made (Loladze 2014) and are anticipated to negatively impact human nutrition (Beach et al. 2019). The effects of high CO_2 on nutrients other than protein and minerals that are crucial for humans, certain nutrients and carotenoids, are now becoming clearer (Loladze et al. 2019). According to variations in physiology, decreased mineral levels are more pronounced in C_3 plants than in C_4 plants (Loladze 2014), while both C_3 and C_4 plants exhibit decreased transpiration. The modeling of carbohydrate synthesis by increased CO_2 is greater in C_3 plants. The effects of the above-mentioned climate change factors on plants generally differ according to plant varieties and where they are grown. There is a clear relationship between radiation utilization efficiency (RUE) and atmospheric CO_2 assimilation for C_3

plants. RUE is the rate of dry matter accumulation and accumulated captured global radiation. In this context, the RUE value is maximum at high levels of photosynthesis (Singh et al. 2012).

The atmospheric CO_2 effect (g MJ^{-1}) on RUE is the increase of RUE for C_3 plants following each ppm of CO_2 increase in the atmosphere. Therefore, the increase in CO_2 level directly affects the yield of C_3 plants. High efficiencies have been calculated at high CO_2 concentration levels due to the increase in RUE. Optimum start, end of optimum and maximum temperature values for each plant are effective temperatures that affect RUE. However, the situation is different for C_4 plants such as maize. According to Leakey et al. (2006), the increase in CO_2 level did not affect the basic enzymes of photosynthesis such as Rubisco, Peptidase C (PEPc) and Pyruvate, orthophosphate dikinase (PPDK). In addition, they also stated that CO_2 concentration did not affect corn yield in this study (Leakey et al. 2006). Atmospheric CO_2 is not a limiting factor for C_4 plants (Leakey et al. 2006). Therefore, the RUE value for C_4 plants is not limiting either.

5.2 *Effect of High Atmospheric Temperatures*

The atmospheric temperature has risen by 0.85 °C (IPCC 2014b) over the previous two decades. In reality, yearly average lowest temperatures in Spain have risen by 1.5 °C during the previous century and by 0.6 °C over the last 25 years (Lopez-Bustins et al. 2014). Climate ensemble model estimates predict rises of 1.8–4 °C close of the twenty-first century (2090–2099) compared to decade of the 1980–1999. Such figures are from the best estimation of the time series for greenhouse gas emissions based only on the six marker scenarios (Solomon et al. 2007). Effective cereal productivity is severely hampered by heat stress, which has been linked to lower grain yields in wheat (Lizana and Calderini 2013) and rice (Chaturvedi et al. 2017) throughout their reproductive or grain-filling phases. Depending on time (days after anthesis) and duration, high- temperature effects on grain filling might vary greatly. In wheat (Guzmán et al. 2016) and rice (Chaturvedi et al. 2017) both persistent fairly high temperatures (25–35 °C) and heat shocks (>35 °C) while filling the grain period are usually linked to a rise in grain protein content. In fact, a higher quantity of protein is found in the grain as a result of high temperature, because starch accumulates in wheat grains more quickly than it does at lower temperatures and over shorter periods of time. Additionally, whereas the pace of protein accumulation has significantly risen, the length of protein accumulation has decreased. Furthermore, leaves senescence prior to the maturation of the heads, which suggests that maximum temperatures may facilitate N remobilization from leaves and stalks (Savill et al. 2018). Furthermore, it has been demonstrated that key variables in the dough characteristics of wheat include the duration and timing of heat stress throughout grains filling (Wardlaw 2002). High temperatures can alter the protein and mineral content of grains, and they do so in response to alterations in the starch- and protein-synthesizing enzymes. Yang et al. (2018) concluded that the heat stress increased the activity of glutamate synthase while dramatically decreasing the activity of sucrose phosphate synthase, sucrose synthase, and soluble starch synthase during grain filling (Yang et al. 2018). Interestingly, Monjardino et al. (2005) discovered that throughout the early stages of endosperm development, heat stress had a deleterious impact on protein content. They discovered that zeins are the protein fractions most altered by heat stress. In reality, zein accumulation was suppressed rather than destroyed throughout the early phases of development. Higher temperature also changes the protein and mineral nutritional content of rice (Chaturvedi et al. 2017). According to Ferreira et al. (2012), the overall amount of N per grain in wheat is relatively unaffected by increasing temperature, but because to reduced yield of grain, the proportion of N on a dry weight basis climb at higher temperatures. Wheat has shown similar increases in % dry weight (Gooding et al. 2003).

5.3 *Effect of High Ultraviolet Light*

The weakening of the ozone layer in the atmosphere has led to an increase in UV radiation reaching the Earth's surface (Worrest and Häder 1989; Smith et al. 1992). UV light is classified into three classes: UV-A (wavelengths from 320 to 390 nm), UV-B (280 to 320 nm), and UV-C (below 280 nm). Even under low ozone circumstances, the stratosphere quickly absorbs UV-C wavelengths, preventing them from reaching the Earth's surface (Rowland 2006). Ozone does not influence UV-A wavelengths, hence alterations in the ozone layer will not have an impact on their influence. Furthermore, when stratospheric ozone declines, UV-B radiation is more likely to rise, reach the ground, and have an effect on plants (Rowland 2006). Numerous writers have made predictions on how rising levels of UV-B radiation may impact and harm different plant systems (Ayalew et al. 2022). Everyone agrees that it is possible for both mutagenic and physiological harm to happen on plants by increased UV-B (Ha et al. 2022). There are five reactions that occur in desert vegetation respond to UV-B radiation: photomorphogenic alterations, auxin inactivation, ATPase destruction, photosynthetic damage, and flavonoid induction (Stapleton 1992). UV-B light can decrease phytoplankton photosynthesis in marine algae (Gao and Häder 2020), however the production of amino acids similar to mycosporine can prevent UV-B radiation (Raven 1991).

Terrestrial species may be harmed by excessive UV-B radiation. However only wavelengths longer than 290 nm may reach the Earth's crust, the UV-B wavelength band spans 280 to 320 nm. UV-B to photosynthetically active radiation (PAR; 400–700 nm) ratio in sunlight varies, mainly due to variations in solar angular position and ozone layer thickness. Emissions of halogenated compounds, such as CFCs (also called halocarbons), cause the ozone layer to thin. This depletion reduces UV-B protection (Fuentes et al. 2022). Current UV-B radiation levels are believed to be close to its maximum since stratospheric ozone levels are close to their lowest point since monitoring started (Fuentes et al. 2022). Additionally, when the ozone layer gets thinner, the spectral UV-composition shifts toward lower wavelengths those are more harmful to living things (Jansen 1998). Based on estimates of halocarbon loading in the atmosphere under various scenarios, calculations of ozone and UV radiation (for latitudes 45° N and S) show that maximum levels may have already been achieved or will be reached soon. Furthermore, such levels will probably continue to be high for several decades (Fuentes et al. 2022). Additionally, seasonal variations are likely to inflict considerable biological harm and may substantially outweigh the average rise in UV radiation.

Land species, such as plants and microorganisms, can be harmed by an increase in UV-B radiation, although these organisms also have defense and recovery systems. The performance of species interactions, rather than the performance of individual species, is where terrestrial ecosystem responses to increasing UV-B (Watermann et al. 2020). However, the conversations that could occur in terms of plant form changes, timing of the life cycle, and the production of chemicals that are not directly related to primary metabolism are extremely sophisticated. Numerous studies have been done on how UV-B affects the morphological, physiological, and biochemical characteristics of higher plants. According to, there have been reduces in leaf growth (Fuentes et al. 2022), dry and fresh weight, photosynthetic capacity, and total biomass. Studies suggest that elevated UV-B has impacts more frequently through changed gene operational patterns than through gene depredation (Fuentes et al. 2022). Although UV-B reactions are instantaneous, impacts in long-lived perennial plants, like trees, can build up over time (Bornman et al. 2019). Improvements in UV-absorbing molecules appear to be a general reaction to elevated UV-B radiation, even though reports of plant reactions to such UV-B modifications have varied depending to species or variety susceptibility as well as the experimental design utilized (Wong et al. 2022). The buildup of these substances decreases the penetration of UV light into plants and other living things. In nectarine fruit, for example, the production of yellow and red pigments considerably decreases UV radiation penetration (Blanke 1996). According to research by Jansen (1998), UV radiation has been shown to increase levels of key antioxidants glutathione and ascorbate as well as some key enzymes involved in the phenyl-propanoid pathway and flavonoid biosynthesis (Schultz et al. 1998). The plant hormone auxin, which absorbs in the UV-B range and may play a substantial part in the production of an off-flavor in white wines, is another molecular property of UV-B that may damage peptides and lipids in addition to being photo-degraded (Vanhaelewyn et al. 2020).

5.4 *Effect of Drought*

In the future decades, rising greenhouse gas emissions may also result in decreased rainfall that would exacerbate already-present drought conditions in the Mediterranean area (McKersie 2015). According to climate change predictions, the Mediterranean area will have 25–30% less precipitation in the latter half of the twenty-first century (Giorgi and Lionello 2008). Furthermore, the periodicity of rainfall is far more significant. In reality, the anticipated Mediterranean rainfall deficit should have a significantly greater influence on summer rainfall than on winter precipitation. But the biggest factor influencing crop growth in the Mediterranean is winter rain. Additionally, it is believed that one of the key elements restricting food output globally is drought. Agricultural output sensitivities to water constraint are influenced by a number of variables, including genetics, stress duration and severity, and plant developmental stage (Altenbach et al. 2003). Recent researches have shown that winter droughts may occur during the first stages of the crop cycle (Russo et al. 2015), despite the fact that winter precipitation has historically been ample and having the lowest levels of evapotranspiration.

6 Drought Factor Interaction with Other Climate-Change Related Factors

Factors related to climate change will affect plants collectively rather than separately. According to Gray and Brady (2016), the combined effect of these variables is typically non-additive and can differ from what single factor studies would predict in terms of magnitude (synergistic and antagonistic, respectively). However, there are still a limited number of researches on the interactions between environmental variables and grapevine physiology. According to research by Martínez-Lüscher et al. (2015, 2016), a combination of high temperature and high CO_2 concentration increases the grapevine's capacity for photosynthetic growth. This expedites the grape's development. Short-term trials have also examined the interaction effects of water deficit and elevated temperature on grapevine physiology, demonstrating a more severe impact of high-temperature occurrences on grapevine undergoing water stress (Edwards et al. 2011; Galat Giorgi et al. 2019). Due to its intricacy, investigations on how the grapevine reacts physiologically and grows in response to several stress elements connected with climate change (CO_2, temperature, and water deficit) are still few. According to results obtained, water deficit abolished the advantages of higher air CO_2 and higher temperatures on the photosynthetic activity and development of Tempranillo (Leibar et al. 2015; Kizildeniz et al. 2018).

In recent decades, drought occurrences have increased, and average rainfall has decreased in Algeria and other Mediterranean nations, according to several studies, with the observable change commencing in the 1970s (Achour et al. 2020; Zkhiri et al. 2019). In Algeria, the intensity of the drought reduces from west to east (Achour et al. 2020). According to regional climate model (RCM) projections for the future climate up to 2100, northwestern Algeria is the region of the country most susceptible to climate change (Zeroual et al. 2019). When Zeroual et al. (2019) looked at the probable change in climatic zone classes between 1951 and 2100; they discovered that the northwest of Algeria had a humid environment in 1951, a semi-arid climate after 1973, and an arid climate by 2100. Moreover, relatively few researches has examined the relationship between temperature and the danger of drought in this area experiencing rapid changes in climate. According to Bouabdelli et al. (2020a), all hydrological drought occurrences in the watersheds of northwest Algeria have return periods that are less than 25 years. Researchers also discovered that the representative concentration pathway (RCP) 8.5 predicts a 40% rise in the danger of hydrological dryness during the next years. Furthermore, according to Bouabdelli et al. (2020b), the direst scenario calls for an increase in the average length of meteorological drought occurrences of 36%. (RCP8.5). The best method for estimating the risk of drought is multivariate frequency analysis since the number of occurrences is underestimated by unvarying return periods when considering their length or intensity. According to the two future scenarios, drought occurrences in the plains are projected to be more severe and stay longer, especially during the hot season (May through September) between 2021 and 2071. Under the RCP4.5 scenario, a spring agricultural drought (between February and April) threatens agricultural productivity and might have a significant impact on agricultural revenue and food security.

6.1 Drought and High Atmospheric CO_2 Interaction

By the end of the twenty-first century, it is expected that the average global surface temperature would be 1.1–6.4 °C higher than it was between 1980 and 1999. This might result in more intense climatic occurrences (IPCC 2007, 2012). Increasing temperature has the ability to grow evapotranspiration, which might worsen water shortages in arid and semiarid regions, increase ecosystem sensitivity, and hasten severe aridification and desertification (Maestre et al. 2012). Numerous reviews and meta-analyses have shown reactions to increased CO_2 at molecular, biochemical, particular plant, ecological, and systemic levels (Smith and Dukes 2013).

Plant interactions to CO_2 elevation have been described mostly in regular and review studies based on a corpus of experimental research. According to evidence, drought can limit plant growth and productivity responses to high CO_2 depending on the plant species, the intensity and length of the drought, and other factors. CO_2 levels can improve plant variety and production across an ecosystem by lowering stomatal conductance and so boosting water use efficiency and soil water availability. According to Poorter and Pérez-Soba (2001), 13 of 14 herbaceous species displayed stronger relative stimulation owing to increased CO_2 during drought circumstances than under well-watered conditions. C_4 plants do not respond as favorably to increase atmospheric CO_2 as do C_3 plants, which have a unique photosynthetic process known as a photosynthetic CO_2 concentrating pump (Long et al. 2006). As example, CO_2 enrichment increased biomass in a C_3 wheat plant under well-watered circumstances by 8 and 15% (Ewert et al. 2002), but in a C_4 sorghum plant under well-watered conditions by 0 and 13% (Conley et al. 2001). According to data from Morgan et al. (2011), the alleviation of drought stress only occurs when there is a water deficit, which shows that for C_4 plants, the stimulation of CO_2 enrichment is highly dependent on water circumstances.

The photosynthetic down regulation phenomena demonstrate how photosynthetic ability may diminish over time under conditions of increased CO_2 owing to acclimatization (Komatsu et al. 2013). According to Yu et al. (2012), under drought stress circumstances, increased CO_2 did not cause photosynthesis to downregulate. Therefore, the species and environmental factors have an impact on how much downregulation occurs.

6.2 Drought and High Temperature Interaction

Numerous species have been examined for changes in physiological responses to heat and drought. In general, when plants are exposed to abiotic stressors, photosynthetic activity is suppressed through Rubisco instability and PSII degradation (Nishiyama and Murata 2014). As per Wang et al. (2010), the combined effects of drought and heat stress on wheat photosynthetic rates were more detrimental than the effects of the separate stressors alone. Drought and heat treatment of *Nicotiana tabacum* plants resulted in reduced photosynthesis, increased respiration, stomatal closure, and increased leaf temperature (Rizhsky et al. 2002). Rubisco activity elevated in *Cicer arietinum* leaves in reaction to heat stress but reduced in reaction to drought. Drought-sensitive genotypes showed higher alterations in such characteristics, with reduced Rubisco activity compared to drought-tolerant cultivars under the stress combination (Carmo-Silva et al. 2012). Drought and heat-induced stomatal responses provide a difficult condition that plants must balance between guarding against overheating and stopping water loss. Typically, heat induces an increase in stomatal conductance as the plant attempts to cool its leaves through transpiration. However, dry conditions have the opposite effect, reducing stomatal conductance to minimize water loss (Mittler and Blumwald 2010). In contrast to plants merely treated to heat stress, tobacco plants exposed to this stress combination showed greater leaf temperatures. This is likely because the plants under this stress combination are unable to enhance transpiration since their stomata are closed (Rizhsky et al. 2002).

6.3 Drought and UV-Light Interaction

The response of a plant to stress, such as ultraviolet radiation or drought, varies significantly depending on factors such as species, cultivar, plant organ, developmental stage, and is also influenced by interactions within eco-physiological systems. Drought stress is a major environmental stressor that has a negative impact on agricultural output and quality. Water scarcity is the principal limiting factor of plant productivity in many environments as a result of global climate change (Tesař et al. 2007). Stress due to drought may change how secondary plant chemicals are produced. For instance, under drought stress conditions, it was discovered that the content of phenolic compounds like caffeic acid was decreased in *Ipomoea batatas* roots (Mao et al. 2004); whereas total soluble phenols, quercetin, and betulinic acid were elevated in *Hypericum brasiliense* (shoots) (De Abreu and Mazzafera 2005). Drought and UV-B exposure have both been shown to increase the activity of phenol metabolism-related enzymes including phenylalanine ammonia-lyase (Treutter 2010). It is known that plant reactions to UV-B depend on the amount of water available to them; hence, a lack of water is often thought to reduce plants' sensitivity to UV-B (Gwynn-Jones et al. 1999). Moreover, plant sensitivity to UV-B or drought varies among species, populations, and variations and depends on the plant's physiological stage and the length of the stress' impact (De Abreu and Mazzafera 2005). In compared to the single stress element, the combination of UV radiation and drought stress applied to the lettuce plants in a study carried out by Rajabbeigi et al. (2013) demonstrated a significant increase in proline concentrations which were increased by up to 1.5 times compared to control and UV-B treatment-only plants. In comparison to control plants, drought or UV-B treatment alone as well as the two stressors together had no effect on the total phenolic content of lettuce plants. When UV-B and drought stress were combined, Phenylalanine ammonialyase activity was enhanced compared to control plants (Rajabbeigi et al. 2013). Researchers have looked at the positive and negative impacts of UV radiation and drought on various plants. In response to both drought and UV radiation, *Populus cathayana* and *Hippophae rhamnoides* showed decreased plant height and leaf area, whereas *Glycine max* did not exhibit any comparable efficacy (Hock et al. 2020). According to Jan et al. (2022) UV-B radiation and drought both slow down photosynthesis in rice.

6.4 Drought and High Atmospheric CO_2, and High Temperature Interactions

In several primary crop producing zones of arid and semiarid areas, relative high drought and heat frequently occur throughout the summer, a peak period for plant development (Dreesen et al. 2012). Temperature may counter the beneficial effects of CO_2 enrichment on soil and plant water status, which may be increased by drought (Zeppel et al. 2012). Thus, the combination of warming and drought may counteract the beneficial effects of CO_2 enrichment on soil and plant water status. Therefore, extreme heat and dry circumstances may concurrently offset the improvement in plant growth and survival caused by CO_2 enrichment, especially over the long term (Norby and Luo 2004). Alfalfa, a forage legume, did not exhibit photosynthetic adaptation 30 days after regrowth under conditions of normal water and temperature, while plants cultivated under drought with extreme temperatures displayed photosynthetic downregulation, including a decrease in Rubisco activity (Erice et al. 2006). Conversely, plants with higher drought tolerance have higher heat stress resistance (Suzuki et al. 2005). In a grassland environment, dominating species responded more strongly to warming than they did to either drought or increased CO_2, but not to each one alone (Bloor et al. 2010). Plants with increased drought resistance are significantly more tolerant of heat stress (Suzuki et al. 2005). Elevated CO_2 with a 2 °C can boost wheat grain production and biomass regardless of watering regimes in Australia, but when temperatures were more than 2 °C higher, the positive effects vanished owing to dryness (De Oliveira et al. 2012). However, the responses of plants to abiotic stressors in combination are special and cannot be understood just from the reactions at the molecular, physiological, whole-plant, or global scales (Yu et al. 2012; Smith and Dukes 2013). As a result, we must properly research this complicated problem.

7 Conclusion

Drought is the world's leading natural catastrophe, with low rainfall compared to typical amounts for subterranean and surface water. Human influence is causing the global climate to alter including the burning of fossil fuels, clearing forests, and rapid industrialization, which raise "greenhouse gases" like carbon dioxide, methane, ozone, and dinitrogen. Drought is the worst hazard to every living creature since it might make it harder for natural resources to feed people, which could result in starvation killing millions of people. In order to prevent future drought, particularly that brought on by global warming, from becoming a significant issue for all organisms, and research is needed to identify plant species that are tolerant to drought, identify the mechanisms of tolerance, and conserve and modify the gene resources of the resistant plants. The multimodal contexts of biotic and abiotic elements, as well as how ecosystems and plants adapt to changes in temperatures and other external factors, are complicated issues that cover many facets of environmental mechanisms at the molecular, biochemical, individual, and ecosystem levels. A plant's ability to positively and plastically respond to one stress may be constrained by another stress or by multifactor stressors. In addition, other elements, such as land use, may change how the ecosystem reacts to higher CO_2 levels. However, we should properly evaluate the effects of upcoming rises in CO_2 levels in the atmosphere, heat, and other external factors on terrestrial ecosystems. Effects of elevated CO_2 levels, warming, and watering studies at the ecosystem scale across a variety of biomass should be studied and incorporated, as well as greater connections among investigations and concepts.

References

Abbas S (2022) Climate change and major crop production: evidence from Pakistan. Environ Sci Pollut Res 29(4):5406–5414

Achour K, Meddi M, Zeroual A, Bouabdelli S, Maccioni P, Moramarco T (2020) Spatio-temporal analysis and forecasting of drought in the plains of northwestern Algeria using the standardized precipitation index. J Earth Syst Sci 129(1):1–22

AghaKouchak A, Cheng L, Mazdiyasni O, Farahmand A (2014) Global warming and changes in risk of concurrent climate extremes: insights from the 2014 California drought. Geophys Res Lett 41(24):8847–8852

Ahmed T, Scholz M, Al-Faraj F, Niaz W (2016) Water-related impacts of climate change on agriculture and subsequently on public health: A review for generalists with particular reference to Pakistan. Int J Environ Res Public Health 13(11):1051

Ainsworth EA, Long SP (2005) What have we learned from 15 years of free-air CO_2 enrichment (FACE)? A meta-analytic review of the responses of photosynthesis, canopy properties and plant production to rising CO_2. New Phytol 165(2):351–372

Altenbach SB, DuPont FM, Kothari KM, Chan R, Johnson EL, Lieu D (2003) Temperature, water and fertilizer influence the timing of key events during grain development in a US spring wheat. J Cereal Sci 37(1):9–20

Amthor JS (1995) Terrestrial higher-plant response to increasing atmospheric [CO_2] in relation to the global carbon cycle. Glob Chang Biol 1(4):243–274

Appiah DO, Akondoh AC, Tabiri RK, Donkor AA (2018) Smallholder farmers' insight on climate change in rural Ghana. Cogent Food Agric 4(1):1436211

Arora A, Sairam RK, Srivastava GC (2002) Oxidative stress and antioxidative system in plants. Curr Sci 82:1227–1238

Ayalew T, Yoseph T, Cadisch G (2022) Elevated UV-B radiation depressed biomass yield and symbiotic N_2-fixation in *Bradyrhizobium* inoculated cowpea varieties. Symbiosis 87:201–211

Beach RH, Sulser TB, Crimmins A, Cenacchi N, Cole J, Fukagawa NK et al (2019) Combining the effects of increased atmospheric carbon dioxide on protein, iron, and zinc availability and projected climate change on global diets: a modelling study. Lancet Planet Health 3(7):e307–e317

Belkhodja R, Morales F, Abadia A, Gomez-Aparisi J, Abadia J (1994) Chlorophyll fluorescence as a possible tool for salinity tolerance screening in barley (*Hordeum vulgare* L.). Plant Physiol 104(2):667–673

Bhardwaj A, Devi P, Chaudhary S, Rani A, Jha UC, Kumar S et al (2021) 'Omics' approaches in developing combined drought and heat tolerance in food crops. Plant Cell Rep 41(3):699–739

Blanke MM (1996) Transmission of UV-irradiance into nectarine fruit. Angew Bot 70:76–77

Bloor JM, Pichon P, Falcimagne R, Leadley P, Soussana JF (2010) Effects of warming, summer drought, and CO_2 enrichment on aboveground biomass production, flowering phenology, and community structure in an upland grassland ecosystem. Ecosystems 13(6):888–900

Bornman JF, Barnes PW, Robson TM, Robinson SA, Jansen MA, Ballare CL, Flint SD (2019) Linkages between stratospheric ozone, UV radiation and climate change and their implications for terrestrial ecosystems. Photochem Photobiol Sci 18(3):681–716

Bouabdelli S, Meddi M, Zeroual A, Alkama R (2020a) Hydrological drought risk recurrence under climate change in the karst area of northwestern Algeria. J Water Clim Change 11(S1):164–188

Bouabdelli S, Zeroual A, Meddi M, Djelloul F, Alkama R (2020b) Past and future drought in northwestern Algeria: the Beni Bahdel dam catchment. Proc Int Assoc Hydrol Sci 383:315–318

Božič B, Zemljič Jokhadar Š, Kristanc L, Gomišček G (2020) Cell volume changes and membrane ruptures induced by hypotonic electrolyte and sugar solutions. Front Physiol 11:582781

Broberg MC, Högy P, Pleijel H (2017) CO_2-induced changes in wheat grain composition: meta-analysis and response functions. Agronomy 7(2):32

Brown ME, Funk CC (2008) Food security under climate change. Science 319(5863):580–581

Carmo-Silva AE, Gore MA, Andrade-Sanchez P, French AN, Hunsaker DJ, Salvucci ME (2012) Decreased CO_2 availability and inactivation of rubisco limit photosynthesis in cotton plants under heat and drought stress in the field. Environ Exp Bot 83:1–11

Chandio AA, Jiang Y, Gessesse AT, Dunya R (2019) The nexus of agricultural credit, farm size and technical efficiency in Sindh, Pakistan: A stochastic production frontier approach. J Saudi Soc Agric Sci 18(3):348–354

Chandio AA, Jiang Y, Akram W, Adeel S, Irfan M, Jan I (2021) Addressing the effect of climate change in the framework of financial and technological development on cereal production in Pakistan. J Clean Prod 288:125637

Changnon SA Jr, Easterling WE (1989) Measuring drought impacts: the Illinois case 1. JAWRA J Am Water Resour Assoc 25(1):27–42

Chaturvedi AK, Bahuguna RN, Pal M, Shah D, Maurya S, Jagadish KS (2017) Elevated CO_2 and heat stress interactions affect grain yield, quality and mineral nutrient composition in rice under field conditions. Field Crop Res 206:149–157

Chen YE, Liu WJ, Su YQ, Cui JM, Zhang ZW, Yuan M et al (2016) Different response of photosystem II to short and long-term drought stress in *Arabidopsis thaliana*. Physiol Plant 158(2):225–235

Clemmensen KE, Durling MB, Michelsen A, Hallin S, Finlay RD, Lindahl BD (2021) A tipping point in carbon storage when forest expands into tundra is related to mycorrhizal recycling of nitrogen. Ecol Lett 24(6):1193–1204

Collins S, Schaum CE (2021) Growth strategies of a model picoplankter depend on social milieu and p CO_2. Proc R Soc B 288(1955):20211154

Conley MM, Kimball BA, Brooks TJ, Pinter PJ Jr, Hunsaker DJ, Wall GW et al (2001) CO_2 enrichment increases water-use efficiency in sorghum. New Phytol 151(2):407–412

Cook BI, Williams AP, Mankin JS, Seager R, Smerdon JE, Singh D (2018) Revisiting the leading drivers of Pacific coastal drought variability in the contiguous United States. J Clim 31(1):25–43

De Abreu IN, Mazzafera P (2005) Effect of water and temperature stress on the content of active constituents of *Hypericum brasiliense* Choisy. Plant Physiol Biochem 43(3):241–248

De Oliveira ED, Bramley H, Siddique KH, Henty S, Berger J, Palta JA (2012) Can elevated CO_2 combined with high temperature ameliorate the effect of terminal drought in wheat? Funct Plant Biol 40(2):160–171

Douville H, Raghavan K, Renwick J, Allan RP, Arias PA, Barlow M et al (2021) Water cycle changes. In: Masson-Delmotte VP, Zhai P, Pirani A, Connors SL, Péan C, Berger S, Caud N, Chen Y, Goldfarb L, Gomis MI, Huang M, Leitzell K, Lonno E, Matthews JBR, Maycock TK, Waterfield T, Yelekçi O, Yu R, Zhou B (eds) Climate change 2021: the physical science basis. Contribution of working group I to 45 the sixth assessment report of the intergovernmental panel on climate change. Cambridge University Press, Cambridge, UK, pp 1055–1210

Drake BG (1992) A field-study of the effects of elevated CO_2 on ecosystem processes in a Chesapeake Bay wetland. Aust J Bot 40(5):579–595

Dreesen FE, De Boeck HJ, Janssens IA, Nijs I (2012) Summer heat and drought extremes trigger unexpected changes in productivity of a temperate annual/biannual plant community. Environ Exp Bot 79:21–30

Edwards EJ, Smithson L, Graham DC, Clingeleffer PR (2011) Grapevine canopy response to a high-temperature event during deficit irrigation. Aust J Grape Wine Res 17:153–161

Engelbrecht FA, McGregor JL, Engelbrecht CJ (2009) Dynamics of the conformal-cubic atmospheric model projected climate-change signal over southern Africa. Int J Climatol 29(7):1013–1033

Erice G, Irigoyen JJ, Pérez P, Martínez-Carrasco R, Sánchez-Díaz M (2006) Effect of elevated CO_2, temperature and drought on photosynthesis of nodulated alfalfa during a cutting regrowth cycle. Physiol Plant 126(3):458–468

Ewert F, Rodriguez D, Jamieson P, Semenov MA, Mitchell RAC, Goudriaan J et al (2002) Effects of elevated CO_2 and drought on wheat: testing crop simulation models for different experimental and climatic conditions. Agric Ecosyst Environ 93(1–3):249–266

FAO (2009) 2050: A third more mouths to feed. https://www.fao.org/news/story/en/item/35571/icode/.pdf Retrieved on 19.09.2022

FAO (2020) Desert locust. Rome: FAO. Retrieved on August 20, 2022 from http://www.fao.org/locusts/en/

Fernando N, Panozzo J, Tausz M, Norton R, Fitzgerald G, Seneweera S (2012) Rising atmospheric CO_2 concentration affects mineral nutrient and protein concentration of wheat grain. Food Chem 133(4):1307–1311

Ferreira MS, Martre P, Mangavel C, Girousse C, Rosa NN, Samson MF, Morel MH (2012) Physicochemical control of durum wheat grain filling and gluten in polymer assembly under different temperature regimes. J Cereal Sci 56(1):58–66

Forootan E, Khaki M, Schumacher M, Wulfmeyer V, Mehrnegar N, van Dijk AI et al (2019) Understanding the global hydrological droughts of 2003-2016 and their relationships with teleconnections. Sci Total Environ 650:2587–2604

Forzieri G, Miralles DG, Ciais P, Alkama R, Ryu Y, Duveiller G et al (2020) Increased control of vegetation on global terrestrial energy fluxes. Nat Clim Chang 10(4):356–362

Fuentes E, Pérez-Velón D, Prieto B (2022) Effects of changes in UV-B radiation levels on biofilm-forming organisms commonly found in cultural heritage surfaces. Environ Res 214(14):114061

Galat Giorgi E, Sadras VO, Keller M, Perez Peña J (2019) Interactive effects of high temperature and water deficit on Malbec grapevines. Aust J Grape Wine Res 25:345–356

Gao K, Häder DP (2020) Photosynthetic performances of marine microalgae and solar UV under radiation influences of rising CO2. In: Wang Q (ed) Microbial photosynthesis. Springer, Singapore, pp 139–150

García-Herrera R, Garrido-Perez JM, Barriopedro D, Ordonez C, Vicente-Serrano SM, Nieto R et al (2019) The European 2016/17 drought. J Clim 32(11):3169–3187

Ghannoum O, Caemmerer SV, Ziska LH, Conroy JP (2000) The growth response of C4 plants to rising atmospheric CO_2 partial pressure: a reassessment. Plant Cell Environ 23(9):931–942

Giorgi F, Lionello P (2008) Climate change projections for the Mediterranean region. Glob Planet Chang 63(2–3):90–104

Giorgi F, Jones C, Asrar GR (2009) Addressing climate information needs at the regional level: the CORDEX framework. World Meteorol Organiz (WMO) Bull 58(3):175

Gooding MJ, Ellis RH, Shewry PR, Schofield JD (2003) Effects of restricted water availability and increased temperature on the grain filling, drying and quality of winter wheat. J Cereal Sci 37(3):295–309

Gray SB, Brady SM (2016) Plant developmental responses to climate change. Dev Biol 419:64–77

Greve P, Roderick ML, Ukkola AM, Wada Y (2019) The aridity index under global warming. Environ Res Lett 14(12):124006

Grolle J (2015) Historical case studies of famines and migrations in the West African Sahel and their possible relevance now and in the future. Popul Environ 37(2):181–206

Guzmán C, Autrique JE, Mondal S, Singh RP, Govindan V, Morales-Dorantes A et al (2016) Response to drought and heat stress on wheat quality, with special emphasis on bread-making quality, in durum wheat. Field Crop Res 186:157–165

Gwynn-Jones D, Johanson U, Phoenix GK, Gehrke C, Callaghan TV, Björn LO, Sonesson M (1999) UV-B impacts and interactions with other co-occuring variables of environmental change: an Arctic perspective. In: Rozema J (ed) Stratospheric ozone depletion. The effects of enhanced UV-B radiation on terrestrial ecosystems. Backhuys Publishers, Leiden, pp 187–201

Ha PTT, Tran NTB, Tram NTN, Duy LT, Nha MT (2022) Effects of ultraviolet light and methyl salicylate on some phenological and agronomic and morphological characteristics of common bean (Phaseolus vulgaris L.). J Plant Biochem Biotechnol 31:1–8

Häder DP, Barnes PW (2019) Comparing the impacts of climate change on the responses and linkages between terrestrial and aquatic ecosystems. Sci Total Environ 682:239–246

Hock M, Plos C, Sporbert M, Erfmeier A (2020) Combined effects of UV-B and drought on native and exotic populations of *Verbascum thapsus* L. Plan Theory 9(2):269

Hulme M (1995) Estimating global changes in precipitation. Weather 50(2):34–42

Intergovernmental Panel on Climate Change IPCC (2007) Technical summary. In: Solomon S, Qin D, Manning M, Chen Z, Marquis M, Averyt KB, Tignor M, Miller HL (eds) Climate change 2007: the physical science basis. Contribution of working group I to the fourth assessment report of the intergovernmental panel on climate change. Cambridge University Press, Cambridge

Intergovernmental Panel on Climate Change IPCC (2012) Summary for policymakers. In: Field CB, Barros V, Stocker TF et al (eds) Managing the risks of extreme events and disasters to advance climate change adaptation. A special report of working groups I and II of the intergovernmental panel on climate change. Cambridge University Press, Cambridge/New York, pp 1–19

Intergovernmental Panel on Climate Change IPCC (2014a) Climate change 2014-synthesis report summary for policymakers. IPCC, Geneva

Intergovernmental Panel on Climate Change IPCC (2014b) In: Pachauri RK, Meyer LA (eds) Climate change 2014: synthesis report. Contribution of working groups 1, 2, and 3 to the fifth assessment report of the intergovernmental panel on climate change. IPCC, Geneva

Intergovernmental Panel on Climate Change IPCC (2018) Global warming of 1.5°C. Intergovernmental panel on climate change special report, Geneva. http://www.ipcc.ch/report/sr15/

Jan R, Khan MA, Asaf S, Waqas M, Park JR, Asif S et al (2022) Drought and UV radiation stress tolerance in rice is improved by over accumulation of non-enzymatic antioxidant flavonoids. Antioxidants 11(5):917

Jansen WJ (1998) Interpreting saving-investment correlations. Open Econ Rev 9(3):207–219

Jiang H, Tang B, Xie Z, Nolan T, Ye H, Song GY et al (2019) GSK 3-like kinase BIN 2 phosphorylates RD 26 to potentiate drought signaling in Arabidopsis. Plant J 100(5):923–937

Kalefetoğlu T, Ekmekci Y (2005) The effects of drought on plants and tolerance mechanisms. Gazi University J Sci 18(4):723–740

Kayabaşı S (2011) Kuraklık stresinde yetiştirilen soyada (Glycine max L.) bazı fizyolojik parametreler ile profilin birikiminin araştırılması. Harran Üniversitesi Fen Bilimleri Enstitüsü Biyoloji Anabilim Dalı Yüksek Lisans Tezi, Şanlıurfa, 40s

Kizildeniz T (2024) Assessing the growth dynamics of alfalfa varieties (*Medicago sativa* cv. Bilensoy 80 and Nimet) response to varied carbon dioxide (CO_2) concentrations. Heliyon 10(7)

Kizildeniz T, Irigoyen JJ, Pascual I, Morales F (2018) Simulating the impact of climate change (elevated CO_2 and temperature, and water deficit) on the growth of red and white Tempranillo grapevine in three consecutive growing seasons (2013–2015). Agric Water Manag 202:220–230

Komatsu M, Tobita H, Yazaki K, Kitao M, Watanabe M, Koike T (2013) Photosynthetic downregulation in leaves of the Japanese white birch grown under elevated CO_2 concentration does not change their temperature-dependent susceptibility to photoinhibition. Physiol Plant 147(2):159–168

Kotschi J (2007) Agricultural biodiversity is essential for adapting to climate change. GAIA-Ecological Perspectives for Science and Society 16(2):98–101

Leakey AD, Uribelarrea M, Ainsworth EA, Naidu SL, Rogers A, Ort DR, Long SP (2006) Photosynthesis, productivity, and yield of maize are not affected by open-air elevation of CO_2 concentration in the absence of drought. Plant Physiol 140(2):779–790

Leibar U, Aizpurua A, Unamunzaga O, Pascual I, Morales F (2015) How will climate change influence grapevine cv. Tempranillo photosynthesis under different soil textures? Photosynth Res 124(2):199–215

Lima ALS, DaMatta FM, Pinheiro HA, Totola MR, Loureiro ME (2002) Photochemical responses and oxidative stress in two clones of *Coffea canephora* under water deficit conditions. Environ Exp Bot 47:239–247

Lizana XC, Calderini DF (2013) Yield and grain quality of wheat in response to increased temperatures at key periods for grain number and grain weight determination: considerations for the climatic change scenarios of Chile. J Agric Sci 151(2):209–221

Lloyd-Hughes B (2014) The impracticality of a universal drought definition. Theor Appl Climatol 117(3):607–611

Lobell DB, Burke MB, Tebaldi C, Mastrandrea MD, Falcon WP, Naylor RL (2008) Prioritizing climate change adaptation needs for food security in 2030. Science 319(5863):607–610

Loladze I (2002) Rising atmospheric CO_2 and human nutrition: toward globally imbalanced plant stoichiometry? Trends Ecol Evol 17(10):457–461

Loladze I (2014) Hidden shift of the ionome of plants exposed to elevated CO_2 depletes minerals at the base of human nutrition. elife 3:e02245

Loladze I, Nolan JM, Ziska LH, Knobbe AR (2019) Rising atmospheric CO_2 lowers concentrations of plant carotenoids essential to human health: a meta-analysis. Mol Nutr Food Res 63(15):1801047

Long SP, Ainsworth EA, Leakey AD, Nosberger J, Ort DR (2006) Food for thought: lower-than-expected crop yield stimulation with rising CO_2 concentrations. Science 312(5782):1918–1921

Lopez-Bustins JA, Pla E, Nadal M, de Herralde F, Savé R (2014) Global change and viticulture in the Mediterranean region: a case of study in North-Eastern Spain. Span J Agric Res 12(1):78–88

Maestre FT, Salguero-Gómez R, Quero JL (2012) It is getting hotter in here: determining and projecting the impacts of global environmental change on drylands. Philos Trans R Soc B 367(1606):3062–3075

Maestro T, Barnett BJ, Coble KH, Garrido A, Bielza M (2016) Drought index insurance for the Central Valley project in California. Appl Econ Perspect Policy 38(3):521–545

Mao L, Jett LE, Story RN, Hammond AM, Peterson JK, Labonte DR (2004) Influence of drought stress on sweet potato resistance to sweet potato weevil, *Cylas formicarius* (Coleoptera: Apoinidae), and storage root chemistry. Fla Entomol 87:261–267

Martínez-Lüscher J, Morales F, Sánchez-Díaz M, Delrot S, Aguirreolea J, Gomès E, Pascual I (2015) Climate change conditions (elevated CO_2 and temperature) and UV-B radiation affect grapevine (Vitis vinifera cv. Tempranillo) leaf carbon assimilation, altering fruit ripening rates. Plant Sci 236:168–176

Martínez-Lüscher J, Kizildeniz T, Vucetic V, Dai Z, Luedeling E, van Leeuwen C, Gomès E et al (2016) Sensitivity of grapevine phenology to water availability, temperature and CO_2 concentration. Front Environ Sci 4:1–14

McKersie B (2015) Planning for food security in a changing climate. J Exp Bot 66(12):3435–3450

Medek DE, Schwartz J, Myers SS (2017) Estimated effects of future atmospheric CO_2 concentrations on protein intake and the risk of protein deficiency by country and region. Environ Health Perspect 125(8):087002

Meinshausen M, Smith SJ, Calvin K, Daniel JS, Kainuma ML, Lamarque JF et al (2011) The RCP greenhouse gas concentrations and their extensions from 1765 to 2300. Clim Chang 109:213–241

Mittler R, Blumwald E (2010) Genetic engineering for modern agriculture: challenges and perspectives. Annu Rev Plant Biol 61:443–462

Monjardino P, Smith AG, Jones RJ (2005) Heat stress effects on protein accumulation of maize endosperm. Crop Sci 45(4):1203–1210

Morales F, Pascual I, Sánchez-Díaz M, Aguirreolea J, Irigoyen JJ, Goicoechea N et al (2014) Methodological advances: using greenhouses to simulate climate change scenarios. Plant Sci 226:30–40

Morgan JA, LeCain DR, Pendall E, Blumenthal DM, Kimball BA, Carrillo Y et al (2011) C_4 grasses prosper as carbon dioxide eliminates desiccation in warmed semi-arid grassland. Nature 476(7359):202–205

Morton JF (2007) The impact of climate change on smallholder and subsistence agriculture. Proc Natl Acad Sci 104(50):19680–19685

Mundree SG, Baker B, Mowla S, Peters S, Marais S, Vander Willigen C et al (2002) Physiological and molecular insights into drought tolerance. Afr J Biotechnol 1(2):28–38

Nishiyama Y, Murata N (2014) Revised scheme for the mechanism of photoinhibition and its application to enhance the abiotic stress tolerance of the photosynthetic machinery. Appl Microbiol Biotechnol 98:8777–8796

Norby RJ, Luo YQ (2004) Evaluating ecosystem responses to rising atmospheric CO_2 and global warming in a multifactor world. New Phytol 162:281–293

Orimoloye IR, Belle JA, Orimoloye YM, Olusola AO, Ololade OO (2022) Drought: A common environmental disaster. Atmos 13(1):111

Ozcelik S (2021) Ozone diplomacy, ozone regime negotiations and Arctic. Int J Polit Secur 3(1):74–98

Perera F (2018) Pollution from fossil-fuel combustion is the leading environmental threat to global pediatric health and equity: solutions exist. Int J Environ Res Public Health 15(1):16

Poorter H, Pérez-Soba M (2001) The growth response of plants to elevated CO_2 under non-optimal environmental conditions. Oecologia 129(1):1–20

Porter JR, Xie L, Challinor AJ, Cochrane K, Howden SM, Iqbal MM et al (2014) Food security and food production systems. Methods 7:1–1

Rajabbeigi E, Eichholz I, Beesk N, Ulrichs C, Kroh LW, Rohn S, Huyskens-Keil S (2013) Interaction of drought stress and UV-B radiation-impact on biomass production and flavonoid metabolism in lettuce (Lactuca sativa L.). J Appl Bot Food Qual 86(1):190–197

Raven JA (1991) Physiology of inorganic C acquisition and implications for resource use efficiency by marine phytoplankton: relation to increased CO_2 and temperature. Plant Cell Environ 14(8):779–794

Raymond F, Ullmann A, Tramblay Y, Drobinski P, Camberlin P (2018) Evolution of Mediterranean extreme dry spells during the wet season under climate change. Reg Environ Chang 19(8):2339–2351

Rizhsky L, Liang H, Mittler R (2002) The combined effect of drought stress and heat shock on gene expression in tobacco. Plant Physiol 130(3):1143–1151

Roul MK, Mishra R, Das A (2017) Impact of increase in greenhouse gases on global warming and its remedies. J Emerg Technol Innov Res 4(12):981–986

Rowland FS (2006) Stratospheric ozone depletion. Philos Trans R Soc B 361(1469):769–790

Russo AC, Gouveia CM, Trigo RM, Liberato ML, DaCamara CC (2015) The influence of circulation weather patterns at different spatial scales on drought variability in the Iberian Peninsula. Front Environ Sci 3:1

Salazar-Parra C, Aguirreolea J, Sánchez-Díaz M, Irigoyen JJ, Morales F (2012) Photosynthetic response of Tempranillo grapevine to climate change scenarios. Ann Appl Biol 161(3):277–292

Salisbury FB, Ross CW (1992) Plant physiology. Wadsworth Pub. Com., Inc., Belmont

Savill GP, Michalski A, Powers SJ, Wan Y, Tosi P, Buchner P, Hawkesford MJ (2018) Temperature and nitrogen supply interact to determine protein distribution gradients in the wheat grain endosperm. J Exp Bot 69(12):3117–3126

Schultz HR, Löhnertz O, Bettner W, Bálo B, Linsenmeier A, Jähnisch A (1998) Is grape composition affected by current levels of UV-B radiation? Vitis 37(4):191–192

Schumacher DL, Keune J, Van Heerwaarden CC, Vilà-Guerau de Arellano J, Teuling AJ, Miralles DG (2019) Amplification of mega-heatwaves through heat torrents fuelled by upwind drought. Nat Geosci 12(9):712–717

Siddiqui R, Samad G, Nasir M, Jalil HH (2012) The impact of climate change on major agricultural crops: evidence from Punjab, Pakistan. Pakistan Dev Rev 51:261–274

Singh S, Gupta AK, Kaur N (2012) Influence of drought and sowing time on protein composition, antinutrients, and mineral contents of wheat. Sci World J 2012:485751

Smirnoff N (1993) The role of active oxygen in the response of plants to water deficit and desiccation. New Phytol 125:27–58

Smith NG, Dukes JS (2013) Plant respiration and photosynthesis in global-scale models: incorporating acclimation to temperature and CO_2. Glob Chang Biol 19(1):45–63

Smith RC, Prezelin BB, Baker KEA, Bidigare RR, Boucher NP, Coley T et al (1992) Ozone depletion: ultraviolet radiation and phytoplankton biology in Antarctic waters. Science 255(5047):952–959

Solomon S, Manning M, Marquis M, Qin D (2007) Climate change 2007-the physical science basis: working group I contribution to the fourth assessment report of the IPCC, vol 4. Cambridge University Press
Spinoni J, Naumann G, Carrao H, Barbosa P, Vogt J (2014) World drought frequency, duration, and severity for 1951–2010. Int J Climatol 34(8):2792–2804
Stapleton AE (1992) Ultraviolet radiation and plants: burning questions. Plant Cell 4(11):1353
Suzuki N, Rizhsky L, Liang H, Shuman J, Shulaev V, Mittler R (2005) Enhanced tolerance to environmental stress in transgenic plants expressing the transcriptional coactivator multiprotein bridging factor 1c. Plant Physiol 139(3):1313–1322
Sweet SK, Wolfe DW, DeGaetano A, Benner R (2017) Anatomy of the 2016 drought in the northeastern United States: implications for agriculture and water resources in humid climates. Agric For Meteorol 247:571–581
Taub DR, Miller B, Allen H (2008) Effects of elevated CO_2 on the protein concentration of food crops: a meta-analysis. Glob Chang Biol 14(3):565–575
Tesař M, Šír M, Lichner Ľ, Čermák J (2007) Plant transpiration and net entropy exchange on the Earth's surface in a Czech watershed. Biologia 62(5):547–551
Tian L, Yuan S, Quiring SM (2018) Evaluation of six indices for monitoring agricultural drought in the south-Central United States. Agric For Meteorol 249:107–119
Time A, Garrido M, Acevedo E (2018) Water relations and growth response to drought stress of *Prosopis tamarugo* Phil. A review. J Soil Sci Plant Nutr 18(2):329–343
Tramblay Y, Hertig E (2018) Modelling extreme dry spells in the Mediterranean region in connection with atmospheric circulation. Atmos Res 202:40–48
Tramblay Y, Koutroulis A, Samaniego L, Vicente-Serrano SM, Volaire F, Boone A et al (2020) Challenges for drought assessment in the Mediterranean region under future climate scenarios. Earth Sci Rev 210:103348
Treutter D (2010) Managing phenol contents in crop plants by phytochemical farming and breeding-visions and constraints. Int J Mol Sci 11(3):807–857
Tripathi A, Prasad AR (2009) Estimation of agricultural supply response by cointegration approach. Indian Econ J 57(1):106–131
Vanhaelewyn L, Van Der Straeten D, De Coninck B, Vandenbussche F (2020) Ultraviolet radiation from a plant perspective: the plant-microorganism context. Front Plant Sci 11:1984
Vidal-Macua JJ, Ninyerola M, Zabala A, Domingo-Marimon C, Gonzalez-Guerrero O, Pons X (2018) Environmental and socioeconomic factors of abandonment of rainfed and irrigated crops in Northeast Spain. Appl Geogr 90:155–174
Vyas SP, Garg BK, Kathju S, Lahiri AN (2003) Relative drought tolerance of moth bean genotypes. Adv Arid Legumes Res:393–397
Wang GP, Hui Z, Li F, Zhao MR, Zhang J, Wang W (2010) Improvement of heat and drought photosynthetic tolerance in wheat by over accumulation of glycine betaine. Plant Biotechnol Rep 4(3):213–222
Wardlaw IF (2002) Interaction between drought and chronic high temperature during kernel filling in wheat in a controlled environment. Ann Bot 90(4):469–476
Watermann LY, Hock M, Blake C, Erfmeier A (2020) Plant invasion into high elevations implies adaptation to high UV-B environments: a multi-species experiment. Biol Invasions 22(3):1203–1218
Weinhouse C (2021) The roles of inducible chromatin and transcriptional memory in cellular defense system responses to redox-active pollutants. Free Radic Biol Med 170:85–108
Wilhite DA, Glantz MH (1985) Understanding: the drought phenomenon: the role of definitions. Water Int 10(3):111–120
Wilhite D, Pulwarty RS (eds) (2017) Drought and water crises: integrating science, management, and policy. CRC Press
Wong TM, Sullivan JH, Eisenstein E (2022) Acclimation and compensating metabolite responses to UV-B radiation in natural and transgenic Populus spp. Defective in lignin biosynthesis. Metabolites 12(8):767

Worrest RC, Häder DP (1989) Effects of stratospheric ozone depletion on marine organisms. Environ Conserv 16(3):261–263

Wu J, Chen X, Yuan X, Yao H, Zhao Y, AghaKouchak A (2021) The interactions between hydrological drought evolution and precipitation-streamflow relationship. J Hydrol 597:126210

Yang H, Gu X, Ding M, Lu W, Lu D (2018) Heat stress during grain filling affects activities of enzymes involved in grain protein and starch synthesis in waxy maize. Sci Rep 8(1):1–9

Yu J, Chen L, Xu M, Huang B (2012) Effects of elevated CO_2 on physiological responses of tall fescue to elevated temperature, drought stress, and the combined stresses. Crop Sci 52(4):1848–1858

Zeppel MJ, Lewis JD, Chaszar B, Smith RA, Medlyn BE, Huxman TE, Tissue DT (2012) Nocturnal stomatal conductance responses to rising [CO_2], temperature and drought. New Phytol 193(4):929–938

Zeroual A, Assani AA, Meddi M, Alkama R (2019) Assessment of climate change in Algeria from 1951 to 2098 using the Köppen-Geiger climate classification scheme. Clim Dyn 52(1):227–243

Zeroual A, Assani AA, Meddi H, Bouabdelli S, Zeroual S, Alkama R (2020) Assessment of projected precipitations and temperatures change signals over Algeria based on regional climate model: RCA4 simulations. Water Res Algeria-Part I:135–159

Zhang G, Sakai H, Tokida T, Usui Y, Zhu C, Nakamura H et al (2013a) The effects of free-air CO_2 enrichment (FACE) on carbon and nitrogen accumulation in grains of rice (*Oryza sativa* L.). J Exp Bot 64(11):3179–3188

Zhang X, Cai J, Wollenweber B, Liu F, Dai T, Cao W, Jiang D (2013b) Multiple heat and drought events affect grain yield and accumulations of high molecular weight glutenin subunits and glutenin macropolymers in wheat. J Cereal Sci 57(1):134–140

Zhao T, Wang QJ, Schepen A, Griffiths M (2019) Ensemble forecasting of monthly and seasonal reference crop evapotranspiration based on global climate model outputs. Agric For Meteorol 264:114–124

Ziska LH (2022) Rising carbon dioxide and global nutrition: evidence and action needed. Plan Theory 11(7):1000

Zkhiri W, Tramblay Y, Hanich L, Jarlan L, Ruelland D (2019) Spatiotemporal characterization of current and future droughts in the high atlas basins (Morocco). Theor Appl Climatol 135(1):593–605

Drought Stress Responses in Plants and Biotechnological Interventions for Its Mitigation

Muhammad Daniyal Junaid, Shajiha Naeem, Atiq ur Rehman Aziz, and Muhammad Arshad Javed

Abstract Drought stress is a formidable challenge to agriculture and threatens food security and ecosystems. This chapter systematically examines the physiological, biochemical, and molecular responses of plants under water deficit conditions. Among the molecular mechanisms governing drought stress responses, we surveyed plants innate mechanisms against water scarcity in soil, moreover the complex signaling pathways, gene expression patterns, and metabolic adjustments that allow plants to adapt and survive in water-deficient environments were discussed in detail. Recent advancements in the identification and characterization of drought-responsive genes and proteins and potential targets for genetic engineering to enhance drought tolerance have been summed up. This chapter also summarized a range of mitigation approaches to alleviate the impact of drought on plant growth and productivity. Including traditional practices for cutting-edge biotechnological interventions, we assessed the potential of various strategies for enhancing water use efficiency and drought resilience in diverse plant species. We also highlight the promising prospects including omics and other molecular approaches and challenges associated with the implementation of these mitigation measures in global agricultural settings. This chapter aims to provide a holistic understanding of the complex responses of plants to drought stress and advancements in landscape mitigation strategies. In doing so, it contributes to the ongoing discourse on sustainable agriculture and encourages the development of innovative approaches to ensure global food security in the face of changing climatic conditions.

Keywords Drought stress · Genomics · Osmotic adjustments · Transcriptomics · Phytohormones

M. D. Junaid (✉) · S. Naeem · A. u. R. Aziz · M. A. Javed
Department of Plant Breeding and Genetics, Faculty of Agricultural Sciences, University of the Punjab, Lahore, Pakistan

U. K. Chaudhry et al. (eds.), *Drought Stress*,
https://doi.org/10.1007/978-3-031-80610-0_3

1 Introduction

Drought is characterized by extended dry spells or periods of low precipitation that lead to hydrological imbalances (Haile et al. 2020). Water scarcity has detrimental impact on plant productivity globally and droughts are also linked to natural disasters (Lee et al. 2020). Plant water linkages are mostly disrupted, and water usage efficiency is reduced during droughts. Water shortage affects nearly every plant process, including phenology, growth rates, photosynthesis, respiration rates, and decreased leaf and stem size (Anjum et al. 2017). Drought affects every geographical area on Earth (Mera 2018). Every year, one-half of the earth is becoming vulnerable to water scarcity. The severity of drought cannot be predicted properly because it is influenced by several factors, such as precipitation, nutrient distribution, the capacity of soils to absorb moisture and evaporation ability (Mukherjee et al. 2018). A 40% yield decrease was observed in maize, and a 21% yield decrease was observed in wheat caused by a 40% reduction in soil moisture (Akhtar et al. 2018) moreover, drought causes 34–68% yield loss in cowpea plants (Siraj et al. 2022).

Drought causes impaired growth and inability to reach crop stands and strongly affects plants at the seedling and germination stages (Seleiman et al. 2021). Drought stress reduces stomatal efficiency to conserve water. As a result, photosynthesis is affected, and CO_2 fixation is hindered, causing lower yields (Morales et al. 2020). Under extreme drought, the stomata close, resulting in a reduction in photosynthesis. Acute drought conditions minimize the chlorophyll content in leaves (Shahzad et al. 2016). Reactive oxygen species cause damage to chloroplasts under drought stress, hence lowering the chlorophyll content (Sehar et al. 2021).

In Pakistan, 65% of labor is directly or indirectly involved in agriculture (Iqbal et al. 2015); similarly, Morocco has a 40% labor force indulged in agriculture. In developed countries, drought stress and floods minimized crop production by 50% during 2001–02 (Tiwari et al. 2018). The global population increase is exponential, and the water supply is limited; these conditions also adds to the effects of drought. To date, no systematic technology has been developed that can expedite plant growth under water scarcity (Ferreira et al. 2017). To meet future food needs, the development of drought-tolerant cultivars is essential. The development of cultivars for abiotic stress resistance requires comprehensive knowledge of the genetic and physiological mechanisms involved (Dresselhaus and Hückelhoven 2018) and their associations at different plant developmental stages. Furthermore, it is immensely important to understand the abiotic factors and their relationship with plant growth to sustain biodiversity and develop better crop varieties (Junaid et al. 2021).

1.1 Drought

Plants are sessile and are frequently subjected to abiotic and biotic stresses (Khare et al. 2020). Among the abiotic stresses, drought is the major factor in reducing agricultural productivity, and it mostly occurs in arid and semiarid areas (Zargar et al. 2017). Drought can be defined in hydrological, agricultural, meteorological and socioeconomic terms. From an agricultural perspective, drought stress occurs when there is inadequate moisture available for plant growth in soil (Ojuederie et al. 2019). Physiologically, water deficiency is characterized by transpiration greater than water uptake by the plant (Siddique and Helen 2020). This difference may be due to differences in water potential between plant roots and soil, which can lead to drought stress. This process is also characterized by decreased atmospheric and soil humidity with high ambient air temperature (Liu et al. 2020). Water deficiency not only is caused by less moisture but also occurs when there is nonuniform precipitation.

1.2 Causes of Drought Stress in Plants

As a result of rising temperatures and atmospheric CO_2 levels caused by global climate change, many regions are experiencing changes in rainfall patterns. Global drought stress is caused primarily by changes in climate (Salehi-Lisar and Bakhshayeshan-Agdam 2016). Droughts can also be caused by a variety of other factors, including high temperatures, intense light, and dry winds, all of which accelerate the evaporation of water from soil. Furthermore, these elements cause plants to lose more water, which increases their vulnerability to water stress. In certain situations, environmental water deficits prevent drought from occurring in the true sense. In certain situations, there may be an adequate amount of water in the soil, but several soil characteristics, including salinity, low soil temperatures, and flooding, hinder or limit the ability of roots to absorb water, which can cause water stress in plants. Pseudo-drought, also known as physiological drought, is a type of drought in which the weather has no influence (Seleiman et al. 2021). Reduced turgor results in inhibited cell growth and expansion. For the first stage of plant growth and establishment, cell growth and expansion are essential phenomena. Under water deficit conditions, the following variables are critical: the degree of stress, duration of stress, and plant reaction.

2 Drought Symptoms in Plants

Plant specie, developmental stage, growth conditions, and other environmental factors all affect the symptoms of drought in plants. Other factors influencing drought symptoms in plants include plant vigor, soil physiochemical conditions, duration of drought, and severity of drought. Loss of leaf turgor, drooping, wilting, etiolation, yellowing, and early leaf fall are typical signs of drought (Yadav and Sharma 2016). In addition, necrosis, branch dieback, tree and shrub canopy thinning, bark cracking, and poor and stunted growth are uncommon symptoms. All these factors lead to plant death in harsh environments.

3 Plant Responses to Drought

As multicellular organisms, plants have complex mechanisms for coping with abiotic stresses (Ullah et al. 2018). Stress protection and tolerance are the two main strategies plants use to combat environmental stress. Plants adapt to water deficit conditions through molecular, biochemical, physiological, and phenological responses. These responses can occur at the whole-plant level or within a specific organ. Three basic responses that a plant shows against abiotic stress are explained in the next sections.

3.1 Escape

Plants that exhibit a shortened life cycle or complete their reproductive cycle before water scarcity are said to exhibit an "escape" response (Volaire 2018). During escape, plants are not exposed to drought, whereas they flower early, which is also a characteristic of stress avoidance (Basu et al. 2016). Early maturity helps plants avoid stress; thus, the development of early maturing varieties can be fruitful for reducing stress-related losses. Plant genotype and environmental conditions determine the fate of a plant's life cycle, and they also determine the ability of a plant to escape drought (Fang and Xiong 2015). Furthermore, if phenological development is synchronized with moisture availability in soil, it results in high seed yields (Hadebe et al. 2017); therefore, plants may escape from drought stress, although "escape" causes yield reduction, but early flowering and early maturity assist plants in avoiding stress (Maphosa et al. 2020). A previous study suggested that drought causes leaf shedding to continue from older to younger leaves, which means that older plant parts are more prone to water deficiency (Hussain and Ali 2015).

3.2 Avoidance

Drought protection includes stomatal control by plants to minimize water loss and conserve water uptake through extensive deep root systems (Meng 2018). As a result, plants maintain high water potential against stress. A deep root system helps plants extract water from considerable depths and in the vicinity of nearby soils (Yang et al. 2017). The presence of cuticles and hairy leaves enables plants to maintain water potential, and is regarded as a key trait in drought tolerance. Plants that exhibit "avoidance" exhibit lower yields and are smaller in size due to the energy consumed on hairy leaves and cuticle production (Salehi-Lisar and Bakhshayeshan-Agdam 2020). Waxy bloom (glaucousness) is a desirable trait for tolerance against drought and helps maintain high water tissue potential on leaves (Chourasia 2017); additionally, glaucousness helps lower leaf senescence and leaf temperature (Hunt et al. 2018).

3.3 Tolerance

Several plant genotypes exhibit xeromorphic (ability to store water in leaves in stems) characteristics, including the production of trichomes and hairiness of leaves in response to drought. Even though it reduces plant yield, such plants also decrease their area and number of leaves. With leaf hairiness, water loss is minimized, and leaf temperature decreases; moreover, plant root architecture is one of the main tolerance traits against drought stress. Root traits, including growth rate, biomass, proliferation capacity and root density, are key factors that protect plants from drought stress (Farooq et al. 2017). The escape, avoidance and tolerance mechanisms of the plants are shown in Fig. 1.

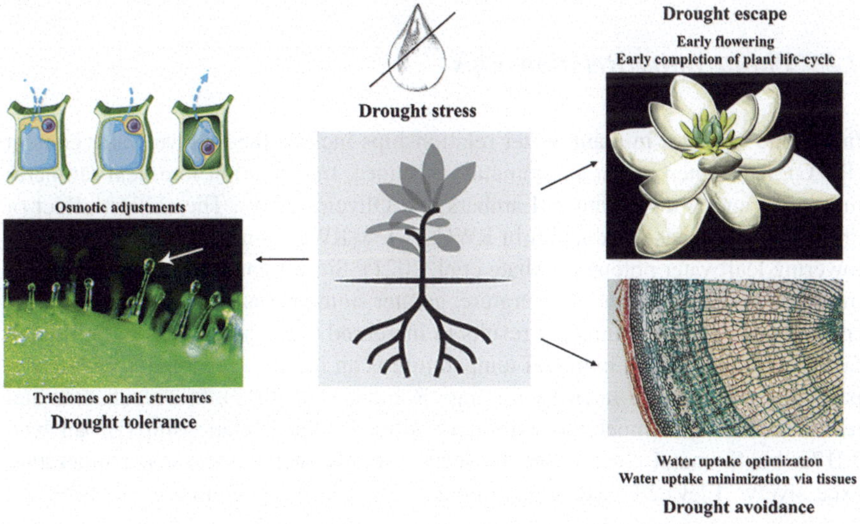

Fig. 1 Plant's drought escape, avoidance and tolerance mechanisms

4 Drought Effects on Plants

4.1 Morphological and Anatomical Characteristics

Numerous facets of a plant's morphology and anatomy can be impacted by drought. Water stress modifies the ultrastructure and anatomy of a leaf. Plants exposed to drought undergo a variety of changes, including a reduction in leaf size, a lower aperture and fewer stomata (Seleiman et al. 2021), thickening of the cell wall, cutinization of the leaf surface and development of a conductive system (an increase in the number of large vessels), submersion of stomata in succulent and xerophyte plants, and the formation of tube leaves in cereals. Furthermore, under water deficit conditions, early leaf senescence increases (Huang et al. 2018). For plants to engage in photosynthesis at its best, stomatal opening and leaf area development must occur at their ideal levels. Therefore, a low leaf area, increased resistance to gas exchange in stomata, and increased leaf senescence result in decreased net photosynthesis in water deficit plants. Size reduction is the primary way that drought stress affects plant morphology (Kaur et al. 2021). Main causes of the decrease in plant size and biomass production is due to low rate of photosynthesis (Postma et al. 2021). A common sign of drought stress that may alter plant morphology is a decrease in chlorophyll content (Salehi-Lisar and Bakhshayeshan-Agdam 2016; Yang et al. 2021). Plants that are dehydrated grow longer roots and take in more water, creating a ramified root system. A greater capacity for water uptake is typically caused by an expansion of the plant's root system and an increase in biomass allocated to the roots during drought conditions. As a result, under moderate water deficit, root growth is not significantly reduced even though shoot growth is. Hence, when plants are dehydrated, they typically have higher root-to-shoot ratios, but their overall biomass is significantly lower (Carvajal et al. 2015; Chang et al. 2016).

4.2 Plant–Water Relationships

Important variables in plant–water relationships include the relative water content (RWC), leaf water potential, stomatal resistance, transpiration rate, leaf temperature, and canopy temperature (Lambers and Oliveira 2019). The earliest effect of drought on plants is a reduction in RWC. A low RWC causes stomatal closure by lowering leaf water potential (Abate et al. 2021). Since transpiration is the primary mechanism regulating leaf temperature, greater stomatal resistance lowers the transpiration rate, which ultimately results in increased leaf temperature. A significant factor contributing to a high leaf temperature is an increase in stomatal resistance, particularly under high-intensity lighting (Rehman et al. 2017). As a result, stomatal resistance and leaf temperature exhibit a positive feedback relationship (Urban et al. 2017a, b). Stomatal closure did, however, initially increase the leaf temperature excessively. Elevated leaf temperatures can cause denaturation of proteins,

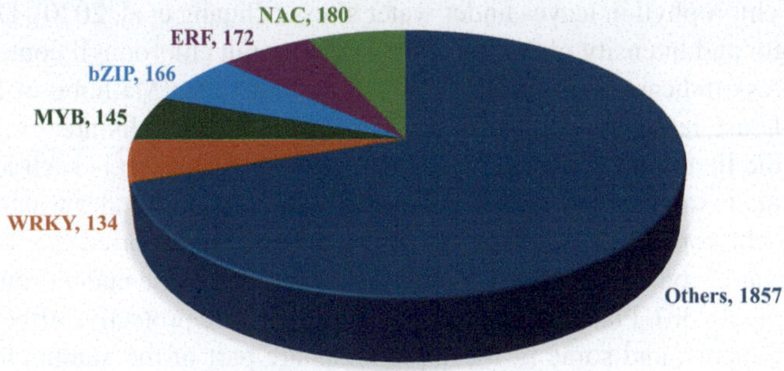

Fig. 2 Pie chart showing the numbers of transcription factor (TF) genes in the NAC, MYB, ERF, bZIP, WRKY, and other families in sorghum

particularly enzymes (Scafaro et al. 2021). Furthermore, variations in membrane permeability are an additional consequence of elevated temperatures, potentially impacting various facets of metabolic processes. The most significant causes of disruption in the metabolic processes of cells, including photosynthesis, respiration, ion uptake, mineral nutrition, and the synthesis of vital macromolecules such as proteins and amino acids, are these changes. Figure 2 shows different transcription factors which are regulated under stress in sorghum.

4.3 *Photosynthesis*

In higher plants, one of the main effects of drought is a reduction in or inhibition of photosynthesis (Siddique et al. 2016). This effect can be caused by a variety of factors, such as a low leaf surface temperature, a decreased leaf expansion rate, elevated leaf temperature, compromised photosynthetic machinery, and early leaf senescence. Under water deficit conditions, both stomatal and nonstomatal factors can effectively inhibit photosynthesis (Urban et al. 2017a, b). Decreased photosynthetic electron transport components accumulate because of carbon dioxide limitations caused by protracted stomatal closure, particularly under light saturation conditions. The buildup of these substances can result in a decrease in molecular oxygen and the generation of reactive oxygen species (ROS) in the form of superoxide, hydroxyl radicals, and H_2O_2, which can damage chloroplasts oxidatively. Furthermore, the main stomatal-dependent factor that lowers the rate of photosynthesis due to decreased activity of CO_2 reduction-related enzymes (the Calvin cycle and dark reactions) is low CO_2 uptake caused by stomatal closure. Inequalities between the light and dark reactions of photosynthesis and the buildup of ROS in chloroplasts may result from the decreased activity of dark reactions. Enzymes, photosynthetic pigments, and thylakoid membranes are among the components of the photosynthetic apparatus that can be harmed by ROS (Yang et al. 2020). Another factor contributing to the decrease in the rate of photosynthesis is a decrease in the amount of chlorophyll in leaves under water stress (Zhuang et al. 2020). Depending on the length and intensity of the drought, a decrease in chlorophyll content during drought stress indicates a reduced ability to harvest light (Mathobo et al. 2017). Carotenoids are less sensitive to water stress than chlorophylls are, according to reports in the literature. On the other hand, xanthophyll pigments such as zeaxanthin and antheraxanthin have been found to increase in plants under water stress, in contrast to chlorophylls (Simkin et al. 2022). It was also reported that carotenoid and anthocyanin contents in carrot plants significantly decrease under drought stress (Junaid et al. 2023b). Plants under stress benefit from the protective effects of xanthophyll pigments, and some of these pigments are part of the xanthophyll cycle, which aids in the detoxification of reactive oxygen species (ROS). Rubisco, also known as ribulose-bisphosphate carboxylase/oxygenase, is the primary enzyme involved in carbon metabolism in the Calvin cycle (Sharma et al. 2020b). The rate of rubisco biosynthesis and degradation determines how much rubisco is present in leaves. Rubisco concentration and activity drop quickly in water-deficient environments. This impact is seen in every plant that has been researched; species-specific variations exist in the extent of the reduction. The factors that cause a decrease in rubisco activity include the acidification of the chloroplast stroma, the absence of CO2 and ribulose-bisphosphate as the carboxylation substrate, a decrease in the amount and/or activity of the coupling factor (ATPase, ATP synthase), structural alterations to the chloroplasts and rubisco, and the release of rubisco from damaged plastids. In addition to rubisco, under drought conditions, the activities of some other carbon metabolism-related enzymes decrease linearly with decreased leaf water potential. These enzymes include fructose-1,6-bisphosphatase, phosphoenolpyruvate carboxylase, NADP-malic enzyme, phosphor-ribulokinase, sucrose phosphate synthase, and pyruvate orthophosphate di-kinase. The cyclic and noncyclic forms of electron transport involved in the light reactions of photosynthesis are also disrupted by drought stress. Both the NADPH/H+ reduction and the photophosphorylation process (ATP biosynthesis) are adversely impacted by a reduced electron transport rate (Salehi-Lisar and Bakhshayeshan-Agdam 2020). When there is water stress, these modifications collectively cause disturbances in the photosynthetic machinery. Both the PSI and PSII photosystems in chloroplasts are impacted by water shortage, mainly due to a reduction in electron transport rate and the accumulation of ROS. Drought-tolerant plants utilize a variety of defense mechanisms to prevent damage to their photosynthetic apparatus, such as thermal dissipation of light energy, photodegradation of PSII D1 protein, water-water cycle activation, xanthophyll cycle triggering, and the dissociation of light-harvesting complexes from photosynthetic reaction centers.

4.4 Protein, Amino Acid, and Mineral Contents

Plants produce compounds like proteins and amino acids and store certain minerals in response to drought stress. According to Aswathi et al. (2022), drought stress modifies gene expression, which in turn influences the synthesis of proteins and mRNAs associated to drought. Proteases, LEA proteins (Li et al. 2018), enzymes required for the biosynthesis of various osmotic-compatible compounds (osmo-protectants), enzymes involved in the detoxification of ROS (e.g., catalase (CAT), peroxidase (POD), ascorbate peroxidase (APX), and glutathione reductase (GR), and protein factors involved in the regulation of signal transduction and gene expression, on the other hand, increase under drought stress. The known mechanism for plant resistance to drought and other stresses, such as salinity, is the accumulation of compatible solutes, called osmo-protectants in some studies, to support osmotic adaptation (osmotic regulation and osmotic adjustment). Low-molecular-weight compatible solutes can accumulate at high concentrations without adversely affecting cell components or metabolism (Salehi-Lisar and Bakhshayeshan-Agdam 2016). Water uptake from the soil is triggered by the accumulation of compatible solutes, which also raises the cellular osmotic pressure. Furthermore, because of their high lipophilicity, compatible solutes prevent water loss from plants (Salehi-Lisar and Bakhshayeshan-Agdam 2020), maintain the turgor pressure and water content of cells, and control the osmotic balance between the cytosol and the vacuole. Compatible solutes may replace the water molecules surrounding ions and guard against the destabilization of critical macromolecules, thereby preventing interactions between ions (at high concentrations) and cellular components (Somero 2022). One of the common amino acids referred to as osmo-protectants is proline. Two mechanisms occur during drought that increase cell proline levels: increased synthesis of proline and decreased activity of the enzymes that degrade proline (Yang et al. 2021). The primary cause of proline accumulation during drought stress is low turgor pressure. Proline has been shown by numerous studies to play a significant role in osmotic regulation. Under drought conditions, proline accumulation and the accumulation of other osmo-protectants cause the water potential of cells to decrease, which facilitates water uptake from the soil. Proline also protects against oxidative stress on cell components, and both its production and breakdown processes are crucial for maintaining the proper energy balance between mitochondria and chloroplasts (Shafi et al. 2019). NADPH/H+ oxidizes to NADP+ in chloroplasts and reduces to NADH/H in mitochondria during the pathways that produce and destroy proline. Because more electrons are consumed during the NADPH/H+ oxidation process in chloroplasts (Sun et al. 2018), the production of ROS decreases. Furthermore, the reduction in NADH/H+ in mitochondria is essential for the cellular energy supply and for recuperation following a stressful event.

In addition to proline, there are other compatible solutes or osmo-protectants whose production and accumulation are stimulated in water-deficient environments. There are six main categories into which compatible solutes fall: (1) sugars, such as glucose and fructose; (2) amino acids; (3) onium compounds; (4) polyols and sugar alcohols; (5) sorbitol, mannitol, pinitol, glycerol, and tertiary and quaternary ammonium; (5) sulfonium compounds; and (6) glycine-betaine and 3-dimethylsulfoniopropionate. To maintain the intracellular water potential, plants occasionally accumulate specific minerals, such as $NaCl$, in addition to compatible compounds. Although mineral accumulation and metabolism are not always compatible, certain plants such as halophytes accumulate minerals via unique mechanisms that protect them from damage. To maintain a balance between the water potentials of the two compartments, plants typically store minerals in the vacuole and compatible compounds in the cytosol.

One significant impact of water deficiency on plants is altering the uptake of nutrients by roots and their transportation to shoots (Bárzana and Carvajal 2020). Drought stress typically results in a decrease in P concentration, an increase in N, and no detrimental effects on plant K concentration (Seleiman et al. 2021). Numerous researchers have also noted a decrease in the Ca content of plants. In response to many stresses, such as drought, plants first target the cell membrane; the stability of the membrane in the roots is crucial for proper plant mineral nutrition. Thus, a key component of plant resistance to drought is maintaining membrane stability. One of the main causes of ion homeostasis disruption in plants is damage to cell membranes in water-deficient environments.

4.5 *Hormones*

The regulation of plant processes is significantly influenced by hormones. Certain hormones play a role in how plants respond to environmental stressors such as drought. When plants are stressed by drought, one of the most useful hormones is abscisic acid (ABA). When plants experience drought, ABA is produced in the roots and moves to the shoots, particularly the leaves. Moreover, chloroplasts undergo ABA synthesis in response to water stress. Furthermore, under water deficit conditions, the activity of PM-ATPase (plasma membrane ATPase) decreases because photosynthesis and respiration result in the production of less ATP. Reduced PM-ATPase increases the pH of the apo-plastic (cell wall) and causes ABA to change to its anionic form (ABA^-). ABA is unable to pass through the leaf cell plasma membrane and instead moves via transpiration stream in the leaf apoplast toward the gourd cells of the stomata (Maurel et al. 2015). The stomatal conductance capacity is reduced, and stomatal closure is induced by ABA translocation to stomata. One of the earliest plant responses to water stress is increased stomatal resistance (Qi et al. 2018), which results in decreased water loss from the leaf surface. Nevertheless, low stomatal CO_2 uptake lowers the rate of photosynthesis in leaves. Additionally, ABA is essential for controlling the activity of aquaporins. It is commonly known that when there is drought, ABA accumulation lowers the amount of ethylene produced (Yao et al. 2021). On the other hand, because indole-3-acetic acid (IAA) downregulation promotes the accumulation of late embryogenesis abundant (LEA) mRNA, auxins function as negative regulators of drought tolerance in plants. The drought stress adaptation of plants, particularly seeds, is mediated by the accumulation of LEA proteins, which are induced by ABA (Ali et al. 2020). Under conditions of water stress, the levels of endogenous cytokinins (zeatin) and gibberellins (GA3) in plants rapidly decrease. Studies have demonstrated that cytokinins can postpone senescence, which may improve plant adaptation by preventing senescence induced by drought (Hönig et al. 2018). Drought generally causes plants to accumulate more brassinosteroids (BRs). Under drought stress, brassinosteroids can decrease ion leakage from membranes by increasing water uptake and stabilizing cell membranes (Ahmad Lone et al. 2022).

4.6 Lipids

Since lipids make up the majority of cell membranes, they are crucial to a plant's ability to withstand environmental stresses (Rawat et al. 2021). Drought stress generally results in disruption of the connection between membrane lipids and proteins, as well as a reduction in the activity of membrane-bound enzymes and in the bilayer's transport capacity. The major glycolipid found in leaves, monogalactosyldiacylglycerol (MGDG), decreases when plants are exposed to drought (Chen et al. 2018). Since MGDG is the most crucial component of the chloroplast membrane, a decrease in its content has a detrimental effect on photosynthesis and causes the membrane to break down. The well-known effect of drought and many other environmental stresses on plants is lipid peroxidation as a result of oxidative damage (Garcia-Caparros et al. 2021).

4.7 Oxidative Stress as a Secondary Stress

When plants are subjected to several environmental stresses, such drought, they create reactive oxygen species (ROS), such as superoxide radicals ($O2^-$), hydroxyl radicals (OH), hydrogen peroxide (H_2O_2), alkoxy radicals (RO), and singlet oxygen (Nieves-Cordones et al. 2019). The structure and functioning of proteins, lipids, and other important macromolecules may be denatured when they interact with ROS. Under drought stress, many cell compartments, including peroxisomes, mitochondria, and chloroplasts, release reactive oxygen species (ROS). In biological systems, ROS generation is mediated by both enzymatic and nonenzymatic mechanisms (Sachdev et al. 2021), which rely on various factors, such as the concentration of oxygen within cells. ROS buildup typically results in lipid peroxidation, protein and photosynthetic pigment oxidation, DNA nicking, and other reactions. Plants have evolved defenses against ROS damage. Together, these enzymatic and nonenzymatic defense mechanisms constitute an antioxidant defense system. Among the enzymes involved in plant antioxidant responses are superoxide dismutase, catalase, peroxidase, etc. (Junaid et al. 2023a). Several compounds, including glutathione, ascorbic acid, carotenoids, and α-tocopherol, are part of the plant antioxidant defense system.

4.8 Respiration

Plants need a sufficient amount of energy to resist water scarcity (DeNicola et al. 2015). Plant roots consume a major portion of fixed carbon for maintenance and dry matter production (Collalti et al. 2020). As photosynthesis and active respiration are negatively affected by water scarcity, the carbon metabolism of plants may be affected. The photosynthetic CO_2 assimilation and respiration ratio determines the plant growth rate (Xing et al. 2021). It is evident from previous literature that under water scarcity, plants spend more energy absorbing moisture from their surroundings. The repair rate is adversely affected by disruption of ATP biosynthesis and the Tri carboxyl acid cycle under drought stress (Salehi-Lisar and Bakhshayeshan-Agdam 2016). Moreover, plants can improve their physiological activity and growth by reducing their root respiration rate and root biomass production (Ma et al. 2018). In plants, ROS are produced in mitochondria during water deficit conditions and can deteriorate the cellular machinery (Chaudhry and Sidhu 2022). There are usually two pathways for mitochondrial electron transport for ubiquinone to oxygen. However, in other pathways, the cytochrome pathway is altered, and electrons are directly transferred to oxygen by alternative oxidases (Simons and Lambers 2018). Alternative oxidase activity transfers electrons and reduces H_2O_2 to O_2, maintaining normal metabolite levels and ROS production (Hou et al. 2021). This alternative pathway can be induced in response to drought stress or through inhibition of the main electron transfer pathway (Salehi-Lisar and Bakhshayeshan-Agdam 2016).

Drought causes a decrease in root respiration, the photosynthesis rate, and root and shoot biomass. Previous studies concluded that severe water scarcity in soil results in improved physiological activity and growth in drought-tolerant wheat cultivars (Ahmad et al. 2018). Drought stress increases the root respiration rate, which leads to disparities in carbon resources, declines in ATP (adenosine triphosphate) production and high levels of ROS production.

4.9 Drought Stress Effects at the Molecular Level

Plants exposed to drought show changed gene expression patterns, they express change in the expression of genes linked to transcription, translation factors, and signal transduction that are involved in early reactions. Moreover, several genes associated with late responses such as those pertaining to oxidative stress, water transport, osmotic balance, and the damage-repair mechanism cause altered expressions (Baldoni et al. 2015). In general, ABA signal transduction and drought signaling are closely related. Drought-induced genes produced in plants are largely regulated by ABA. The transcriptional, post-transcriptional, translational, and post-translational phases are among the stages at which plant gene expression is regulated (Zhang et al. 2020).

4.10 Growth and Yield

Impaired germination and poor stand establishment are the primary effects of drought (Chandrasekaran et al. 2017). A polyethylene glycol-induced water deficit in alfalfa (*Medicago sativa*) resulted in decreased germination potential, hypocotyl length, and fresh and dry weights of the shoots and roots but increased root length (Singh et al. 2016). Cell division, enlargement, and differentiation are the three processes that make up growth. Complex interactions between genetic, physiological, ecological, and morphological events also play a role. These occurrences, which are impacted by a water deficit, determine the type and amount of plant growth. Because of the decrease in turgor pressure, cell growth is one of the physiological processes most susceptible to drought (Anjum et al. 2017).

5 Physiological Responses to Drought Stress

Plants adapt to a number of physiological adjustments in preparation for escaping water deficit, including reduced leaf area, increased leaf thickness, leaf folding, a deeper root system, and reduced stomatal number and conductance, to reduce evapo-transpiration (Peguero-Pina et al. 2020). According to Lee and Suh, the biosynthesis of cuticular wax on aerial plant parts is also associated with responses to drought. Some plants undergo cellular and biochemical modifications to maintain tissue hydrostatic pressure (Negin et al. 2021). Physiological adjustments play a key role under fluctuating environmental conditions as they maintain turgor pressure. Moreover, water availability and root anatomy affect water movement in plants (Brunner et al. 2015). Studies have shown that if photosynthetic components, including pigments, the electron transport system, CO_2 pathways and photosystems, are subjected to drought stress, this results in decreased photosynthetic activity (Sharma et al. 2020a). This may be due to the reason that under drought stress, plant undergoes stomatal closure which results in leaf senescence, lower cell expansion and decreased photosynthetic efficiency (Seleiman et al. 2021). Eventually, it inhibits ATP synthases and key photosynthesis-related enzymes. Water use efficiency (WUE) is the main plant response to osmotic imbalances and is directly associated with photosynthetic activity (Kang et al. 2021). It has also been reported that drought in combination with heat deteriorates proteins and inhibits electron transport in plants (Qaseem et al. 2019). Carbon isotope discrimination" can be utilized for measuring transpiration efficiency, which provides information about CO_2 exchange and the water economy (Wang et al. 2016).

Plants recognize water deficiency via response signaling sensors, which play important roles in several signaling pathways and transcription factors (Kaur et al. 2021). Some plants exhibit stress responses via the activation of chemical, hydraulic or electrical long-distance signals via innate processes. During water deficit conditions, plants also reduce their root hydraulic conductivity to reduce water loss

(Siddique and Helen 2020). Aquaporins govern the above mentioned functions; they are membrane proteins that act as channels for the movement of water and small solutes (Subileau and Vittet 2021). Osmotic adjustment and cell wall elasticity can improve the tissue water status in plants. In extended periods of drought, this process helps maintain turgor and other physiological activities (Ilyas et al. 2021). Previously, dehydrins, abscisic acid (ABA) and osmotic adjustment were shown to increase plant tolerance against drought stress by maintaining high tissue water potential. Osmotic adjustment helps in the accumulation of solutes in the cytoplasm, thus reducing the adverse effects of drought stress (Singh et al. 2015a, b). Dehydrative damage is delayed when plants undergo osmotic adjustment due to underwater scarcity; moreover osmo-protectants decrease osmotic potential and enhance water retention in cells of other vegetables and carrots during drought conditions (Junaid et al. 2023c).

5.1 Biochemical Responses

Enzymatic and nonenzymatic components make up plants' antioxidant defense mechanism. As controls, we employed glutathione reductase, ascorbate peroxidase (APX), peroxidase (POD), catalase (CAT), and superoxide dismutase (SOD). Among the nonenzymatic components include reduced glutathione, cysteine, and ascorbic acid (Das et al. 2018). Many antioxidant enzymes as well as compounds that are soluble in lipids and water scavenge reactive oxygen species (ROS) in plants (Mansoor et al. 2022). The ascorbate glutathione pathway helps in scavenging H_2O_2 and superoxide radicals and includes ascorbate peroxidase, monodehydroascorbate reductase, glutathione reductase and dehydro-ascorbate reductase (Hasanuzzaman et al. 2019). Most of the enzymes involved in the aforementioned phenomenon are located in the cytosol, chloroplast stroma, peroxisomes and mitochondria (Singh et al. 2022). Glutathione reductase plays a key role in sustaining the reduced glutathione pool under drought stress, whereas ascorbate peroxidase acts as an important antioxidant enzyme (Nahar et al. 2015).

The fission of superoxide into O_2 and H_2O_2 is performed by superoxide dismutase, which is a basic enzymatic mechanism (Azadmanesh and Borgstahl 2018) and is also regarded as the first step in the scavenging of reactive oxygen species (ROS). Increased concentrations of SOD also lead to oxidative stress tolerance (Junaid et al. 2023a). A study on *Coffea canephora* suggested that enhanced antioxidant activity is positively correlated with drought tolerance (Pinheiro 2023). In another study, opposing results were obtained under long-term drought conditions for the same plant species. Although carotenoids can scavenge ROS under water scarcity, little attention has been given to understanding their drought tolerance mechanism (Ilyas et al. 2021). ROS significantly damage the cellular machinery during drought, and ascorbate peroxidase and glutathione reductase concentrations increase and reverse cell damage in response to drought stress (Sarker and Oba 2018). A previous study reported that sensitive carrot cultivars exhibit decreased biochemical and antioxidant gene expression levels under drought stress (Junaid et al. 2023a).

5.2 ROS

When plants are exposed to abiotic stress, ROS are generated, including hydrogen peroxide (H_2O_2), alkoxy radicals (ROs), hydroxyl radicals (OHs) and singlet oxygen (O_2), which ultimately cause oxidative stress. ROS disrupts the normal functioning of plant cells by interacting with lipids, DNA and proteins. Pigments present in the thylakoid membrane of chloroplasts may interact with O_2 and produce the harmful oxidant O^{-2} (Wang and Fu 2016). Moreover, ROS production may also increase via the interaction of O_2 with the electron transport chain in mitochondria (Zhao et al. 2019). It is also known that during photorespiration, glycolate oxidizes into glyoxylic acid, and peroxisomes produce H_2O_2.

In biological systems, enzymatic and nonenzymatic reactions generate reactive oxygen species depending on the concentration of oxygen. A reduction in O_2 occurs at higher oxygen concentrations in nonenzymatic reactions. ROS generation continues even at lower O_2 concentrations through the mitochondrial electron transport chain (Hernansanz-Agustín and Enríquez 2021). The deterioration of superoxide anions by superoxide dismutase (SOD) is another enzymatic procedure that generates H_2O_2 (Batinic-Haberle et al. 2015). Moreover, CAT and POD also play key roles in the activation and deactivation of H_2O_2 (Nandi et al. 2019). Studies have shown that under normal and stress conditions, apo-plastic enzymes may lead to the generation of ROS. Glucose oxidase and amino acid oxidase may also accumulate H_2O_2, and ROS are also produced as byproducts in chloroplast electron transport chain, moreover ROS may produce as a result of electron leakage in mitochondria at ubiquinone, the cytochrome *b* region (Sarewicz and Osyczka 2015) or the matrix side of complex 1 (NADH dehydrogenase). The literature suggests that, under water deficit conditions, ROS have deleterious effects on plant growth and productivity (Nxele et al. 2017).

Along with ROS formation, plants also exhibit increased concentrations of methylglyoxal during abiotic stress (Hasanuzzaman et al. 2017). It is an extremely reactive organic compound that can deteriorate DNA and cellular processes. Methylglyoxal may be detoxified via glyoxalase I and II (the glyoxalase system), which catalyze the conversion of methylglyoxal into D-lactate (Wang et al. 2019). To combat the toxic effects of reactive oxygen species (ROS), two common methods of preventing oxidative stress, prevention or avoidance of ROS formation, include the use of detoxifying/scavenging enzymes and multiple antioxidants (Lamaoui et al. 2018).

5.3 Cell Membrane Stability (CMS)

Abiotic stress primarily damages biological membranes, and plants that exhibit drought tolerance are considered to be able to maintain membranous integrity and stability under scarce water conditions (Rawat et al. 2021). The membrane stability of leaf segments is the main characteristic of germplasm screening for drought tolerance. Drought and heat stress considerably deteriorate cell membrane stability in Kentucky bluegrass (Sheikh-Mohamadi et al. 2018). K nutrition enhances drought tolerance in maize, mainly through the disruption of cell membrane stability (Ahmad et al. 2015). Cultivars that have better cell membrane stability are considered more tolerant to fungal infection and vice versa. Any variation in cell membrane stability, stomatal regulation or root capacity is negatively associated with osmotic adjustment. In holm oak (*Quercus ilex*) plants, drought tolerance was increased by hardening, as it reduces osmotic potential and stomatal regulation and enhances cell membrane stability (Ilyas et al. 2021). Abiotic stress-induced membrane disruptions raise the likelihood of molecular interactions leading to membrane fusion and protein denaturation; nevertheless, several drugs have been found to block these detrimental molecular interactions. According to Khan et al. (2020), these substances include oligosaccharides, sorbitol, mannitol, glycine betaine, polyols, trehalose, proline, sucrose, and carnitine. Ion leakage may result from the thermally induced inhibition of the enzymes that keep chemical gradients in cells intact (Chourasia 2017).

5.4 Compatible Solutes and Osmotic Adjustment

The overproduction of various compatible organic solutes is a widely used stress tolerance mechanism in plants. They defend plants against stress in a variety of ways, including regulation of osmotic pressure, detoxifying reactive oxygen species, stabilizing membranes, and preserving the native structures of proteins and enzymes (Fig. 3). One way to maintain stable water conditions in the face of osmotic stress is through osmotic adjustment. A variety of osmotically active molecules and/or ions, such as organic acids, soluble sugars, sugar alcohols, proline, glycine betaine, calcium, potassium, chloride ions, etc., accumulate. The osmotic potential of the cell decreases in a situation of water deficit and solute accumulation. This draws water into the cell and aids in maintaining turgor. The organelles and cytoplasm of plants function at a roughly normal pace through osmotic adjustment, which aids in improved growth, photosynthesis, assimilate partitioning, and grain filling (Hossain et al. 2021).

In settings with a restricted water supply, osmotic adjustment has been suggested as a critical feature in postponing the stress of dehydration (Ndlovu et al. 2021). The degree of osmotic adjustment was shown to relate to the chickpea seed production when cultivated in the field under a line-source irrigation system. One of the most

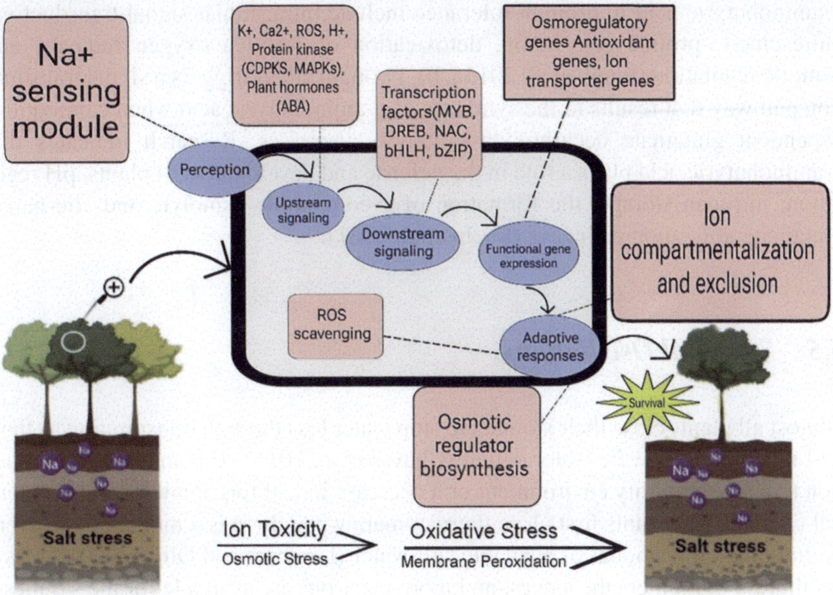

Fig. 3 Role of osmotic adjustment under drought stress

important cytosolic compounds is proline, which is readily accumulated by bacteria, mammals, algae, and higher plants in response to reduced water potential (Zulfiqar et al. 2023). Slow oxidation in mitochondria and enhanced biosynthesis are the two factors that lead to its synthesis in leaves at low water potential. Free proline has been linked to numerous physiological functions, including the stabilization of macromolecules, the storage of carbon and nitrogen for later use in the event of a water deficit, and the sinking of excess reductants, despite some disagreement on this point. It has been reported that under drought conditions, drought-tolerant petunia (*Petunia hybrida*) plants accumulate free proline, which functions as an osmo-protectant and induces drought tolerance. One of the most researched quaternary ammonium compounds, glycine betaine, is a compatible solute found in bacteria, plants, and animals (Pade et al. 2016). It is already known that glycine betaine is crucial for improving plant resistance to a variety of abiotic stresses, such as drought. Apart from its direct protective functions, glycine betaine can also act as an osmo-protectant, improve enzyme and membrane integrity, or indirectly shield cells from environmental stressors by engaging in signal transduction pathways (Zulfiqar et al. 2020). Despite engaging in regular C3-type photosynthesis, which appears to be related to citrulline accumulation, wild watermelon (*Citrullus lanatus*) can adapt to extreme drought stress. In addition to proline and glycine-betaine, citrulline, glutamate, and arginine are the main nutrients that accumulate in wild watermelon (Joshi and Fernie 2017). Citrulline can successfully shield enzymes and DNA from oxidative damage (He et al. 2017). The physiological functions of γ-aminobutyric acid in drought tolerance include intracellular signal transduction, putrescine-to-proline conversion, detoxication of reactive oxygen radicals, and osmotic regulation (Singh et al. 2015a, b). Drought stress triggers a signal transduction pathway that results in the synthesis of γ-aminobutyric acid when calmodulin-dependent glutamate decarboxylase activity increases. Research indicates that γ-aminobutyric acid plays a role in the defense and development of plants, pH regulation, nitrogen storage, the formation of a compatible osmolyte and alternative glutamate utilization pathways (Shelp et al. 2021).

5.5 Stomatal Oscillations

Almost all plants close their stomata to stop water loss through transpiration as their first reaction to a severe water deficit (Chaves et al. 2016). This may occur in reaction to a low-humidity environment or a decrease in leaf turgor and/or water potential (Tufa 2022). Plants first close their stomata when there is a moderate or severe limitation on the amount of available soil water (Lambers and Oliveira 2019). As a result, less CO_2 enters the leaves, and more electrons are available for the synthesis of active oxygen species. The amount of heat that can be dissipated increases as the rate of transpiration decreases (Ballinas and Barradas 2016). Numerous studies have demonstrated that stomatal responses are frequently more directly related to the amount of moisture in the soil than to the level of leaf water. This indicated that while leaf water status is maintained constant, stomata react to chemical cues, such as abscisic acid produced by dehydrating roots (Fig. 4). Higher levels of cytokinin in the xylem sap influence stomatal sensitivity to abscisic acid and directly encourage stomatal opening (Pirasteh-Anosheh et al. 2016).

6 Mitigation of Drought Stress

In reaction to stress, trehalose is biosynthesized by a wide variety of species, including bacteria, fungus, algae, insects, invertebrates, and lower plants (John et al. 2017; Kosar et al. 2019). Certain drought-tolerant ferns, the desiccation-tolerant angiosperm *Myrothamnus flabellifolia* (Gaff and Oliver 2013), and the revival plant *Selaginella lepidophylla* may all collect large levels of trehalose. Trehalose is present in trace levels in many higher plants, including tobacco plants (John et al. 2017). According to previous research, improved water status and the expression of heterologous trehalose biosynthesis genes during Arabidopsis root development are prerequisites for enhanced trehalose drought tolerance (Shim et al. 2019). When trehalose accumulating organisms are under stress, the enzymes trehalose-6-phosphate synthase and phosphatase work together to produce sugar in two steps. Transgenic plants that overexpress trehalose-6-phosphate synthase have been shown to have improved drought tolerance despite the minute accumulation of trehalose

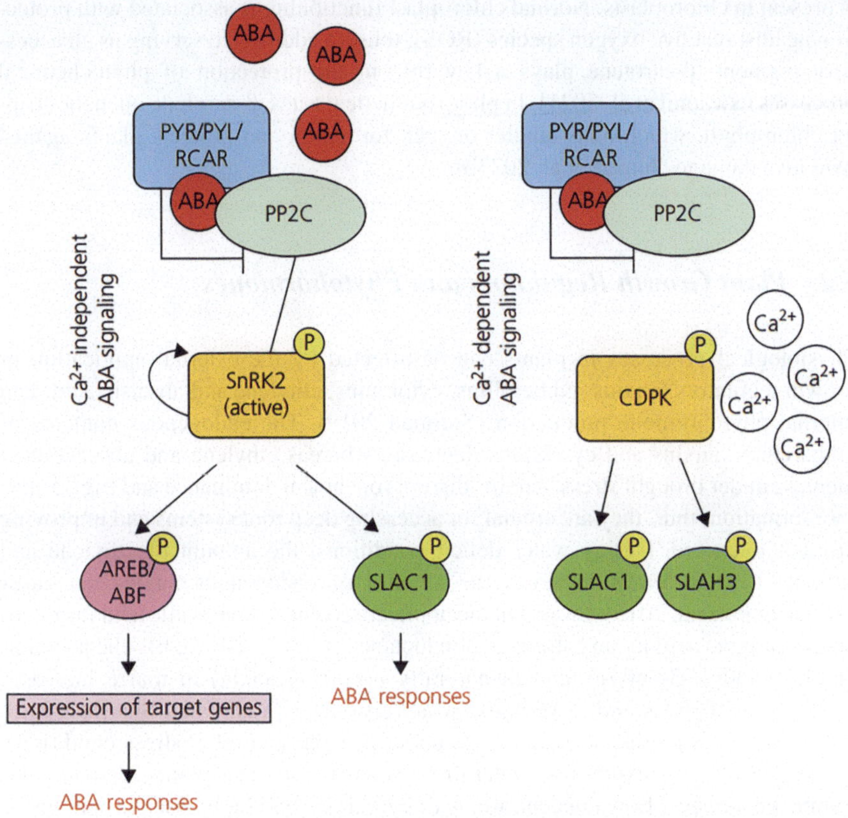

Fig. 4 Schematic representation of abscisic acid signaling under drought stress. ABA levels were increased by environmental stimuli

(Abdi et al. 2022). Plants can resist drought stress by maintaining the antioxidant defense system to scavenge reactive oxygen species, stabilize cell membranes, and primarily adjust their osmotic balance to conserve cell and tissue water (Saxena et al. 2019). Drought tolerance is also greatly aided by plant growth regulators, polyamines, γ-aminobutyric acid, free amino acids, and carbohydrates, which scavenge reactive oxygen species, regulate stomatal function, protect essential macromolecules, and maintain the cell water balance.

6.1 Carotenes

Carotenes play important roles in the plant antioxidant defense system against abiotic stress (Hasanuzzaman et al. 2020), although they are prone to oxidative damage. In green plants, β-carotene is attached to photosystem I and photosystem II and is present in chloroplasts. Normal chloroplast functioning is associated with protection against reactive oxygen species (ROS); thus, in addition to serving as an accessory pigment, β-carotene plays a key role in the protection of phytochemical processes (Swapnil et al. 2021). In photosynthetic tissues, β-carotene quenches triplet chlorophyll, which halts singlet oxygen formation and protects plants against oxidative damage (Junaid et al. 2023b).

6.2 *Plant Growth Regulators and Phytohormones*

Physiological processes in plants can be affected by the external application of growth regulators (auxins, gibberellins, cytokinins, ethylene and abscisic acid) and internal phytohormone production (Soliman 2019). The endogenous contents of gibberellins, auxins and cytokinins decrease, whereas ethylene and abscisic acid increase under drought stress. Auxins disrupt root apical dominance and trigger new root formation; thus, they are crucial for accessing deep root systems and improving drought tolerance. During water deficit conditions, the amount of ethylene and abscisic acid increases, whereas the amount of endogenous auxins decreases (Romanenko et al. 2015). In cotton, the application of indole-3-acetic acid improves photosynthetic activity and stomatal conductance (Li et al. 2021a, b). The accumulation of indole-3-butyric acid (a naturally occurring auxin) in maize increases under drought stress and in response to abscisic acid (Yadav and Sharma 2016). Abscisic acid is produced in all plants under drought and other stress conditions, and it controls gene expression under drought stress and helps plants adapt to environmental stresses. Low concentrations of cytokinins and high abscisic acid levels help in stomatal closure, reducing water loss under drought stress (Pirasteh-Anosheh et al. 2016). Moreover, abscisic acid (ABA) causes changes that maintain cellular turgor in response to water deficiency. Studies suggest that ethylene is a key player in leaf senescence and leaf function throughout the life span of a plant, it also induces premature leaf shedding as a result of water scarcity (Jalil et al. 2022). Drought stress hinders hypocotyl growth, and foliar application of gibberellic acid has been proven to reverse this negative effect. It has been observed that it increases the water status of plants and positively affects protein synthesis (Wojtyla et al. 2016). In cotton, gibberellic acid also increases photosynthesis, the transpiration rate and stomatal conductance (Nimir et al. 2016). Moreover, under water deficit conditions, it encourages pollen formation and seed cone production in *Picea sitchensis* (Li et al. 2021a, b). Previous studies suggested that a biosynthetic mutant of gibberellic acid, "*ga5*", increases drought rhizogenesis. Under drought stress, drought rhizogenesis occurs, as characterized by the formation of short, tuberized and hairless roots. These roots can tolerate water scarcity, and when rehydrated, new functional roots can also be produced. This phenomenon was first reported in the Brassicaceae family (Tekle and Alemu 2016).

A remarkable increase in water potential and chlorophyll content was observed in soybean plants under drought conditions following the application of exogenous uniconazole, brassinolide and abscisic acid (ABA) (Jiang et al. 2021). Another study suggested that the exogenous application of brassinolide, methyl jasmonate (MJ) and uniconazole to maize increases the activities of ascorbate peroxidase (APX), catalase, carotenoids, abscisic acid (ABA) and superoxide dismutase (SOD), which ultimately increase drought tolerance (Ahanger et al. 2018). In a study on pear, it was observed that drought tolerance increases with increasing betaine levels following the exogenous application of jasmonic acid, hence delaying senescence (Yousefi et al. 2019). Salicylic acid can also increase drought tolerance. Salicyclic acid is directly associated with improved catalase activity in winter wheat (Sharma et al. 2017).

6.3 Molecular Approaches

Traits associated with drought tolerance are complex and controlled by multiple genes. Therefore, identification of abiotic stress-related genes is difficult because they are present at multiple quantitative trait loci (QTLs) (Soriano et al. 2021). Conventional plant breeding has played a significant role in improving the traits associated with drought tolerance. The use of marker-assisted breeding and quantitative trait loci (QTLs) has proven to be beneficial in breeding stress-resilient crops (Sandhu et al. 2021). Drought resistance in elite durum (*T. turgidum* ssp. durum), and *T. aestivum* was improved by introgression of QTLs from wild emmer wheat (Merchuk-Ovnat et al. 2016). Although several plants have extremely large genome sizes and ploidy levels, significant efforts have been expended in the genome sequencing, annotation and characterization of important genes (Sun et al. 2022). Several miRNAs have been shown to be preserved in different species and are regulated by drought. In wild emmer wheat (drought resistant), miR166 was downregulated under drought stress (Ferdous et al. 2015). In drought tolerance, b-diketone wax plays an important role in minimizing water loss, and recently, a noncoding miRNA was identified that controls wax formation (Furlan et al. 2021). From these studies, we can assume that miRNA modifications can potentially improve drought tolerance in cereal crops.

To develop drought-resistant crops, genomic studies of highly drought-tolerant plants are important because they can be used as models for drought resilience (Hrmova and Hussain 2021). The study of the changes caused by modifications in DNA structure without alterations in nucleotide sequence is known as "epigenetics". These include the modification of histones, methylation of DNA and noncoding RNA expression; thus, epigenetics can be a potential source for trait improvement in plant breeding (Summanwar et al. 2020). Epigenetic modifications occur not only under stress but also within several generations of a species (Kawakatsu and Ecker 2019). Molecular and biochemical studies have identified several ABA- and stress-responsive genes; moreover, few transcription factors have been found to be responsible for ABA production (Singh and Laxmi 2015).

6.4 Functional Genomics Approaches

Since drought tolerance is a genetically regulated phenomena, bioinformatics technologies have been used to characterize many quantitative trait loci associated with membrane stability and other functionally relevant genes (Magwanga et al. 2020). The ability to identify important genes that react to drought stress and connect their regulation to adaptation activities that take place during stress has been made possible by the development of whole-genome sequencing and similar technologies (Langridge and Reynolds 2015). One of the first techniques for concurrently searching for variations in the quantities of complementary DNA fragments produced from messenger RNA extracted from samples between experimental treatments was differential display (Klein et al. 2020). According to Debbarma et al. (2019), there is evidence linking stress-tolerant quantitative trait loci to several stress-related genes and responsive elements that have been gradually cloned. These genes might potentially serve as the molecular underpinnings of stress tolerance. on a recent work, mapping of quantitative trait loci for grain yield and its components using a simple sequence repeat/expressed sequence tag marker map explained substantial variance on chromosome 4A of wheat (Subbaramamma et al. 2017). It is evident that combining genetic engineering, marker-assisted selection, and conventional and molecular breeding may make it possible to increase crop resistance to abiotic stress efficiently (Younis et al. 2020).

6.5 Agronomic Strategies

The following crop management techniques, when used in conjunction with irrigation, crop residues and mulching, soil management and culture techniques, and crop variety selection, have the potential to mitigate the negative effects of drought and heat stress. It has also been reported that applying nutrients such as N, K, Ca, and Mg can decrease ROS toxicity by increasing the concentration of antioxidant enzymes in plant cells (Ahanger et al. 2017). However, a number of studies have demonstrated that applying fertilizer has no discernible impact on drought stress and that an ideal soil moisture content is necessary because water is essential for nutrient mobility and metabolism (Tavanti et al. 2021). Exogenous silicon (Si) has been applied to mitigate the effects of drought on rice and wheat plants. Higher levels of photosynthetic pigments, increased antioxidant activities, and altered gene expression in the ascorbate-induced glutathione cycle, flavonoid biosynthesis, and antioxidant response were all observed in plants treated with Si (Sun et al. 2020). Salicylic acid, cytokinin, and ABA are among the plant growth substances that have been shown to be important for drought tolerance. Treatment with plant growth regulators can greatly increase the water potential and chlorophyll content in situations where there is water stress (Leufen et al. 2016). In situations where there is a water deficit, the exogenous application of ABA increases soybean yields (Hu et al. 2022). To preserve the marketability of horticulture crops by mitigating the symptoms of drought stress, recent studies have evaluated the use of concentrated ABA or ABA analogs (Lamaoui et al. 2018).

It is well known that polyamines have a significant impact on the growth and development of plants. As polyamines are cationic, they can bind to anionic membrane constituents such as phospholipids, shielding the lipid bilayer from the damaging effects of stress (Zhang et al. 2021). Over the past 20 years, a great deal of research has been conducted on the role that polyamines play in plant defense responses against environmental stressors, as interest in this topic has increased (Tyagi et al. 2023). Numerous genes encoding polyamine metabolism-related enzymes have been cloned from various species, and their expression under various stress scenarios has been examined. For instance, overexpression of the apple spermidine synthase gene results in high levels of spermidine synthase, which significantly increases resistance to abiotic stresses, such as drought (Wang et al. 2022).

6.6 Seed Priming

One of the most effective but temporary ways to counteract the impacts of drought stress is by seed priming. According to Corbineau et al. (2023), seed priming is a technique where seeds are partially moistened to the point where germination-related metabolic processes begin but radicle emergence does not. According to Singh et al. (2015a, b), primed seeds often show better germination rates, improved germination uniformity, and occasionally higher total germination percentages. This tactic has been employed by several crop species to counteract the effects of drought stress. There is currently interest in improving rice and other crops for growing in areas that are water scarce. Drought frequency and intensity may increase significantly in recently introduced aerobic rice culture systems. It was reported that osmo-priming with 4% KCl solution and saturated CaHPO4 solution was successful at improving seedling emergence, crop stand establishment (Subbaramamma et al. 2017), and yield under stress while testing the efficacy of various osmotica for improving the performance of direct-seeded rice plants. Primed rice seeds germinate well, and seedlings emerge more quickly and uniformly in drought-prone areas, increasing yield (Zheng et al. 2016). Germination was tested for eleven types of highland rice plants with restricted soil moisture. Compared to unprimed seeds, seed priming significantly increased the germination and water-use efficiency of drought-stressed plants, improving their performance by 44% (Rai-Kalal et al. 2021).

6.7 Aquaporins

Passive water exchange across membranes can be facilitated and regulated by aquaporins. These proteins are members of a family of major intrinsic membrane proteins that are highly conserved (Li and Wang 2017). Aquaporins are widely distributed in the plasma and vacuolar membranes of plants. The general mechanism of protein-mediated membrane water transport has been revealed by structural analysis of aquaporins. The identification of plant aquaporins has led to a paradigm shift in the knowledge of plant water relations (Putpeerawit et al. 2017), but the connection between aquaporins and drought resistance in plants has not been determined. Moreover, mercury may inhibit aquaporins. This was made clear by several studies on the reduction in root hydraulic conductivity caused by mercury, which demonstrated that aquaporins are important for overall root water uptake and for the cellular osmoregulation of highly compartmented root cells. An elegant method for investigating the roles of aquaporins in plant water relations is reverse genetics (Maurel et al. 2015). The important factors influencing aquaporin activity include phosphorylation calcium, and pH (Qiu et al. 2020). A specific function of aquaporins is to regulate transcellular water transport. For example, they mediate soil water uptake and are widely expressed in roots (Maurel et al. 2015); similarly, transgenic plants that downregulate one or more genes encoding prolactin-inducible proteins have reduced capacities for root water uptake (Tekle and Alemu 2016).

6.8 Stress Proteins

Stress protein synthesis is a common reaction to address stressful conditions, such as water deficit. Since most stress proteins are soluble in water, they hydrate cellular structures, which helps to explain the phenomenon of stress tolerance. Drought tolerance is linked to the synthesis of several transcription factors and stress proteins (Manna et al. 2021). The abiotic stress signaling pathway involves dehydration-responsive element-binding genes. By adjusting the expression of dehydration-responsive element binding genes, stress tolerance could be engineered in transgenic plants (Manna et al. 2021). The ability of rice and groundnuts to withstand drought was significantly enhanced by the introduction of a novel dehydration-responsive element-binding gene transcription factor (Sarkar et al. 2016; Sarkar et al. 2019). Many Capsellia bursa-pastoris like genes have been reported to be synthesized in response to drought stress in various plant species, including tomato some monocots and brassica (Tekle and Alemu 2016). The drought resistance and high proline accumulation of the transgenic tall fescue (*Festuca arundinacea*), in which the dehydration-responsive element-binding gene 1A gene was inserted into the plant, increased. The expression levels of genes abundant in late embryogenesis and in dehydrin type plants, as well as those of molecular chaperones that shield cellular proteins from denaturation, are altered by drought stress in numerous ways

(Vazquez-Hernandez et al. 2021). Heat shock proteins are a part of the broader class of molecules known as chaperones. These proteins play a role in maintaining the stability of the structures of other proteins. Low-molecular-weight heat shock proteins are typically generated only in response to environmental stress, especially high temperatures. However, a variety of stresses, including low temperatures, anaerobic environments, and drought, can cause a large number of heat shock proteins (Zhang et al. 2020). Another significant group of proteins that contribute to drought tolerance is membrane-stabilizing proteins, also known as late embryogenic abundant proteins. By providing a protective environment for other proteins or structures known as dehydrins, these proteins or structures improve the capacity to bind water. Additionally, they are essential for the sequestration of concentrated ions that occur during dehydration of cells (Smith and Graether 2022). Dehydrins are a class of abundant late embryogenesis proteins that accumulate in response to low temperature and dehydration (Hanin et al. 2011). These compounds are synthesized during the desiccation stage of seeds and accumulate in vegetative tissues when there is a water deficit. Different transcription factors linked to abiotic stress identified in different plant species are given in Table 1.

Table 1 Transcription factors linked to abiotic stress identified in different plant species

TF family	DNA-binding domain	Cis-acting element	Structural features
NAC	NAC domain	NACRS (TCNACACGCATGT)	NAC domain consist of 150 amino acids residues in N-terminal and variable transcription regulatory in their C-terminal.
MYB	MYB domain	MYBR (TAACNA/G)	MYB domain composed of multiple repeats each repeat approximately 52 amino acids which forming a helix–turn–helix (HTH) structure.
WRKY	WRKYGQK domain	W-box (TTGACT/C)	WRKY domain is ~60 amino acid residues in length, and has a zinc-finger structure at the C-terminus which either Cx4-5Cx22-23HxH or Cx7Cx23HxC.
ERF/ DREB	AP2/ERF domain	GCC box (AGCCGCC) and (TACCGACAT)	Composed of 60 amino acids with conserved domain consist of Three parallel β-sheets and putative amphiphilic α-helix.
bZIP	bZIP domain	C-box (GACGTC), A-box (TACGTA), G-box (CACGTG), PB-like (TGAAAA), and GLM (GTGAGTCAT)	bZIP domain consist of ~16 amino acid residues which containing nuclear localization signal followed by an invariant N-x7-R/K motif contacts the DNA.

7 Future Perspectives

To survive a wide range of intricate abiotic stresses, plants have evolved sophisticated adaptive mechanisms. Significant advancements in our understanding of these intricate features in higher plants have been made possible by the development of new technologies such as genetic transformation and genomics. However, additional field validation of products or prototypes is necessary for the commercial application of successful research outputs. New frontiers in the study of heat and drought tolerance in plants include epigenetics, genetic engineering, and plant–soil microbiome interactions. For instance, there are chances to improve biotic and abiotic stress in crops through the study of soil microbiomes. The microbes around a plant, especially roots, can mediate plant mechanisms to escape drought and/or heat stress. These processes are associated with various developmental phases of plants, physiological cascades, and molecular and biochemical events occurring at the cellular, tissue, or whole plant level. The use of new molecular and genomic tools and methodologies has opened up new research pathways into plant microbiome. These promising advancements make it possible to investigate the biological roles of a large variety of microorganisms both inside and outside the tissues of plant hosts. Thanks to significant advancements in crop genome characterization and the optimization of genome editing technologies in crops, our understanding of how to produce stress-tolerant crops has evolved and will continue to do so. In the end, conventional and marker-assisted breeding approaches must be combined with genome editing or transgenic procedures to get the desired enhanced cultivars. Furthermore, it is critical to consider spatially different climate change models to guide breeding programs in target trait identification for the selection and identification of novel adaptive germplasms. The implementation of these programs is expected to yield tangible and practical outcomes that have the potential to mitigate the impacts of climate change, specifically about heat waves and droughts, and to improve agricultural production and food security.

8 Conclusion

Drought significantly impacts plant growth, development, and metabolism, especially in arid and semi-arid regions. It affects plants from the molecular level to the whole plant and can be triggered by factors such as high temperatures, light intensity, and low rainfall. Plants employ various drought resistance mechanisms, categorized as avoidance and tolerance tactics. Drought tolerance involves physiological, morphological, ecological, biochemical, and molecular processes. Water deficit impacts plant growth, photosynthesis, and water-use efficiency, with complex regulation of stomatal conductance and enzyme activity. Reactive oxygen species damage cellular components, while growth chemicals like salicylic acid, cytokinin, and abscisic acid contribute to drought tolerance. Effective strategies to mitigate drought stress include developing drought-resistant genotypes, seed priming, and using plant growth regulators and osmo-protectants. Future research should focus on assimilate partitioning, plant stress responses, root signaling, and osmotic adjustment. Understanding the genetic basis of drought tolerance through genomics could lead to the development of drought-resistant plants with improved yields.

References

Abate E, Nardini A, Petruzzellis F, Trifilò P (2021) Too dry to survive: Leaf hydraulic failure in two Salvia species can be predicted on the basis of water content. Plant Physio Biochem 166:215–224

Abdi G, Wahab A, Khurram MF et al (2022) Trehalose-6-phosphate: Biosynthesis, plant metabolism, and crop yields. Research Square

Ahanger MA, Tomar NS, Tittal M, Argal S, Agarwal R (2017) Plant growth under water/salt stress: ROS production; antioxidants and significance of added potassium under such conditions. Physio Mol Biol Plants 23:731–744

Ahanger MA, Ashraf M, Bajguz A, Ahmad P (2018) Brassinosteroids regulate growth in plants under stressful environments and crosstalk with other potential phytohormones. J Plant Growth Reg 37:1007–1024

Ahmad Lone W, Majeed N, Yaqoob U, John R (2022) Exogenous brassinosteroid and jasmonic acid improve drought tolerance in *Brassica rapa* L. genotypes by modulating osmolytes, antioxidants and photosynthetic system. Plant Cell Rep 41(3):603–617

Ahmad N, Khan MB, Farooq S, Shahzad M, Farooq M, Hussain M (2015) Potassium nutrition improves the maize productivity under water deficit conditions. Soil Environ 34(1):15–26

Ahmad Z, Waraich EA, Akhtar S, Anjum S, Ahmad T, Mahboob W, Hafeez OBA, Tapera T, Labuschagne M, Rizwan M (2018) Physiological responses of wheat to drought stress and its mitigation approaches. Acta Physiol Plant 40:1–13

Akhtar K, Wang W, Khan A, Ren G, Afridi MZ, Feng Y, Yang G (2018) Wheat straw mulching with fertilizer nitrogen: An approach for improving soil water storage and maize crop productivity. Plant Soil Environ 64(7):330–337

Ali S, Hayat K, Iqbal A, Xie L (2020) Implications of abscisic acid in the drought stress tolerance of plants. Agronomy 10(9):1323

Anjum SA, Ashraf U, Zohaib A, Tanveer M, Naeem M, Ali I, Tabassum T, Nazir U (2017) Growth and developmental responses of crop plants under drought stress: a review. Zemdirbyste-Agriculture 104(3):267–276

Aswathi KR, Kalaji HM, Puthur JT (2022) Seed priming of plants aiding in drought stress tolerance and faster recovery: a review. Plant Growth Regul 97:1–19

Azadmanesh J, Borgstahl GE (2018) A review of the catalytic mechanism of human manganese superoxide dismutase. Antioxidants 7(2):25

Baldoni E, Genga A, Cominelli E (2015) Plant MYB transcription factors: their role in drought response mechanisms. Int J Mol Sci 16(7):15811–15851

Ballinas M, Barradas VL (2016) Transpiration and stomatal conductance as potential mechanisms to mitigate the heat load in Mexico City. Urban For Urban Green 20:152–159

Bárzana G, Carvajal M (2020) Genetic regulation of water and nutrient transport in water stress tolerance in roots. J Biotech 324:134–142

Basu S, Ramegowda V, Kumar A, Pereira A (2016) Plant adaptation to drought stress. F1000 Res 5

Batinic-Haberle I, Tovmasyan A, Spasojevic I (2015) An educational overview of the chemistry, biochemistry and therapeutic aspects of Mn porphyrins–from superoxide dismutation to H_2O_2-driven pathways. Redox Biol 5:43–65

Bertolino LT, Caine RS, Gray JE (2019) Impact of stomatal density and morphology on water-use efficiency in a changing world. Front Plant Sci 10:225

Brunner I, Herzog C, Dawes MA, Arend M, Sperisen C (2015) How tree roots respond to drought. Front Plant Sci 6:547

Carvajal DE, Loayza AP, Squeo FA (2015) Contrasting responses to water-deficit among *Encelia canescens* populations distributed along an aridity gradient. Am J Bot 102(9):1552–1557

Chandrasekaran P, Sivakumar R, Nandhitha G, Vishnuveni M, Boominathan P, Senthilkumar M (2017) Impact of PPFM and PGRs on seed germination, stress tolerant index and catalase activity in tomato (*Solanum lycopersicum* L) under drought. Int. J. Curr. Microbiol. App. Sci. 6(6):540–549

Chang H, Chen D, Kam J, Richardson T, Drenth J, Guo X, McIntyre CL, Chai S, Rae AL, Xue GP (2016) Abiotic stress upregulated TaZFP34 represses the expression of type-B response regulator and SHY2 genes and enhances root to shoot ratio in wheat. Plant Sci 252:88–102

Chaudhry S, Sidhu GPS (2022) Climate change regulated abiotic stress mechanisms in plants: A comprehensive review. Plant Cell Rep 41(1):1–31

Chaves M, Costa JM, Zarrouk O, Pinheiro C, Lopes C, Pereira JS (2016) Controlling stomatal aperture in semiarid regions—The dilemma of saving water or being cool? Plant Sci 251:54–64

Chen D, Wang S, Qi L, Yin L, Deng X (2018) Galactolipid remodeling is involved in drought-induced leaf senescence in maize. Environ Exp Bot 150:57–68

Chourasia KN (2017) Resistance/Tolerance mechanism under water deficit (Drought) condition in plants. Int J Curr Microbiol App Sci 6(4):66–78

Collalti A, Tjoelker MG, Hoch G, Mäkelä A, Guidolotti G, Heskel M, Petit G, Ryan MG, Battipaglia G, Matteucci G (2020) Plant respiration: controlled by photosynthesis or biomass? Glo Change Biol 26(3):1739–1753

Corbineau F, Taskiran-Özbingöl N, El-Maarouf-Bouteau H (2023) Improvement of seed quality by priming: concept and biological basis. Seeds 2(1):101–115

Das P, Manna I, Biswas AK, Bandyopadhyay M (2018) Exogenous silicon alters ascorbate-glutathione cycle in two salt-stressed indica rice cultivars (MTU 1010 and Nonabokra). Environ Sci Pollut Res 25:26625–26642

Debbarma J, Sarki YN, Saikia B, Boruah HPD, Singha DL, Chikkaputtaiah C (2019) Ethylene response factor (ERF) family proteins in abiotic stresses and CRISPR–Cas9 genome editing of ERFs for multiple abiotic stress tolerance in crop plants: a review. Mol Biotechnol 61(2):153–172

DeNicola E, Aburizaiza OS, Siddique A, Khwaja H, Carpenter DO (2015) Climate change and water scarcity: the case of Saudi Arabia. Ann Glob Health 81(3):342–353

Dresselhaus T, Hückelhoven R (2018) Biotic and abiotic stress responses in crop plants. Agronomy 8:267

Fang Y, Xiong L (2015) General mechanisms of drought response and their application in drought resistance improvement in plants. Cell Mol Life Sci 72:673–689

Farooq M, Gogoi N, Hussain M, Barthakur S, Paul S, Bharadwaj N, Migdadi HM, Alghamdi SS, Siddique KH (2017) Effects, tolerance mechanisms and management of salt stress in grain legumes. Plant Physiol Biochem 118:199–217

Ferdous J, Hussain SS, Shi BJ (2015) Role of micro RNA s in plant drought tolerance. Plant Biotechnol J 13(3):293–305

Ferreira TH, Tsunada MS, Bassi D, Araújo P, Mattiello L, Guidelli GV, Righetto GL, Gonçalves VR, Lakshmanan P, Menossi M (2017) Sugarcane water stress tolerance mechanisms and its implications on developing biotechnology solutions. Front Plant Sci 8:1077

Furlan A, Bianucci E, Llanes A, Peralta JM, Castro S (2021) Abiotic stress tolerance including salt, drought and metal (loid) s in legumes. In: Sustainable agriculture reviews legume agriculture and biotechnology, vol 2, pp 135–180

Gaff DF, Oliver M (2013) The evolution of desiccation tolerance in angiosperm plants: a rare yet common phenomenon. Funct Plant Biol 40(4):315–328

Garcia-Caparros P, De Filippis L, Gul A, Hasanuzzaman M, Ozturk M, Altay V, Lao MT (2021) Oxidative stress and antioxidant metabolism under adverse environmental conditions: a review. Bot Rev 87:421–466
Hadebe ST, Mabhaudhi T, Modi AT (2017) Water use of sorghum (*Sorghum bicolor* L. Moench) in response to varying planting dates evaluated under rainfed conditions. Water SA 43(1):91–103
Haile GG, Tang Q, Li W, Liu X, Zhang X (2020) Drought: Progress in broadening its understanding. Wiley Interdiscip Rev Water 7(2):e1407
Hanin M, Brini F, Ebel C, Toda Y, Takeda S, Masmoudi K (2011) Plant dehydrins and stress tolerance: versatile proteins for complex mechanisms. Plant Signal Behav 10:1503–1509
Hasanuzzaman M, Nahar K, Hossain MS, Mahmud JA, Rahman A, Inafuku M, Oku H, Fujita M (2017) Coordinated actions of glyoxalase and antioxidant defense systems in conferring abiotic stress tolerance in plants. Int J Mol Sci 18(1):200
Hasanuzzaman M, Bhuyan MB, Anee TI, Parvin K, Nahar K, Mahmud JA, Fujita M (2019) Regulation of ascorbate-glutathione pathway in mitigating oxidative damage in plants under abiotic stress. Antioxidants 8(9):384
Hasanuzzaman M, Bhuyan MB, Zulfiqar F, Raza A, Mohsin SM, Mahmud JA, Fujita M, Fotopoulos V (2020) Reactive oxygen species and antioxidant defense in plants under abiotic stress: revisiting the crucial role of a universal defense regulator. Antioxidants 9(8):681
He L, He T, Farrar S, Ji L, Liu T, Ma X (2017) Antioxidants maintain cellular redox homeostasis by elimination of reactive oxygen species. Cell Physiol Biochem 44(2):532–553
Hernansanz-Agustín P, Enríquez JA (2021) Generation of reactive oxygen species by mitochondria. Antioxidants 10(3):415
Hönig M, Plíhalová L, Husičková A, Nisler J, Doležal K (2018) Role of cytokinins in senescence, antioxidant defense and photosynthesis. Int J Mol Sci 19(12):4045
Hossain A, Skalicky M, Brestic M, Maitra S, Alam MA, Syed MA, Hossain J, Sarkar S, Saha S, Bhadra P (2021) Consequences and mitigation strategies of abiotic stresses in wheat (*Triticum aestivum* L.) under the changing climate. Agronomy *11*(2):241
Hou L, Zhao M, Huang C, He Q, Zhang L, Zhang J (2021) Alternative oxidase gene induced by nitric oxide is involved in the regulation of ROS and enhances the resistance of *Pleurotus ostreatus* to heat stress. Microb Cell Fact 20(1):1–20
Hrmova M, Hussain SS (2021) Plant transcription factors involved in drought and associated stresses. Int J Mol Sci 22(11):5662
Hu W, Zhang J, Wu Z, Loka DA, Zhao W, Chen B, Wang Y, Meng Y, Zhou Z, Gao L (2022) Effects of single and combined exogenous application of abscisic acid and melatonin on cotton carbohydrate metabolism and yield under drought stress. Ind Crop Prod 176:114302
Huang Y, Guo Y, Liu Y, Zhang F, Wang Z, Wang H, Wang F, Li D, Mao D, Luan S (2018) 9-cis-epoxycarotenoid dioxygenase 3 regulates plant growth and enhances multiabiotic stress tolerance in rice. Front Plant Sci 9:162
Hunt JR, Hayman PT, Richards RA, Passioura JB (2018) Opportunities to reduce heat damage in rain-fed wheat crops based on plant breeding and agronomic management. Field Crop Res 224:126–138
Hussain B, Ali B (2015) Leaf longevity in plants under water stress—a review. Indian J Plant Sci 4(4):2319–3824
Ilyas M, Nisar M, Khan N, Hazrat A, Khan AH, Hayat K, Fahad S, Khan A, Ullah A (2021) Drought tolerance strategies in plants: a mechanistic approach. J. Plant Growth Regul 40:926–944
Iqbal MA, Iqbal A, Afzal S, Akbar N, Abbas RN, Khan HZ (2015) In Pakistan, agricultural mechanization status and future prospects. Am Eurasian J Agric Environ Sci 15(1):122–128
Jalil SU, Ansari SA, Ansari MI (2022) Role of environment stress leaf senescence and crop productivity. In: Augmenting crop productivity in stress environment. Springer, pp 13–31
Jiang Y, Sun Y, Zheng D, Han C, Cao K, Xu L, Liu S, Cao Y, Feng N (2021) Physiological and transcriptome analyses for assessing the effects of exogenous uniconazole on drought tolerance in hemp (*Cannabis sativa* L.). Sci Rep 11(1):14476

John R, Raja V, Ahmad M, Jan N, Majeed U, Ahmad S, Yaqoob U, Kaul T (2017) Trehalose: metabolism and role in stress signaling in plants. Stress signaling in plants, genomics and proteomics perspective 2:261–275

Joshi V, Fernie AR (2017) Citrulline metabolism in plants. Amino Acid 49:1543–1559

Junaid MD, Chaudhry UK, Gökçe AF (2021) Climate change and plant growth–South Asian perspective. In: Climate change plants. CRC Press, Boca Raton, pp 37–53

Junaid MD, Öztürk ZN, Gökçe AF (2023a) Investigation of drought induced biochemical and gene expression changes in carrot cultivars. Mol Biol Rep 50(1):349–359

Junaid MD, Öztürk ZN, Gökçe AF (2023b) Drought stress effects on morphophysiological and quality characteristics of commercial carrot cultivars. Turk J Bot 47(2):111–126

Junaid MD, Öztürk ZN, Gökçe AF (2023c) Exploitation of tolerance to drought stress in carrot (*Daucus carota* L.): an overview. Stress Biol 3(1):55

Kang J, Hao X, Zhou H, Ding R (2021) An integrated strategy for improving water use efficiency by understanding physiological mechanisms of crops responding to water deficit: present and prospect. Agric Water Manag 255:107008

Kaur H, Kohli SK, Khanna K, Bhardwaj R (2021) Scrutinizing the impact of water deficit in plants: transcriptional regulation, signaling, photosynthetic efficacy, and management. Physiol Plant 172(2):935–962

Kawakatsu T, Ecker JR (2019) Diversity and dynamics of DNA methylation: Epigenomic resources and tools for crop breeding. Breed Sci 69(2):191–204

Khan N, Ali S, Tariq H, Latif S, Yasmin H, Mehmood A, Shahid MA (2020) Water conservation and plant survival strategies of rhizobacteria under drought stress. Agronomy 10(11):1683

Khare S, Singh N, Singh A, Hussain I, Niharika K, Yadav V, Bano C, Yadav RK, Amist N (2020) Plant secondary metabolites synthesis and their regulations under biotic and abiotic constraints. J Plant Biol 63:203–216

Klein JC, Agarwal V, Inoue F, Keith A, Martin B, Kircher M, Ahituv N, Shendure J (2020) A systematic evaluation of the design and context dependencies of massively parallel reporter assays. Nat Methods 17(11):1083–1091

Kosar F, Akram NA, Sadiq M, Al-Qurainy F, Ashraf M (2019) Trehalose: a key organic osmolyte effectively involved in plant abiotic stress tolerance. J Plant Growth Regul 38:606–618

Lamaoui M, Jemo M, Datla R, Bekkaoui F (2018) Heat and drought stresses in crops and approaches for their mitigation. Front Chem 6:26

Lambers H, Oliveira RS (2019) Plant water relations. In: Plant Physiological Ecology. Springer, pp 187–263

Langridge P, Reynolds MP (2015) Genomic tools to assist breeding for drought tolerance. Curr Opin Biotechnol 32:130–135

Lee J, Perera D, Glickman T, Taing L (2020) Water-related disasters and their health impacts: a global review. Prog Disaster Sci 8:100123

Leufen G, Noga G, Hunsche M (2016) Drought stress memory in sugar beet: mismatch between biochemical and physiological parameters. J Plant Growth Regul 35:680–689

Li C, Wang W (2017) Molecular biology of aquaporins. Aqua 969:1–34

Li N, Zhang S, Liang Y, Qi Y, Chen J Zhu W, Zhang L (2018) Label-free quantitative proteomic analysis of drought stress-responsive late embryogenesis abundant proteins in the seedling leaves of two wheat (*Triticum aestivum* L.) genotypes. J Proteome 172:122–142

Li C, Kong X, Luo Z, Li W, Tang W, Zhang D, Ma C, Dong H (2021a) Exogenous application of acetic acid improves the survival rate of cotton by increasing abscisic acid and jasmonic acid contents under drought stress. Acta Physiol Plant 43:1–10

Li Y, Li X, Zhao M-H, Pang Z-Y, Wei J-T, Tigabu M, Chiang VL, Sederoff H, Sederoff R, Zhao X-Y (2021b) An overview of the practices and management methods for enhancing seed production in conifer plantations for commercial use. Horticulturae 7(8):252

Liu L, Gudmundsson L, Hauser M, Qin D, Li S, Seneviratne SI (2020) Soil moisture dominates dryness stress on ecosystem production globally. Nat Commun 11(1):4892

Ma S-C, Wang T-C, Guan X-K, Zhang X (2018) Effect of sowing time and seeding rate on yield components and water use efficiency of winter wheat by regulating the growth redundancy and physiological traits of root and shoot. Field Crop Res 221:166–174

Magwanga RO, Lu P, Kirungu JN, Cai X, Zhou Z, Agong SG, Wang K, Liu F (2020) Identification of QTLs and candidate genes for physiological traits associated with drought tolerance in cotton. J Cotton Res 3:1–33

Manna M, Thakur T, Chirom O, Mandlik R, Deshmukh R, Salvi P (2021) Transcription factors as key molecular target to strengthen the drought stress tolerance in plants. Physiol Plant 172(2):847–868

Mansoor S, Ali O, Lone JK, Manhas S, Kour N, Alam P, Ahmad A, Ahmad P (2022) Reactive oxygen species in plants: from source to sink. Antioxidants 11(2):225

Maphosa L, Richards MF, Norton SL, Nguyen GN (2020) Breeding for abiotic stress adaptation in chickpea (*Cicer arietinum* L.): a comprehensive review. Crop Breed App Biotech 4(3):e200015

Mathobo R, Marais D, Steyn JM (2017) The effect of drought stress on yield, leaf gaseous exchange and chlorophyll fluorescence of dry beans (*Phaseolus vulgaris* L.). Agric Water Manage 180:118–125

Maurel C, Boursiac Y, Luu D-T, Santoni V, Shahzad Z, Verdoucq L (2015) Aquaporins in plants. Physiol Rev 95(4):1321–1358

Meng LS (2018) Compound synthesis or growth and development of roots/stomata regulate plant drought tolerance or water use efficiency/water uptake efficiency. J Agric Food Chem 66(14):3595–3604

Mera GA (2018) Drought and its impacts in Ethiopia. Weather Clim Extremes 22:24–35

Merchuk-Ovnat L, Barak V, Fahima T, Ordon F, Lidzbarsky GA, Krugman T, Saranga Y (2016) Ancestral QTL alleles from wild emmer wheat improve drought resistance and productivity in modern wheat cultivars. Front Plant Sci 7:452

Morales F, Ancín M, Fakhet D, González-Torralba J, Gámez AL, Seminario A, Soba D, Ben-Mariem S, Garriga M, Aranjuelo I (2020) Photosynthetic metabolism under stressful growth conditions as a bases for crop breeding and yield improvement. Plan Theory 9(1):88

Mukherjee S, Mishra A, Trenberth KE (2018) Climate change and drought: a perspective on drought indices. Curr Clim Chang Rep 4:145–163

Nahar K, Hasanuzzaman M, Alam MM, Fujita M (2015) Glutathione-induced drought stress tolerance in mung bean: coordinated roles of the antioxidant defense and methylglyoxal detoxification systems. AoB Plants 7:plv069

Nandi A, Yan LJ, Jana CK, Das N (2019) Role of catalase in oxidative stress- and age-associated degenerative diseases. Oxid Med Cell Longev 2019:9613090–9613019

Ndlovu E, Van-Staden J, Maphosa M (2021) Morpho-physiological effects of moisture, heat and combined stresses on *Sorghum bicolor* [Moench (L.)] and its acclimation mechanisms. Plant Stress 2:100018

Negin B, Hen-Avivi S, Almekias-Siegl E, Shachar L, Aharoni A (2021) Cuticular wax composition is essential for plant recovery following drought with little effect under optimal conditions. bioRxiv 447:487

Nieves-Cordones M, García-Sánchez F, Pérez-Pérez JG, Colmenero-Flores JM, Rubio F, Rosales MA (2019) Coping with water shortage: an update on the role of K+, Cl-, and water membrane transport mechanisms on drought resistance. Front Plant Sci 10, 1619

Nimir NEA, Zhou G, Guo W, Ma B, Lu S, Wang Y (2016) Effect of foliar application of GA3, kinetin, and salicylic acid on ions content, membrane permeability, and photosynthesis under salt stress of sweet sorghum [*Sorghum bicolor* (L.) Moench]. Can J Plant Sci 97(3):525–535

Nxele X, Klein A, Ndimba BK (2017) Drought and salinity stress alters ROS accumulation, water retention, and osmolyte content in sorghum plants. S Afr J Bot 108:261–266

Ojuederie OB, Olanrewaju OS, Babalola OO (2019) Plant growth promoting rhizobacterial mitigation of drought stress in crop plants: implications for sustainable agriculture. Agronomy 9(11):712

Pade N, Michalik D, Ruth W, Belkin N, Hess WR, Berman-Frank I, Hagemann M (2016) Trimethylated homoserine functions as the major compatible solute in the globally significant oceanic cyanobacterium Trichodesmium. Pro Natl Acad Sci 113(46):13191–13196

Peguero-Pina JJ, Vilagrosa A, Alonso-Forn D, Ferrio JP, Sancho-Knapik D, Gil-Pelegrín E (2020) Living in drylands: Functional adaptations of trees and shrubs to cope with high temperatures and water scarcity. Forests 11(10):1028

Pinheiro HA (2023) Physiological and morphological adaptations as associated with drought tolerance in Robusta coffee (*Coffea canephora* Pierre var. *kouillou*). UFRA/Campus Belém

Pirasteh-Anosheh H, Saed-Moucheshi A, Pakniyat H, Pessarakli M (2016) Stomatal responses to drought stress. In: Water stress and crop plants, A sustainable approach, vol 1, pp 24–40

Postma JA, Hecht VL, Hikosaka K, Nord EA, Pons TL, Poorter H (2021) Dividing the pie: a quantitative review on plant density responses. Plant Cell Environ 44(4):1072–1094

Putpeerawit P, Sojikul P, Thitamadee S, Narangajavana J (2017) Genome-wide analysis of aquaporin gene family and their responses to water-deficit stress conditions in cassava. Plant Physiol Biochem. 121:118–127

Qaseem MF, Qureshi R, Shaheen H (2019) Effects of preanthesis drought, heat and their combination on the growth, yield and physiology of diverse wheat (*Triticum aestivum* L.) genotypes varying in sensitivity to heat and drought stress. Sci Rep 9(1):6955

Qi J, Song CP, Wang B, Zhou J, Kangasjärvi J, Zhu JK, Gong Z (2018) Reactive oxygen species signaling and stomatal movement in plant responses to drought stress and pathogen attack. J Integr Plant Biol 60(9):805–826

Qiu J, McGaughey SA, Groszmann M, Tyerman SD, Byrt CS (2020) Phosphorylation influences water and ion channel function of AtPIP2; 1. Plant Cell Environ 43(10):2428–2442

Rai-Kalal P, Tomar RS, Jajoo A (2021) Seed nanopriming by silicon oxide improves drought stress alleviation potential in wheat plants. Funct Plant Biol 48(9):905–915

Rawat N, Singla-Pareek SL, Pareek A (2021) Membrane dynamics during individual and combined abiotic stresses in plants and tools to study the same. Physiol Plant 171(4):653–676

Rehman M, Ullah S, Bao Y, Wang B, Peng D, Liu L (2017) Light-emitting diodes: whether an efficient source of light for indoor plants? Environ Sci Pollut Res 24:24743–24752

Romanenko E, Kosakovskaya I, Romanenko P (2015) Phytohormones of microalgae: biological role and involvement in the regulation of physiological processes. Pt I. auxins, abscisic acid, ethylene. Int J Algae 17(3):275–289

Sachdev S, Ansari SA, Ansari MI, Fujita M, Hasanuzzaman M (2021) Abiotic stress and reactive oxygen species: Generation, signaling, and defense mechanisms. Antioxidants 10(2):277

Salehi-Lisar SY, Bakhshayeshan-Agdam H (2016) Drought stress in plants: causes, consequences, and tolerance. In: Drought stress tolerance in plants, physiology and biochemistry, pp 1–16

Salehi-Lisar SY, Bakhshayeshan-Agdam H (2020) Agronomic crop responses and tolerance to drought stress. In: Agronomic crops, stress responses and tolerance, pp 63–91

Sandhu N, Yadav S, Catolos M, Cruz MTS, Kumar A (2021) Developing climate-resilient, direct-seeded, adapted multiple-stress-tolerant rice applying genomics-assisted breeding. Front Plant Sci 12:637488

Sarewicz M, Osyczka A (2015) Electronic connection between the quinone and cytochrome c redox pools and its role in regulation of mitochondrial electron transport and redox signaling. Physiol Rev 95(1):219–243

Sarkar T, Thankappan R, Kumar A, Mishra GP, Dobaria JR (2016) Stress inducible expression of AtDREB1A transcription factor in transgenic peanut (*Arachis hypogaea* L.) conferred tolerance to soil-moisture deficit stress. Front Plant Sci 7:935

Sarkar T, Thankappan R, Mishra GP, Nawade BD (2019) Advances in the development and use of DREB for improved abiotic stress tolerance in transgenic crop plants. Physiol Mol Bio Plant 25:1323–1334

Sarker U, Oba S (2018) Catalase, superoxide dismutase and ascorbate-glutathione cycle enzymes confer drought tolerance of *Amaranthus tricolor*. Sci Rep 8(1):16496

Saxena R, Kumar M, Tomar RS (2019) Plant responses and resilience toward drought and salinity stress. Plant Arch 19:50–58

Scafaro AP, Fan Y, Posch BC, Garcia A, Coast O, Atkin OK (2021) Responses of leaf respiration to heatwaves. Plant Cell Environ 44(7):2090–2101

Sehar Z, Jahan B, Masood A, Anjum NA, Khan NA (2021) Hydrogen peroxide potentiates defense system in presence of sulfur to protect chloroplast damage and photosynthesis of wheat under drought stress. Physiol Plant 172(2):922–934

Seleiman MF, Al-Suhaibani N, Ali N, Akmal M, Alotaibi M, Refay Y, Dindaroglu T, Abdul-Wajid HH, Battaglia ML (2021) Drought stress impacts on plants and different approaches to alleviate its adverse effects. Plan Theory 10(2):259

Shafi A, Zahoor I, Mushtaq U (2019) Proline accumulation and oxidative stress: diverse roles and mechanism of tolerance and adaptation under salinity stress. In: Salt stress, microbes, and plant interactions, mechanisms and molecular approaches, pp 269–300

Shahzad MA, Jan SU, Afzal F, Khalid M, Gul A, Sharma I, Sofo A, Ahmad P (2016) Drought stress and morphophysiological responses in plants. In: Water stress and crop plants, A sustainable approach, pp 452–467

Sharma M, Gupta SK, Majumder B, Maurya VK, Deeba F, Alam A, Pandey V (2017) Salicylic acid mediated growth, physiological and proteomic responses in two wheat varieties under drought stress. J Proteome 163:28–51

Sharma A, Wang J, Xu D, Tao S, Chong S, Yan D, Li Z, Yuan H, Zheng B (2020a) Melatonin regulates the functional components of photosynthesis, antioxidant system, gene expression, and metabolic pathways to induce drought resistance in grafted *Carya cathayensis* plants. Sci Total Environ 713:136675

Sharma S, Joshi J, Kataria S, Verma SK, Chatterjee S, Jain M, Pathak K, Rastogi A, Brestic M (2020b) Regulation of the Calvin cycle under abiotic stresses: an overview. In: Plant life under changing environment, pp 681–717

Sheikh-Mohamadi MH, Etemadi N, Arab MM, Aalifar M, Arab M (2018) Physiological and ascorbate-glutathione pathway-related genes responses under drought and heat stress in crested wheatgrass. Sci Hortic 242:195–206

Shelp BJ, Aghdam MS, Flaherty EJ (2021) γ-Aminobutyrate (GABA) regulated plant defense: mechanisms and opportunities. Plan Theory 10(9):1939

Shim JS, Seo JS, Kim Y, Koo Y, Do-Choi Y, Jung C (2019) Heterologous expression of bacterial trehalose biosynthetic genes enhances trehalose accumulation in potato plants without adverse growth effects. Plant Biotechnol Rep 13:409–418

Siddique KH, Helen B (2020) Water deficits: development. In: Fresh water and watersheds. CRC Press, pp 215–219

Siddique Z, Jan S, Imadi SR, Gul A, Ahmad P (2016) Drought stress and photosynthesis in plants. In: Water stress and crop plants: a sustainable approach, pp 1–11

Simkin AJ, Kapoor L, Doss CGP, Hofmann TA, Lawson T, Ramamoorthy S (2022) The role of photosynthesis related pigments in light harvesting, photoprotection and enhancement of photosynthetic yield in planta. Photosynth Res 152(1):23–42

Simons BH, Lambers H (2018) The alternative oxidase: is it a respiratory pathway allowing a plant to cope with stress? In: Plant responses to environmental stresses, pp 265–286

Singh D, Laxmi A (2015) Transcriptional regulation of drought response: a tortuous network of transcription factors. Front Plant Sci 6:895

Singh H, Jassal RK, Kang J, Sandhu S, Kang H, Grewal K (2015a) Seed priming techniques in field crops-A review. Agric Rev 36(4):251–264

Singh M, Kumar J, Singh S, Singh VP, Prasad SM (2015b) Roles of osmoprotectants in improving salinity and drought tolerance in plants: a review. Rev Environ Sci Biotechnol 14:407–426

Singh A, Srivastava J, Lal J (2016) Effect of PEG-6000 induced osmotic stress on germination, growth and nutrient uptake of two lentil [*Lens culinaris* (Medikus)] genotypes. J Food Leg 29(3and4):188–194

Singh A, Mehta S, Yadav S, Nagar G, Ghosh R, Roy A, Chakraborty A, Singh IK (2022) How to cope with the challenges of environmental stresses in the era of global climate change: an update on ROS stave off in plants. Int J Mol Sci 23(4):1995

Siraj S, Khan MA, Hamayun M, Ali S, Khan SA, Hussain A, Iqbal A, Khan H, Kang SM, Lee IJ (2022) Microbacterium oxydans regulates physio-hormonal and molecular attributes of *Solanum lycopersicum* under drought stress. Agronomy 12(12):3224

Smith MA, Graether SP (2022) The disordered dehydrin and its role in plant protection: a biochemical perspective. Biomol Ther 12(2):294

Soliman AS (2019) Plant growth hormones. Cell Growth 1

Somero GN (2022) Solutions: how adaptive changes in cellular fluids enable marine life to cope with abiotic stressors. Mar Life Sci Technol 4(3):389–413

Soriano JM, Colasuonno P, Marcotuli I, Gadaleta A (2021) Meta-QTL analysis and identification of candidate genes for quality, abiotic and biotic stress in durum wheat. Sci Rep 11(1):11877

Subbaramamma P, Sangamitra M, Manjusha D (2017) Mitigation of drought stress in production of pulses. Int J Multidiscip Adv Res Trends 4:41–62

Subileau M, Vittet D (2021) Lymphatics in eye fluid homeostasis: minor contributors or significant actors? Biology 10(7):582

Summanwar A, Basu U, Rahman H, Kav NN (2020) Noncoding RNAs as emerging targets for crop improvement. Plant Sci 297:110521

Sun M, Jiang F, Cen B, Wen J, Zhou Y, Wu Z (2018) Respiratory burst oxidase homolog-dependent H_2O_2 and chloroplast H_2O_2 are essential for the maintenance of acquired thermotolerance during recovery after acclimation. Plant Cell Environ 41(10):2373–2389

Sun J, Qiu C, Ding Y, Wang Y, Sun L, Fan K, Gai Z, Dong G, Wang J, Li X (2020) Fulvic acid ameliorates drought stress-induced damage in tea plants by regulating the ascorbate metabolism and flavonoids biosynthesis. BMC Genomics 21(1):1–13

Sun Y, Shang L, Zhu QH, Fan L, Guo L (2022) Twenty years of plant genome sequencing: achievements and challenges. Trends Plant Sci 27(4):391–401

Swapnil P, Meena M, Singh SK, Dhuldhaj UP, Marwal A (2021) Vital roles of carotenoids in plants and humans to deteriorate stress with its structure, biosynthesis, metabolic engineering and functional aspects. Curr Plant Biol 26:100203

Tavanti TR, De-Melo AAR, Moreira LDK, Sanchez DEJ, Dos-Santos SR, Da-Silva RM, Dos-Reis AR (2021) Micronutrient fertilization enhances ROS scavenging system for alleviation of abiotic stresses in plants. Plant Physiol Biochem 160:386–396

Tekle AT, Alemu MA (2016) Drought tolerance mechanisms in field crops. World J Bio Med Sci 3(2):15–39

Tiwari GS, Pareek N, Praverkar K (2018) Increased heat and drought stress under climate change and their impact on physiological growth and development of crops: a review. Ind J Agric Sci 88(12):1818–1825

Tufa KN (2022) Review on effects, mechanisms and managements of plants water stress. Irrig Drain Syst 11:2–7

Tyagi A, Ali S, Ramakrishna G, Singh A, Park S, Mahmoudi H, Bae H (2023) Revisiting the role of polyamines in plant growth and abiotic stress resilience: mechanisms, crosstalk, and future perspectives. J Plant Growth Reg 42(8):5074–5098

Ullah A, Hussain A, Shaban M, Khan AH, Alariqi M, Gul S, Jun Z, Lin S, Li J, Jin S (2018) Osmotin: a plant defense tool against biotic and abiotic stresses. Plant Physiol Biochem 123:149–159

Urban J, Ingwers M, McGuire MA, Teskey RO (2017a) Stomatal conductance increases with rising temperature. Plant Signal Behav 12(8):e1356534

Urban L, Aarrouf J, Bidel LP (2017b) Assessing the effects of water deficit on photosynthesis using parameters derived from measurements of leaf gas exchange and of chlorophyll a fluorescence. Front Plant Sci 8:2068

Vazquez-Hernandez M, Romero I, Sanchez-Ballesta MT, Merodio C, Escribano MI (2021) Functional characterization of VviDHN2 and VviDHN4 dehydrin isoforms from *Vitis vinifera* (L.): an in silico and in vitro approach. Plant Physiol Biochem 158:146–157

Volaire F (2018) A unified framework of plant adaptive strategies to drought: crossing scales and disciplines. Glob Chang Biol 24(7):2929–2938

Wang D, Fu A (2016) The plastid terminal oxidase is a key factor balancing the redox state of thylakoid membrane. The Enzymes 40:143–171

Wang Y, Zhang X, Zhang X, Shao L, Chen S, Liu X (2016) Soil water regime affecting correlation of carbon isotope discrimination with yield and water-use efficiency of winter wheat. Crop Sci 56(2):760–772

Wang Y, Ye XY, Qiu XM, Li ZG (2019) Methylglyoxal triggers the heat tolerance in maize seedlings by driving AsA-GSH cycle and reactive oxygen species–/methylglyoxal-scavenging system. Plant Physiol Biochem 138:91–99

Wang P, Xu Z, Zhang Y, Ma Y, Yang J, Zhou F, Gao Y, Li G, Hu X (2022) Over-expression of spermidine synthase 2 (SlSPDS2) in tomato plants improves saline-alkali stress tolerance by increasing endogenous polyamines content to regulate antioxidant enzyme system and ionic homeostasis. Plant Physiol Biochem 192:172–185

Wojtyla Ł, Lechowska K, Kubala S, Garnczarska M (2016) Molecular processes induced in primed seeds—increasing the potential to stabilize crop yields under drought conditions. J Plant Physiol 203:116–126

Xing H, Zhou W, Wang C, Li L, Li X, Cui N, Hao W, Liu F, Wang Y (2021) Excessive nitrogen application under moderate soil water deficit decreases photosynthesis, respiration, carbon gain and water use efficiency of maize. Plant Physiol. Biochem. 166:1065–1075

Yadav S, Sharma KD (2016) Molecular and morphophysiological analysis of drought stress in plants. Plant Growth 10(5772):65246

Yang F, Feng Z, Wang H, Dai X, Fu X (2017) Deep soil water extraction helps to drought avoidance but shallow soil water uptake during dry season controls the interannual variation in tree growth in four subtropical plantations. Agric Meteorol 234:106–114

Yang W, Wang F, Liu LN, Sui N (2020) Responses of membranes and the photosynthetic apparatus to salt stress in cyanobacteria. Front Plant Sci 11:713

Yang X, Lu M, Wang Y, Wang Y, Liu Z, Chen S (2021) Response mechanism of plants to drought stress. Horticulturae 7(3):50

Yao GQ, Li FP, Nie ZF, Bi MH, Jiang H, Liu XD, Wei Y, Fang XW (2021) Ethylene, not ABA, is closely linked to the recovery of gas exchange after drought in four *Caragana* species. Plant Cell Environ 44(2):399–411

Younis A, Ramzan F, Ramzan Y, Zulfiqar F, Ahsan M, Lim KB (2020) Molecular markers improve abiotic stress tolerance in crops: a review. Plan Theory 9(10):1374

Yousefi M, Naseri L, Zaare-Nahandi F (2019) Nitric oxide ameliorates salinity tolerance in Pyrodwarf pear (Pyrus communis) rootstocks by regulating polyamine content. Iran J Plant Physiol 10(1):3023–3033

Zandalinas SI, Mittler R, Balfagón D, Arbona V, Gómez-Cadenas A (2018) Plant adaptations to the combination of drought and high temperatures. Physiol Plant 162(1):2–12

Zargar SM, Gupta N, Nazir M, Mahajan R, Malik FA, Sofi NR, Shikari AB, Salgotra R (2017) Impact of drought on photosynthesis: molecular perspective. Plant Gene 11:154–159

Zhang Z, Pei X, Zhang R, Lu Y, Zheng J, Zheng Y (2020) Molecular characterization and expression analysis of small heat shock protein 17.3 gene from *Sorbus pohuashanensis* (Hance) Hedl. in response to abiotic stress. Mol Biol Rep 47:9325–9335

Zhang Y, Sun C, Wang C, Jankovic KE, Dong Y (2021) Lipids and lipid derivatives for RNA delivery. Chem Rev 121(20):12181–12277

Zhao RZ, Jiang S, Zhang L, Yu ZB (2019) Mitochondrial electron transport chain, ROS generation and uncoupling. Int J Mol Med 44(1):3–15

Zheng M, Tao Y, Hussain S, Jiang Q, Peng S, Huang J, Cui K, Nie L (2016) Seed priming in dry direct-seeded rice: consequences for emergence, seedling growth and associated metabolic events under drought stress. Plant Growth Regul 78:167–178

Zhuang J, Wang Y, Chi Y, Zhou L, Chen J, Zhou W, Song J, Zhao N, Ding J (2020) Drought stress strengthens the link between chlorophyll fluorescence parameters and photosynthetic traits. PeerJ 8:e10046

Zulfiqar F, Akram NA, Ashraf M (2020) Osmoprotection in plants under abiotic stresses: new insights into a classical phenomenon. Planta 251:1–17

Zulfiqar U, Haider FU, Ahmad M, Hussain S, Maqsood MF, Ishfaq M, Shahzad B, Waqas MM, Ali B, Tayyab MN (2023) Chromium toxicity, speciation, and remediation strategies in soil–plant interface: a critical review. Front Plant Sci 13:1081624

The Impact of Drought Stress on Vegetables

Muhammad Daniyal Junaid, Zahide Neslihan Öztürk, and Ali Fuat Gökçe

Abstract Climate change has triggered several abiotic factors which are negatively affecting plant growth, among them water scarcity is most important. Negative effects on vegetables production caused by water shortage are way higher than other abiotic factors and it is most detrimental for adequate plant functioning and growth. Due to burgeoning human population this increase in production is still not enough and diverse, as we are depending only on 200 major plants although, there are more than 400 thousand plant species in the world. Most of the studies in literature have been done on major cereals and vegetables like potato and tomato, however little emphasis has been given to the studies on neglected vegetables like onion, carrot, pepper, and leafy vegetables against abiotic stresses especially drought. Various molecular approaches including RNA-seq, QTLs, GWAS, NGS and others have the potential to help understand drought tolerance in vegetables and incorporate drought resistance in vegetables. In this chapter, we overviewed the drought related studies in major vegetables, furthermore we also highlighted the progress made in drought stress related studies in neglected vegetable plants.

Keywords Water scarcity · Major vegetable crops · Abiotic stress · Vegetable production

1 Introduction

Vegetables are alterably nutrition source and for small farmers they are suitable source of income and healthy diet (Schreinemachers et al. 2018). According to FAO, vegetable average yields in South Asia, Southeast Asia and Africa are 64, 48 and 36% respectively (FAO 2022). There are several reasons of low productivity

M. D. Junaid (✉) · Z. N. Öztürk · A. F. Gökçe
Department of Agricultural Genetic Engineering, Ayhan Şahenk Faculty of Agricultural Sciences and Technologies, Niğde Ömer Halisdemir University, Niğde, Türkiye
e-mail: afgokce@ohu.edu.tr

U. K. Chaudhry et al. (eds.), *Drought Stress*,
https://doi.org/10.1007/978-3-031-80610-0_4

and other socioeconomic factors that influence vegetable production in these areas. These factors are further combined with climate change to decrease vegetable production in the world. Climate change alters weather patterns, precipitations, and cause extreme events (IPCC 2019). It is estimated that until 2100 average vegetable yields could decrease by 35% or more if climate change continues to increase worlds temperature and it will give rise to severe soil salinity, droughts, and other abiotic factors (Scheelbeek et al. 2018). Here, we primarily focus on drought stress and its influence on vegetables with respect to neglected vegetables.

Drought can be defined as prolonged absence or deficiency of precipitation or periods of dry weathers causing hydrological imbalances (Trenberth et al. 2014). Negative effects on plant production caused by water shortage are way higher than other abiotic factors. Drought aligns at first in natural catastrophes in terms of total number of affected people worldwide. It mainly disturbs plant water association and diminishes water-use efficiency. Almost all the plant's mechanisms get affected by water scarcity e.g., phenology, growth rates, photosynthesis, respiration rates, reduced leaves and stem size (Farooq et al. 2009). All geographic parts of our planet are affected by drought, it is also reported that every year, one-half of the earth is becoming vulnerable to water scarcity. Severity of drought cannot be predicted properly because it is influenced by several factors such as precipitation, its distribution, capacity of soils to absorb moisture and their evaporative abilities (Farooq et al. 2009).

Drought causes impaired growth, heavily affects the plant at seedling and germination stage. Water deficit is responsible for low production and poor quality of the plants; plant cell growth is also dependent on moisture availability (Chai et al. 2016). Drought stress reduces stomatal efficiency to conserve water. As a result, photosynthesis is affected, and CO2 fixation is hindered causing lower yields (Shahzad et al. 2017). Under extreme drought, the stomata closedown resulting into reduced photosynthetic process, acute drought conditions minimize chlorophyll content in the leaves (Ommen et al. 1999). Reactive oxygen species cause detrimental effects to chloroplasts under water scarcity, hence lowering chlorophyll content (Li et al. 2020). Drought resilience can be briefly explained as the ability of a plant in maintaining its economic yield under limited water supply (Beck et al. 2007).

Green vegetables usually contain more than 85% water and are succulent, this makes them more susceptible to drought stress especially during critical growth periods including bulb, leafy head, tuber and curd, and flowering to seed development stages (Nakanwagi et al. 2020). Vegetables usually cope with water deficit by adapting different morpho-physiological and biochemical mechanisms (Nemeskéri and Helyes 2019). Drought tolerance has complex nature, so development of drought resilient vegetable genotypes is not an easy task, but it can be achieved with exploitation of diverse germplasm, wild relatives, deep long extensive root systems and indigenous varieties (Wasaya et al. 2018; Kuo et al. 2020). Here we have discussed briefly about drought stress effects on vegetables and some recent advancements in major and neglected vegetables, with respect to their response to drought stress.

2 Climate Change

Climate change is the change in weather patterns of our planet, it affects agro-ecosystems via changes over period of time in key factors affecting plant growth (e.g., rising temperatures, low moisture availability etc.) (Lorenzoni et al. 2006). Global warming is constantly increasing CO_2 levels in atmosphere, resulting into decreased available land and quality (Peters et al. 2011). Agricultural yields and productivity are estimated to fall by 3–16% till 2080, under-developed poor countries that have already higher average crop temperatures speculated to face 10 to 25% decrease in agricultural productivity up to 2080, well developed countries having lower temperatures are expected to face lower decline in yields varying between 8% expansion of productivity to 6% decline till 2080 (Mahato 2014). Various effects are caused by climate changes i.e., changes in CO_2 level, variability in sea levels, fluctuation in precipitations and continuous higher temperatures (Wheeler et al. 2013).

Fluctuations in precipitations extend the occurrence of short-term vegetable/crop failures and it leads to decline in production. Change in climate aggravates temperatures, fluctuation in precipitations and cause extreme weather conditions, the collective effects of changing climate on plants and overall agriculture will be gloomy, which will endanger global food security (IPCC 2019). Increased decline in underground aquifers, extensive waste of cultivable land to urbanization and elevated water scarcity and soil flooding occur as a result of climate change (Schuster and Torero 2016). The impact of global warming on the crop yields will decrease productivity immediately, as weather changes in most of the geographic regions will increase their average temperature hence reducing the proper sowing and maturing timings of crops.

3 Drought Stress

Plants are sessile, and commonly subjected to abiotic and biotic stresses (Zlatev and Lidon 2012). Out of all abiotic stresses, water scarcity is the major factor in reducing agricultural productivity and it mainly occurs in rainfed and semiarid areas (Rana et al. 2013). Drought can be defined in hydrological, agricultural, meteorological, and socio-economic terms. From an agricultural perspective, drought stress is when there is inadequate moisture for plants growth in soil (Keyvan 2010). Physiologically, water deficiency is characterized when there is higher transpiration than water uptake by the plant (Salehi-Lisar and Bakhshayeshan-Agdam 2016). This may occur due to differences between water potential in plant root and soil. An imbalance between soil water intake and evapotranspiration flux leads to drought stress. This is also characterized by decreased atmospheric and soil humidity with high ambient air temperature (Lipiec et al. 2013). Water deficiency is not only caused by less moisture, but it also occurs when there is non-uniform precipitation, various factors which give rise to water deficit in plants are briefly discussed in next section.

3.1 Factors Triggering Drought Stress

Climate change causes alterations in rainfall patterns which influence increasing temperatures, variation in CO_2 (Arbona et al. 2013; Dai 2012). These factors are playing key role in triggering drought stress. Furthermore, several other factors including high light intensity, dry winds, low rainfalls, high evaporation of soil moisture contribute to occurrence of severe water scarcity in plants (Nezhadahmadi et al. 2013; Salehi-Lisar and Bakhshayeshan-Agdam 2016; Trenberth et al. 2014). In some cases, even there is adequate soil moisture available in soil but due to salinity, low temperature, and excessive water in soil may also cause drought condition in plants and make it unable for plants to uptake required moisture (Arbona et al. 2013; Salehi-Lisar and Bakhshayeshan-Agdam 2016).

3.2 Plant Responses to Drought Stress

Drought stress is genetic linked trait, so plants respond to it in various ways, literature described that species resilient to drought stress lower their moisture loss by decreasing leaf area and controlling stomatal opening (Lazaridou and Koutroubas 2004). This leads to lower biomass production, moreover in plants that don't adopt this mechanism reduce their water use efficiency. Water deficit leads to provocation of osmotic, oxidative and temperature stresses (Landi et al. 2017). Furthermore, drought couples with other abiotic factors and may induce reduction in transpiration, stomatal conductance, and other physiological functioning of plant (Król and Weidner 2017). Drought tolerance can be described as the ability of a plant to produce satisfactory economic yield under low moisture condition. In tropical and subtropical environment drought occurs simultaneously with high temperatures and cause reduction in plant yields (Zandalinas et al. 2017). Drought stress disturbs plant water relation at cell levels, tissues and whole plants which causes specific and unspecific reactions (Beck et al. 2007). Water deficit causes anatomical and morphological changes to plant; leaves which develop during drought stress possess small cell size and elevated stomatal density (Tisne et al. 2010). To survive and cope with water scarcity plants show various responses at morpho-physiological, and biochemical levels (Beck et al. 2007; Goisser et al. 2016).

3.3 Drought Stress Signaling

As plant senses any stress, it activates its signaling cascades, this phenomenon may depend on plant or stress type. Stress signaling is carried out by cell surface receptors and is a common phenomenon in all plants (Shulaev et al. 2008). Environmental stimuli are perceived by different receptors. First receptor in plants was decribed in

1990s, the Receptor-Like Kinase (RLK) (He et al. 1996). A sub-family of genes known as RLKs which are also named as WAKs (Wall Associated Kinases) receive signals coming from external environment like drought stress and activates signaling cascades in vascular plants (Shulaev et al. 2008). Studies have concluded that aquaporins play an important role in hydraulic functions are regulated by environmental factors at biochemical level (Lamaoui et al. 2018). Plant hormones play important role in hydraulic functioning, its conductivity and root, shoot signaling, which is a key phenomenon under water deficit condition. Abscisic acid (ABA) is the one of the primary hormones which plays an important role in regulation of resilience to abiotic stresses such as drought stress, salinity stress, cold stress, heat and waterlogging (Lata and Prasad 2011). ABA is an important chemical involved in root/shoot stress signaling (Schachtman and Goodger 2008), leaf inhibition and stomatal closure. ABA is also responsible for a systematic response again drought stress before occurrence of any visible changes in water in plant leaves and nutrient concentrations (Suzuki et al. 2013). It is evident from previous literature that drought causes osmotic stress which ultimately leads to production or catabolism of numerous plant growth regulators, which include ethylene, cytokinins, gibberellins, auxins and jasmonic acid with other factors; these factors are involved in physiological mechanisms as they act as signaling compounds (Verma et al. 2016).

Scientists believe that there are more than 400,000 species of plants on our planet, and among them 50% are edible. Albeit it is unfortunate that human rely on only 200 plant species, among them rice, maize and wheat account for staple food of worlds 60% population. In last 100 years, 75% of genetic diversity including plants and animals have been vanished because of human's focus on small number of food sources. According to FAO, 75% of worlds food is produced from only 5 animals and 12 plant species (FAO 2022).

4 Drought Stress Studies in Major Vegetables

In vegetables, drought stress causes physiological changes which result in poor yields and quality. Tomato (*Solanum lycopersicum* L.) is one of the important vegetables which belongs to Solanaceae family, moreover it is ranks second vegetable crop after potato (*Solanum tuberosum* L.) (Quinet et al. 2019). Time period and severity of water scarcity negatively influences growth, yield, quality, and plant stand in tomato plant (Gerszberg and Hnatuszko-Konka 2017). Drought deteriorates chloroplast, disturbs photosynthesis, this results in inactivity of electron transport chains which in turn produces reactive oxygen species in plants (Asada 1999). Furthermore, drought stress causes loss of turgor by affecting protoplast, disturbing membrane potentials and osmotic processes. Loss of water from plant cells causes disintegrated membranes, which causes lower fruit quality in tomato. According to Sibomana et al. (2013), water stress leads to 24% decreased plant height, 18% decreased stem size, 32% lower chlorophyll concentration and 69% yield loss in tomato plants. Tomato plant needs adequate water availability in early critical stages of its growth (first 20 days), if exposed to drought stress at early stages tomato plant may undergo decline in leaf water potential, lower nitrate reductase activity, lower photosynthesis rate, decreased RWC and decreased membrane stability. Tomato fruit quality highly depends on the presence of sugars and organic acid contents in it including citric, malic, and ascorbic acids (Nahar and Gretzmacher 2002; Albert et al. 2016), however drought stress, heat and salinity significantly decreases ascorbic acid content in tomato (Hernandez-Espinoza and Barrios-Masias 2020). It was previously described that high CO_2 levels in tomato may increase yield, but they negatively affect tomato quality (Mamatha et al. 2014). Tomato plants also have the ability adjust their osmotic levels against water stress by increasing fructose, glucose and sucrose levels, moreover high ascorbate concentration is also in association with water deficit in tomato plant (Favati et al. 2009). Studies have also revealed that concentrations of all forementioned compounds in tomato plants may also depend on its genetic, agronomic, environmental, and postharvest properties (Flores et al. 2016). Tomato plant also undergoes metabolic changes in retaliation to abiotic stresses including water deficit, however it is presumed that these drought responsive metabolic changes may attribute to cultivar specificity (Flores et al. 2016; Marsic et al. 2018). Conclusively, tomato plant has widely studied via molecular and physiological approaches to know defence mechanisms against water deficit and other abiotic stresses (AbuQamar et al. 2009).

Several studies on other important vegetable plants to understand drought responsive mechanisms. Recently it is reported that U-Box Ligase proteins are involved in drought tolerance in potato (Tang et al. 2022). RNA-seq studies are used to investigate genes that are linked with drought resilience in plants. Various drought related transcriptomic studies were done on major vegetable plants including potato (Zhang et al. 2014), and common beans (Müller et al. 2014). Similar studies have been done on other horticultural plants including poplars (Tang et al. 2013), chrysanthemum (Xu et al. 2013) and in eucalyptus (Villar et al. 2011). Molecular studies showed that drought stress is critically dangerous for growth, grain filling and cause severe economic loss in common beans (Müller et al. 2014).

5 Recent Advancements in Drought Resilience in Some Neglected Vegetables

5.1 Carrot

Carrot ranks tenth in global vegetable production and is widely used in the world. Here we discuss some recent developments in carrot drought related studies. Water scarcity is associated with up-regulation or down regulation of *dcTLP* gene which encodes for thaumatin loke protein (TLP), moreover it is independent of plant growth stage (Jung et al. 2005). In another study, heat shock protein *DcHsp17* was reported, which is also related with osmotic stress resistance (Ahn and Song 2012).

This protein plays an important role in tolerance to abiotic stresses in carrots. Genes belonging to ethylene responsive family ERF in carrot plants express altered expression under drought, cold and heat stress (Li et al. 2015). Moreover, basic helix-loop-helix transcription factors also play an important role in abiotic stress response in carrots however their expression changes depending upon the stress type (Chen et al. 2015). Furthermore, another study described that zinc finger family is a DNA-binding family consisting of 46 genes, it is associated with its specific expression under abiotic stress condition in carrot (Huang et al. 2016). Similarly, WRKY family consists of transcription factors which play important role in abiotic stress related response in Arabidopsis (Li et al. 2016). Another RNA-seq study on purple carrot lines showed that increased expression of MYB-48 transcription factor, MAPK mitogen-activated protein kinase, geraniol 8-hydroxylase, ABA-induced in somatic embryo 3, putative F-box protein, ferric reduction oxidase, and probable disease resistance protein play important role in drought stress tolerance in carrot (Öztürk Gökçe et al. 2022). Still there is far great research gap, which needs to be filled, similar studies which are mentioned above can be further implied in carrot plant to investigate its response to abiotic stress including drought.

5.2 *Onion*

Onion (*Allium cepa* L.) is one of the most important and diverse vegetable crop, which is diversely used in the world. Desirable onion traits include its shape, size, total soluble solids, pungency, firmness, resistance to abiotic and biotic stresses (Gökçe 2018). It is reported previously that proline contents, total soluble solids and antioxidant enzyme activity increases in onion plant when exposed to drought stress, moreover drought stress may cause up-to 65% yield loss in onion plant, furthermore decreased quality traits including flavonoids, phenol and pyruvic acid were also observed in onion plant when exposed to drought stress (Ghodke et al. 2018). Moreover, different onion genotypes were exposed to drought and salinity stress, genotypes showed differential response for drought related traits including proline, chlorophyll-a, leaf water potential and plant morphological traits (Hanci and Cebeci 2015). Another study on onion cultivars showed that tolerant onion cultivars have the ability to accumulate high levels of CAT gene and they also exhibit low degradation of PSII mRNA (Chaudhry et al. 2021). Moreover, it is also reported that malondialdehyde accumulation decreases in tolerant onion genotypes and increases in susceptible onion genotypes under drought stress (Öztürk Gökçe et al. 2022).

5.3 Leafy Vegetables

Different plant species which belong to leafy vegetables category have natural antioxidant activity and can detoxify reactive oxygen species in human body, so they are important for food industry. Hence it is important to understand drought responsive mechanisms in leafy vegetables, as they also require more water than other vegetable plants. It is reported that drought stress in leafy vegetables causes an increase in concentration of vitamins, phenolics, flavonoids, bioactive compounds, and antioxidant activity (Sarker and Oba 2018). Leafy vegetables are prone to drought stress especially during vegetative growth, they may accumulate significantly higher amounts of secondary metabolites under drought stress at this stage. Some leafy vegetables including plants belonging to Brassicaceae family may not attain full growth size, proper quality and biomass under water scarce conditions (Dietz et al. 2021). Some leafy vegetables like Amaranths including A. tricolor are multi-purpose plants and can show resilience to drought stress (Liu and Stutzel 2002; Hura et al. 2007; Sarker and Oba 2018). Unfortunately, there is scarce data available related to the effects of water deficit on metabolites (primary and secondary) in leafy vegetables. In some studies, it is described that drought stress enhanced metabolite contents in Choysum (Hanson et al. 2011), tomato (Stagnari et al. 2016), buckwheat (Siracusa et al. 2017), Achillea species (Gharibi et al. 2016) and herbaceous plants (Hillova et al. 2014).

5.4 Pepper

It is known that pepper (*Capsicum annuum* L.) is severely affected when exposed to water scarcity (Jaimez et al. 2000). Limited literature is reported on effects of environment including drought stress on pepper plant. It is previously known that drought stress causes 46% reduction in dry weight and reduces yield by 25% in pepper. Moreover, with severity of drought stress this percentage loss might increase upto 34 and 58% respectively. Furthermore, pepper plants when subjected to drought stress exhibit significantly higher concentration of potassium ions in their leaves, hence it might be a response to water deficit to maintain turgor in plant cells (De Pascale et al. 2003). Capsanthin is a carotenoid which is mainly present in *Capsicum annuum* L. plants, it was reported that drought stress significantly decreases capsanthin contents however in well-watered plants its concentration is higher. It was due to the reason that mainly two genes *Lcyb* and *Crtz* are involved in capsanthin production, drought stress negatively affects their regulation, this in turn may cause lower quality of pepper plant (Tian et al. 2014). Drought stress also negatively affects yield and physiological properties of pepper (Mardani et al. 2017). Drought stress effects in different vegetables are mentioned (Table 1).

Table 1 Drought stress effects on some vegetable species

Vegetable name	Drought stress effects	Reference
Tomato	Flower shedding, lack of fertilization, split fruit, unviable seeds	Mishra (2021)
Potato	Poor tuber quality and growth, potato splitting, brown spots inside	Levy et al. (2013)
Carrots	Poor taproot growth, pungent taste and smell, accumulation of unwanted solutes in root	Bahadur et al. (2011)
Onion	Multi-centred bulb, doubling and splitting of bulb, poor storage life	Mahdl (2015)
Pepper	Flower shedding, low dry matter, unviable seeds	Chatterjee and Solankey (2015)
Leafy vegetables	Distorted leaves, poor- and low-quality foliage, high accumulation of nitrates	Kumar et al. (2012)

6 Prospects

It is a need to develop drought stress tolerant vegetable genotypes. Plant breeders are making constant efforts to develop resilient vegetable genotypes which can withstand environmental adversaries including drought stress. Agricultural crops including wheat, maize, cotton, rice and major vegetables including tomato and potato have always been given priority in drought stress related studies over neglected vegetables which are briefly described in above section. Climate change has severely disrupted weather patterns and crop cycles, it is needed to combine traditional plant breeding approaches and modern molecular and biotechnological tools to develop resilient vegetable cultivars. Some traditional and modern biotechnological plant development strategies which are previously used in major crops and have the potential to be used for vegetable development are discussed below.

It is believed, with the understanding of mechanisms associated with drought stress tolerance, including physiological and molecular mechanisms in vegetable plants, it would be easier to develop/breed high yielding and more tolerant vegetable cultivars (Cattivelli et al. 2008). Identification of gene regions responsible for yield and resilience is an important factor to understand physiological processes in plants (Kirigwi et al. 2007). It is impracticable to characterize enormous number of plant genotypes in natural environment for drought stress resilience because of ununiform and variable response of plants to water deficit, however it is manageable and easier under controlled environment. In such conditions, it is suggested to correlate selection response in target plant population and compare it under natural and controlled stress environment (Venuprasad et al. 2007). Classical breeding in combination with molecular markers and quantitative trait loci QTLs provides more robust and reliable vegetable cultivar development path. The developed population which is to be examined for drought tolerance, may be subjected to gene mapping and QTL analysis, this helps in sorting out the population and discarding unwanted plant material in plant material. For development of drought tolerant vegetable varieties marker assisted selection (MAS) and gene pyramiding (GP) are beneficial approaches. Furthermore, progeny testing of population for drought tolerant at multi-locations is also beneficial to identify drought tolerant genotypes (Farooq et al. 2009).

Incorporation of indigenous vegetables into breeding programs might be crucial role in development of climate-resilient vegetables because they contain a broad gene-pool for vegetable plant improvement (Keatinge et al. 2016). Traits which are associated with drought tolerance are complex and are controlled by multiple genes. Vegetable plants usually respond to water deficit at morphological, physiological or molecular level, it is due to complex nature of drought stress. This makes drought resilient cultivar development difficult (Wasaya et al. 2018). So, identification of stress related genes is difficult as they are present on multiple quantitative trait loci (QTL). Conventional plant breeding played a significant role in improvement of the traits related to drought resilience. A drought toleant wheat cultivar "Ripper" was developed two decade ago, it exhibited better grain yield and baking quality even grown under non-irrigated conditions in Colorado (Haley et al. 2007). Superior maize varieties under drought in terms of yield were developed by Badu-Apraku and Yallou, (2009). The use of marker assisted breeding and QTLs (quantitative trait loci) has proven to be beneficial in breeding stress resilient crops (Mir et al. 2012). Drought resilience in elite durum (*T. turgidum* ssp. durum) specie and *T. aestivum* was improved by introgressing QTLs from specie belongs to wild emmer wheat (Merchuk-Ovnat et al. 2016). In a recent study, a QTL was mapped in wheat for 2 years under drought and heat conditions. It was concluded that identified regions could be used for further validation in other cultivars for the identification of superior genotypes which can tolerate drought and high temperature stress (Tahmasebi et al. 2016). Though several plants have extremely big genome size and ploidy levels, significant efforts are done for genome sequencing, its annotation and characterization (Uauy et al. 2017). Several miRNAs are shown to be preserved along different species and are regulated by water deficit. In drought stress resistant wheat plant (wild emmer wheat), miR166 was observed to be down-regulated under water deficit (Kantar et al. 2011). In drought tolerance, b-diketone wax plays important role in minimizing water loss, recently a non-coding miRNA has identified which controls wax formation (Furlan et al. 2021). From these mentioned studies, we can assume that miRNA modifications can potentially improve drought tolerance in cereal crops.

To develop drought resistant vegetable crops, genomic studies of highly drought tolerant plants are important as they can be used as model for drought resilience (Giarola et al. 2017). A monocotyledonous plant: *Xerophyta viscosa* is drought tolerant plant closely related to cereals, thus it can be used as model plant for monocots (Costa et al. 2017). Study of the changes occurred by modifications in DNA structure without alteration in nucleotide sequence is known as "Epigenetics". It includes, modification of histones, methylation of DNA and non-coding RNA expression so epigenetics can be a potential source for trait improvement in vegetable plant breeding (Mirouze and Paszkowski 2011). Epigenetic modifications do not occur only under stress but also occur along several generations of a specie (Boyko and Kovalchuk 2008). Hirayama and Shinozaki (2010) suggested that ABA along with abiotic stress signaling modifies development and gene expression profiles via epigenetic modifications to resist against abiotic stress. This knowledge might be used in vegetables to develop drought resistant cultivars. miRNA technology is a recent technique which is now widely used by researchers to identify abiotic stress related miRNAs. In tomato, several miRNAs related to desirable traits are identified (Cao et al. 2014).

Similarly, environmental responses of plants can also be analyzed by examining their metabolic profiles under stress condition. Furthermore, Next Generation Sequencing (NGS) technology has enabled researchers to examine billions of nucleotides at same time and it helps in understanding complex genetic traits including drought stress tolerance. NGS provides information about variation in genomes, insertions or deletions in genome including mutations. Furthermore, NGS technology has the potential to identify stress tolerant gene regions at much reduced cost (Poland and Rife 2012).

The genome of many vegetable species is yet to be mapped/sequenced, to devise appropriate breeding strategies for such vegetables it would be essential to have genetic knowledge. Such information will clarify the pathways or genetic associations in vegetable plants which are directly or indirectly linked with drought tolerance in cost effective way. These mentioned tools/techniques have the potential to be used in the development of drought tolerant vegetable cultivars for future.

References

AbuQamar S, Luo H, Laluk K, Mickelbart MV, Mengiste T (2009) Crosstalk between biotic and abiotic stress responses in tomato is mediated by the AIM1 transcription factor. Plant J 58(2):347–360

Ahn YJ, Song NH (2012) A cytosolic heat shock protein expressed in carrot (*Daucus carota* L.) enhances cell viability under oxidative and osmotic stress conditions. Hort Science 47(1):143–148

Albert E, Segura V, Gricourt J, Bonnefoi J, Derivot L, Causse M (2016) Association mapping reveals the genetic architecture of tomato response to water deficit: focus on major fruit quality traits. J Exp Bot 67(22):6413–6430

Arbona V, Manzi M, de Ollas C, Gómez-Cadenas A (2013) Metabolomics as a tool to investigate abiotic stress tolerance in plants. Int J Mol Sci 14(3):4885–4911

Asada K (1999) The water-water cycle in chloroplasts: scavenging of active oxygens and dissipation of excess photons. Annu Rev Plant Biol 50(1):601–639

Bahadur A, Chatterjee A, Kumar R, Singh M, Naik PS (2011) Physiological and biochemical basis of drought tolerance in vegetables. Int J Veg Sci:1–16

Beck EH, Fettig S, Knake C, Hartig K, Bhattarai T (2007) Specific and unspecific responses of plants to cold and drought stress. J Biosci 32:501–510

Boyko A, Kovalchuk I (2008) Epigenetic control of plant stress response. Environ Mol Mutagen 49(1):61–72

Cao X, Wu Z, Jiang F, Zhou R, Yang Z (2014) Identification of chilling stress-responsive tomato microRNAs and their target genes by high-throughput sequencing and degradome analysis. BMC Genomics 15:1–16

Cattivelli L, Rizza F, Badeck FW, Mazzucotelli E, Mastrangelo AM, Francia E, Marè C, Tondelli A, Stanca AM (2008) Drought tolerance improvement in crop plants: an integrated view from breeding to genomics. Field Crop Res 105(1-2):1–14

Chai Q, Gan Y, Zhao C, Xu H-L, Waskom RM, Niu Y, Siddique KHM (2016) Regulated deficit irrigation for crop production under drought stress. A review. Agron Sustain Develop 36(1)

Chatterjee A, Solankey SS (2015) Functional physiology in drought tolerance of vegetable crops: an approach to mitigate climate change impact. In: Chaudhary ML, Patel VB, Siddiqui MW, Mahdi SS (eds) Climate change: The principles and applications in horticultural science. CRC Press, pp 149–171

Chaudhry UK, Gökçe ZNÖ, Gökçe AF (2021) Drought and salt stress effects on biochemical changes and gene expression of photosystem II and catalase genes in selected onion cultivars. Biologia 76:3107–3121

Chen YY, Li MY, Wu XJ, Huang Y, Ma J, Xiong AS (2015) Genome-wide analysis of basic helix–loop– helix family transcription factors and their role in responses to abiotic stress in carrot. Mol Breed 35:1–12

Costa MCD, Artur MA, Maia J et al (2017) A footprint of desiccation tolerance in the genome of *Xerophyta viscosa*. Nat Plants 3(4):1–10

Dai A (2012) Drought under global warming: a review. WIREs Clim Change 3:45–65

De Pascale S, Ruggiero C, Barbieri G, Maggio A (2003) Physiological responses of pepper to salinity and drought. J Am Soc Hortic Sci 128:48–54

Dietz KJ, Zörb C, Geilfus CM (2021) Drought and crop yield. Plant Biol 23:881–893

FAO (2022) https://www.fao.org/3/y5609e/y5609e02.htm. Accessed on 1 June 2022

Farooq A, Wahid A, Kobayashi N, Fujita D, Basra SMA et al (2009) Plant drought stress: effects mechanisms and management. In: Sustainable agriculture. Springer, Netherlands, Dordrecht, pp 153–188

Favati F, Lovelli S, Galgano F, Miccolis V, Di Tommaso T, Candido V (2009) Processing tomato quality as affected by irrigation scheduling. Sci Hortic 122:562–571

Flores P, Hernández V, Hellín P, Fenoll J, Cava J, Mestre T, Martínez V (2016) Metabolite profile of the tomato dwarf cultivar micro-tom and comparative response to saline and nutritional stresses with regard to a commercial cultivar. J Sci Food Agric 96:1562–1570

Furlan A, Bianucci E, Llanes A, Peralta JM, Castro S (2021) Abiotic stress tolerance including salt, drought and metal(loid)s in legumes. Sustainable Agriculture Reviews 51: Legume Agriculture and Biotechnology Vol 2:135–180

Gerszberg A, Hnatuszko-Konka K (2017) Tomato tolerance to abiotic stress: a review of most often engineered target sequences. Plant Growth Regul 83:175–198

Gharibi S, Tabatabaei BES, Saeidi G, Goli SAH (2016) Effect of drought stress on total phenolic, lipid peroxidation, and antioxidant activity of Achillea species. Appl Biochem Biotechnol 178:796–809

Ghodke PH, Andhale PS, Gijare UM, Thangasamy A, Khade YP, Mahajan V, Singh M (2018) Physiological and biochemical responses in onion crop to drought stress. Int J Curr Microbiol App Sci 7:2054–2062

Giarola V, Hou Q, Bartels D (2017) Angiosperm plant desiccation tolerance: hints from transcriptomics and genome sequencing. Trend Plant Sci 22(8):705–717

Goisser M, Geppert U, Rötzer T, Paya A, Huber A, Kerner R, Bauerle T, Pretzsch H, Pritsch K, Häberle KH, Matyssek R (2016) Does belowground interaction with *Fagus sylvatica* increase drought susceptibility of photosynthesis and stem growth in *Picea abies*? For Ecol Manag 375:268–278

Gökçe AF (2018) Classical genetics on gene mapping. In: The Allium Genomes. Springer, pp 53–66

Haley SD, Johnson JJ, Peairs FB et al (2007) Registration of 'Ripper' wheat. J Plant Reg 1(1):1–6

Hanci F, Cebeci E (2015) Comparison of salinity and drought stress effects on some morphological and physiological parameters in onion (*Allium cepa* L.) during early growth phase. Bulg J Agric Sci 21:1204–1210

Hanson P, Yang RY, Chang LC, Ledesma L, Ledesma D (2011) Carotenoids, vitamin C, minerals, and total glucosinolates in choysum (*Brassica rapa* cv *g. parachinensis*) and kailaan (*B. oleracea* Alboglabra group) as affected by variety and wet and dry season production. J Food Compos Anal 24:950–962

He ZH, Fujiki M, Kohorn BD (1996) A cell wall-associated, receptor-like protein kinase. J Biol Chem 271(33):19789–19793

Hernandez-Espinoza LH, Barrios-Masias FH (2020) Physiological and anatomical changes in tomato roots in response to low water stress. Sci Hortic 265:109208

Hillova D, Takacsova M, Lichtnerova H (2014) Stomatal response to water stress in herbaceous perennials. In: Plants in Urban Areas and Landscape, pp 52–56

Hirayama T, Shinozaki K (2010) Research on plant abiotic stress responses in the post-genome era: past, present and future. Plant J 61(6):1041–1052

Huang W, Huang Y, Li MY, Wang F, Xu ZS, Xiong AS (2016) Dof transcription factors in carrot: genome-wide analysis and their response to abiotic stress. Biotechnol Lett 38:145–155

Hura T, Hura K, Grzesiak M, Rzepka A (2007) Effect of long-term drought stress on leaf gas exchange and fluorescence parameters in C3 and C4 plants. Acta Physiol Plant 29:103–113

IPCC (Intergovernmental Panel on Climate Change). (2019) Food Secur

Jaimez RE, Vielma O, Rada F, García-Núñez C (2000) Effects of water deficit on the dynamics of flowering and fruit production in *Capsicum chinense* Jacq in a tropical semiarid region of Venezuela. J Agron Crop Sci 185:113–119

Jung YC, Lee HJ, Yum SS, Soh WY, Cho DY, Auh CK, Lee TK, Soh HC, Kim YS, Lee SC (2005) Drought-inducible—but ABA-independent—thaumatin-like protein from carrot (*Daucus carota* L.). Plant Cell Rep 24(6):366–373

Kantar M, Lucas SJ, Budak H (2011) miRNA expression patterns of *Triticum dicoccoides* in response to shock drought stress. Planta 233:471–484

Keatinge JDH, Ledesma DR, Hughes JA et al (2016) How future climatic uncertainty and biotic stressors might influence the sustainability of African vegetable production. Acta Hortic 1225:23–42

Keyvan S (2010) The effects of drought stress on yield, relative water content, proline, soluble carbohydrates and chlorophyll of bread wheat cultivars. J Anim Plant Sci 8(3):1051–1060

Kirigwi FM, Van Ginkel M, Brown-Guedira G et al (2007) Markers associated with a QTL for grain yield in wheat under drought. Mol Breed 20:401–413

Król A, Weidner S (2017) Changes in the proteome of grapevine leaves (*Vitis vinifera* L.) during long-term drought stress. J Plant Physiol 211:114–126

Kumar R, Solankey SS, Singh M (2012) Breeding for drought tolerance in vegetables. Veg Sci 39(1):1–15

Kuo CG, Schafleitner R, Schreinemachers P Wopereis M (2020) Vegetables and climate change: pathways to resilience (No. OTHER)

Lamaoui M, Jemo M, Datla R, Bekkaoui F (2018) Heat and drought stresses in crops and approaches for their mitigation. Front Chem 6:26

Landi S, De Lillo A, Nurcato R, Grillo S, Esposito S (2017) In-field study on traditional Italian tomato landraces: the constitutive activation of the ROS scavenging machinery reduces effects of drought stress. Plant Physiol Biochem 118:150–160

Lata C, Prasad M (2011) Role of DREBs in regulation of abiotic stress responses in plants. J Exp Bot 62(14):4731–4748

Lazaridou M, Koutroubas SD (2004, September) Drought effect on water use efficiency of berseem clover at various growth stages. In: New directions for a diverse planet: proceedings of the 4th international crop science congress Brisbane, pp 467–475

Levy D, Coleman WK, Veilleux RE (2013) Adaptation of potato to water shortage: irrigation management and enhancement of tolerance to drought and salinity. Am J Potato Res 90:186–206

Li MY, Xu ZS, Huang Y, Tian C, Wang F, Xiong AS (2015) Genome-wide analysis of AP2/ERF transcription factors in carrot (*Daucus carota* L.) reveals evolution and expression profiles under abiotic stress. Mol Genet Genom 290(6):2049–2061

Li W, Wang H, Yu D (2016) Arabidopsis WRKY transcription factors WRKY12 and WRKY13 oppositely regulate flowering under short-day conditions. Mol Plant 9:1492–1503

Li S, Zhang N, Zhu X, Ma R, Yang J, Tang X, Si H (2020) Enhanced drought tolerance with artificial microRNA-mediated StProDH1 gene silencing in potato. Crop Sci 60:1462–1471

Lipiec J, Doussan C, Nosalewicz A, Kondracka K (2013) Effect of drought and heat stresses on plant growth and yield: a review. Int Agrophys 27:463–477

Liu F, Stutzel H (2002) Leaf water relations of vegetable amaranth (*Amaranthus* spp.) in response to soil drying. Eur J Agron 16:137–150

Lorenzoni I, Leiserowitz A, de Franca Doria M et al (2006) Cross-national comparisons of image associations with "global warming" and "climate change" among laypeople in the United States of America and Great Britain. J Risk Res 9(03):265–281

Mahato A (2014) Climate change and its impact on agriculture. Int J Sci Res Pub 4(4):1–6

Mahdl SS (2015) Climate dynamics in horticultural science, volume 1: the principles and applications. Apple Academic Press

Mamatha H, Srinivasa Rao NK, Laxman RH, Shivashankara KS, Bhatt RM, Pavithra KC (2014) Impact of elevated CO_2 on growth, physiology, yield, and quality of tomato (*Lycopersicon esculentum* Mill) cv. Arka Ashish. Photosynthetica 52:519–528

Mardani S, Tabatabaei SH, Pessarakli M, Zareabyaneh H (2017) Physiological responses of pepper plant (*Capsicum annuum* L.) to drought stress. J Plant Nutr 40:1453–1464

Marsic NK, Vodnik D, Mikulic-Petkovsek M, Veberic R, Sircelj H (2018) Photosynthetic traits of plants and the biochemical profile of tomato fruits are influenced by grafting, salinity stress, and growing season. J Agric Food Chem 66:5439–5450

Merchuk-Ovnat L, Barak V, Fahima T, Ordon F, Lidzbarsky GA, Krugman T, Saranga Y (2016) Ancestral QTL alleles from wild emmer wheat improve drought resistance and productivity in modern wheat cultivars. Front Plant Sci 7:452

Mir RR, Zaman-Allah M, Sreenivasulu N, Trethowan R, Varshney RK (2012) Integrated genomics, physiology and breeding approaches for improving drought tolerance in crops. Theor Appl Genet 125:625–645

Mirouze M, Paszkowski J (2011) Epigenetic contribution to stress adaptation in plants. Curr Opin Plant Biol 14(3):267–274

Mishra A (2021) Chapter-4 breeding for resistance to abiotic stress in vegetable crops. Chief Editor Manoj Kumar Ahirwar 55

Müller BSDF, Sakamoto T, Silveira RDD, Zambussi-Carvalho PF, Pereira M, Pappas GJ, do Carmo Costa MM, Guimarães CM, Pereira WJ, Brondani C, Vianello-Brondani RP (2014) Differentially expressed genes during flowering and grain filling in common bean (*Phaseolus vulgaris*) grown under drought stress conditions. Plant Mol Biol Rep 32:438–451

Nahar K, Gretzmacher R (2002) Effect of water stress on nutrient uptake, yield and quality of tomato (*Lycopersicon esculentum* mill.) under subtropical conditions. Die Bodenkultur J Land Manage Food Environ 53(1):45–51

Nakanwagi MJ, Sseremba G, Kabod NP, Masanza M, Kizito EB (2020) Identification of growth stage-specific watering thresholds for drought screening in *Solanum aethiopicum* Shum. Sci Rep 10:1–11

Nemeskéri E, Helyes L (2019) Physiological responses of selected vegetable crop species to water stress. Agronomy 9(8):447

Nezhadahmadi A, Prodhan ZH, Faruq G (2013) Drought tolerance in wheat. Sci World J 2013:610721

Ommen A, Donnelly S, Vanhoutvin M, van Oijen R, Manderscheid (1999) Chlorophyll content of spring wheat flag leaves grown under elevated CO_2 concentrations and other environmental stresses within the 'ESPACE-wheat' project. Eur J Agron 10:197–203

Öztürk Gökçe ZN, Gökçe AF, Junaid MD et al (2022) Comparative transcriptomics of drought stress response of taproot meristem region of contrasting purple carrot breeding lines supported by physio-biochemical parameters. Funct Integr Genomics 22:697–710

Peters GP, Aamaas BT, Lund M, Solli C, Fuglestvedt JS (2011) Alternative "global warming" metrics in life cycle assessment: a case study with existing transportation data. Environ Sci Technol 45(20):8633–8641

Poland JA, Rife TW (2012) Genotyping-by-sequencing for plant breeding and genetics. The Plant Genome 5:92–102

Quinet M, Angosto T, Yuste-Lisbona FJ, Blanchard-Gros R, Bigot S, Martinez JP, Lutts S (2019) Tomato fruit development and metabolism. Front Plant Sci 10:1554

Rana RM, Rehman SU, Ahmed J, Bilal M (2013) A comprehensive overview of recent advances in drought stress tolerance research in wheat (*Triticum aestivum* L.). Asian J Agric Biol 1(1):29–37

Salehi-Lisar SY, Bakhshayeshan-Agdam H (2016) Drought stress in plants: causes, consequences, and tolerance. In: Drought stress tolerance in plants, vol 1. Springer, Cham, pp 1–16

Sarker U, Oba S (2018) Drought stress enhances nutritional and bioactive compounds, phenolic acids and antioxidant capacity of Amaranthus leafy vegetable. BMC Plant Biol 18(1):258

Schachtman DP, Goodger JQ (2008) Chemical root to shoot signaling under drought. Trends Plant Sci 13(6):281–287

Scheelbeek PF, Bird FA, Tuomisto HL, Green R, Harris FB, Joy EJ, Chalabi Z, Allen E, Haines A, Dangour AD (2018) Effect of environmental changes on vegetable and legume yields and nutritional quality. Proceed Nat Acad Sci 115:6804–6809

Schreinemachers P, Simmons EB, Wopereis MC (2018) Tapping the economic and nutritional power of vegetables. Glob Food Secur:1636–1645

Schuster M, Torero M (2016) Toward a sustainable food system: Reducing food loss and waste. In: 2016 Global Food Policy Report. Chapter 3. International Food Policy Research Institute (IFPRI). PP 22–31. Washington, D.C. https://ebrary.ifpri.org/cdm/ref/collection/p15738coll2/id/130211

Shahzad S, Khan MY, Zahir ZA, Asghar HN, Chaudhry UK (2017) Comparative effectiveness of different carriers to improve the efficacy of bacterial consortium for enhancing wheat production under salt affected field conditions. Pak J Bot 49:1523–1530

Shulaev V, Cortes D, Miller G, Mittler R (2008) Metabolomics for plant stress response. Physiol Plant 132(2):199–208

Sibomana IC, Aguyoh JN, Opiyo AM (2013) Water stress affects growth and yield of container grown tomato (*Lycopersicon esculentum* Mill) plants. Global J Bio Sci Biotechnol 2(4):461–466

Siracusa L, Gresta F, Sperlinga E, Ruberto G (2017) Effect of sowing time and soil water content on grain yield and phenolic profile of four buckwheat (*Fagopyrum esculentum* Moench.) varieties in a Mediterranean environment. J Food Compos Anal 62:1–7

Stagnari F, Galieni A, Pisante M (2016) Drought stress effects on crop quality. In: Parvaiz A (ed) Water stress and crop plants: a sustainable approach, vol 2. Wiley, pp 375–392

Suzuki N, Miller G, Salazar C et al (2013) Temporal-spatial interaction between reactive oxygen species and abscisic acid regulates rapid systemic acclimation in plants. Plant Cell 25(9):3553–3569

Tahmasebi S, Heidari B, Pakniyat H, McIntyre CL (2016) Mapping QTLs associated with agronomic and physiological traits under terminal drought and heat stress conditions in wheat (Triticum aestivum L.). Genome 60(1):26–45

Tang S, Liang H, Yan D, Zhao Y, Han X, Carlson JE, Xia X, Yin W (2013) *Populus euphratica*: the transcriptomic response to drought stress. Plant Mol Biol 83:539–557

Tang X, Ghimire S, Liu W, Fu X, Zhang H, Sun F, Zhang N, Si H (2022) Genome-wide identification of U-box genes and protein ubiquitination under PEG-induced drought stress in potato. Physiol Plant 174(1):13475

Tian SL, Lu BY, Gong ZH, Shah SNM (2014) Effects of drought stress on capsanthin during fruit development and ripening in pepper (*Capsicum annuum* L.). Agric Water Manag 137:46–51

Tisne S, Schmalenbach I, Reymond M, Dauzat M, Pervent M, Vile D, Granier C (2010) Keep on growing under drought: genetic and developmental bases of the response of rosette area using a recombinant inbred line population. Plant Cell Environ 33(11):1875–1887

Trenberth KE, Dai A, van der Schrier G, Jones PD, Barichivich J, Briffa KR, Sheffield J (2014) Global warming and changes in drought. Nat Clim Chang 4:17–22

Uauy C, Wulff BB, Dubcovsky J (2017) Combining traditional mutagenesis with new high-throughput sequencing and genome editing to reveal hidden variation in polyploid wheat. Annu Rev Genet 51(1):435–454

Venuprasad R, Lafitte HR, Atlin GN (2007) Response to direct selection for grain yield under drought stress in rice. Crop Sci 47(1):285–293

Verma V, Ravindran P, Kumar PP (2016) Plant hormone-mediated regulation of stress responses. BMC Plant Biol 16:1–10

Villar E, Klopp C, Noirot C, Novaes E, Kirst M, Plomion C, Gion JM (2011) RNA-Seq reveals genotype-specific molecular responses to water deficit in eucalyptus. BMC Genomics 12:1–18

Wasaya A, Zhang X, Fang Q, Yan Z (2018) Root phenotyping for drought tolerance: a review. Agronomy 8(11):241

Wheeler S, Zuo A, Bjornlund H (2013) Farmers' climate change beliefs and adaptation strategies for a water scarce future in Australia. Glob Environ Change 23(2):537–547

Xu Y, Gao S, Yang Y, Huang M, Cheng L, Wei Q, Fei Z, Gao J, Hong B (2013) Transcriptome sequencing and whole genome expression profiling of chrysanthemum under dehydration stress. BMC Genomics 14:1–15

Zandalinas SI, Balfagón D, Arbona V, Gómez-Cadenas A (2017) Modulation of antioxidant defense system is associated with combined drought and heat stress tolerance in citrus. Front Plant Sci 8:266741

Zhang N, Yang J, Wang Z et al (2014) Identification of novel and conserved microRNAs related to drought stress in potato by deep sequencing. PloS one 9(4):95489

Zlatev Z, Lidon FC (2012) An overview on drought induced changes in plant growth, water relations and photosynthesis. Emirates J Food Agric 24(1)

The Impact of Drought Stress on the Nutritional Quality of Vegetables

Muhammed Yasir Naeem

Abstract Human health, food productivity, food security, and the nutritional quality of vegetables and fruits are all expected to be jeopardized because of climate change. Drought is a natural part of the climatic cycle. Drought activates plant defense mechanisms, causing multiple physiological stress responses that lead to a major shift in metabolite synthesis, affecting the nutritional and health benefits of crops. Due to rising abiotic stresses, especially drought stress, the quantity and concentration of bioactive components in vegetables is deteriorating slowly but surely. The basic secondary metabolites that are nutritionally vital for human health against a series of illnesses have a major impact because of water stress, causing either desirable or undesirable changes in the chemistry and organoleptic qualities of vegetables. This chapter emphasizes the detrimental impact of water stress on agricultural production and its nutritional effect, as the negative impact of drought has been widely studied, but its influence on nutritional quality has received less attention. These findings might be valuable to both crop breeders and agronomists in developing techniques for obtaining highly nutritive and quality vegetables and appropriate genotypes in drought environments, with just a minor impact on productivity.

Keywords Vegetables · Water stress · Defense mechanisms · Secondary metabolites · Nutritional quality · Human health

1 Introduction

Today there is compelling evidence that human-caused disturbances to our Earth's natural biosphere are heating up to the point that they are endangering human health by disrupting global systems. Agriculture production and food nutritional contents are threatened by changes like rising carbon dioxide (CO_2) concentrations, altering

M. Y. Naeem (✉)
Department of Plant Productions and Technologies, Niğde Ömer Halisdemir University, Niğde, Türkiye

U. K. Chaudhry et al. (eds.), *Drought Stress*,
https://doi.org/10.1007/978-3-031-80610-0_5

hydrological cycle, variations in global temperature, and stratosphere ozone (O_3) degradation. Such changes have the potential to have a negative influence on food security when they're not addressed with appropriate adaptation measures. Worldwide research initiatives have already shown reduced agricultural production and nutritional quality in response to environmental conditions, focusing mostly on staple food crops (Willett et al. 2019).

Fruits and vegetables are full of nutrients and bioactive substances that are beneficial to one's health and help in illness protection. As in global burden of disease 2017 models mentioned, that insufficient fruit and vegetable intake is one of the three leading nutritional risk variables for mortalities and disability-adjusted existence, and modelled projections indicate that weather patterns variations in fruit and vegetable intake will be one of the world's biggest linked sources of fatalities from climate change by 2050 (Afshin et al. 2019).

Based on rising apprehension on the effects of animal source production of food on climate change, global healthy and sustainable reference diets now recommend minimal quantities of animal products and encourage higher intake of plant-based food like fruits and vegetables (Willett et al. 2019).

Vegetables are susceptible to seasonal variations in environmental stress. The frequency of hot days, the overall climate throughout the growth period, especially during fruit growth and development, are significantly impacted by variations in minimum and maximum daily temperatures as well as other factors. High temperatures during the latent period, and an insufficient water during ripening process in perennial fruit and vegetables might cause considerable harm to fruit output and nutritional quality (Hatfield and Prueger 2015).

Abiotic and biotic stress issues generate significant yield losses in agriculture and are becoming more prevalent as a result of the effects of global warming. Drought is the most prominent abiotic factor: it impacts growth and development of plants, reducing field crops output (Stefanelli et al. 2010; Ferrara et al. 2011). Owing to increase agricultural water demands, contamination of groundwater supplies, and global warming scenario, scarcity of water and reduced irrigation availability of water may become a major concern in the near future (Rebey et al. 2012).

Plant water relationships research has mostly focused on gaining a better understanding of the roles that water plays in crop growth and yield. Furthermore, the influence of water on quality and plant reaction at the biochemical level has received less attention. Drought causes the activation of numerous signal transduction in plant, led to a major shift in expression of genes and the plant metabolic activity; one essential modulator of such networks is the phytohormone abscisic acid (ABA), that develops in consequence to osmotic stress (Clauw et al. 2015). As a result, the production of certain metabolites (amino acids, sugars, indoles, phenolics, and glucosinolates) is significantly increased. Some phytochemicals responses in plants are mentioned in Fig. 1.

In order to minimize nutritional deficits, it became extremely necessary to enhance food crops' dietary value and healthfulness. As a result, generating mild

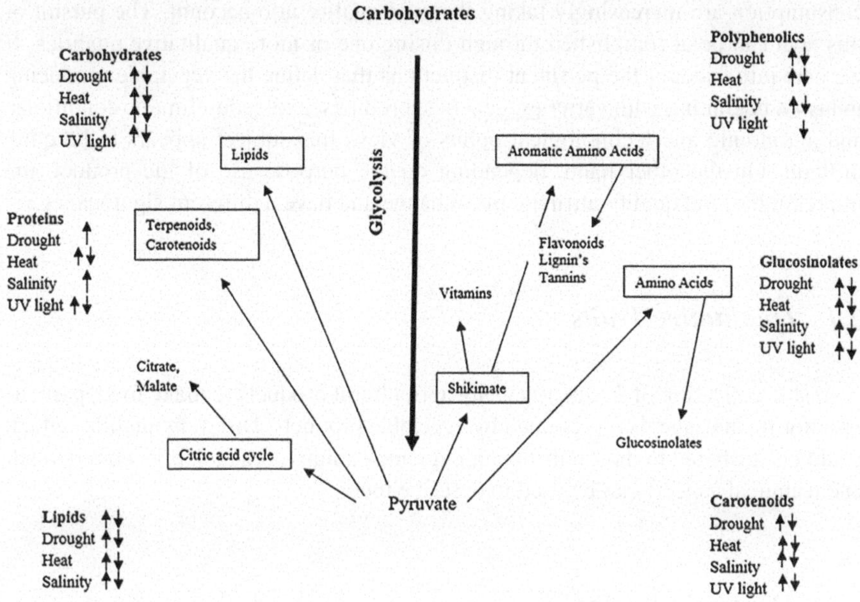

Fig. 1 A brief description of the biosynthetic process and the stress-induced metabolite synthesis. The primary and secondary metabolites' responses to various biotic and abiotic stressful circumstances (increased; reduced) are given outside the circle in which the fundamental metabolic pathway is depicted. (Modified from Jahangir et al. (2009) and Wang and Frei (2011))

environmental stress throughout the plant growth phase could be a good way to boost certain phytonutrients while also improving plant nutritive value without lowering yields.

Therefore, the goal of this section is to emphasize plant reactions to drought in order to determine the significance quality characteristics. The impacts of various stresses on the accumulation of metabolites responsible for nutritive benefits (sugars, proteins, lipids, and mineral elements) and those relevant to health value in vegetable crops are also extensively covered in this chapter (secondary metabolites).

2 Vegetables and Quality

Quality is a broad and complicated topic that involves a number of variables for most of the horticultural products. The consumer's quality perception includes elements derived from sensory properties stimuli (such as flavor, touch), as well as real or imagined health expectations connected to healthiness (such as sustainable farming, or the ingredients in nutritional supplements), in addition to fundamental quality attributes like external aspect, and integrity. Productions intended for conscious consumption are increasingly taking the goal quality into account. The pursuit of this goal can be accomplished through raising one or more qualitative qualities. If we take into account the pertinent distinctions that define the vegetable producing industry, beginning with varieties, genetics, products, and pedo-climatic conditions, and agronomic and technological points of view, this subject appears to be quite difficult. On the other hand, depending on the purpose use of the product, the impression of the quality attribute may change and have a different significance.

2.1 Qualitative Traits

A sizable collection of frequently using agricultural products to make food, particularly for fresh usage, is represented by vegetable products. Different qualities which could be attributed to their nutritional, hygienic-sanitary, organoleptic, commercial, and technical aspects can be used to describe them.

2.2 Organoleptic Traits

It comprises all of our senses' observable characteristics. The flavor and texture of vegetable products are the most crucial factors. The first one is mostly controlled by the sugar and organic acid levels in fruits (such as tomato and melon). The taste also encapsulates the element of the distinctive scents of each product as they are experienced through the nose and, in the case of volatile chemicals, the sense of smell. Some substances may impart disagreeable flavors and odors, which would negatively impact quality. Cucurbitacin, that is accountable for the sour taste of cucumbers, and capsicin, which is responsible for the spiciness of peppers, are two noteworthy examples.

2.3 Nutritional Characteristics

It explains the nutritional value and chemical make-up of the items. Since the buyer cannot immediately detect them, they are seldom evaluable at the moment of purchase. For the evaluation of items designed for industrial processing, composition has historically been more significant. Prior until now, only fresh food consumption has been impacted by real nutritional qualities (such as the amount of carotene in carrots) or apprehensive ones (such as the amount of readily iron content of spinach or vitamin C in sweet peppers) (Montoya and Raffaelli 2010).

2.4 *Marketable Traits*

Include all the factors that were considered for classifying the product in accordance with the rule set out by the European Control Agency. With a few rare exceptions (such the soluble solids in melons), these are external criteria used to classify the commodity into more or less uniform groups that are taken into account in commerce. Even though they don't always change the organoleptic properties, they are crucial for customer choice. Such qualities specify "quality requirements" for horticultural goods' presentation and packaging to enable them to sell inside the European Union, either imported from or exported to other states (Montoya and Raffaelli 2010).

3 Stress Components and Vegetables

According to prior research, a stress state can cause exceptionally unfavorable production phenomenon, which restricts the capacity to create enough food for the entire world's population. As plants are generally farmed in situations with several stressors that are frequently present at the same time, such as open fields, it can be challenging to comprehend how stress works. Like to certain other farmed species, vegetables are revealed to a variety of abiotic and biotic stressors (Fig. 2). Climate change and the expansion of agriculture practices that may impact the availability of natural resources have increased the frequency of unfavorable growth and production circumstances in recent years (Montoya and Raffaelli 2010). Numerous authors are becoming interested in studying this context. Tobacco (Li et al. 2014; Dimlioğlu et al. 2015), cotton (Abdelraheem et al. 2017; He et al. 2017), wheat (Huang et al. 2016), and other widely dispersed crops have been the focus of the majority of research undertaken thus far. Although there is less information accessible for vegetables' nutritional quality, it is still feasible to discover enough information to characterize the stress-related issue. Abiotic stress is often mentioned in the most studies on vegetables, with over 3000 study publications. The quantity of articles on biotic stressors has been drastically cut down to about 1200. Brassicas, legumes, and tomatoes are some of the horticultural crops that have been researched the most. The majority of these works focus on physiologically identifying plant response pathways and biotechnological research.

3.1 *Stress and Quality of Vegetables*

Variations in morphological and metabolic features are involved in a large portion of biotic assaults on plants, and as a result, also on vegetables. Biotic stress could cause development to slow down, reduced items as a consequence, changes in color

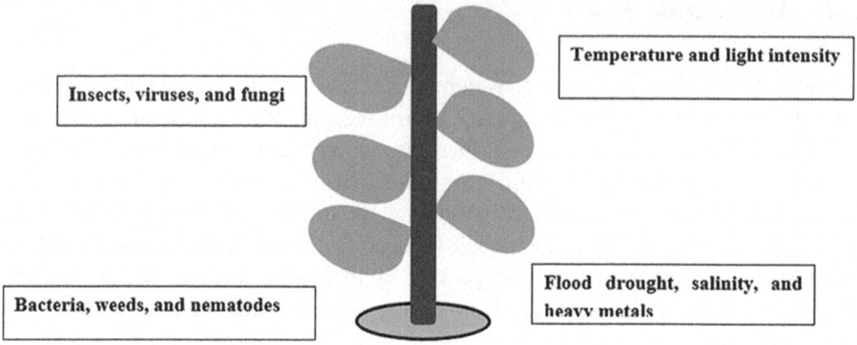

Fig. 2 A schematic diagram of important stresses affecting plants

and form, low yield, and, in the end, a shorter shelf life. In general, poor growing conditions such infections and abiotic stressors are linked to decreased crop growth. According to certain scientific studies in this field that assessed how organic farming affected the output of horticultural crops, organic crops typically yield between 5.0% and 50.0% fewer than traditional crops (Benbrook 2009; Quirós et al. 2014). There are a wide variety of chemical substances that the plant can produce (Suzuki et al. 2013), including carotenoid (Barański et al. 2014), glucosinolates (Nicoletto et al. 2016), proline (Sperdouli and Moustakas 2012), poly-amines (Hussain et al. 2011), and tocopherol (Benbrook 2009).

4 Mineral Compounds

Mineral content in crops is connected to genotypes, growth factors, soil conditions, and crop maturity during time of harvest (Martinez-Ballesta et al. 2010). Drought has a significant impact as availability of water in soil is directly related to mineral intake by plants; on the other hand, it can reduce flourish translocation of nutrients due to decreased rates of transpiration and degrade active transport and membrane permeability (Misle et al. 2014). Generally, (i) Drought lowers N absorption in the soil, that might lower the availability of N; (ii) High P absorption is seen in dry environments that are mostly calcareous, have high pH values (e.g., Mediterranean regions), or have low pH and rich in aluminum and iron oxides, and (iii) Soil K^+ mobility is closely correlated with water holding capacity. Due to such factors, the majority of research have focused on changing fertilization tactics during drought to enhance mineral availability in soil. Deficit irrigation, or withdrawing or lowering irrigation input for fixed intervals throughout crop cycles, improves quality of the crop while also allowing for more effective managing irrigation water (Martinez-Ballesta et al. 2010). The irrigation volume and grafting combination had a favorable effect on K^+ and Mg^{2+} levels in water melon, but P and Ca^{2+} levels did not alter substantially (Rouphael et al. 2008).

Water stress appears to favor K^+ deposition in sunflower (Esmaeilian et al. 2012), Ocimum gratissimum L. (Osuagwu et al. 2010), Lupinus albus, and Lupinus mutabilis (De Carvalho 2005), likely due to its essential factor in osmoregulation in plant organs, whereas it reduces K^+ content in maize (Kaya et al. 2006). Water deficit has conflicting results on Ca^{2+} and Mg^{2+} concentrations: some research shows an upsurge (De Carvalho 2005), while others show a reduction (Kaya et al. 2006).

K^+ is a significant mineral nutrient which decreases in concentration on water stress in banana plants (Martinez-Ballesta et al. 2010), while, in contrast proportions various fundamentals, like Fe, Ca^{2+}, Na^+, and Zn levels might rise, resulting in an increase in enhancement of fruit organoleptic properties (Mahouachi 2007).

4.1 Lipids

Lipids serve both as the structural and storing aspects of cell membranes. Reactive oxygen species (ROS) are harmful to lipids under drought circumstances because they trigger the generation of different oxidative radicals which enable oxidation (Ahmad et al. 2014). As a result, lipid content of the plasma membrane fluctuates significantly, particularly in terms of changes in membrane fatty acid (FA) composition (Laribi et al. 2009). As a result of changes in the lipid matrix context in which they are immersed, these alterations influence permeability and inherent membrane protein activities. Most research on the relationship among water deficit and plant oil yield has been conducted on plants that also generate essential oils (Wang and Frei 2011). Drought lessens the lipid content in the harvesting fractions. Lipolytic and peroxidative activity promotion, in addition to restriction of lipid production, are examples of degradative activities (Labusch et al. 2013), are linked to a reduction in membranes concentration; nevertheless, the reported effects are highly dependent on crop species and genotypes (i.e., environmental conditions, such as mean air temperature) (Wang and Frei 2011). Drought during the ripening stage has an effect on the production of oil in spring cultivars of the Brassicaceae family. Drought appears to lower oil accumulation and lipid content in lupine (Carvalho et al. 2004), corn (Ali et al. 2013), peanuts (Zhao and Bartley 2014), and wheat (Chaiyadee et al. 2013). It also lowers yield and oil production in sunflower (Esmaeilian et al. 2012). Furthermore, temperature, geography, and soil type are among environmental factors that influence oil content (Kaya and Kolsarici 2011).

Water stress timing and genotype play critical roles (Shahsavari et al. 2014). Scarcity of water before and during blooming does not produce variations in FA profiles in rapeseed, but during maturity, it decreases stearic (C 18:0) and oleic acid (C 18:1) concentrations, but appears to be significant for oil linoleic acid (C 18:2) content (Latif et al. 2016). After being exposed to mild to extreme water scarcity, monounsaturated and polyunsaturated FAs both drop (on average by 21%), whereas saturated FAs and the proportion of main saturated FA, palmitic acid, both rise significantly (Rebey et al. 2012).

Water availability effects fruit development in addition to the content of olive oil (Martinelli et al. 2012). Water scarcity consequences a significant rise in palmitic and arachidic acids, whereas pelargonic and lauric acid concentrations are modestly reduced (Martinelli et al. 2013).

4.2 *Proteins*

Protein contributes considerably in terms of nutritional value of numerous plants and generally, crop protein content varies between 10% and 30% on total dry matter (Wang and Frei 2011).

As per various research, water stress raises the protein concentration of grains, particularly gluten (Zhang et al. 2014; Bouacha et al. 2015). In addition, the gliadin/glutenin ratio declines linearly in the days following anthesis, since gliadin production is greater in the early stages of maturation, but glutenin's accumulates relatively fast in the latter stages (Fois et al. 2011). As a result, period of stress imposed (late water stress) may play an essential role in glutenin production. According to Flagella et al. (2010), the gliadin/glutenin ratio is nearly unaltered in both availability of water and drought, while the fraction of highly polymeric protein is larger in drought.

Several horticultural and commercial crops, such as onion and sunflower, when water is low, there is a rise in crude protein (Esmaeilian et al. 2012). Water scarcity during blooming associated with maximum Seed protein content in the later. Water stress, on the other hand, has been proven to diminish both protein and oil content in soybean seeds (Kirnak et al. 2010).

4.3 *Total Soluble Solids and Soluble Sugars*

Many agricultural plants have been shown to accumulate total soluble solids (TSS) as a key adaptation mechanism in response to water stress (Babita et al. 2010). A lesser amount of irrigation water increases TSS levels (Patanè et al. 2011) an average TSS value, reported as °Brix, was 5.78 for low water irrigation level and 4.30 for maximum water application (Favati et al. 2009). Similarly, studies on irrigation constraints have found a higher in peaches, citrus fruit (Navarro et al. 2010), grapes, watermelon, and cantaloupe (Ahmadi-Mirabad et al. 2014). In a number of crop species, the sugar content rises in low moisture circumstances, including tomatoes, cucumbers, and grapes (Deluc et al. 2009). Several researchers assert that the amount of sugar of sugar beet storage organs has increased, while some suggest that it has decreased (Tsialtasa et al. 2009). Under moderate and severe stressful circumstances in red beet, a drop in total sugar content exhibited (Stagnari et al. 2014).

4.4 Starch

Starch content in cereal grains is high (approximately 65%), that shown to be considerably affected by water deficiency, particularly once exposed to adverse circumstances as the grain is being filled (Wang and Frei 2011). Water deficit throughout growing season minimizes the level of carbohydrate acquired in stems and leaves, resulting in a seed mass constraint. Because the biomass produced in these tissues during the transition from the reproductive to vegetative phases is necessary to accumulate adequate resources for relocation (Marcaida et al. 2014); Reduced water availability during the reproductive periods directly inhibits the growth of seeds, causing early grain dehydration (Fábián et al. 2011). Starch development is greatly influenced by the sucrose content of the seeds as well as the activity of enzymes involved in starch synthesis (e.g., sucrose synthase) (Saeedipour and Moradi 2011). This shows that under drought condition, starch reduction is caused by both a low sucrose concentration and a decrease in enzyme activity. This pattern has been reported in barley (Jiang et al. 2012), corn (Ali et al. 2013), and wheat (Thitisaksakul et al. 2015). Cassava and potato (Ballmer et al. 2012) are two starchy tuber crops that have shown a drop in starch content because of the drought.

4.5 Pigments

Because of the protective role against ROS, there has been much research on the impact of water limitation on carotenoid deposition (Esteban et al. 2015). Water scarcity increases fruits' lycopene and total carotenoids content, but has little influence on β-carotene or xanthophyll levels. Extended watering periods combined with a lower irrigation schedule considerably improve the levels of lycopene and β-carotene in processed tomatoes. The most of water stress research in carrots are focused on the development of undesired chemicals in roots, Zeipiņa et al. (2014), on the other hand, found no greater carotene content in water deprived plants. The fruit pericarp of pepper contains a lot of carotenoids, primarily carotene, capsanthin, and capsorubin, which collectively account for 60.0% of the total carotenoids. Because of decreased watering rate, some authors observed an overall rise only in red fruits carotenoids (by 30%) appear. (Marin et al. 2009). Esteban et al. (2014) evaluated several ways for increasing the carotenoid zeaxanthin content of rocket and spinach using abiotic stress (freezing, maximum light concentration, and dryness). Authors discovered that a mild water deficit was particularly successful in boosting the violaxanthin, antheraxanthin, and zeaxanthin content, as well as the overall carotenoid level, in rocket but not in spinach. On a dry weight basis, the lutein and carotene contents of two years old rosemary (Rosmarinus officinalis L.) plants enduring a protracted drought during a Mediterranean summer decreased; however, on a chlorophyll basis, the level of lutein climbed (by 80%), while β-carotene did not change (Munné-Bosch and Alegre 2000). Under water stress, both betaxanthin and betacyanin concentrations increased in red beet, but the betacyanin/betaxanthin ratio remained same (Stagnari et al. 2014).

4.6 Glucosinolates

Glucosinolates are one of the most significant Brassicaceae metabolites, derived from amino acid metabolism (Podsedek 2007). Several research on Nasturtium officinale L. Brassica oleracea L. var. capitata, Brassica rapa ssp. rapifera L., and Brassica carinata L. (Zhang et al. 2011) have shown that water constraint enhances glucosinolates accumulation in (Schreiner et al. 2009). Water stress causes greater values of each individual and total glucosinolates in cabbage plants during head development (Radovich et al. 2005). Regardless of the fact that the duration of water deficit imposition (i.e., phenological phases) has the greatest impact on glucosinolates accumulation in Brassica napus L, a general rise was detected (Latif et al. 2016).

4.7 Ascorbic Acid

In comparison to phenolics, there are less studies on the water restriction have an influence on ascorbate (vitamin C) crops accumulation. Numerous research in tomato have reported a rise in ascorbate content owing to water scarcity (Patanè et al. 2011). Vitamin C content rises when watering levels are reduced to 20 and 40 mm. Moreover, this should be cultivar dependant, as prior research has shown that during drought conditions, there is a reduced amount of vitamin C and soluble solids and a larger produce (Dumas et al. 2003). In table grapes, partially root zone drying increases vitamin C content by 15–42%, while in strawberries, it increases vitamin C and ellagic acid concentrations (Dodds et al. 2007). In contrast, hot pepper plants cultivated in soilless culture with insufficient watering lose vitamin C in fruits (Ahmad et al. 2014), and Ocimum gratissimum L. and Vigna unguiculata L. in seeds (Osuagwu et al. 2010). This decline may be caused by oxygen and its derivative species destroying more directly during water deficit (Fang and Xiong 2015). Tocopherol levels have also been shown to be higher in sage and lemon balm, as well as soybeans (Britz and Kremer 2002), peanuts, and tobacco (Liu et al. 2008). Meanwhile, investigations on Catharanthus roseus have shown that the rosea variety has a greater amount of tocopherol than the Alba variety, indicating that genotype has no secondary impact (Jaleel et al. 2008).

4.8 Polyphenols

The most common antioxidants in human nutrition are polyphenols. Phenolics are secondary metabolites found in all plants that are essential to their survival. They are particularly noticeable when judging the color, look, flavor, and taste of fruits and vegetables. Even though the effects of genotype, harvest maturity, environment, culture, and handling and storage are significant, the few known field studies on broccoli demonstrate a significant increase in polyphenolic contents of 35.5% due to dryness (Fortier et al. 2006; Podsedek 2007). In contrast, low soil water content appears to reduce the concentration of phenolic chemicals in greenhouse conditions (Cogo et al. 2011). Young lettuce seedlings (4 weeks old) exposed to moderate dehydration increased significantly in total phenolic content and antioxidant activities (Oh et al. 2010). At time of harvesting, this has also seen for a variety of lettuce genotypes (Eichholz et al. 2014). In farm tomatoes, restricted root zone irrigation raises total antioxidant content by 2% in whole fruits and 7% in fresh fruit sap (Jensen et al. 2010). When compared to unstressed plants, antiradical activity in farm potatoes rises by 10% (Jensen et al. 2010; Jovanovic et al. 2010). Moreover, under supervised circumstances (50% and 30% of water holding capacity), Beta vulgaris var. conditiva Alef. storage roots showed a rise in overall phenolic content and corresponding antioxidant activity. (Stagnari et al. 2014). The same trend of activity was seen in potatoes (Wegener et al. 2015). A rise in oil polyphenol content may be accomplished in olive trees (Gómez-Rico et al. 2009), and Withania somnifera Dun. (Kannan and Kulandaivelu 2011) by providing a water deficit throughout the season. Apigenin, a flavonoid variant essential for the plant's adaptation, is unaffected by soil moisture loss in chamomile (Baghalian et al. 2011). *Cucumis sativus* L. plants subjected to varied water regimes had considerably higher anthocyanin content (Castellarin et al. 2007; Donavon et al. 2013) as a result of enhanced early sugar buildup or promotion of anthocyanin hydroxylation (Castellarin et al. 2007).

5 Conclusions and Future Opportunities

Drought has an impact on crop quality features through interacting with genetic, physiological, and environmental variables. Furthermore, a multifactorial interaction occurs in all of the experimental trials, since water constraint is frequently associated with greater levels of salt or heightened temperatures. As a result, the accompanying outcomes are frequently inconclusive. Nonetheless, conclusive results about the development of primary and secondary metabolites may be drawn as drought-induced increase of soluble sugars, proteins, and secondary metabolites, or decreased starch and fat levels, is quite obvious. Water-stressed plants may be a feasible source of bioactive compounds, especially some confined in tissues and with restricted biomass yield, given the increased interest in functional meals. Therefore, additional research is needed to highlight how water restriction affects the several interrelated elements that affect metabolite levels in various field settings.

References

Abdelraheem A, Liu F, Song M, Zhang JF (2017) A meta-analysis of quantitative trait loci for abiotic and biotic stress resistance in tetraploid cotton. Mol Gen Genomics 292(6):1221–1235

Afshin A, Sur PJ, Fay KA, Cornaby L, Ferrara G, Salama JS et al (2019) Health effects of dietary risks in 195 countries, 1990–2017: a systematic analysis for the Global Burden of Disease Study 2017. Lancet 393(10184):1958–1972

Ahmad P, Jamsheed S, Hameed A, Rasool S, Sharma I, Azooz MM, Hasanuzzaman M (2014) Drought stress induced oxidative damage and antioxidants in plants. In: Oxidative damage to plants. Academic, Amsterdam, pp 345–367

Ahmadi-Mirabad A, Lotfi M, Roozban MR (2014) Growth, yield, yield components and water-use efficiency in irrigated cantaloupes under full and deficit irrigation. Electron J Biol 10(3):79–84

Ali Q, Anwar F, Ashraf M, Saari N, Perveen R (2013) Ameliorating effects of exogenously applied proline on seed composition, seed oil quality and oil antioxidant activity of maize (*Zea mays* L.) under drought stress. Int J Mol Sci 14(1):818–835

Babita M, Maheswari M, Rao LM, Shanker AK, Rao DG (2010) Osmotic adjustment, drought tolerance and yield in castor (*Ricinus communis* L.) hybrids. Environ Exp Bot 69(3):243–249

Baghalian K, Abdoshah S, Khalighi-Sigaroodi F, Paknejad F (2011) Physiological and phytochemical response to drought stress of German chamomile (*Matricaria recutita* L.). Plant Physiol Biochem 49(2):201–207

Ballmer T, Hebeisen T, Wüthrich R, Gut F (2012) Potential for drip irrigation in potato production under changing climatic conditions. Rech Agron Suisse 5(2012):244–251

Barański M, Średnicka-Tober D, Volakakis N, Seal C, Sanderson R, Stewart GB et al (2014) Higher antioxidant and lower cadmium concentrations and lower incidence of pesticide residues in organically grown crops: a systematic literature review and meta-analyses. Br J Nutr 112(5):794–811

Benbrook C (2009) Impacts of genetically engineered crops on pesticide use in the United States: the first thirteen years. The Organic Center 24:880–882

Bouacha OD, Rhazi L, Aussenac T, Rezgui S, Nouaigui S (2015) Molecular characterization of storage proteins for selected durum wheat varieties grown in different environments. J Cereal Sci 61:97–104

Britz SJ, Kremer DF (2002) Warm temperatures or drought during seed maturation increase free α-tocopherol in seeds of soybean (*Glycine max* [L.] Merr.). J Agric Food Chem 50(21):6058–6063

Carvalho ISD, Ricardo CP, Chaves M (2004) Quality and distribution of assimilates within the whole plant of lupines (*L. albus* and *L. mutabilis*) influenced by water stress. J Agron Crop Sci 190(3):205–210

Castellarin SD, Pfeiffer A, Sivilotti P, Degan M, Peterlunger E, Di Gaspero G (2007) Transcriptional regulation of anthocyanin biosynthesis in ripening fruits of grapevine under seasonal water deficit. Plant Cell Environ 30(11):1381–1399

Chaiyadee S, Jogloy S, Songsri P, Singkham N, Vorasoot N, Sawatsitang P et al (2013) Soil moisture affects fatty acids and oil quality parameters in peanut. Int J Plant Prod 7(1):81–95

Clauw P, Coppens F, De Beuf K, Dhondt S, Van Daele T, Maleux K et al (2015) Leaf responses to mild drought stress in natural variants of Arabidopsis. Plant Physiol 167(3):800–816

Cogo SL, Chaves FC, Schirmer MA, Zambiazi RC, Nora L, Silva JA, Rombaldi CV (2011) Low soil water content during growth contributes to preservation of green colour and bioactive compounds of cold-stored broccoli (*Brassica oleraceae* L.) florets. Postharvest Biol Technol 60(2):158–163

De Carvalho IMMS (2005) Effects of water stress on the proximate composition and mineral contents of seeds of two lupins (*Lupinus albus* and *Lupinus mutabilis*). J Food Qual 28(4):325–332

Deluc LG, Quilici DR, Decendit A, Grimplet J, Wheatley MD, Schlauch KA et al (2009) Water deficit alters differentially metabolic pathways affecting important flavor and quality traits in grape berries of Cabernet Sauvignon and Chardonnay. BMC Genomics 10(1):1–33

Dimlioğlu G, Daş ZA, Bor M, Özdemir F, Türkan İ (2015) The impact of GABA in harpin-elicited biotic stress responses in *Nicotiana tabaccum*. J Plant Physiol 188:51–57
Dodds PAA, Taylor JM, Else MA, Atkinson CJ, Davies WJ (2007) Partial rootzone drying increases antioxidant activity in strawberries. Acta Hortic 744:295–302
Donavon S, Patrick AN, Charles L (2013) The effects of various drip fertigated water quantities on flavonoid and anthocyanin content on hydroponically cultivated *Cucumis sativa* L. Int J Phys Sci 8(19):1012–1016
Dumas Y, Dadomo M, Di Lucca G, Grolier P (2003) Effects of environmental factors and agricultural techniques on antioxidant content of tomatoes. J Sci Food Agric 83(5):369–382
Eichholz I, Förster N, Ulrichs C, Schreiner M, Huyskens-Keil S (2014) Survey of bioactive metabolites in selected cultivars and varieties of *Lactuca sativa* L. under water stress. J Appl Bot Food Qual 87:265–273
Esmaeilian Y, Sirousmehr AR, Asghripour MR, Amiri E (2012) Comparison of sole and combined nutrient application on yield and biochemical composition of sunflower under water stress. Int J Appl 2(3):214–220
Esteban R, Fleta-Soriano E, Buezo J, Míguez F, Becerril JM, García-Plazaola JI (2014) Enhancement of zeaxanthin in two-steps by environmental stress induction in rocket and spinach. Food Res Int 65:207–214
Esteban R, Moran JF, Becerril JM, García-Plazaola JI (2015) Versatility of carotenoids: an integrated view on diversity, evolution, functional roles and environmental interactions. Environ Exp Bot 119:63–75
Fábián A, Jäger K, Rakszegi M, Barnabás B (2011) Embryo and endosperm development in wheat (*Triticum aestivum* L.) kernels subjected to drought stress. Plant Cell Rep 30(4):551–563
Fang Y, Xiong L (2015) General mechanisms of drought response and their application in drought resistance improvement in plants. Cell Mol Life Sci 72(4):673–689
Favati F, Lovelli S, Galgano F, Miccolis V, Di Tommaso T, Candido V (2009) Processing tomato quality as affected by irrigation scheduling. Sci Hortic 122(4):562–571
Ferrara A, Lovelli S, Di Tommaso T, Perniola M (2011) Flowering, growth and fruit setting in greenhouse bell pepper under water stress. J Agron 10(1):12–19
Flagella Z, Giuliani MM, Giuzio L, Volpi C, Masci S (2010) Influence of water deficit on durum wheat storage protein composition and technological quality. Eur J Agron 33(3):197–207
Fois S, Schlichting L, Marchylo B, Dexter J, Motzo R, Giunta F (2011) Environmental conditions affect semolina quality in durum wheat (*Triticum turgidum* ssp. *durum* L.) cultivars with different gluten strength and gluten protein composition. J Sci Food Agric 91(14):2664–2673
Fortier E, Desjardins Y, Tremblay N, Bélec C, Côté M (2006) Influence of irrigation and nitrogen fertilization on broccoli polyphenolics concentration. Int Symp Veg Safety Hum Health 856:55–62
Gómez-Rico A, Salvador MD, Fregapane G (2009) Virgin olive oil and olive fruit minor constituents as affected by irrigation management based on SWP and TDF as compared to ETc in medium-density young olive orchards (*Olea europaea* L. cv. Cornicabra and Morisca). Food Res Int 42(8):1067–1076
Hatfield JL, Prueger JH (2015) Temperature extremes: effect on plant growth and development. Weather Clim Extremes 10:4–10
He Z, Zhang H, Tewolde H, Shankle M (2017) Chemical characterization of cotton plant parts for multiple uses. Agri Environ Lett 2(1):110044
Huang J, Sedano F, Huang Y, Ma H, Li X, Liang S et al (2016) Assimilating a synthetic Kalman filter leaf area index series into the WOFOST model to improve regional winter wheat yield estimation. Agri For Meteorol 216:188–202
Hussain MI, González L, Souto C, Reigosa MJ (2011) Ecophysiological responses of three native herbs to phytotoxic potential of invasive *Acacia melanoxylon* R. Br Agrofor Syst 83(2):149–166
Jahangir M, Abdel-Farid IB, Kim HK, Choi YH, Verpoorte R (2009) Healthy and unhealthy plants: the effect of stress on the metabolism of Brassicaceae. Environ Exp Bot 67(1):23–33

Jaleel CA, Gopi R, Manivannan P, Gomathinayagam M, Sridharan R, Panneerselvam R (2008) Antioxidant potential and indole alkaloid profile variations with water deficits along different parts of two varieties of *Catharanthus roseus*. Colloids Surf B Biointerfaces 62(3):312–318

Jensen CR, Battilani A, Plauborg F, Psarras G, Chartzoulakis K, Janowiak F et al (2010) Deficit irrigation based on drought tolerance and root signalling in potatoes and tomatoes. Agric Water Manag 98(3):403–413

Jiang SS, Liang XN, Li X, Wang SL, Lv DW, Ma CY et al (2012) Wheat drought-responsive grain proteome analysis by linear and nonlinear 2-DE and MALDI-TOF mass spectrometry. Int J Mol Sci 13(12):16065–16083

Jovanovic Z, Stikic R, Vucelic-Radovic B, Paukovic M, Brocic Z, Matovic G et al (2010) Partial root-zone drying increases WUE, N and antioxidant content in field potatoes. Eur J Agron 33(2):124–131

Kannan ND, Kulandaivelu G (2011) Drought induced changes in physiological, biochemical and phytochemical properties of *Withania somnifera* Dun. J Med Plants Res 5(16):3929–3935

Kaya MD, Kolsarici O (2011) Seed yield and oil content of some sunflower (*Helianthus annuus* L.) hybrids irrigated at different growth stages. Afr J Biotechnol 10(22):4591–4595

Kaya C, Tuna L, Higgs D (2006) Effect of silicon on plant growth and mineral nutrition of maize grown under water-stress conditions. J Plant Nutr 29(8):1469–1480

Kirnak H, Dogan E, Türkoğlu H (2010) Effect of drip irrigation intensity on soybean seed yield and quality in the semiarid Harran plain. Turk Span J Agric Res 8(4):1208–1217

Labusch C, Shishova M, Effendi Y, Li M, Wang X, Scherer GF (2013) Patterns and timing in expression of early auxin-induced genes imply involvement of phospholipases A (pPLAs) in the regulation of auxin responses. Mol Plant 6(5):1473–1486

Laribi B, Bettaieb I, Kouki K, Sahli A, Mougou A, Marzouk B (2009) Water deficit effects on caraway (*Carum carvi* L.) growth, essential oil and fatty acid composition. Ind Crop Prod 30(3):372–379

Latif F, Ullah F, Mehmood S, Khattak A, Khan AU, Khan S, Husain I (2016) Effects of salicylic acid on growth and accumulation of phenolics in *Zea mays* L. under drought stress. Acta Agric Scand B Soil Plant Sci 66(4):325–332

Li JF, Zhang D, Sheen J (2014) Cas9-based genome editing in Arabidopsis and tobacco. Methods Enzymol 546:459–472

Liu X, Hua X, Guo J, Qi D, Wang L, Liu Z, Jin Z, Chen S, Liu G (2008) Enhanced tolerance to drought stress in transgenic tobacco plants overexpressing VTE1 for increased tocopherol production from *Arabidopsis thaliana*. Biotechnol Lett 30(7):1275–1280

Mahouachi J (2007) Growth and mineral nutrient content of developing fruit on banana plants (*Musa acuminata* AAA, 'Grand Nain') subjected to water stress and recovery. J Hort Sci Biotechnol 82(6):839–844

Marcaida M III, Li T, Angeles O, Evangelista GK, Fontanilla MA, Xu J et al (2014) Biomass accumulation and partitioning of newly developed Green Super Rice (GSR) cultivars under drought stress during the reproductive stage. Field Crop Res 162:30–38

Marin A, Rubio JS, Martinez V, Gil MI (2009) Antioxidant compounds in green and red peppers as affected by irrigation frequency, salinity and nutrient solution composition. J Sci Food Agric 89(8):1352–1359

Martinelli F, Basile B, Morelli G, d'Andria R, Tonutti P (2012) Effects of irrigation on fruit ripening behavior and metabolic changes in olive. Sci Hortic 144:201–207

Martinelli F, Remorini D, Saia S, Massai R, Tonutti P (2013) Metabolic profiling of ripe olive fruit in response to moderate water stress. Sci Hortic 159:52–58

Martinez-Ballesta MC, Dominguez-Perles R, Moreno DA, Muries B, Alcaraz-López C, Bastías E et al (2010) Minerals in plant food: effect of agricultural practices and role in human health. Rev Agron Sustain Dev 30(2):295–309

Misle E, Kahlaoui B, Garrido E, Hachicha M (2014) Using an allometric model for the accumulation of mineral nutrients in crops under saline and water stress: a field experience in fertigation. In: Improvement of crops in the era of climatic changes. Springer, New York, pp 81–106

Montoya JM, Raffaelli D (2010) Climate change, biotic interactions and ecosystem services. Philos Trans R Soc B 365(1549):2013–2018

Munné-Bosch S, Alegre L (2000) Changes in carotenoids, tocopherols and diterpenes during drought and recovery, and the biological significance of chlorophyll loss in *Rosmarinus officinalis* plants. Planta 210(6):925–931

Navarro JM, Pérez-Pérez JG, Romero P, Botía P (2010) Analysis of the changes in quality in mandarin fruit, produced by deficit irrigation treatments. Food Chem 119(4):1591–1596

Nicoletto C, Santagata S, Pino S, Sambo P (2016) Antioxidant characterization of different Italian broccoli landraces. Hortic Bras 34:74–79

Oh MM, Carey EE, Rajashekar CB (2010) Regulated water deficits improve phytochemical concentration in lettuce. J Am Soc Hortic Sci 2135(3):23–229

Osuagwu GGE, Edeoga HO, Osuagwu AN (2010) The influence of water stress (drought) on the mineral and vitamin potential of the leaves of *Ocimum gratissimum* L. Recent Res Sci Technol 2(2):27–33

Patanè C, Tringali S, Sortino O (2011) Effects of deficit irrigation on biomass, yield, water productivity and fruit quality of processing tomato under semi-arid Mediterranean climate conditions. Sci Hortic 129(4):590–596

Podsedek A (2007) Natural antioxidants and antioxidant capacity of Brassica vegetables: a review. LWT Food Sci Technol 40(1):1–11

Quirós R, Villalba G, Muñoz P, Font X, Gabarrell X (2014) Environmental and agronomical assessment of three fertilization treatments applied in horticultural open field crops. J Clean Prod 67:147–158

Radovich TJ, Kleinhenz MD, Streeter JG (2005) Irrigation timing relative to head development influences yield components, sugar levels, and glucosinolate concentrations in cabbage. J Am Soc Hortic Sci 130(6):943–949

Rebey IB, Jabri-Karoui I, Hamrouni-Sellami I, Bourgou S, Limam F, Marzouk B (2012) Effect of drought on the biochemical composition and antioxidant activities of cumin (*Cuminum cyminum* L.) seeds. Ind Crop Prod 36(1):238–245

Rouphael Y, Cardarelli M, Colla G, Rea E (2008) Yield, mineral composition, water relations, and water use efficiency of grafted mini-watermelon plants under deficit irrigation. HortScience 43(3):730–736

Saeedipour S, Moradi F (2011) Comparison of the drought stress responses of tolerant and sensitive wheat cultivars during grain filling: impact of invertase activity on carbon metabolism during kernel development. J Agric Sci 3(2):1916–9760

Schreiner M, Beyene B, Krumbein A, Stutzel H (2009) Ontogenetic changes of 2-propenyl and 3-indolylmethyl glucosinolates in *Brassica carinata* leaves as affected by water supply. J Agric Food Chem 57(16):7259–7263

Shahsavari N, Jais HM, Shirani Rad AH (2014) Responses of canola oil quality characteristics and fatty acid composition to zeolite and zinc fertilization under drought stress. Int J Agri Sci 4(1):49–59

Sperdouli I, Moustakas M (2012) Interaction of proline, sugars, and anthocyanins during photosynthetic acclimation of *Arabidopsis thaliana* to drought stress. J Plant Physiol 169(6):577–585

Stagnari F, Galieni A, Speca S, Pisante M (2014) Water stress effects on growth, yield and quality traits of red beet. Sci Hortic 165:13–22

Stefanelli D, Goodwin I, Jones R (2010) Minimal nitrogen and water use in horticulture: effects on quality and content of selected nutrients. Food Res Int 43(7):1833–1843

Suzuki K, Golaz JC, Stephens GL (2013) Evaluating cloud tuning in a climate model with satellite observations. Geophys Res Lett 40(16):4464–4468

Tsialtasa JT, Soulioti E, Maslaris N, Papakosta D (2009) Genotypic response to re-growth of defoliated sugar beets after re-watering in a water-limited environment: effects on yield and quality

Thitisaksakul M, Jiménez RC, Arias MC, Beckles DM (2015) Effects of environmental factors on cereal starch biosynthesis and composition. J Cereal Sci 65:2–11

Wang Y, Frei M (2011) Stressed food–the impact of abiotic environmental stresses on crop quality. Agric Ecosyst Environ 141(4):271–286
Wegener CB, Jansen G, Jürgens HU (2015) Bioactive compounds in potatoes: Accumulation under drought stress conditions. Funct Food in Health Disease 5(3):108–116
Willett W, Rockström J, Loken B, Springmann M, Lang T, Vermeulen S, Murray CJ (2019) Food in the Anthropocene: the EAT–Lancet Commission on healthy diets from sustainable food systems. Lancet 393(10170):447–492
Zhang C, Wang L, Zhang Y, Dai H, Wan S, Yang Y, Tian J (2011) Effects of shading and nitrogen nutrition on the quality, free amino acids and sugar metabolism of tea. Food Chem 125(1):44–49
Zhang D, Li Z, Wang F, Zhang Y, Yao Y (2014) Long-term effects of irrigation amount and frequency on soil and plant water status, vegetative growth and yields of pear-jujube trees in the semiarid Loess Plateau of China. Agric Water Manag 135:61–70
Zhao Y, Bartley LE (2014) Comparative genomic analysis of the R2R3 MYB secondary cell wall regulators of Arabidopsis, poplar, rice, maize, and switchgrass. BMC Plant Biol 14(1):135
Zeipiņa S, Alsiņa I, Lepse L (2014) Stinging nettle–the source of biologically active compounds as sustainable daily diet supplement. Res Rural Dev 20:34–38

The Impact of Drought Stress on Plant Disease Dynamics

Najma Majeed, Ahmad Iqbal, Sajeela Sehar, Mehwish Sanaullah, Minahil Athar, Ayesha Ali, Muhammad Imran, and Ahmad Mahmood

Abstract Drought stress poses a significant challenge to plant health and productivity worldwide. On one hand, it can enhance plant defense mechanisms against certain diseases by triggering systemic acquired resistance (SAR) and inducing the production of protective compounds. On the other hand, drought can also compromise the plant's immune system, making it more vulnerable to infection and disease spread. Therefore, here we explore the physiological and biochemical changes that plants undergo in response to water deficit, discussing the growth patterns, nutrient uptake, and photosynthetic efficiency, followed by exploration of how these drought-induced modifications can influence plant susceptibility to pathogens. We also examine the role of environmental factors and plant genotype in modulating these effects. Practical implications for agriculture and disease management are considered, with a focus on strategies to mitigate the adverse effects of drought on plant disease. In conclusion, understanding the effects of drought stress on plant disease is crucial for developing resilient agricultural systems.

Keywords Drought stress · Plant diseases · Water scarcity · Disease susceptibility · Plant defense · Agriculture resilience

---

N. Majeed · S. Sehar · M. Sanaullah · M. Athar · A. Ali · M. Imran
Department of Soil and Environmental Sciences, MNS-University of Agriculture, Multan, Pakistan

A. Iqbal
Institute of Plant Protection, MNS-University of Agriculture, Multan, Pakistan

A. Mahmood (✉)
Department of Soil and Environmental Sciences, MNS-University of Agriculture, Multan, Pakistan

Department of Climate Change, MNS-University of Agriculture, Multan, Pakistan

School of Interdisciplinary Engineering & Science (SINES), National University of Science and Technology (NUST), Islamabad, Pakistan
e-mail: ahmad.mahmood@mnsuam.edu.pk

U. K. Chaudhry et al. (eds.), *Drought Stress*,
https://doi.org/10.1007/978-3-031-80610-0_6

1 Introduction

In agriculture, plant diseases are a major problem that affects crop yields and income. Numerous pathogens, including bacteria, viruses, fungi, and other microorganisms, can be the cause of these illnesses (Sharma et al. 2017). Plant diseases have an influence on social, cultural, and environmental factors in addition to financial losses (Almeida and Nunney 2015). Wilting, rots, damping-off, and leaf spots are just a few of the symptoms that can be caused by the pathogens that cause plant diseases (Qurrahmah et al. 2022; Srinivas et al. 2019). Plant disease management and detection are essential to maintaining crop health while guaranteeing food security. Timely responses to prevent significant crop damage can be facilitated by the early detection of diseases using techniques such as image analysis and machine learning (Ra et al. 2021; Suljović et al. 2022). The technological developments like the application of nanomaterials like silver nanoparticles have demonstrated promise in the fight against pathogen-caused plant diseases (Gupta and Chauhan 2017).

Disease development is also influenced by nutritional factors; nutrients such as silicon, phosphorus, potassium, sulfur, nitrogen, and sulfur affect the growth of pathogens and the course of disease (Walters and Bingham 2007). The significance of efficient soil management techniques in disease prevention is highlighted by the ubiquity of soilborne pathogens (Panth et al. 2020).

Reduced subsurface water reserves cause continuous water loss through transpiration, which disrupts physiological and biochemical processes and ultimately lowers plant growth and yield (Samarah 2005). This is known as drought stress in plants. Particularly in light of the warming climate, this stress is thought to be a major abiotic factor restricting plant growth and yield worldwide (Chen et al. 2022). Reduced chlorophyll content, changed antioxidant potential, and increased osmolyte contents are some of the effects of drought stress on plants, which also affects photosynthesis, antioxidant systems, and osmotic adjustment (Arivalagan and Somasundaram 2017; Dong et al. 2019). The floral characteristics, pollinator attraction, and seed set can all be impacted by drought stress, which lowers plant reproductive success (Rering et al. 2020).

It is crucial to understand how drought stress affects plants to formulate strategies that will increase plant resilience. According to studies, drought stress can lower transpiration rates, net leaf photosynthetic rates, and leaf water potential, which can have an impact on plant growth and yield (Kumar 2019). To improve crop productivity and guarantee food security under changing climates, researchers are examining the responses of plants to drought stress and identifying mechanisms to alleviate its harmful effects. In the context of global warming and climate change, the effects of changing climate on drought are a crucial concern. According to studies, the frequency and intensity of drought episodes are predicted to climb dramatically as Earth's temperature rises, creating the possibility of longer and more severe drought conditions (Jiang et al. 2019a; Tabari and Willems 2022). The increased likelihood of drought presents a significant concern to many areas across the world, impacting soil moisture content, hydrological systems, and crop yield (Alahacoon and Edirisinghe 2021).

The economic effects as well as environmental ones are anticipated to result from climate change's effects on drought. In locations where drought damages are expected to increase significantly, the intensification of droughts under scenarios of global warming may result in increased economic damages (Miao et al. 2020; Sam et al. 2023). Proactive adaptation and mitigation efforts are necessary since these impacts can result in ecosystem vulnerabilities, infrastructure failures, and worsened socioeconomic implications (Chen et al. 2018). Moreover, it has been shown that the effects of global warming are asymmetrical with respect to the types of drought that occur, with intense droughts being significantly more affected than typical drought conditions (Xu et al. 2019). The complex relationship between drought and climate change and the need for specialized strategies to address the various effects of shifting climatic patterns on the distribution and intensity of drought.

A large area of the Earth's surface is affected by drought stress, which has a substantial impact on agricultural land worldwide. About 45% of the world's agricultural land area experiences regular or persistent drought (Per et al. 2017). Drought stress affects over 40% of the world's wheat fields and as much as 42% of the global wheat production area (Acuña-Galindo et al. 2015; Ma et al. 2023). This stress has serious consequences for crop production. Furthermore, affecting 64% of the world's land area, drought stress is thought to be the single biggest abiotic factor limiting food production globally (Rabara et al. 2014). Furthermore, drought-induced yield losses have been observed on a large scale, with 75% of global cultivated land experiencing such losses between 1983 and 2005. This resulted in substantial economic repercussions, with drought-induced yield losses amounting to about 166 billion United States dollars during that period (Nguyen et al. 2021).

As global climate change progresses, the frequency and severity of drought stress are expected to increase, posing a serious challenge to world food security and sustainability. With rising temperatures and changing precipitation patterns, the risk of drought stress in many areas globally is anticipated to escalate, leading to even higher levels of water scarcity and agricultural land affected by drought (Akinroluyo et al. 2020; Wang et al. 2022). Drought stress is a significant global concern, with various studies highlighting the potential areas that could be affected in the near future. The combination of drought and high temperatures is expected to increase, emphasizing the need to study this stress combination to enhance crop tolerance (Zandalinas et al. 2018). Wheat and barley, crops vital for food and feed in many regions, are particularly vulnerable to drought stress (Sallam et al. 2019). Climate models predict future increases in drought stress for over 70% of the global land area, indicating a widespread impact (Swann et al. 2016). Plants in their natural habitat often face a combination of stresses like drought and heat, further emphasizing the need to understand and mitigate these challenges (Rizhsky et al. 2004).

Large tracts of maize fields are under danger as a result of growing drought stress brought on by population increase and climate change. Maize is a staple crop in many locations (Fan 2019). According to forecasts, the effects of climate change and population expansion will likely make drought stress much more of a problem for plant growth and agricultural output worldwide (Jha et al. 2014). Due to sporadic and frequent droughts, maize-growing regions in Africa are already experiencing yield losses; this trend is predicted to worsen in the upcoming years (Singh et al. 2022). As evidenced by recent years, there is cause for concern regarding the global worsening of drought stress brought on by irregular rainfall patterns (Jaldhani et al. 2021). Early on in a drought, soil moisture depletion can cause vegetation to experience drought stress more quickly and severe (Sungmin and Park 2023). Predictions related to climate change suggest that drought conditions would worsen, creating difficulties for agriculture globally (Hone et al. 2021). Reactive oxygen species overproduction can cause oxidative damage to plants under drought stress (Alabdallah et al. 2021).

Global crop output is severely limited by drought stress, which highlights the need for studies and programs to increase crop resilience to drought (Zou et al. 2023). Droughts related to global change, which are marked by warmer and longer drought episodes, may result in a widespread die-off of trees, which could have an effect on ecosystems (Breshears et al. 2009). Given that climate change is still affecting temperatures and precipitation patterns, it is imperative to invest in breeding crops like maize to withstand drought (Nurmberg et al. 2022).

1.1 Importance of Understanding the Link Between Drought and Plant Disease

Plant diseases can be greatly impacted by drought stress through a number of different methods. Abscisic acid (ABA) synthesis in plants can spike in response to drought stress, which can boost microbial virulence by reducing plant immunity and help certain infections cause water-soaking lesions (Choudhary and Senthil-Kumar 2024). Furthermore, drought stress triggers a general reaction of important plant hormones that react to water stress as well as be vital in plants' defenses against infections (Li et al. 2021). There may be a connection between drought acclimation and plant defense mechanisms because the levels of reactive oxygen species (ROS) in drought-acclimated plants are highly connected with disease resistance (Ramegowda et al. 2013). Dryness can alter a plant's physiological state, altering its susceptibility to disease and promoting pathogen invasions (Barradas et al. 2018). The composition and diversity of phyllosphere-associated microorganisms, which are important in plant-pathogen interactions and impact disease resistance, can also be impacted by drought stress (Lin et al. 2023). The potential linkages between pathogen stress and drought at the plant-pathogen interface, which could modify later responses to abiotic or biotic challenges, are suggested by the crosstalk and convergence of processes in hormone production and signaling pathways (Gupta et al. 2016a).

Additionally, the length and intensity of the stress can affect how plant diseases are affected by drought stress. Both mild and severe droughts cause different plant response mechanisms to be activated, which changes the plant's defenses to pathogen infections (Gupta et al. 2016a). The intricate interaction between abiotic and biotic stresses is shown by the fact that concurrent pathogen infection and drought stress can cause both similar and unique transcriptome responses in plants (Sinha et al. 2017). It is anticipated that climate change will make it more common for plants to experience disease and drought stress at the same time, underscoring the importance of comprehending and reducing the combined impacts of these stressors on plant health (Heyman et al. 2020).

In summary, a variety of factors, including alterations in plant physiology, hormone responses, microbial populations, and defense mechanisms, interact with drought stress and plant diseases. In order to improve plant resilience to pathogen stress and drought in the face of shifting environmental conditions, it is imperative to comprehend these interactions.

1.1.1 Drought Stress and Bacterial Diseases

Plant bacterial infections can be impacted by drought stress in a number of ways. Research has indicated that drought stress might modify the transcriptional response and plant growth-promoting activity of bacteria such as Pseudomonas fluorescens, hence impacting the interactions between plants and bacteria (Nishu et al. 2022). There may be a connection between drought stress and bacterial illnesses since drought stress can promote bacterial multiplication while decreasing plant growth and biomass (Gupta et al. 2016a). The rhizosphere of crop plants is home to microbial populations that can impact plant responses to various stresses, such as drought. This underscores the significance of microbiome modification in improving plant stress tolerance (Xie et al. 2021).

Additionally, it has been demonstrated that bacteria like Azospirillum sp. promote wheat development in drought-stressed environments, highlighting the significance of rhizosphere bacteria in plant stress responses (Arzanesh et al. 2011). It has been discovered that plants that have acclimated to drought stress can confer resistance to bacterial pathogens such as Pseudomonas syringae. This finding suggests that there may be a cross-tolerance mechanism between bacterial diseases and drought stress (Ramegowda et al. 2013). A dynamic link between drought stress, bacteria, and plant health has been shown by correlations between changes in the rhizosphere bacterial community during drought stress and soil characteristics and nutrient levels (Liu et al. 2021).

Further underlining the potential of bacteria in boosting plant resilience to stress conditions, studies have also shown that bacterial consortia can promote plant growth and secondary metabolite accumulation in plants under drought stress (Moreno-Galván et al. 2020). According to Veach et al. (2020), bacterial endophytes outperform endophytic fungi in promoting plant development under drought stress, highlighting the significance of microbial interactions in plant stress responses. Furthermore, it has been demonstrated that seed endophyte bacteria improve plants' ability to withstand drought stress by controlling physiological traits, antioxidants, and mineral uptake (Pan et al. 2022).

1.1.2 Drought Stress and Fungal Diseases

Through a variety of ways, drought stress can have a substantial impact on fungal infections in plants. Studies have indicated that the dynamics of fungal communities in the rhizosphere and plant root microbiome can be influenced by drought conditions, resulting in changes to the fungal composition and biodiversity (Carbone et al. 2021). Changes in the fungal community associated with roots have been connected to drought stress, which may have an impact on plant health and disease susceptibility (Andreo-Jimenez et al. 2019). It has been discovered that the plant host and drought circumstances have an impact on the makeup of the root-associated fungal microbiome in rice, highlighting the complex interaction between fungal communities and drought stress (Andreo-Jimenez et al. 2019).

It has been demonstrated that fungi endophytes increase plant resistance to drought stress and encourage plant growth in situations where water is scarce (Javed et al. 2022). According to Morales-Quintana et al. (2022), these endophytic fungi can cause changes in plant physiology and biochemistry, which can lead to enhanced photosynthetic efficiency, modification of antioxidant enzyme activity, and increased proline content in plants under drought stress. Furthermore, by modifying growth-defense regulating proteins and phytohormone-dependent pathways, fungal endophytes can give plants persistent resistance to biotic and abiotic stressors (Javed et al. 2022). Furthermore, boosting plant resilience to drought stress is largely dependent on the symbiotic connection between plants and arbuscular mycorrhizal fungus (AMF) (Chialva et al. 2020). According to Fan et al. (2023), AMF can improve plant fitness and stress tolerance by increasing nutrition and water absorption. AMF has been shown to positively affect plant development and drought tolerance, suggesting that these fungi may be able to lessen the negative impacts of drought stress on plants (Fan et al. 2023).

1.1.3 Drought Stress and Viral Diseases

In fact, plant-virus interactions and disease progression can be significantly impacted by drought stress in relation to viral infections in plants. Research has indicated that the effects of drought stress on many physiological and molecular processes in plants can alter their susceptibility to viral infections. For example, it has been shown that drought stress in tomato plants lowers the titer of the tomato yellow leaf curl virus (TYLCV) by triggering genes related to autophagy, which in turn causes the viral coat proteins to break down, especially in the roots (Choudhary and Senthil-Kumar 2024). This suggests that drought stress may affect the way viruses replicate and propagate within plant tissues, which may change the course of illness. The complicated molecular processes are involved in how plants react to compounded stresses like viral infections and drought. Plant responses to drought stress may be mediated by the MYB transcription factor family, which is known to be involved in plant responses to a variety of pressures, including biotic challenges such viral infections (Yao et al. 2022). The interaction of viral illnesses with drought stress can alter plant defense mechanisms, gene expression, and physiological processes, which in turn can affect how viral infections in plants turn out.

The effects of drought stress on defense mechanisms and antioxidant systems in plants can affect how those plants react to viral infections. Antioxidant enzyme activity has been demonstrated to rise in plants under drought stress, which may have consequences for plant defense against viral infections (Cui et al. 2016). The plant's capacity to fend against viral infections and slow the spread of disease may be impacted by the modification of antioxidant systems during drought stress. Overall, there are many different facets to the association between viral infections and drought stress in plants, including nuanced interactions at the molecular, physiological, and ecological levels. For the purpose of creating methods to increase plant resistance to viral infections in the face of changing climatic conditions, it is essential to comprehend how drought stress affects plant-virus interactions.

1.1.4 Drought Stress and Nematodal Diseases

Through a variety of ways, drought stress can have a major impact on plant nematode diseases. Research has indicated that drought stress might affect a plant's vulnerability to nematode infection; in fact, the consequences of drought-induced stress can obscure the spectral signatures that indicate nematode attack (Susič et al. 2018). Complex molecular responses are involved in the interaction between drought stress and nematode diseases. This is demonstrated by the induction of certain genes, such as RAPID ALKALINIZATION FACTOR-LIKE8 (AtRALFL8), in roots under joint stresses, which can confer susceptibility to both nematode infection and drought stress (Atkinson et al. 2013). Plant development and stress responses may be impacted by the connection between nematode infections and drought stress. According to observations, nematode-infected plants lose soil moisture more slowly than healthy plants, which results in less severe drought circumstances for the infected plants (Haverkort et al. 1992).

In drought-stressed plants, plant responses that enhance root allocation relative to leaves may be overwhelmed by the quantity of root-feeding nematodes, which could affect how above- and below-ground biomass is partitioned (Franco et al. 2020). Additionally, there may be a connection between the physiological reactions of plants to nematode infection and drought stress. Plant responses to these stresses can be impacted by nematode invasion, which can upset hormonal homeostasis in plants. This can activate defensive mechanisms and lower pathogen infection (Li et al. 2018). The combined stresses of drought and nematode infection can lead to changes in plant gene expression, defense mechanisms, and physiological processes, affecting plant health and productivity under challenging environmental conditions.

1.1.5 Drought Stress and Protozoal Diseases

Plant protozoal infections can be affected by drought stress through a number of different methods. Although the studies given do not specifically address the relationship between plant protozoal infections and drought stress, we can draw some conclusions about possible relationships based on our general understanding of plant stress responses and pathogen interactions. Plant diseases caused by protozoa are less frequent than those caused by bacteria, viruses, and fungus. Drought stress, however, can change physiological processes and impair plant defenses, thereby increasing a plant's susceptibility to a variety of diseases, including protozoa. Plant health can be harmed by drought stress, which can result in diminished vitality and immune system weakness, which can foster an atmosphere that is favorable to protozoal infections.

The physiological changes brought on by stress can affect a plant's resistance to protozoal infections. These changes can include adjustments to hormone levels, nutrition availability, and defense mechanisms. Furthermore, drought stress has the ability to upset the delicate equilibrium of the plant microbiome, which may have an impact on how plants interact with helpful microbes like protozoa that aid in disease defense. Furthermore, plants' susceptibility to protozoal infections may be indirectly influenced by the effects of drought stress on their root systems and general health. Plants may be more susceptible to protozoal infections as a result of changes in root growth, nitrogen uptake, and water stress responses brought on by drought. In the general understanding of plant stress responses suggests that drought stress can potentially influence plants' susceptibility to protozoal infections by compromising plant health and altering physiological processes. Specific references directly linking drought stress to protozoal diseases in plants were not provided. Additional studies in this field may provide additional light on the precise mechanisms behind the interactions between plant protozoal infections and drought stress.

2 Physiological Effects of Drought Stress on Plants

Drought stress has a substantial physiological impact on plants, influencing many elements of their growth and development. Plants respond to drought stress by undergoing physiological, biochemical, and molecular changes (Wang et al. 2016). Drought stress modifies plant metabolism, causing changes in physiological and biochemical pathways (Farooq et al. 2009). These changes affect water and nutrient interactions, photosynthesis, assimilate partitioning, and respiration (Afshari et al. 2020). Furthermore processes, which impair plant growth and development, drought stress causes alterations in essential physiological and biochemical (Muhammad Rashid and Seema Mahmood 2018). Plants respond to drought stress by activating defensive systems that minimize its negative consequences. These processes include physiological changes such as stomatal conductance, osmotic adjustment, and enhanced antioxidant activity (Saud et al. 2014).

Drought stress also alters the antioxidant defense system and photosynthetic activity in plants, with treatments such as salicylic acid enhancing these responses (González-Villagra et al. 2022). Drought stress also impacts the activity of antioxidant metabolism enzymes such superoxide dismutase, peroxidase, catalase, and glutathione reductase (Sudiana et al. 2022). Drought stress has a morphological impact on plants, resulting in reductions in root and shoot length, chlorophyll content, and photosynthetic activity (Mohi-Ud-Din et al. 2021). Furthermore, drought stress causes a decrease in photosynthesis, transpiration rate, stomatal conductance, leaf water content, relative growth rate, and water consumption (Sallam et al. 2019).

2.1 Disruption of Plant Water Relations

Drought stress disturbs plant water interactions, causing substantial effects on plant physiology and growth. Drought stress disrupts plant-water connections, which has a significant impact on different physiological and biochemical processes in plants (Hussain et al. 2019). Drought stress inhibits plant growth primarily by interfering with water relations and physiological processes (Liu et al. 2019). This disruption in water relations is evident in the reduction of water use efficiency and the disturbance of cell membrane stability, which are essential for maintaining plant health under stress conditions (Maurel and Chrispeels 2001). Additionally, drought stress affects stomatal conductance, resulting in alterations in water loss regulation and plant water status.

Drought stress disrupts plant water relations, resulting in changes in physiological processes such as stomatal closure, osmotic adjustment, and antioxidant activity (Jiang et al. 2019b). Stomatal closure is an important physiological process that plants use to withstand drought stress by limiting water loss via transpiration. Moreover, drought stress-induced disruptions in plant water relations can result in changes in leaf water content, relative water content, and cell membrane stability, compromising overall plant health and production (González-Villagra et al. 2022).

2.2 Impacts on Photosynthesis and Nutrient Uptake

Drought stress has a profound impact on plant physiology, specifically photosynthesis and nutrient uptake. Stress-induced limits in water availability and stomatal conductance reduce photosynthetic activity (Yang et al. 2023). This decrease in photosynthesis has a negative impact on plant growth and development, resulting in lower productivity (Chen et al. 2021). Additionally, drought stress alters nutrient uptake in plants, resulting in imbalances and deficits in critical elements such as nitrogen and phosphorus (Bista et al. 2018). Drought stress can cause lower nutrient uptake and utilization, exacerbating the impact on plant growth (Fiorilli et al. 2022).

The interruption of photosynthesis and nitrogen uptake during drought stress is an important part of plant responses to water constraint. Drought stress not only impacts physiological processes associated to photosynthesis and nutrient intake, but it also causes alterations in hormone balance, membrane stability, and general plant metabolism (Abdelaal et al. 2021). These changes in plant physiology under drought can have a knock-on effect on growth, development, and stress tolerance mechanisms (Bista et al. 2020). Moreover, the interplay between drought stress, photosynthesis, and nutrient uptake illustrates the complex responses of plants to environmental stressors and emphasizes the necessity of understanding these interactions in creating ways to promote plant resistance to drought (Pereira et al. 2023).

2.3 *Weakening of Plant Cell Walls*

The alteration of plant cell walls during drought stress is an important part of plant responses to water constraint. Drought stress alters the composition and structure of the cell wall, which affects plant growth and development. Plant cell walls respond to drought stress by increasing the production of cellulose, xyloglucan, and pectin in order to maintain cell wall integrity and turgor pressure, allowing for continuous cell development under low water potential (Morcillo et al. 2021). Expansions, which bind to cellulose and mediate acid-induced loosening of plant cell walls, are elevated under abiotic stress conditions, contributing to cell wall changes associated with drought tolerance (Lenk et al. 2019).

Drought stress has been demonstrated in studies to reduce cell wall content, potentially due to thinner cell walls in plants subjected to water shortage circumstances (van der Weijde et al. 2017). Additionally, tolerant plant cultivars respond to drought stress by increasing the expression of lignin biosynthesis genes, which strengthens the plant cell wall and reduces water loss, emphasizing the relevance of cell wall reinforcement in drought tolerance processes (Ponce et al. 2022). While drought-induced alterations in cell wall composition may differ between tissues and the time of drought induction, these changes generally do not severely impact biomass conversion in plants like Sorghum bicolor (Scavuzzo-Duggan et al. 2021).

Furthermore, disruption of plant cell walls and cell wall enzymes can have a major impact on plant growth and development under drought stress circumstances (Hu et al. 2020). Expansins, which are important regulators of cell wall extension during plant growth, have a function in cell growth mediated by abscisic acid and indole-3-acetic acid during drought stress, indicating their importance in plant responses to water shortage (Zhao et al. 2012). Adjusting cell wall composition and structure is critical for plant adaptation to drought stress, since it influences growth processes and responses to environmental cues (Yan et al. 2018).

3 Molecular Responses to Drought Stress

Plants undergo a series of biochemical reactions in response to drought stress that allow them to adapt and endure harsh environments. Plants that are stressed by drought show distinct patterns of gene expression, according to transcriptome analysis. For example, sorghum plants tend to have more transcripts regulated by heat and drought stress combined than by each stress alone, suggesting a more intricate molecular response to the stress combination (Zandalinas et al. 2018). One typical method for comprehending the molecular mechanisms of drought stress tolerance in plants is to identify and characterize genes activated under drought stress (RS and Petrović 2019).

Research has demonstrated that plants activate particular genes and pathways in response to drought stress. For instance, it has been discovered that, in plants experiencing progressive mild drought, the phosphate starvation response occurs before the abscisic acid response, demonstrating the connection between drought stress responses and nutrition (Nagatoshi et al. 2021). In order to respond and adapt to abiotic stresses such as drought, plants have complex molecular systems that include the control of stress-related proteins, metabolites, and gene expression (Shinozaki and Yamaguchi-Shinozaki 2022).

Plants respond molecularly to drought stress through intricate signaling networks and regulatory components. Through a variety of sensors, plants detect drought stress, which triggers downstream reactions like stomatal closure, the production of stress-related proteins, and the production of metabolites (Kuromori et al. 2022). To improve agricultural resilience to changing environmental circumstances and develop crops that are drought-tolerant, it is imperative to comprehend the molecular underpinnings of plant responses to drought stress (Rabara et al. 2015). In conclusion, plants respond molecularly to drought stress by activating particular genes, regulatory pathways, and physiological processes to deal with conditions of water deficiency. In order to improve agricultural sustainability and drought tolerance in crops, researchers can pinpoint important targets by deciphering the molecular mechanisms underlying plant responses to drought stress.

3.1 Gene Expression Changes

Many studies that clarify the molecular reactions involved in plant adaptation to water shortage circumstances have been carried out by researchers in an effort to understand the changes in gene expression in plants under drought stress. Certain gene expression patterns and regulatory networks that are triggered in response to drought stress have been identified by transcriptomic analysis. Studies have, for instance, highlighted the importance of transcription factors and cis-acting regions in stress-inducible promoters for controlling the expression of genes during cold and drought stressors (Yamaguchi-Shinozaki and Shinozaki 2006). Studies have shown that plants respond differently to drought stress in terms of their gene expression profiles. Specifically, more transcripts are modified under drought conditions than under heat stress or in combination with other stresses (Zandalinas et al. 2018).

In order to increase stress tolerance, plants' molecular responses to drought stress include the up-regulation of stress-related genes, such as those involved in the production of proline and oxylipin (Kumar et al. 2021). In order to withstand drought stress, plants employ a variety of physiological and molecular defense mechanisms, including the control of stress-related proteins, metabolites, and gene expression (Hussain et al. 2019). Furthermore, research has identified particular genes and pathways in plants that react negatively to drought stress. For example, it has been discovered that the MYB44-ENAP1/2 complex controls the expression of genes that respond to drought and histone acetylation (Zhao et al. 2022). Furthermore, via up-regulating stress-responsive genes, the overexpression of MYB family genes, such as OsMYB6, has been linked to improved resistance to salinity and drought stress in transgenic rice (Tang et al. 2019).

Plants respond molecularly to drought stress by activating pathways that are sensitive to stress, regulating gene expression dynamically, and modifying physiological processes to increase their ability to withstand conditions of water scarcity. It is essential to comprehend the complex molecular mechanisms that underlie plant responses to drought stress in order to develop strategies that improve crop tolerance to drought and maintain agricultural sustainability in the face of shifting environmental conditions.

3.2 Hormonal Regulation

Hormonal controls are essential for plants' reactions to water deficits under drought stress. A crucial hormone involved in the plant's reaction to drought stress is abscisic acid (ABA). Under drought stress, plants quickly store and synthesise ABA in their roots. The ABA is subsequently transferred to the aboveground portions of the plant through the xylem to control leaf stomatal closure, lower transpiration loss, and increase water retention capacity, all of which improve plant water use efficiency (Liu and Wei 2019). Complex signaling pathways and interactions are involved in the control of hormones during drought stress. Stress from drought causes ABA to build up, and ABA is essential for controlling stomatal closure and gene expression to support adaptive reactions in situations where there is a water shortage (Gamboa-Tuz et al. 2018). Furthermore, ABA-dependent and ABA-independent signaling pathways have been shown to be involved in controlling the expression of genes linked to abiotic stress tolerance in plants under drought stress, according to the hormonal profiling of these plants (Gamboa-Tuz et al. 2018).

Moreover, it has been shown that the interaction of hormones such as ABA and jasmonic acid (JA) controls how plants react to drought stress. Under drought stress,

JA, a crucial regulator of herbivore-induced plant defenses, builds up in plants, suggesting its function in mediating plant responses to circumstances of water deficit (Nguyen et al. 2016a, b). The intricate hormonal interplay that regulates plant responses to a variety of stressors, including drought, is highlighted by the connection between ABA and JA. In summary, hormone controls—especially those pertaining to ABA—are essential in coordinating how plants react to drought stress. The plant's capacity to adapt and endure in environments with water scarcity is facilitated by the complex interactions and balance between several hormones. Developing ways to improve plant resilience and drought tolerance in crops requires an understanding of the hormonal regulation during drought stress.

4 Influence of Drought Stress on Plant Defense Mechanisms

Drought stress in plants can lead to a reduction in aphid fitness, plant growth, and vigor, along with an increase in the concentration of defensive compounds within the plant tissues (Leybourne et al. 2022, 2021). These defensive compounds, such as glucosinolates, can impact aphid performance on drought-stressed plants (Mewis et al. 2012). Furthermore, drought stress induces various physiological and biochemical modifications in plants, including the excessive production of reactive oxygen species (ROS) and alterations in antioxidant defense systems (Gowtham et al. 2022; Zhang et al. 2019). The presence of drought stress can impair plant growth, photosynthesis, yield, nitrogen metabolism, and antioxidant defense while increasing oxidative stress (Akin 2023). Additionally, drought stress can lead to a decrease in the levels of certain phytochemicals and an increase in others, affecting the overall defense mechanisms of plants (Adebusola et al. 2019).

Plants respond to drought stress by activating defense mechanisms, such as the production of specialized metabolic compounds and antioxidant defense systems (Nardino et al. 2022). Exogenous applications of compounds like melatonin and probiotics have been shown to mitigate the oxidative damages induced by drought stress in plants (Sharma and Zheng 2019; Hossain et al. 2020). Moreover, treatments with fulvic acid have been found to enhance the drought tolerance of tea plants by improving antioxidant defense mechanisms (Sun et al. 2020). Additionally, the application of silicon and proline can upregulate antioxidant enzymes and mitigate the effects of drought stress in sugar beet plants.

(i) **Jasmonic Acid (JA)**

Jasmonic acid is responsible for regulating the plant response in abiotic stresses. During the drought stress the biosynthesis gene of Jasmonic acid activates which results in increasing the level of JA. Like abscise acid, JA is also transferred to upper part through both xylem and phloem (Yoon et al. 2020). Under the drought conditions, JA participates in the regulation of stomatal closing. Stomatal gaps are also indirectly regulated by it, enhancing the drought resistance in plants. Moreover, it also plays a key role in conductivity of water under drought conditions. It was also observed that the application of exogeneous JA enhanced the antioxidant activity of plants under water stress (Iqbal et al. 2022).

(ii) **Salicylic Acid (SA)**

Salicylic acid is a phenolic compound that is produced through the secondary metabolism that is responsible for various physiological functions such as stomatal regulation (Mir et al. 2022). Salicylic acid biosynthesis occurs in the chloroplast of the cell and then conjugates with cytoplasm with segregates in the vacuole. SA is moved to its target place through xylem or by cell-to-cell transport (Ullah et al. 2018). It showed a key role in drought stress tolerance by regulating the closing of stomata and transcriptional modification of drought-related genes (Yadav et al. 2021).

(iii) **Abscisic Acid (ABA)**

It is a plant's natural stress hormone that controls many physiological functions. Water stress or drought triggers the production of abscise acid (Ali et al. 2020). When the plasma membrane receives the stress signal, it starts to produce abscise acid in plastids. Most of the ABA is produced in the root that is transferred through xylem tissue to the upper part of the plant (Fraudentali et al. 2020). It is very crucial for plants under the water stress. ABA performs two main functions during drought: water balance and dehydration tolerance of cell. By controlling the guard cells, the water balance is maintained in all cells and the production of dehydration tolerance (Hsu et al. 2021).

4.1 Impaired Stomatal Function

Under the drought conditions, stomata reduce their frequency and promote closing of stomata as a result transpirational water loss is reduced. Sulfate concentration is increased in the xylem that promotes the stomatal closing. Water stress also modulates the phytohormones especially ABA known as "stress hormone". Under drought conditions, Hydraulic signal emerges from root which transmit to shoot and trigger the closing of stomata (Agurla et al. 2018).

4.2 Suppression of Plant Immune Response

The manipulation of plant immune responses by pathogens involves intricate mechanisms that target the plant's defense signaling network to suppress the host immune response. Pathogens can influence phytohormone levels, such as salicylic acid (SA), to hinder the plant's immune system Pieterse et al. (2009). This suppression can result in compromised antibacterial defense, decreased SA accumulation, and reduced SA-dependent autophagy, ultimately promoting bacterial growth on virus-infected plants (Zvereva et al. 2016).

The interaction between the Phytophthora effector Avh241 and host NDR1-like proteins showcases an effective counter-defense mechanism where the effector interferes with NDR1 function to suppress plant immune responses during infection (Yang et al. 2021). The balance between biotic and abiotic stress responses in plants involves intricate interactions of stress hormones like abscisic acid and salicylic acid, which can modulate immune responses in a leaf age-dependent manner (Berens et al. 2019). Furthermore, the crosstalk between plant hormones, such as cytokinin and salicylic acid, plays a role in regulating plant immunity, with cytokinin enhancing SA-dependent defense responses while SA inhibits cytokinin signaling (Argueso et al. 2012).

4.3 Activation of Defense Signaling Pathways

Mitogen-activated protein kinase (MAPK) induces OsMAP1 gene when biotic and abiotic stress like drought occurs (Xiong et al. 2018). Overexpression of OsMAP1 is associated with kinase activity and increased drought resistance (Bai and Matton 2018). Ca-dependent protein kinases (CDPKs) are activated by Ca^{2+} and target Ca signaling in plants. CDPKs play key role in stress signaling transduction pathway by stimulus-dependent activation or by directed functional target protein phosphorylation (Shi et al. 2018). Some CDPKs genes are responsible for response to abiotic stress and others are linked with ABA signaling. CPK4 and CPK11 phosphorylate are two ABA responsive transcription factors named ABF1 and ABF4 which facilitate ABA signaling pathway. Overexpression of CPK6 shows improved drought resistance. CPK3 and CPK6 are also responsible for regulation of ABA stomatal signaling and ion channel of guard signaling (Aslam et al. 2022).

4.4 Crosstalk Between Drought and Disease Resistance Pathways

The interaction between drought and disease resistance pathways in plants involves complex interactions among various signaling molecules, transcription factors, and defense responses. Studies have demonstrated the intricate interplay between abiotic stress responses, such as drought, and biotic stress responses, including disease resistance mechanisms Suzuki et al. (2014). This interconnection can result in physiological outcomes such as drought avoidance, tolerance, or resistance in plants (Ollas and Dodd 2016). The regulatory networks associated with the response to drought stress often overlap with pathways reacting to other abiotic stresses like cold and heat, indicating a broader crosstalk in abiotic stress responses (Nakashima et al. 2014). Negative interactions between disease resistance and abiotic stress tolerance signaling pathways have been recognized, indicating a delicate balance between these pathways in plants (Zuluaga et al. 2020). Hormonal interactions, particularly between abscisic acid (ABA) and jasmonic acid (JA), are crucial in mediating responses to both abiotic and biotic stresses (Nguyen et al. 2016a, b). The convergence of mechanisms in hormone biosynthesis and signaling pathways suggests potential interactions between drought and pathogen stress responses in plants (Gupta et al. 2017).

Transcription factors, such as RD26, have been identified as key mediators of crosstalk between drought and brassinosteroid signaling pathways, highlighting the complexity of regulatory networks in stress responses (Ye et al. 2017). Additionally, the BES/BZR family transcription factor TaBZR2 has been found to positively regulate drought responses by activating stress-related genes, underscoring the crosstalk between brassinosteroid and drought signaling pathways (Cui et al. 2019). The identification of crucial transcription factor hubs and protein kinases as central points of crosstalk between cold and drought stress signaling pathways emphasizes the interconnected nature of stress responses in plants. The involvement of signaling molecules like hydrogen sulfide in mitigating drought stress through the ABA signaling pathway indicates a complex crosstalk between different stress signaling pathways.

5 Interaction Between Drought Stress and Plant Pathogens

The interaction between drought stress and plant pathogens is a complex phenomenon that involves intricate physiological and molecular responses in plants. Drought stress can have significant impacts on plant-pathogen interactions, influencing the susceptibility of plants to various pathogens. Studies have highlighted the multifaceted nature of these interactions and the diverse responses exhibited by plants under combined drought and pathogen stress conditions (Gupta et al. 2016a, b; Sinha et al. 2017). Drought stress can modulate plant responses to pathogen infections by altering physiological and biochemical processes within the host plant. For instance, drought stress has been shown to enhance microbial virulence by suppressing plant immunity, leading to increased pathogen growth and the development of water-soaking lesions by specific pathogens. Additionally, the co-occurrence of drought stress and pathogen infection can result in reduced plant growth and biomass, indicating a complex interplay between stressors (Gupta et al. 2016a, b).

5.1 Mechanisms of Pathogen Infection Under Drought Conditions

The occurrence of plant infections caused by pathogens during drought stress circumstances is influenced by a complex interaction of molecular and physiological pathways (Sunarti et al. 2022). The presence of drought stress can affect the way plants respond to pathogen infections through many routes. For example, severe drought stress can elicit a whole-plant response guided by essential phytohormones, which not only respond to water stress but also play an important role in the plant's response to pathogen-induced biotic stress (Suzuki et al. 2014). This reaction can affect the plant's susceptibility to pathogen infections during drought circumstances. Pathogens can exploit plants' reduced defenses during drought stress to produce infections (Sinha et al. 2019). When plant basal defense systems are already activated due to drought stress, pathogens may decrease them by releasing effectors into the plant cell, promoting infection (Pandey et al. 2017). The modification of plant defenses by pathogens during drought stress demonstrates the intricate relationship between stress responses and pathogen virulence methods.

The interaction of phytohormones such as abscisic acid (ABA) and jasmonic acid (JA) during combined drought and pathogen stress can influence plant responses at the abiotic-biotic stress interface (Gupta et al. 2017). The confluence of hormone production mechanisms and signalling pathways shows that drought and pathogen stress may cooperate to influence plant responses. Moreover, alterations in the phyllosphere microbiome under drought conditions might influence the dynamics of plant-pathogen interactions and modify disease outcomes (Sinha et al. 2017). Drought stress can cause changes in plant chemical composition, affecting nutrient concentration and balance, altering levels of defensive allelochemicals, and changing levels of volatile stress metabolites, all of which can affect plant-pathogen interactions (Szczepaniec and Finke 2019).

5.1.1 Changes in Plant Physiology

Pathogen infections in plants cause a series of physiological changes that affect many elements of plant life. These adjustments include changes in phytochemical status, such as changes in photosynthetic pigments, phenolic compounds, starch content, and antioxidative enzyme regulation (Yousaf et al. 2016). Furthermore, pathogen-induced metabolic changes might cause changes in plant metabolism, influencing cellular reproduction, function, and differentiation (Berger et al. 2007). Plant responses to pathogen infections are characterized by metabolic alterations that coincide with symptom development or defense reactions (El-Dougdoug 2020). Furthermore, pathogen infections can induce changes in plant volatiles, affecting airborne communication between plants and potentially altering interactions with insect vectors (Moreira et al. 2021). Plant-pathogen interactions can also lead to differential expression of genes and physiological responses. For example, alterations in water relations that do not physiologically support the pathogen or prepare the plant for strong defence mechanisms may be the cause of the reduction in pathogen multiplication in plants (Gupta et al. 2016b). Physiological changes resulting from drought and recovery are also likely to be involved in plant-pathogen interactions (Ponce et al. 2022). These interactions may lead to complex regulation of biochemical and physiological process in plants (Wojakowska et al. 2015) (Table 1).

Table 1 Pathogen-induced changes in plant growth and development with known molecular mechanism

Pathogens	Molecular mechanism	References
Meloidogyne Incognita and Heterodera Schachii	Synthesis harmone auxin and induce plant cell reprograming; induce a feeding cell complex or form a giant cell.	Lee et al. (2011)
Heterodera Schachii	Secret effector protein 19007; target the Arabidopsis LAX) auxin import protein trigger the cell wall hydrolysis to stimulate syncytium development and lateral root emerge.	Lee et al. (2011)
Meloidogyne hapla and Rotylenchulus reniformis	Secret the CEP-like peptide effector mimics and mediate the nitrogen-demand signaling, nodulation, and lateral root development.	Tabata and Sawa (2014)
Aster Yellow phytoplasma	Secret effector SAP54; degradation of floral homeotic proteins and MTFs; promotes insect colonization with the RAD23 to alter flower development, cause phyllody. Secret effector TENGU and down-regulate the auxin-responsive genes to alter plants development.	MacLean et al. (2011)
Pseudomonas syringae	Arabidopsis shed cauline leaves, HAESA/HAESA-like 2. INFLORESCENCE DEFICIENT IN ABSCISSION, and NEVERSHED were required for the leaf abscission.	Patharkar et al. (2017)

5.1.2 Altered Host-Pathogen Interactions

Significant changes in host-pathogen interactions occur during drought stress, which may have an effect on the course and results of disease. By interfering with physiological and biochemical processes, drought stress can modify a plant's sensitivity to pathogen infections, potentially increasing or decreasing a plant's vulnerability to pathogens (Gupta et al. 2016b). The intricacy of these relationships is additionally enhanced by the interplay and convergence of mechanisms in hormone biosynthesis and signaling pathways, suggesting potential interplays between pathogen stress and drought that may influence later reactions to biotic or abiotic stressors (Gupta et al. 2017). Research has demonstrated that drought stress can alter the way genes pertaining to plant-pathogen interactions are expressed, with the plant-pathogen interaction pathway's genes being overexpressed under drought stress (Li et al. 2021). Additionally, the effect of drought stress on foliar and xylem-invading bacterial pathogens in chickpea has been studied, showing the relevance of varied stress intensities on stress interactions and their overall impact (Sinha et al. 2016). Furthermore, the net outcome of host-pathogen interactions under drought circumstances is determined by a variety of parameters, including plant genotype, pathogen species, and the timing, severity, and duration of the stress (Heyman et al. 2021). Furthermore, drought stress can affect pathogen virulence programs, as demonstrated by the rice blast fungus (Magnaporthe oryzae), which drastically changes its virulence program under drought conditions (Bidzinski et al. 2016). The coupling of drought stress and pathogen infection can have a significant impact on plant immunity, disease development, and host-pathogen interactions, emphasising the complex link between abiotic and biotic stress in plant systems.

5.2 *Effect of Drought Stress on Pathogen Growth and Development*

Drought stress has the potential to significantly affect pathogen proliferation and development, hence affecting disease dynamics. Abiotic stressors, such as dryness, might influence plant-pest interactions by increasing host plant vulnerability to pathogenic organisms and decreasing competitive ability with weeds (Pandey et al. 2017). Furthermore, drought stress can increase aflatoxin production by Aspergillus flavus, an opportunistic disease of plants such as maize and peanut (Fountain et al. 2018). Drought has a deleterious impact on infections, however fungal pathogens can develop at lower water potentials than their host plants (Desprez-Loustau et al. 2006). Drought stress can also influence plant susceptibility to infections, as evidenced by the increased prevalence of fungal diseases under severe drought stress conditions (Davis et al. 2015). Drought stress has a significant impact on pathogen growth and development, as evidenced by reductions in photosynthetic rates, stomatal conductance, and transpiration, as well as alterations in xylem water potential (Ghanbary et al. 2017). Furthermore, the interaction between drought stress and pathogen infection might result in complex results. Drought stress, for example, can drive conditional outcomes in host-pathogen interactions, influencing plant growth and performance under stress (Ahmad et al. 2016; Davis et al. 2015). Furthermore, drought stress can alter plant pathogen susceptibility, as seen in the instance of grapevines, where drought stress increased sensitivity to Plasmopara viticola (Heyman et al. 2021).

5.2.1 Influence on Pathogen Reproduction

Drought stress can have a substantial effect on pathogen reproduction. Stress may affect pathogen virulence or pathogenicity, potentially resulting in the emergence of previously unknown potential pathogens (Venuprasad et al. 2007). Drought stress can also have an impact on host metabolic composition, potentially affecting interactions with invading diseases and pathogen reproduction (Rering et al. 2020). The interaction of drought stress and pathogen infection can result in diverse results. Drought stress may generate conditional outcomes in host-pathogen interactions, influencing plant growth and performance under stress. Furthermore, drought stress can influence plant responses to pathogens via gene-for-gene interactions, with the severity of stress playing an important role in shaping these interactions (Lauder et al. 2019).

5.2.2 Enhanced Virulence Under Drought Conditions

The phenomenon known as "Enhanced Virulence Under Drought Condition" relates to the fact that certain diseases may exhibit greater virulence or pathogenicity when plants are stressed by a drought. The interplay of drought stress and pathogen virulence can result in more severe disease consequences in plants. According to research, some viruses may adapt and become more aggressive during droughts, potentially leading to increased disease development (Barnawal et al. 2017). Drought stress has been shown in studies to alter plant physiology and metabolism, potentially allowing pathogens to survive and reproduce more efficiently (Cohen et al. 2020). Drought stress can modify endogenous phytohormone levels in plants, potentially influencing the expression of genes involved in stress responses and pathogen defence systems (Shemi et al. 2021). Furthermore, the induction of salicylic acid in virus-infected plants under drought circumstances has been associated to improved drought tolerance, implying a complex interplay between infections and drought responses (Aguilar et al. 2017). In addition, the effects of drought stress on pathogen reproduction might be complex. Drought stress may cause the formation of reactive oxygen species and enhanced lipid peroxidation, fostering pathogen development and reproducti (Ahmad et al. 2021). Genetic changes, such as overexpression of certain genes, can modulate plant responses to drought stress, influencing pathogen reproduction and virulence under stress conditions (Bang et al. 2018).

5.3 Case Studies and Examples of Drought-Pathogen Interactions

Drought stress can significantly impact plant-pathogen interactions, affecting disease dynamics and outcomes. Various case studies shed light on the intricate relationship between drought stress and pathogen responses: (Pandey et al. 2017) conducted a study concentrating on the combination of drought and pathogen stress to better understand the impacts of abiotic and biotic stress interactions on plants. This review examined the drought and pathogen stress interaction model, focusing on the implications for plant development and stress responses (Gupta et al. 2016a). The degree of common responses between combined and individual stresses differs depending on the pathogens infecting drought-stressed plants. This study emphasized the need of taking into account the specific pathogens implicated in drought-induced interactions (Sinha et al. 2019). Drought stress was explored to see how it affected the interaction of soil-borne pathogens with chickpea in natural circumstances. This study shed light on the complicated relationship between drought stress and pathogen infections by investigating several drought-affected diseases (Desprez-Loustau et al. 2006). Maize grown in high-CO_2 and drought conditions was found to be more susceptible to Fusarium verticillioides proliferation and fumonisin contamination. This study demonstrated how coupled pressures might exacerbate pathogen proliferation and mycotoxin contamination (Aguilar et al. 2017). Maize cultivated under high CO_2 and drought circumstances was found to be more prone to Fusarium verticillioides proliferation and fumonisin contamination. This study showed how linked stressors could enhance pathogen proliferation and mycotoxin contamination.

These case studies and examples showcase the diverse impacts of drought-pathogen interactions on plant health, stress responses, and disease outcomes. Understanding these interactions is vital for devising strategies to alleviate the impact of diseases in agricultural systems confronting changing environmental conditions.

6 Impact of Drought Stress on Specific Plant Diseases

To investigate the impact of drought stress on specific plant diseases, it is essential to understand the intricate interactions between drought conditions and pathogen infections. Drought stress has been shown to significantly affect plant susceptibility to various pathogens, thereby influencing disease development and plant health.

6.1 Fungal Diseases Favored by Drought

Drought stress has been shown to have a significant impact on fungal diseases, affecting their development and spread. Research indicates that drought stress can prolong the initial infection phase of fungal pathogens like Fusarium, but it enhances their proliferation and spread after the initial infection phase (Liu and Liu 2016). While the direct effects of drought on pathogens are generally negative, most fungal pathogens exhibit plasticity and can grow at water potentials well below the minimum for the growth of their host plants (Desprez-Loustau et al. 2006). When severe drought occurs, number of fungal diseases including dry root rot and black root rot remarkedly increases in crops such as chickpea (Sinha et al. 2019). The intertwining of drought stress and fungal diseases may result in complicated physiological and biochemical responses in plants. For example, plants have been shown to be more tolerant of both necrotrophic fungi and hemibiotic bacteria when they have previously experienced drought (Ramegowda et al. 2013). The production of certain genes in plants under drought stress can contribute to the plant's tolerance of water scarcity while also increasing its chances for infection (Campo et al. 2012).

Studies have also demonstrated that the response of fungal microbiota associated with diseases like black-foot disease and potential biocontrol agents to drought stress is consistent, with their relative abundances being significantly higher under certain irrigation regimes (Carbone et al. 2021). The interplay of drought stress and fungal infections has the potential to alter plant defense mechanisms. Drought-stressed plants may demonstrate improved defense capacity against fungal infections, possibly through mechanisms like the generation of reactive oxygen species (Barradas et al. 2018). The impact of drought stress on fungal diseases extends beyond individual pathogens and can also influence the overall fungal community dynamics in the soil. Drought-induced shifts in soil fungal communities have been observed, which can be linked to specific root traits across various plant species (Lozano et al. 2021). Additionally, the combined stress of drought and pathogen infection can lead to unique and shared responses in plants, indicating a new state of stress when exposed to both stressors simultaneously (Gupta et al. 2016b). The study by (Santos et al. 2024) highlighted the interaction between drought stress and the fungal pathogen Neofusicoccum kwambonambiense in Eucalyptus seedlings, emphasizing the need to understand how drought affects disease development.

Furthermore, (Vurukonda et al. 2022) demonstrated the effectiveness of Pseudomonas sp. Strain FCBB-2 in promoting plant growth and exhibiting antifungal activity under drought stress conditions. This indicates the potential of certain microbial strains in mitigating the impact of fungal pathogens during periods of water scarcity. Overall, these studies collectively suggest that drought stress can exacerbate fungal diseases in plants by altering the composition of fungal communities, increasing disease incidence, and influencing plant-fungal interactions.

6.2 *Bacterial Diseases and Drought Stress*

Drought stress significantly impacts bacterial diseases by influencing the composition and activity of bacterial communities. Research has shown that drought stress can lead to a restructuring of root-associated microbiomes, enriching certain bacterial taxa such as Actinobacteria and Chloroflexi, while depleting others like Acidobacteria and Deltaproteobacteria (Santos-Medellín et al. 2017). Drought stress affects iron homeostasis in plant roots, influencing microbial community composition and resulting in increased levels of Actinobacteria, a lineage that thrives under drought conditions (Xu et al. 2021). The decrease in soil respiration under drought stress has been associated with reduced fungal and bacterial biomass, indicating heightened physiological and nutritional stress in the bacterial community (Zhou et al. 2020).

The interplay between drought stress and bacterial diseases has been investigated in crops like rice, where the combined stress of bacterial leaf blight and drought impacts gene expression and potential genes for multi-stress tolerance (Pal et al. 2022). In response to low oxygen levels induced by drought stress, metabolic pathways involved in drought stress tolerance, such as pyruvate decarboxylases, contribute to resistance against bacterial wilt disease (Wang et al. 2020). Drought stress can also shape rhizosphere bacterial community patterns, reducing diversity and increasing the abundance of drought-resistant bacteria (Liu et al. 2021).

The impact of drought stress on bacterial diseases extends beyond plants to the entire ecosystem. Drought stress negatively affects microbial activity, with the extent of this response influenced by interactions above and below ground (Veach et al. 2020). Additionally, adaptation processes of bacterial communities to drought have been observed, with a higher abundance of Firmicutes reported under drought stress in soils previously exposed to drought. In conclusion, drought stress plays a crucial role in shaping bacterial communities and their interactions with plants, influencing disease susceptibility and overall ecosystem dynamics. Understanding these impacts is vital for developing strategies to mitigate the effects of drought on bacterial diseases and ecosystem health (Siebielec et al. 2020).

6.3 Interaction of Drought with Viral Diseases

Drought stress has been shown to have a significant impact on plant viral diseases. When plants are subjected to drought stress, their growth, photosynthesis, water relations, pigment content, and membrane integrity are all affected (Kumar et al. 2021). This reduction in plant fitness due to drought stress can lead to an increase in disease development. For instance, studies have reported that plants under environmental stress conditions are more predisposed to diseases such as stem canker caused by Cytospora (Budi and Widodo 2022). The interaction between drought stress and plant viral diseases involves complex defense pathways. Plants subjected to a combination of drought and heat stress have been found to accumulate sugars like sucrose, maltose, and gulose, which can impact their response to pathogens (Rizhsky et al. 2004). Drought stress can also decrease the resistance of plants to diseases, making them more susceptible to pathogen attacks (Li et al. 2020).

Drought stress can modulate plant-virus-vector interactions, affecting symptom expression, plant immunity, and vector behavior. The severity of the interaction between drought and disease can vary depending on factors such as the pathogen, plant genotype, and drought intensity (Vasquez et al. 2022). Additionally, drought stress has been shown to enhance the movement of aphids, increasing the spread of viral diseases among plants. The impact of drought stress on plant viral diseases is multifaceted, affecting plant physiology, defense mechanisms, and interactions with pathogens. Drought stress not only weakens plants, making them more susceptible to diseases, but it also alters their metabolic profiles and immune responses, influencing the spread and severity of viral infections in plants (Nachappa et al. 2016).

7 Indirect Effects of Drought on Plant Disease

Drought stress can have significant indirect effects on plant diseases by altering various physiological and molecular processes within plants. Studies have highlighted the intricate interplay between drought conditions and pathogen infections, shedding light on the complex relationships between abiotic stress and disease development (Rering et al. 2020). Drought stress can indirectly impact plant diseases by affecting host physiology, altering plant metabolism, and modulating plant-microbe interactions. For instance, drought-induced changes in plant physiology can lead to increased susceptibility to pests and pathogens, exacerbating disease development.

7.1 *Increased Insect Vector Activity and Disease Spread*

Drought stress in plants can have indirect effects on increasing insect vector activity and disease spread. Prolonged drought stress can limit insect vector population size and the reservoirs of viruses and vectors, such as weed and volunteer crop plants ("Impact of Temperature and Rainfall Change on Epidemics Caused by Plant Viruses" 2021). Plant viruses can manipulate the behavior of their insect vectors, enhancing transmission efficiency and spread. For instance, aphids infected with Barley yellow dwarf virus (BYDV) exhibit a preference for noninfected plants, contributing to the spread of the virus (Ingwell et al. 2012). Additionally, plant viruses can induce changes in host plants that modify aphid feeding behavior, leading to enhanced virus acquisition from infected plants (Carmo-Sousa et al. 2016). Furthermore, some plant viruses have been found to enhance the tolerance of host plants to environmental stressors like drought, potentially increasing the survival and spread of both the virus and its insect vectors (Porras et al. 2020). Effective management of insect vectors is crucial to minimize the spread of vector-borne diseases and reduce crop damage (Garzo et al. 2020). The process of host-plant selection by insect vectors, influenced by factors like virus infection, affects the success of pathogen acquisition and transmission. The interplay between plant viruses, insect vectors, and environmental stressors like drought can create a complex scenario where drought stress indirectly influences insect vector behavior, leading to increased disease spread (Maluta et al. 2017).

7.2 *Changes in Soil Microbial Communities and Potential Pathogens*

Drought stress has been shown to have significant impacts on soil microbial communities. Studies have demonstrated that drought consistently alters the composition of soil fungal and bacterial communities. This alteration in microbial communities is considered a major driver of soil function in terrestrial ecosystems (Ochoa-Hueso et al. 2018). The response of the bacterial microbiota to drought stress is taxonomically consistent across soils and cultivars, with an enrichment of Actinobacteria and Chloroflexi, and a depletion of Acidobacteria and Deltaproteobacteria (Santos-Medellín et al. 2017). Additionally, under drought stress, the biodiversity in the soil decreases significantly, while it increases in the rhizosphere community. Specific soil parameters seem to play a role in determining the enrichment of bacterial flora in the rhizosphere under drought conditions (Chen et al. 2022).

Furthermore, recurrent drought events have been found to increase the dissimilarity of microbial communities compared to control and single drought events, enhancing soil multifunctionality during drought (Canarini et al. 2021). Extreme and frequent drought can induce a shift to an alternative soil microbial state characterized by altered bacterial and fungal community structures with reduced complexity and functionality (Cordero et al. 2021). Drought stress has also been reported to reduce the water potential of the soil physical matrix, leading to phosphorus deficiency and impacting the acquisition, mobility, and accessibility of phosphorus for soil biota, including microbes (Olatunji et al. 2018).

Drought stress can influence plant-soil feedback effects, altering biomass allocation and plant performance (Wilschut and Van Kleunen 2021). Changes in the microbial soil community caused by drought have consequences on the composition of the root-associated microbiome, affecting plant growth and fitness under stress (Carbone et al. 2021). Additionally, drought can leave lasting impacts on soil and root-associated microbiomes, known as legacy effects, which can further influence plant growth and microbiome assembly under drought stress (Swift et al. 2023). The timing of pathogen infection or drought stress occurrence under combined stress conditions plays a crucial role in determining the overall effect (Pandey et al. 2017). Studies have indicated that drought stress can reduce pathogen multiplication in plants experiencing combined stress (Gupta et al. 2016a). Severe drought triggers a comprehensive response in plants guided by key phytohormones, influencing their reaction not only to water stress but also to biotic stress caused by herbivores and pathogens (Szczepaniec and Finke 2019).

Drought stress may influence pathogen virulence or pathogenicity, potentially leading to the emergence of different potential pathogens that were not previously recognized (Sinha et al. 2016). The combined effect of drought stress and pathogen infection can alter resistance responses in plants, highlighting the role of certain plant hormones like ABA and JA as major regulators in combating host pathogens (Gupta et al. 2017). Additionally, drought stress-induced changes, such as the accumulation of osmolytes and nutrient leakage, can provide enriched nutrient sources for pathogens, affecting their proliferation (Pandey et al. 2015). Observations have shown that drought stress can both increase susceptibility to bacterial pathogens in some cases and enhance tolerance towards bacterial pathogens in others (De Pascali et al. 2019). The interaction between drought stress and pathogen infection can induce common and distinct transcriptomic responses in plants, indicating a close association between drought stress and pathogen stress effects (Sinha et al. 2017). With climate change, more instances of plants facing simultaneous drought and pathogen stress are expected, emphasizing the need to understand the combined impact of these stressors on plant health (Heyman et al. 2021).

8 Management Strategies for Drought and Plant Disease

(i) **Selection of drought-tolerant plant varieties**
Drought is among the major abiotic stresses that impose negative impact on the yield of crops. Drought and plant diseases can be managed by numerous strategies for better crop health and production. It can be managed by appropriate agronomic practices together with adopting drought-resistant varieties (Zia et al. 2021). Researchers have developed various techniques to enhance drought resistance in plants. The development of transgenic plants is the most notable method. Plant genome has broad range of gene which are associated with drought that gives amazing opportunity for the improvement of crops (Salehi-Lisar and Bakhshayeshan-Agdam 2020). Regulation of these stress-related genes plays a significant role in increasing the stress resistance in plants. Large numbers of genes are responsible for drought stress resistance like AtNAC2 gene is crucial for drought stress tolerance. The overexpression of AtNAC2 in transgenics plants boost the drought resistance in plants. (Ilyas et al. 2021). Similarly, root diameter is enlarged by the over expression of OsNAC5 that leads to increase the drought resistance in plant (Anwar and Kim 2020).

(ii) **Agronomic Practices to Reduce Drought Stress Impact**
Agronomic strategies like sowing time, management of fertilizers and plant geometry help to enhance drought resistance. The major impact of these strategies is to reduce water loss by lessening the evapotranspiration. The application of Gypsum improves the soil structure which flocculate the soil particle that improve the growth of roots and water infiltration (Nadeem et al. 2019). During rainy season, other agronomic practices like mulching, ploughing and zero tillage also mitigate the negative impact of water stress (Johnson et al. 2018). Soil health is affected by nutrients management that also improves the soil's water holding capacity. Various inorganics nutrients like N, Zn, P, Se and K were discovered to mitigate the detrimental impacts of drought stress. The application of silicon with foliar also improve the drought tolerance (Nasir and Toth 2022).

(iii) **Integrated Pest and Disease Management Approaches**
Plant physiology is affected by drought stress that attracts pests especially at high temperatures. Prolonged drought stress impacts the growth as well as survival of the plant but also causes outbreak of the pests. Moreover, drought also increases the entry of pests like aphids that disseminate viruses. To overcome these problems, integrated pest management is very crucial to control the damage due to the drought stress (Yihdego et al. 2019).

Abiotic stresses impose serious effects on the productivity of crops globally. Drought stress is gaining serious attention as it retards plant growth and reduce the production. Genetic and molecular breeding techniques especially QLT mapping highlight the complex inheritance of drought resistance in plants (Nadeem et al. 2019).

(iv) **Use of biostimulants and biocontrol agents**

In addition to selection of suitable variety, growth regulators, either natural or synthetic, can also reduce the negative impact of water stress on plants. Foliar application of Osmo protectants and growth regulators can enhance drought tolerance (Khan et al. 2018). The application of growth regulators like abscisic acid (ABA), Auxins (IAA), brassinosteroids (BRs), salicylic acid (SA), gibberellins (GA3), cytokinins (Cks) and jasmonates (JA) efficiently reduce the stress condition. During drought stress, abscisic acid is responsible for signaling between the shoot and root that cause antitranspirant response by closing of stomata and reducing the leaf expansion.

(v) **Future Directions and Research Needs**

In future, researchers should focus on bacteria which are associated with plants that have ability to enhance water use efficiency and tolerance to drought conditions. These species helps plant to enhance their root system and maintain its cellular moisture level which results in the survival of the plant in stressed conditions (Khan et al. 2020). Moreover, research should be carried out to understand the aggregation of polyphenols and polyamines and their effects under drought conditions. This research will help the breeders to develop the drought resistance cultivars (Batool et al. 2023).

9 Conclusion

The relationship between drought and plant disease showcases an important aspect of agricultural science. The complexity between drought and plant diseases is not just a collection of individual effects but a dynamic interaction that can significantly influence plant health and disease severity. Drought conditions can both suppress and ameliorate various plant diseases, asking for a better understanding of the mechanisms behind. Comprehending the importance of considering drought stress in plant disease management is essential for developing effective strategies to mitigate the risks. This includes selecting drought-tolerant crop varieties, employing water-saving irrigation practices, and timing the application of plant protectants to coincide with drought-induced susceptibility windows. As climate changes and water scarcity becomes more prevalent, the need to integrate drought considerations into plant disease management will only grow. Ensuring agricultural resilience and ecosystem health in the face of these challenges is paramount for sustainable food production and the preservation of biodiversity.

References

Abdelaal K, AlKahtani M, Attia K, Hafez Y, Király L, Künstler A (2021) The role of plant growth-promoting bacteria in alleviating the adverse effects of drought on plants. Biology 10(6):Article 6

Acuña-Galindo MA, Mason RE, Subramanian NK, Hays DB (2015) Meta-analysis of wheat QTL regions associated with adaptation to drought and heat stress. Crop Sci 55(2):477–492

Adebusola A, Victor O, Emmanuel I, Alam A (2019) Effect of nitrogen application on the phytochemical production in Jatropha curcas l. (angiosperms: euphorbiaceae) during drought conditions. Plant Sci Today 6(2):183–189

Afshari M, Naderi A, Mojadam M, Lack S, Alavifazel M (2020) Zinc and iron-mediated alleviation water deficiency of maize by modulating antioxidant metabolism. Notulae Botanicae Horti Agrobotanici Cluj-Napoca 48(2):989–1004

Aguilar E, Cutrona C, del Toro FJ, Vallarino JG, Osorio S, Pérez-Bueno ML, Barón M, Chung BN, Canto T, Tenllado F (2017) Virulence determines beneficial trade-offs in the response of virus-infected plants to drought via induction of salicylic acid. Plant Cell Environ 40(12):2909–2930

Agurla S, Gahir S, Munemasa S, Murata Y, Raghavendra AS (2018) Mechanism of stomatal closure in plants exposed to drought and cold stress. In: Survival strategies in extreme cold and desiccation: adaptation mechanisms and their applications. Springer, Singapore, pp 215–232

Ahmad M, Ok YS, Kim B-Y, Ahn J-H, Lee YH, Zhang M, Moon DH, Al-Wabel MI, Lee SS (2016) Impact of soybean stover-and pine needle-derived biochars on Pb and As mobility, microbial community, and carbon stability in a contaminated agricultural soil. J Environ Manag 166:131–139

Ahmad S, Ihsan M, Wang GY, Zeeshan M, Yang L, Ali I, Zhou XB (2021) Ameliorative effect of melatonin improves drought tolerance by regulating growth, photosynthetic traits and leaf ultrastructure of maize seedlings. BMC Plant Biol 21(1):368

Akin S (2023) Asparagine and nitric oxide jointly enhance antioxidant capacity and nitrogen metabolism to improve drought resistance in cotton: evidence from long-term field trials. Food Energy Secur 13(1)

Akinroluyo OK, Jaškūnė K, Kemešytė V, Statkevičiūtė G (2020) Drought stress response of westerwolths ryegrass (Lolium multiflorum ssp. Multiflorum) cultivars differing in their ploidy level. Zemdirbyste-Agriculture 107(2):161–170

Alabdallah NM, Hasan MM, Salih AM, Roushdy SS, Al-Shammari AS, Alsanie SI, El-Zaidy M (2021) Silver nanoparticles improve growth and protect against oxidative damage in eggplant seedlings under drought stress. Plant Soil Environ 67(11):617–624

Alahacoon N, Edirisinghe M (2021) Spatial variability of rainfall trends in Sri Lanka from 1989 to 2019 as an indication of climate change. ISPRS Int J Geo Inf 10(2):84

Ali S, Hayat K, Iqbal A, Xie L (2020) Implications of abscisic acid in the drought stress tolerance of plants. Agronomy 10(9):1323

Almeida RP, Nunney L (2015) How do plant diseases caused by Xylella fastidiosa emerge? Plant Dis 99(11):1457–1467

Andreo-Jimenez B, Vandenkoornhuyse P, Lê Van A, Heutinck A, Duhamel M, Kadam N, Jagadish K, Ruyter-Spira C, Bouwmeester H (2019) Plant host and drought shape the root associated fungal microbiota in rice. PeerJ 7:e7463

Anwar A, Kim J-K (2020) Transgenic breeding approaches for improving abiotic stress tolerance: recent progress and future perspectives. Int J Mol Sci 21(8):Article 8

Argueso C, Ferreira F, Epple P, To, J, Hutchison C, Schäller G et al (2012) Two-component elements mediate interactions between cytokinin and salicylic acid in plant immunity. PLoS Genet 8(1):e1002448

Arivalagan M, Somasundaram R (2017) Alteration of photosynthetic pigments and antioxidant systems in tomato under drought with Tebuconazole and Hexaconazole applications. J Sci Agri 1:146–157

Arzanesh MH, Alikhani HA, Khavazi K, Rahimian HA, Miransari M (2011) Wheat (*Triticum aestivum* L.) growth enhancement by Azospirillum sp. under drought stress. World J Microbiol Biotechnol 27(2):197–205

Aslam MM, Rashid MAR, Siddiqui MA, Khan MT, Farhat F, Yasmeen S et al (2022) Recent insights into signaling responses to cope drought stress in rice. Rice Sci 29(2):105–117

Atkinson NJ, Lilley CJ, Urwin PE (2013) Identification of genes involved in the response of Arabidopsis to simultaneous biotic and abiotic stresses. Plant Physiol 162(4):2028–2041

Bai F, Matton DP (2018) The Arabidopsis mitogen-activated protein kinase kinase kinase 20 (MKKK20) C-terminal domain interacts with MKK3 and harbors a typical DEF mammalian MAP kinase docking site. Plant Signal Behav 13(8):e1503498

Bang SW, Lee D-K, Jung H, Chung PH, Kim YS, Choi YD, Suh J-W, Kim J-K (2018) Overexpression of OsTF1L, a rice HD-Zip transcription factor, promotes lignin biosynthesis and stomatal closure that improves drought tolerance. Plant Biotechnol J 17(1):118–131

Barnawal D, Bharti N, Pandey SS, Pandey A, Chanotiya CS, Kalra A (2017) Plant growth-promoting rhizobacteria enhance wheat salt and drought stress tolerance by altering endogenous phytohormone levels and TaCTR1/TaDREB2 expression. Physiol Plant 161(4):502–514

Barradas C, Pinto G, Correia B, Castro BB, Phillips AJL, Alves A (2018) Drought × disease interaction in *Eucalyptus globulus* under *Neofusicoccum eucalyptorum* infection. Plant Pathol 67(1):87–96

Batool M, El-Badri AM, Hassan MU, Haiyun Y, Chunyun W, Zhenkun Y, Jie K, Wang B, Zhou G (2023) Drought stress in *Brassica napus*: effects, tolerance mechanisms, and management strategies. J Plant Growth Regul 42(1):21–45

Berens M, Wolinska K, Spaepen S, Ziegler J, Nobori T, Nair A et al (2019) Balancing trade-offs between biotic and abiotic stress responses through leaf age-dependent variation in stress hormone cross-talk. Proc Natl Acad Sci 116(6):2364–2373

Berger S, Sinha AK, Roitsch T (2007) Plant physiology meets phytopathology: plant primary metabolism and plant–pathogen interactions. J Exp Bot 58(15–16):4019–4026

Bidzinski P, Ballini E, Ducasse A, Michel C, Zuluaga P, Genga A, Chiozzotto R, Morel J-B (2016) Transcriptional basis of drought-induced susceptibility to the Rice blast fungus *Magnaporthe oryzae*. Front Plant Sci 7

Bista DR, Heckathorn SA, Jayawardena DM, Mishra S, Boldt JK (2018) Effects of drought on nutrient uptake and the levels of nutrient-uptake proteins in roots of drought-sensitive and -tolerant grasses. Plan Theory 7(2):Article 2

Bista DR, Heckathorn SA, Jayawardena DM, Boldt JK (2020) Effect of drought and carbon dioxide on nutrient uptake and levels of nutrient-uptake proteins in roots of barley. Am J Bot 107(10):1401–1409

Breshears DD, Myers OB, Meyer CW, Barnes FJ, Zou CB, Allen CD, McDowell NG, Pockman WT (2009) Tree die-off in response to global change-type drought: mortality insights from a decade of plant water potential measurements. Front Ecol Environ 7(4):185–189

Budi FS, Widodo (2022) Effects of drought stress and plant growth-promoting rhizobacteria on sudden-death disease (*Cytospora euginae*) development of clove. IOP Conf Ser Earth Environ Sci 974(1):012092

Campo S, Peris-Peris C, Montesinos L, Peñas G, Messeguer J, San Segundo B (2012) Expression of the maize ZmGF14-6 gene in rice confers tolerance to drought stress while enhancing susceptibility to pathogen infection. J Exp Bot 63(2):983–999

Canarini A, Schmidt H, Fuchslueger L, Martin V, Herbold CW, Zezula D, Gündler P, Hasibeder R, Jecmenica M, Bahn M, Richter A (2021) Ecological memory of recurrent drought modifies soil processes via changes in soil microbial community. Nat Commun 12(1):5308

Carbone MJ, Alaniz S, Mondino P, Gelabert M, Eichmeier A, Tekielska D, Bujanda R, Gramaje D (2021) Drought influences fungal community dynamics in the grapevine rhizosphere and root microbiome. J Fungi 7(9):686

Carmo-Sousa M, Moreno A, Plaza M, Garzo E, Fereres A (2016) Cucurbit aphid-borne yellows virus (CABYV) modifies the alighting, settling and probing behaviour of its vector *Aphis gos-*

sypii favouring its own spread: persistent transmitted viruses effect on vector behaviour. Ann Appl Biol 169(2):284–297

Chen J, Liu Y, Pan T, Liu Y, Sun F, Ge Q (2018) Population exposure to droughts in China under the 1.5 °C global warming target. Earth Syst Dyn 9(3):1097–1106

Chen S, Li P, Tan S, Pu X, Zhou Y, Hu K, Huang W, Liu L (2021) Combined proteomic and physiological analysis of chloroplasts reveals drought and recovery response mechanisms in *Nicotiana benthamiana*. Plan Theory 10(6):Article 6

Chen Y, Yao Z, Sun Y, Wang E, Tian C, Sun Y, Liu J, Sun C, Tian L (2022) Current studies of the effects of drought stress on root exudates and rhizosphere microbiomes of crop plant species. Int J Mol Sci 23(4):2374

Chialva M, Lanfranco L, Guazzotti G, Santoro V, Novero M, Bonfante P (2020) *Gigaspora margarita* and its E*ndobacterium modulate* symbiotic marker genes in tomato roots under combined water and nutrient stress. Plan Theory 9(7):886

Choudhary A, Senthil-Kumar M (2024) Drought: a context-dependent damper and aggravator of plant diseases. Plant Cell Environ 47(6):2109–2126

Cohen I, Zandalinas SI, Fritschi FB, Sengupta S, Fichman Y, Azad RK, Mittler R (2020) The impact of water deficit and heat stress combination on the molecular response, physiology, and seed production of soybean. Physiol Plant 172(1):41–52

Cordero I, Leizeaga A, Hicks LC, Rousk J, Bardgett RD (2021) Extreme drought triggers transition to an alternative soil microbial state. https://doi.org/10.1101/2021.12.10.472086

Cui Z-H, Bi W-L, Hao X-Y, Xu Y, Li P-M, Walker MA, Wang Q-C (2016) Responses of in vitro-grown plantlets (*Vitis vinifera*) to grapevine leafroll-associated Virus-3 and PEG-induced drought stress. Front Physiol 7

Cui X, Gao Y, Guo J, Yu T, Zheng W, Yongwei L et al (2019) BES/BZR transcription factor TaBZR2 positively regulates drought responses by activation of TaGST1. Plant Physiol 180(1):605–620

Davis TS, Bosque-Pérez NA, Foote NE, Magney TS, Eigenbrode SD (2015) Environmentally dependent host–pathogen and vector–pathogen interactions in the Barley yellow dwarf virus pathosystem. J Appl Ecol 52(5):1392–1401

De Pascali M, Vergine M, Sabella E, Aprile A, Nutricati E, Nicolì F, Buja I, Negro C, Miceli A, Rampino P, De Bellis L, Luvisi A (2019) Molecular effects of *Xylella fastidiosa* and drought combined stress in olive trees. Plan Theory 8(11):437

Desprez-Loustau M-L, Marçais B, Nageleisen L-M, Piou D, Vannini A (2006) Interactive effects of drought and pathogens in forest trees. Ann For Sci 63(6):597–612

Dong S, Jiang Y, Dong Y, Wang L, Wang W, Ma Z, Yan C, Ma C, Liu L (2019) A study on soybean responses to drought stress and rehydration. Saudi J Biol Sci 26(8):2006–2017

El-Dougdoug NK (2020) Physiological and molecular defense level in potato cultivars against potato virus X. Ann Agri Sci Moshtohor 58(4):1079–1088

Fan D (2019) The effect of calcium to maize seedlings under drought stress. Am J Plant Sci 10(08):1391–1396

Fan X, Xie H, Huang X, Zhang S, Nie Y, Chen H, Xie X, Tang M (2023) A module centered on the transcription factor MSN2 from arbuscular mycorrhizal fungus *Rhizophagus irregularis* regulates drought stress tolerance in the host plant. New Phytol 240(4):1497–1518

Farooq M, Wahid A, Kobayashi N, Fujita D, Basra SMA (2009) Plant drought stress: effects, mechanisms and management. In: Lichtfouse E, Navarrete M, Debaeke P, Véronique S, Alberola C (eds) Sustainable Agriculture. Springer, Dordrecht, pp 153–188

Fiorilli V, Maghrebi M, Novero M, Votta C, Mazzarella T, Buffoni B, Astolfi S, Vigani G (2022) Arbuscular mycorrhizal symbiosis differentially affects the nutritional status of two durum wheat genotypes under drought conditions. Plan Theory 11(6):Article 6

Fountain JC, Koh J, Yang L, Pandey MK, Nayak SN, Bajaj P, Zhuang W-J, Chen Z-Y, Kemerait RC, Lee RD, Chen S, Varshney RK, Guo B (2018) Proteome analysis of aspergillus flavus isolate-specific responses to oxidative stress in relationship to aflatoxin production capability. Sci Rep 8(1):3430

Franco ALC, Gherardi LA, De Tomasel CM, Andriuzzi WS, Ankrom KE, Bach EM, Guan P, Sala OE, Wall DH (2020) Root herbivory controls the effects of water availability on the partitioning between above- and below-ground grass biomass. Funct Ecol 34(11):2403–2410

Fraudentali I, Ghuge SA, Carucci A, Tavladoraki P, Angelini R, Rodrigues-Pousada RA, Cona A (2020) Developmental, hormone-and stress-modulated expression profiles of four members of the Arabidopsis copper-amine oxidase gene family. Plant Physiol Biochem 147:141–160

Frontiers | Chilling and Drought Stresses in Crop Plants: Implications, Cross Talk, and Potential Management Opportunities. (n.d.). Retrieved May 10, 2024, from https://www.frontiersin.org/journals/plant-science/articles/10.3389/fpls.2018.00393/full

Gamboa-Tuz SD, Pereira-Santana A, Zamora-Briseño JA, Castano E, Espadas-Gil F, Ayala-Sumuano JT, Keb-Llanes MÁ, Sanchez-Teyer F, Rodríguez-Zapata LC (2018) Transcriptomics and co-expression networks reveal tissue-specific responses and regulatory hubs under mild and severe drought in papaya (*Carica papaya* L.). Sci Rep 8(1):14539

Garzo E, Moreno A, Plaza M, Fereres A (2020) Feeding behavior and virus-transmission ability of insect vectors exposed to systemic insecticides. Plan Theory 9(7):895

Ghanbary E, Kouchaksaraei MT, Mirabolfathy M, Sanavi SAMM, Rahaie M (2017) Growth and physiological responses of *Quercus brantii* seedlings inoculated with Biscogniauxia mediterranea and Obolarina persica under drought stress. For Pathol 47(5)

González-Villagra J, Reyes-Díaz MM, Tighe-Neira R, Inostroza-Blancheteau C, Escobar AL, Bravo LA (2022) Salicylic acid improves antioxidant defense system and photosynthetic performance in *Aristotelia chilensis* plants subjected to moderate drought stress. Plants 11(5):Article 5

Gowtham H, Singh S, Shilpa N, Aiyaz M, Kalegowda N, Udayashankar A et al (2022) Insight into recent progress and perspectives in improvement of antioxidant machinery upon PGPR augmentation in plants under drought stress: a review. Antioxidants 11(9):1763

Gupta D, Chauhan P (2017) In vitro disease burden analysis using green silver nanoparticles. Adv Mater Proc 2(2):113–116

Gupta A, Dixit SK, Senthil-Kumar M (2016a) Drought stress predominantly endures *Arabidopsis thaliana* to P*seudomonas syringae* infection. Front Plant Sci 7

Gupta A, Sarkar AK, Senthil-Kumar M (2016b) Global transcriptional analysis reveals unique and shared responses in *Arabidopsis thaliana* exposed to combined drought and pathogen stress. Front Plant Sci 7

Gupta A, Hisano H, Hojo Y, Matsuura T, Ikeda Y, Mori IC, Senthil-Kumar M (2017) Global profiling of phytohormone dynamics during combined drought and pathogen stress in *Arabidopsis thaliana* reveals ABA and JA as major regulators. Sci Rep 7(1):4017

Haverkort AJ, Boerma M, Velema R, Waart M (1992) The influence of drought and cyst nematodes on potato growth. 4. Effects on crop growth under field conditions of four cultivars differing in tolerance. Neth J Plant Pathol 98(3):179–191

Heyman L, Chrysargyris A, Demeestere K, Tzortzakis N, Höfte M (2020) Responses to drought stress modulate the susceptibility to *Plasmopara viticola* in grapevine. Scite.Ai

Heyman L, Chrysargyris A, Demeestere K, Tzortzakis N, Höfte M (2021) Responses to drought stress modulate the susceptibility to *Plasmopara viticola* in *Vitis vinifera* self-rooted cuttings. Plan Theory 10(2):Article 2

Hone H, Mann R, Yang G, Kaur J, Tannenbaum I, Li T, Spangenberg G, Sawbridge T (2021) Profiling, isolation and characterisation of beneficial microbes from the seed microbiomes of drought tolerant wheat. Sci Rep 11(1):11916

Hossain M, Li J, Sikdar A, Hasanuzzaman M, Uzizerimana F, Muhammad I et al (2020) Exogenous melatonin modulates the physiological and biochemical mechanisms of drought tolerance in tartary buckwheat (F*agopyrum tataricum* (L.) *gaertn*). Molecules 25(12):2828

Hsu PK, Dubeaux G, Takahashi Y, Schroeder JI (2021) Signaling mechanisms in abscisic acid-mediated stomatal closure. Plant J 105(2):307–321

Hu X, Cui Y, Lu X, Song W, Lei L, Zhu J, Lai J, Lizhu E, Zhao H (2020) Maize WI5 encodes an endo-1,4-β-xylanase required for secondary cell wall synthesis and water transport in xylem. J Integr Plant Biol 62(10):1607–1624

Hussain S, Hussain S, Qadir T, Khaliq A, Ashraf U, Parveen A, Saqib M, Rafiq M (2019) Drought stress in plants: an overview on implications, tolerance mechanisms and agronomic mitigation strategies. Plant Science Today 6(4):Article 4

Ilyas M, Nisar M, Khan N, Hazrat A, Khan AH, Hayat K, Fahad S, Khan A, Ullah A (2021) Drought tolerance strategies in plants: a mechanistic approach. J Plant Growth Regul 40(3):926–944

Impact of Temperature and Rainfall Change on Epidemics caused by Plant viruses (2021) J Nat Sci Research. https://doi.org/10.7176/JNSR/12-3-05

Ingwell LL, Eigenbrode SD, Bosque-Pérez NA (2012) Plant viruses alter insect behavior to enhance their spread. Sci Rep 2(1):578. https://doi.org/10.1038/srep00578

Iqbal S, Wang X, Mubeen I, Kamran M, Kanwal I, Díaz GA et al (2022) Phytohormones trigger drought tolerance in crop plants: outlook and future perspectives. Front Plant Sci 12:799318

Jaldhani V, Senguttuvel P, Srikanth B, Raghuveer Rao P, Subrahmanyam D, Sanjeeva Rao D (2021) Present status and future prospects of drought tolerance in Rice. In: Kumar Goyal A (ed) Cereal Grains—Volume 2. IntechOpen. https://doi.org/10.5772/intechopen.97461

Javed J, Rauf M, Arif M, Hamayun M, Gul H, Ud-Din A, Ud-Din J, Sohail M, Rahman MM, Lee I-J (2022) Endophytic fungal consortia enhance basal drought-tolerance in *Moringa oleifera* by upregulating the antioxidant enzyme (APX) through heat shock factors. Antioxidants 11(9):1669

Jha UC, Bohra A, Singh NP (2014) Heat stress in crop plants: its nature, impacts and integrated breeding strategies to improve heat tolerance. Plant Breed 133(6):679–701

Jiang S, Wang M, Ren L, Xu C-Y, Yuan F, Liu Y, Lu Y, Shen H (2019a) A framework for quantifying the impacts of climate change and human activities on hydrological drought in a semiarid basin of northern China. Hydrol Process 33(7):1075–1088

Jiang Y, Cai T, Tang F (2019b) Effect of different drought stress on seedling growth and physiological characteristics of *Cenchrus pauciflorus* Benth. Open access. Libr J 6(12):Article 12

Johnson Y, Ayuke F, Kinama J, Sijali I (2018) Effects of tillage practices on water use efficiency and yield of different drought tolerant common bean varieties in Machakos County, Eastern Kenya. https://doi.org/10.13140/RG.2.2.23410.61126

Khan A, Pan X, Najeeb U, Tan DKY, Fahad S, Zahoor R, Luo H, Khan A, Pan X, Najeeb U, Tan DKY, Fahad S, Zahoor R, Luo H (2018) Coping with drought: stress and adaptive mechanisms, and management through cultural and molecular alternatives in cotton as vital constituents for plant stress resilience and fitness. Biol Res 51:47

Khan N, Ali S, Tariq H, Latif S, Yasmin H, Mehmood A, Shahid MA (2020) Water conservation and plant survival strategies of Rhizobacteria under drought stress. Agronomy 10(11):Article 11

Kumar M (2019) Recent advances in the drought stress tolerance in Rice

Kumar M, Kumar Patel M, Kumar N, Bajpai AB, Siddique KHM (2021) Metabolomics and molecular approaches reveal drought stress tolerance in plants. Int J Mol Sci 22(17):9108

Kuromori T, Fujita M, Takahashi F, Yamaguchi-Shinozaki K, Shinozaki K (2022) Inter-tissue and inter-organ signaling in drought stress response and phenotyping of drought tolerance. Plant J 109:342–358

Lauder J, Moran E, Hart SC (2019) Fight or flight? Potential tradeoffs between drought defense and reproduction in conifers. Tree Physiol 39(7):1071–1085

Lee C, Chronis D, Kenning C, Peret B, Hewezi T, Davis EL, Baum TJ, Hussey R, Bennett M, Mitchum MG (2011) The novel cyst nematode effector protein 19C07 interacts with the Arabidopsis auxin influx transporter LAX3 to control feeding site development. Plant Physiol 155(2):866–880

Lenk I, Fisher LHC, Vickers M, Akinyemi A, Didion T, Swain M, Jensen CS, Mur LAJ, Bosch M (2019) Transcriptional and metabolomic analyses indicate that cell wall properties are associated with drought tolerance in *Brachypodium distachyon*. Int J Mol Sci 20(7):Article 7

Leybourne D, Preedy K, Valentine T, Bos J, Karley A (2021) Drought has negative consequences on aphid fitness and plant vigor: insights from a meta-analysis. Ecol Evol 11(17):11915–11929

Leybourne D, Valentine T, Binnie K, Taylor A, Karley A, Bos J (2022) Drought stress increases the expression of barley defence genes with negative consequences for infesting cereal aphids. J Exp Bot 73(7):2238–2250

Li X, Xing X, Xu S, Zhang M, Wang Y, Wu H, Sun Z, Huo Z, Chen F, Yang T (2018) Genome-wide identification and functional prediction of tobacco lncRNAs responsive to root-knot nematode stress. PLoS One 13(11):e0204506

Li F, Duan T, Li Y (2020) Effects of the fungal endophyte *Epichloë festucae* var. Lolii on growth and physiological responses of perennial ryegrass cv. Fairway to combined drought and pathogen stresses. Microorganisms 8(12):1917

Li Y, Si Y-T, He Y-X, Li J-X (2021) Comparative analysis of drought-responsive and -adaptive genes in Chinese wingnut (*Pterocarya stenoptera* C. DC). BMC Genomics 22(1):155

Lin T, Tang J, Li S, Li S, Han S, Liu Y, Yang C, Chen G, Chen L, Zhu T (2023) Drought stress-mediated differences in phyllosphere microbiome and associated pathogen resistance between male and female poplars. Plant J 115(4):1100–1113

Liu X, Liu C (2016) Effects of drought-stress on Fusarium crown rot development in barley. PLoS One 11(12):e0167304

Liu Y, Wei X (2019) Dark septate endophyte improves drought tolerance of *Ormosia hosiei* Hemsley & E. H. Wilson by modulating root morphology, ultrastructure, and the ratio of root hormones. Forests 10(10):830

Liu F, Ma H, Du Z, Ma B, Liu X, Peng L, Zhang W (2019) Physiological response of North China red elder container seedlings to inoculation with plant growth-promoting rhizobacteria under drought stress. PLoS One 14(12):e0226624

Liu Q, Xie S, Zhao X, Liu Y, Xing Y, Dao J, Wei B, Peng Y, Duan W, Wang Z (2021) Drought sensitivity of sugarcane cultivars shapes rhizosphere bacterial community patterns in response to water stress. Front Microbiol 12:732989

Lozano YM, Aguilar-Trigueros CA, Roy J, Rillig MC (2021) Drought induces shifts in soil fungal communities that can be linked to root traits across 24 plant species. New Phytol 232(5):1917–1929

Ma J, Geng Y, Liu H, Zhang M, Liu S, Hao C, Hou J, Zhang Y, Zhang D, Zhang W (2023) TaTIP41 and TaTAP46 positively regulate drought tolerance in wheat by inhibiting PP2A activity. J Integr Plant Biol 65(9):2056–2070

MacLean AM, Sugio A, Makarova OV, Findlay KC, Grieve VM, Tóth R, Nicolaisen M, Hogenhout SA (2011) Phytoplasma effector SAP54 induces indeterminate leaf-like flower development in Arabidopsis plants. Plant Physiol 157(2):831–841

Maluta NKP, Fereres A, Lopes JRS (2017) Settling preferences of the whitefly vector *Bemisia tabaci* on infected plants varies with virus family and transmission mode. Entomol Exp Appl 165(2–3):138–147

Maurel C, Chrispeels MJ (2001) Aquaporins. A molecular entry into plant water relations. Plant Physiol 125(1):135–138

Mewis I, Khan M, Glawischnig E, Schreiner M, Ulrichs C (2012) Water stress and aphid feeding differentially influence metabolite composition in A*rabidopsis thaliana* (L.). PLoS One 7(11):e48661

Miao L, Li S, Zhang F, Chen T, Shan Y, Zhang Y (2020) Future drought in the dry lands of Asia under the 1.5 and 2.0 °C warming scenarios. Earth's Future 8(6):e2019EF001337

Mir RA, Bhat BA, Yousuf H, Islam ST, Raza A, Rizvi MA et al (2022) Multidimensional role of silicon to activate resilient plant growth and to mitigate abiotic stress. Front Plant Sci 13:819658

Mohi-Ud-Din M, Talukder D, Rohman M, Ahmed JU, Jagadish SVK, Islam T, Hasanuzzaman M (2021) Exogenous application of methyl Jasmonate and salicylic acid mitigates drought-induced oxidative damages in French Bean (*Phaseolus vulgaris* L.). Plan Theory 10(10):Article 10

Morales-Quintana L, Moya M, Santelices-Moya R, Cabrera-Ariza A, Rabert C, Pollmann S, Ramos P (2022) Improvement in the physiological and biochemical performance of strawberries under drought stress through symbiosis with Antarctic fungal endophytes. Front Microbiol 13:939955

Morcillo RJL, Vílchez JI, Zhang S, Kaushal R, He D, Zi H, Liu R, Niehaus K, Handa AK, Zhang H (2021) Plant transcriptome reprograming and bacterial extracellular metabolites underlying tomato drought resistance triggered by a beneficial soil bacteria. Meta 11(6):Article 6

Moreira X, Granjel RR, de la Fuente M, Fernández-Conradi P, Pasch V, Soengas P, Turlings TCJ, Vázquez-González C, Abdala-Roberts L, Rasmann S (2021) Apparent inhibition of induced plant volatiles by a fungal pathogen prevents airborne communication between potato plants. Plant Cell Environ 44(4):1192–1201

Moreno-Galván A, Romero-Perdomo FA, Estrada-Bonilla G, Meneses CHSG, Bonilla RR (2020) Dry-Caribbean *Bacillus* spp. strains ameliorate drought stress in maize by a strain-specific antioxidant response modulation. Microorganisms 8(6):823

Muhammad Aslam M, Okal E, Idris A, Zhang Q, Xu W, Jk K, Wani S, Yuan W (2022) Rhizosphere microbiomes can regulate plant drought tolerance. Pedosphere 32:61–74

Muhammad Rashid MAS, Seema Mahmood KHS (2018) 17. Study of adverse effects of drought stress on two different hybrids of maize (*Zea mays* L.). Pure Appl Biol (PAB) 7(4):Article 4

Nachappa P, Culkin CT, Saya PM, Han J, Nalam VJ (2016) Water stress modulates soybean aphid performance, feeding behavior, and virus transmission in soybean. Front Plant Sci 7

Nadeem M, Li J, Yahya M, Sher A, Ma C, Wang X, Qiu L (2019) Research progress and perspective on drought stress in legumes: a review. Int J Mol Sci 20(10):Article 10

Nagatoshi Y, Ikazaki K, Kobayashi Y, Mizuno N, Sugita R, Takebayashi Y, Kojima M, Sakakibara H, Kobayashi NI, Tanoi K, Fujii K, Baba J, Ogiso-Tanaka E, Ishimoto M, Yasui Y, Oya T, Fujita Y (2021) Phosphate starvation response precedes abscisic acid response under progressive mild drought in plants. https://doi.org/10.1101/2021.07.28.453724

Nakashima K, Yamaguchi-Shinozaki K, Shinozaki K (2014) The transcriptional regulatory network in the drought response and its crosstalk in abiotic stress responses including drought, cold, and heat. Front Plant Sci 5

Nardino M, Perin E, Aranha B, Carpes S, Fontoura B, Sousa D et al (2022) Understanding drought response mechanisms in wheat and multi-trait selection. PLoS One 17(4):e0266368

Nasir MW, Toth Z (2022) Effect of drought stress on potato production: a review. Agronomy 12(3):Article 3

Nguyen D, D'Agostino N, Tytgat TOG, Sun P, Lortzing T, Visser EJW, Cristescu SM, Steppuhn A, Mariani C, Van Dam NM, Rieu I (2016a) Drought and flooding have distinct effects on herbivore-induced responses and resistance in *Solanum dulcamara*. Plant Cell Environ 39(7):1485–1499

Nguyen D, Rieu I, Mariani C, Dam N (2016b) How plants handle multiple stresses: hormonal interactions underlying responses to abiotic stress and insect herbivory. Plant Mol Biol 91(6):727–740

Nguyen LV, Tran NMT, Nguyen LV, Phan HNT, Rumanzi MS, Pham CV, Tang HT (2021) Variation in root development response of napier grass to drought stress. J Ecol Eng 22(8):64–74

Nishu SD, No JH, Lee TK (2022) Transcriptional response and plant growth promoting activity of *Pseudomonas fluorescens* DR397 under drought stress conditions. Microbiol Spectr 10(4):e00979–e00922

Nurmberg PL, Brito A, Zimmer CM, Schwantes DO, Mantovani EE, Bosetti F, Rotundo J, Truong SK, Borras L, De La Vega A, Messina C, McCormick RF (2022) A commercial breeding perspective of maize improvement for drought stress tolerance. Revista Brasileira de Milho e Sorgo 21. https://doi.org/10.18512/rbms2022v21e1277

Ochoa-Hueso R, Collins SL, Delgado-Baquerizo M, Hamonts K, Pockman WT, Sinsabaugh RL, Smith MD, Knapp AK, Power SA (2018) Drought consistently alters the composition of soil fungal and bacterial communities in grasslands from two continents. Glob Chang Biol 24(7):2818–2827

Olatunji OA, Luo H, Pan K, Tariq A, Sun X, Chen W, Wu X, Zhang L, Xiong Q, Li Z, Song D, Zhang A, Sun F (2018) Influence of phosphorus application and water deficit on the soil microbiota of N 2 -fixing and non-N-fixing tree. Ecosphere 9(6):e02276

Ollas C, Dodd I (2016) Physiological impacts of aba–ja interactions under water-limitation. Plant Mol Biol 91(6):641–650

Pal G, Bakade R, Deshpande S, Sureshkumar V, Patil SS, Dawane A, Agarwal S, Niranjan V, PrasannaKumar MK, Vemanna RS (2022) Transcriptomic responses under combined bacterial blight and drought stress in rice reveal potential genes to improve multi-stress tolerance. BMC Plant Biol 22(1):349

Pan Y, Kang P, Tan M, Hu J, Zhang Y, Zhang J, Song N, Li X (2022) Root exudates and rhizosphere soil bacterial relationships of *Nitraria tangutorum* are linked to k-strategists bacterial community under salt stress. Front Plant Sci 13:997292

Pandey P, Sinha R, Mysore KS, Senthil-Kumar M (2015) Impact of concurrent drought stress and pathogen infection on plants. In: Mahalingam R (ed) Combined stresses in plants. Springer International Publishing, pp 203–222

Pandey P, Irulappan V, Bagavathiannan MV, Senthil-Kumar M (2017) Impact of combined abiotic and biotic stresses on plant growth and avenues for crop improvement by exploiting physio-morphological traits. Front Plant Sci 8

Panth M, Hassler SC, Baysal-Gurel F (2020) Methods for management of soilborne diseases in crop production. Agriculture 10(1):16

Patharkar OR, Gassmann W, Walker JC (2017) Leaf shedding as an anti-bacterial defense in Arabidopsis cauline leaves. PLoS Genet 13(12):e1007132

Per TS, Khan NA, Reddy PS, Masood A, Hasanuzzaman M, Khan MIR, Anjum NA (2017) Approaches in modulating proline metabolism in plants for salt and drought stress tolerance: Phytohormones, mineral nutrients and transgenics. Plant Physiol Biochem 115:126–140

Pereira EC, Zabalgogeazcoa I, Arellano JB, Ugalde U, Vázquez de Aldana BR (2023) *Diaporthe atlantica* enhances tomato drought tolerance by improving photosynthesis, nutrient uptake and enzymatic antioxidant response. Front Plant Sci 14

Pieterse C, León-Reyes A, Ent S, Wees S (2009) Networking by small-molecule hormones in plant immunity. Nat Chem Biol 5(5):308–316

Ponce OP, Torres Y, Prashar A, Buell R, Lozano R, Orjeda G, Compton L (2022) Transcriptome profiling shows a rapid variety-specific response in two Andigenum potato varieties under drought stress. Front Plant Sci 13

Porras MF, Navas CA, Marden JH, Mescher MC, De Moraes CM, Pincebourde S, Sandoval-Mojica A, Raygoza-Garay JA, Holguin GA, Rajotte EG, Carlo TA (2020) Enhanced heat tolerance of viral-infected aphids leads to niche expansion and reduced interspecific competition. Nat Commun 11(1):1184

Poudel M, Mendes R, Costa LAS, Bueno CG, Meng Y, Folimonova SY, Garrett KA, Martins SJ (2021) The role of plant-associated bacteria, fungi, and viruses in drought stress mitigation. Front Microbiol 12

Qurrahmah ANF, Mutfidasari U, Aisyah PI, Butar-Butar IPP, Rosadi I, Oktavianingsih L (2022) Severity analysis of fruit plant diseases based on leaf symptoms using ImageJ software. BIOMA Jurnal Ilmiah Biologi 11(2)

Ra R, Rb P, Tc P (2021) Detection and classification of diseases in cucumber leaves. In: Smart Intelligent Computing and Communication Technology, vol 38. IOS Press, Amsterdam, p 30

Rabara RC, Tripathi P, Rushton PJ (2014) The potential of transcription factor-based genetic engineering in improving crop tolerance to drought. Omics J Integr Biol 18(10):601–614

Rabara RC, Tripathi P, Choudhary MK, Timko MP, Shen QJ, Rushton PJ (2015) Transcriptome profiling of tobacco under water deficit conditions. Genomic Data 5:61–63

Ramegowda V, Senthil-Kumar M, Ishiga Y, Kaundal A, Udayakumar M, Mysore K (2013) Drought stress acclimation imparts tolerance to *Sclerotinia sclerotiorum* and P*seudomonas syringae* in *Nicotiana benthamiana*. Int J Mol Sci 14(5):9497–9513

Rering CC, Franco JG, Yeater KM, Mallinger RE (2020) Drought stress alters floral volatiles and reduces floral rewards, pollinator activity, and seed set in a global plant. Ecosphere 11(9)

Rizhsky L, Liang H, Shuman J, Shulaev V, Davletova S, Mittler R (2004) When defense pathways collide. The response of Arabidopsis to a combination of drought and heat stress. Plant Physiol 134(4):1683–1696

Salehi-Lisar SY, Bakhshayeshan-Agdam H (2020) Agronomic crop responses and tolerance to drought stress. In: Hasanuzzaman M (ed) Agronomic crops: volume 3: stress responses and tolerance. Springer, pp 63–91

Sallam A, Alqudah AM, Dawood MFA, Baenziger PS, Börner A (2019) Drought stress tolerance in wheat and barley: advances in physiology, breeding and genetics research. Int J Mol Sci 20(13):Article 13

Sam TT, Nhi PTT, Mai NTH, Linh DQ, Loi PT (2023) The climate change effects on agricultural drought in the be River Basin. IOP Conf Ser Earth Environ Sci 1170(1):012006

Samarah NH (2005) Effects of drought stress on growth and yield of barley. Agron Sustain Dev 25(1):145–149

Santos B, Pinto G, Berenguer H, Correia B, Amaral J, Gómez-Cadenas A, Alves A (2024) Physiological and hormonal responses of drought-stressed eucalyptus seedlings infected with *Neofusicoccum kwambonambiense*. Plant Pathol 73(4):810–823

Santos-Medellín C, Edwards J, Liechty Z, Nguyen B, Sundaresan V (2017) Drought stress results in a compartment-specific restructuring of the Rice root-associated microbiomes. mBio 8(4):e00764–e00717

Saud S, Li X, Chen Y, Zhang L, Fahad S, Hussain S, Sadiq A, Chen Y (2014) Silicon application increases drought tolerance of Kentucky bluegrass by improving plant water relations and morphophysiological functions. Sci World J 2014:1–10

Scavuzzo-Duggan T, Varoquaux N, Madera M, Vogel JP, Dahlberg J, Hutmacher R, Belcher M, Ortega J, Coleman-Derr D, Lemaux P, Purdom E, Scheller HV (2021) Cell wall compositions of *Sorghum bicolor* leaves and roots remain relatively constant under drought conditions. Front Plant Sci 12

Sharma A, Zheng B (2019) Melatonin mediated regulation of drought stress: physiological and molecular aspects. Plan Theory 8(7):190

Sharma P, Sharma A, Sharma M, Bhalla N, Estrela P, Jain A, Thakur P, Thakur A (2017) Nanomaterial fungicides: in vitro and in vivo antimycotic activity of cobalt and nickel nanoferrites on phytopathogenic fungi. Global Chall 1(9):1700041

Shemi R, Wang R, Gheith EMS, Hussain HA, Hussain S, Irfan M, Cholidah L, Zhang K, Zhang S, Wang L (2021) Effects of salicylic acid, zinc and glycine betaine on morpho-physiological growth and yield of maize under drought stress. Sci Rep 11(1):3195

Shi S, Li S, Asim M, Mao J, Xu D, Ullah Z et al (2018) The Arabidopsis calcium-dependent protein kinases (CDPKs) and their roles in plant growth regulation and abiotic stress responses. Int J Mol Sci 19(7):1900

Shinozaki K, Yamaguchi-Shinozaki K (2022) Functional genomics in plant abiotic stress responses and tolerance: from gene discovery to complex regulatory networks and their application in breeding. Proc Jpn Acad Ser B 98(8):470–492

Siebielec S, Siebielec G, Klimkowicz-Pawlas A, Gałązka A, Grządziel J, Stuczyński T (2020) Impact of water stress on microbial community and activity in Sandy and Loamy soils. Agronomy 10(9):1429

Singh GM, Gosavi G, Patel SS, Solanki PS, Zhang F, Xu J, Mishra VK, Sharma S (2022) Impact of drought on maize yield and exploration of in-situ maize crop genetic resources for drought tolerance. https://doi.org/10.20944/preprints202212.0210.v1

Sinha R, Gupta A, Senthil-Kumar M (2016) Understanding the impact of drought on foliar and xylem invading bacterial pathogen stress in chickpea. Front Plant Sci 7

Sinha R, Gupta A, Senthil-Kumar M (2017) Concurrent drought stress and vascular pathogen infection induce common and distinct transcriptomic responses in chickpea. Front Plant Sci 8

Sinha R, Irulappan V, Mohan-Raju B, Suganthi A, Senthil-Kumar M (2019) Impact of drought stress on simultaneously occurring pathogen infection in field-grown chickpea. Sci Rep 9(1):5577

Srinivas C, Devi DN, Murthy KN, Mohan CD, Lakshmeesha TR, Singh B, Kalagatur NK, Niranjana SR, Hashem A, Alqarawi AA (2019) *Fusarium oxysporum* f. Sp. Lycopersici causal agent of vascular wilt disease of tomato: biology to diversity–A review. Saudi J Biol Sci 26(7):1315–1324

Sudiana IM, Chandra ND, Mangunwardoyo W, Kanti A, Napitupulu TP, Idris I, Sumerta IN (2022) Effects of *Funneliformis mosseae* inoculation on chili pepper growth under repeated drought stress. Jurnal Teknologi 84(6):Article 6

Suljović A, Čakić S, Popović T, Šandi S (2022). Detection of Plant Diseases Using Leaf Images and Machine Learning. 2022 21st International Symposium INFOTEH-JAHORINA (INFOTEH), 1–4

Sun J, Qiu C, Ding Y, Wang Y, Sun L, Fan K et al (2020) Fulvic acid ameliorates drought stress-induced damage in tea plants by regulating the ascorbate metabolism and flavonoids biosynthesis. BMC Genomics 21(1):411

Sunarti S, Kissoudis C, Van Der Hoek Y, Van Der Schoot H, Visser RGF, Van Der Linden CG, Van De Wiel C, Bai Y (2022) Drought stress interacts with powdery mildew infection in tomato. Front Plant Sci 13

Sungmin O, Park SK (2023) Flash drought drives rapid vegetation stress in arid regions in Europe. Environ Res Lett 18(1):014028

Susič N, Žibrat U, Širca S, Strajnar P, Razinger J, Knapič M, Vončina A, Urek G, Gerič Stare B (2018) Discrimination between abiotic and biotic drought stress in tomatoes using hyperspectral imaging. Sensors Actuators B Chem 273:842–852

Suzuki N, Rivero RM, Shulaev V, Blumwald E, Mittler R (2014) Abiotic and biotic stress combinations. New Phytol 203(1):32–43

Swann ALS, Hoffman FM, Koven CD, Randerson JT (2016) Plant responses to increasing CO_2 reduce estimates of climate impacts on drought severity. Proc Natl Acad Sci 113(36):10019–10024

Swift JF, Kolp MR, Carmichael A, Ford NE, Hansen PM, Sikes BA, Kleiner M, Wagner MR (2023). Environmental legacy effects impact maize growth and microbiome assembly under drought stress. https://doi.org/10.1101/2023.04.11.536405

Szczepaniec A, Finke D (2019) Plant-vector-pathogen interactions in the context of drought stress. Front Ecol Evol 7. https://doi.org/10.3389/fevo.2019.00262

Tabari H, Willems P (2022) Trivariate analysis of changes in drought characteristics in the CMIP6 multimodel ensemble at global warming levels of 1.5, 2, and 3 C. J Clim 35(18):5823–5837

Tabata R, Sawa S (2014) Maturation processes and structures of small secreted peptides in plants. Front Plant Sci 5. https://doi.org/10.3389/fpls.2014.00311

Tang Y, Bao X, Zhi Y, Wu Q, Guo Y, Yin X, Zeng L, Li J, Zhang J, He W, Liu W, Wang Q, Jia C, Li Z, Liu K (2019) Overexpression of a MYB family gene, OsMYB6, increases drought and salinity stress tolerance in transgenic Rice. Front Plant Sci 10:168

Ullah A, Manghwar H, Shaban M, Khan AH, Akbar A, Ali U et al (2018) Phytohormones enhanced drought tolerance in plants: a coping strategy. Environ Sci Pollut Res 25:33103–33118

University of Belgrade, Faculty of Agriculture, RS, Petrović I (2019) Two-step RT-qPCR analysis of expression of 7 drought-related genes in tomato (Lycopersicon esculentum Mill.). In Faculty of Agriculture, University of Belgrade, RS, B. Vucelić-Radović, D. Lazić, Faculty of Agriculture, University of Belgrade, RS, M. Nikšić, & Faculty of Agriculture, University of Belgrade, RS, Application of Molecular Methods and Raman Microscopy/Spectroscopy in Agricultural Sciences and Food Technology. Ubiquity Press, pp 3–14. https://doi.org/10.5334/bbj.a

van der Weijde T, Huxley LM, Hawkins S, Sembiring EH, Farrar K, Dolstra O, Visser RGF, Trindade LM (2017) Impact of drought stress on growth and quality of miscanthus for biofuel production. GCB Bioenergy 9(4):770–782

Vasquez DF, Hernandez A, Torres D, Borrero-Echeverry F, Zuluaga P, Rincon DF (2022) Drought as a modulator of plant–virus–vector interactions: effects on symptom expression, plant immunity and vector behaviour. Plant Pathol 71(6):1282–1292

Veach AM, Chen H, Yang ZK, Labbe AD, Engle NL, Tschaplinski TJ, Schadt CW, Cregger MA (2020) Plant hosts modify belowground microbial community response to extreme drought. mSystems 5(3):e00092–e00020

Venuprasad R, Lafitte HR, Atlin GN (2007) Response to direct selection for grain yield under drought stress in Rice. Crop Sci 47(1):285–293

Vurukonda SSKP, Vardharajula S, Shaik Zulfikar A (2022) In vitro assessment of pseudomonas sp. strain FCBB-2 for effective plant growth promotion and antifungal activity under drought stress. Asian Plant Res J:13–27

Walters DR, Bingham IJ (2007) Influence of nutrition on disease development caused by fungal pathogens: implications for plant disease control. Ann Appl Biol 151(3):307–324

Wang X, Cai X, Xu C, Wang Q, Dai S (2016) Drought-responsive mechanisms in plant leaves revealed by proteomics. Int J Mol Sci 17(10):Article 10

Wang Y., Morcillo RJL, Yu G, Zhao A, Xue H, Rufian JS, Sang, Y., & Macho, A. P. (2020). A bacterial effector uncovers a metabolic pathway involved in resistance to bacterial wilt disease. https://doi.org/10.1101/2020.09.22.307595

Wang X, Li X, Dong S (2022) Screening and identification of drought tolerance of spring soybean at seedling stage under climate change. Front Sustainable Food Syst 6

Wilschut RA, Van Kleunen M (2021) Drought alters plant-soil feedback effects on biomass allocation but not on plant performance. Plant Soil 462(1–2):285–296

Wojakowska A, Kułak K, Jasiński M, Kachlicki P, Stawiński S, Stobiecki M (2015) Metabolic response of narrow leaf lupine (*Lupinus angustifolius*) plants to elicitation and infection with *Colletotrichum lupini* under field conditions. Acta Physiol Plant 37(8):152

Xie J, Dawwam GE, Sehim AE, Li X, Wu J, Chen S, Zhang D (2021) Drought stress triggers shifts in the root microbial community and alters functional categories in the microbial Gene Pool. Front Microbiol 12:744897

Xiong H, Yu J, Miao J, Li J, Zhang H, Wang X et al (2018) Natural variation in OsLG3 increases drought tolerance in rice by inducing ROS scavenging. Plant Physiol 178(1):451–467

Xu L, Chen N, Zhang X (2019) Global drought trends under 1.5 and 2 °C warming. Int J Climatol 39(4):2375–2385

Xu L, Dong Z, Chiniquy D, Pierroz G, Deng S, Gao C, Diamond S, Simmons T, Wipf HM-L, Caddell D, Varoquaux N, Madera MA, Hutmacher R, Deutschbauer A, Dahlberg JA, Guerinot ML, Purdom E, Banfield JF, Taylor JW et al (2021) Genome-resolved metagenomics reveals role of iron metabolism in drought-induced rhizosphere microbiome dynamics. Nat Commun 12(1):3209

Yadav B, Jogawat A, Gnanasekaran P, Kumari P, Lakra N, Lal SK et al (2021) An overview of recent advancement in phytohormones-mediated stress management and drought tolerance in crop plants. Plant Gene 25:100264

Yamaguchi-Shinozaki K, Shinozaki K (2006) Transcriptional regulatory networks in cellular responses and tolerance to dehydration and cold stresses. Annu Rev Plant Biol 57(1):781–803

Yan J, Aznar A, Chalvin C, Birdseye DS, Baidoo EEK, Eudes A, Shih PM, Loqué D, Zhang A, Scheller HV (2018) Increased drought tolerance in plants engineered for low lignin and low xylan content. Biotechnol Biofuels 11(1):195

Yang B, Yang S, Guo B, Wang Y, Zheng W, Tian M et al (2021) The phytophthora effector AVH241 interacts with host NDR1-like proteins to manipulate plant immunity. J Integr Plant Biol 63(7):1382–1396

Yang Y, Gu M, Lu J, Li X, Liu D, Li X, Wang L (2023) GR24 alleviates the adverse effects of drought stress on physiology and photosystem II function in alfalfa (*Medicago sativa* L.). Grassl Sci 69(2):113–119

Yao C, Li X, Li Y, Yang G, Liu W, Shao B, Zhong J, Huang P, Han D (2022) Overexpression of a *Malus baccata* MYB transcription factor gene MbMYB4 increases cold and drought tolerance in Arabidopsis thaliana. Int J Mol Sci 23(3):1794

Ye H, Liu S, Tang B, Chen J, Xie Z, Nolan T et al (2017) Rd26 mediates crosstalk between drought and brassinosteroid signalling pathways. Nature. Communications 8(1)

Yihdego Y, Salem HS, Muhammed HH (2019) Agricultural Pest management policies during drought: case studies in Australia and the State of Palestine. Nat Hazard Rev 20(1):05018010

Yoon Y, Seo DH, Shin H, Kim HJ, Kim CM, Jang G (2020) The role of stress-responsive transcription factors in modulating abiotic stress tolerance in plants. Agronomy 10(6):788

Yousaf B, Liu G, Wang R, Zia-ur-Rehman M, Rizwan MS, Imtiaz M, Murtaza G, Shakoor A (2016) Investigating the potential influence of biochar and traditional organic amendments on the bioavailability and transfer of Cd in the soil–plant system. Environ Earth Sci 75:1–10

Zandalinas SI, Mittler R, Balfagón D, Arbona V, Gómez-Cadenas A (2018) Plant adaptations to the combination of drought and high temperatures. Physiol Plant 162(1):2–12

Zhang X, Goatley M, Wu W, Ervin E, Shang C (2019) Drought-induced injury is associated with hormonal alteration in Kentucky bluegrass. Plant Signal Behav 14(10):e1651607

Zhao M, Han Y, Feng Y, Li F, Wang W (2012) Expansins are involved in cell growth mediated by abscisic acid and indole-3-acetic acid under drought stress in wheat. Plant Cell Rep 31(4):671–685

Zhao B, Shao Z, Wang L, Zhang F, Chakravarty D, Zong W, Dong J, Song L, Qiao H (2022) MYB44-ENAP1/2 restricts HDT4 to regulate drought tolerance in Arabidopsis. PLoS Genet 18(11):e1010473

Zhou G, Zhou X, Liu R, Du Z, Zhou L, Li S, Liu H, Shao J, Wang J, Nie Y, Gao J, Wang M, Zhang M, Wang X, Bai SH (2020) Soil fungi and fine root biomass mediate drought-induced reductions in soil respiration. Funct Ecol 34(12):2634–2643

Zia R, Nawaz MS, Siddique MJ, Hakim S, Imran A (2021) Plant survival under drought stress: implications, adaptive responses, and integrated rhizosphere management strategy for stress mitigation. Microbiol Res 242:126626

Zou C, Guo Z, Zhao S, Chen J, Zhang C, Han H (2023) Genome-wide analysis of long non-coding RNAs in sugar beet (*Beta vulgaris* L.) under drought stress. Front Plant Sci 14:1118011

Zuluaga A, Bidzinski P, Chanclud E, Ducasse A, Cayrol B, Selvaraj M et al (2020) The rice DNA-binding protein ZBED controls stress regulators and maintains disease resistance after a mild drought. Front Plant Sci 11

Zvereva A, Golyaev V, Turco S, Gubaeva E, Rajeswaran R, Schepetilnikov M, Pooggin M (2016) Viral protein suppresses oxidative burst and salicylic acid-dependent autophagy and facilitates bacterial growth on virus-infected plants. New Phytol 211(3):1020–1034

The Impact of Drought Stress on Nutrient and Heavy Metal Uptake by Vegetables

Hamza Yousaf, Muhammad Ishfaq, Asma Zafar, Sana Yaqoob, Dua E. Zainab, Azra Parveen, Muhammad Sarfraz, and Syed Azhar Hussain

Abstract Drought stress poses a significant threat to agricultural productivity, especially in vegetable production. Given the importance of the severity and duration of drought, it is likely that crop losses from drought will be greater than losses from any other stress. It affects the uptake, transportation, and distribution of nutrients, which in turn hampers plant growth and ultimately, production. Moreover, drought also facilitates the absorption of heavy metals, leading to contamination and devastating yield losses. It is widely recognized that the application of additional nutrients will not promote plant growth if the soil already contains sufficient amounts and the drought is severe, although there is inconsistent data on the impact of nutrient addition on plant growth under drought conditions. A better understanding of the role of mineral nutrients in plant tolerance to drought will contribute to improved fertilizer management in arid and semi-arid environments, as well as in regions experiencing short-term droughts. Understanding the impact of water stress on nutrient uptake and assimilation is crucial, particularly in semi-arid regions where vegetable production is prevalent. This chapter provides an overview of the

H. Yousaf (✉)
College of Resources and Environmental Sciences, China Agricultural University, Beijing, People's Republic of China

M. Ishfaq
College of Life Science and Oceanography, Shenzhen University, Shenzhen, China

A. Zafar · D. E. Zainab
College of Natural Resources and Environment, Northwest A&F University, Yangling, China

S. Yaqoob
Institute of Soil and Environmental Science, University of Agriculture, Faisalabad, Pakistan

A. Parveen
Department of Soil and Environmental Sciences, Faculty of Agricultural Sciences, Ghazi University, Dera Ghazi Khan, Pakistan

M. Sarfraz
Soil Salinity Research Institute, Pindi Bhattian, Punjab, Pakistan

S. A. Hussain
Soil and water testing lab for research, Sargodha, Punjab, Pakistan

U. K. Chaudhry et al. (eds.), *Drought Stress*,
https://doi.org/10.1007/978-3-031-80610-0_7

impact of drought on the nutrition of vegetables in drought-vulnerable regions, with a particular focus on how drought affects nutrient availability, the absorption of heavy metals, transportation, and accumulation in vegetables, as well as the interaction between nutrient supply and drought response. Additionally, this chapter discusses various drought management strategies to enhance nutrient availability for vegetables.

Keywords Drought · Vegetables · Nutrient · Heavy metals

1 Drought Stress and Climatic Changes

Plants grow and develop under a variety of environmental conditions in both natural and agricultural settings. Drought is one of the most severe circumstances that has an adverse effect on plant output. Water is critical to many key plant functions, including the maintenance of physiological function, which is necessary for healthy growth and development (Abbasi and Abbasi 2010). Consequently, drought results from a reduction in the amount of water available, either momentarily or throughout the course of the plant's growth period, especially in arid climatic zones worldwide (Diatta et al. 2020). Drought has historically resulted in terrible famines and constitutes a serious threat to future global food security (Okorie et al. 2019). The worrying global population increase and rising food demand compound the effects of drought on agriculture (O'Connell 2017). Drought is unpredictable due to a number of causes, such as altered rainfall patterns, increased evapotranspiration, and a lack of water in the rhizosphere (Passioura and Angus 2010).

A condition known as physiological drought, often called pseudo-drought, occurs when plants are unable to take in water from the soil. Even in the event that the root zone has enough moisture (Daryanto et al. 2017). Every year, the repercussions of climate change such as increasing temperatures, altered precipitation patterns, and unprecedented natural disasters become more apparent and have a negative impact on significant human endeavors. These climate shifts will unavoidably have an impact on crop yield and quality (Dong et al. 2024). Crop productivity would drastically decrease as a result of these climate disruptions, and drought events are predicted to occur more frequently with higher air evaporation losses (Teuling et al. 2013). Presently, irregular precipitation patterns are causing an annual increase in the frequency of drought occurrences, and temperature increases are also a contributing factor to climate change.

Recent studies by the IPCC (Intergovernmental Panel on Climate Change) have stated that climate change is already affecting both natural and human systems. This includes temperatures rising, patterns of precipitation changes, and an increased occurrence of severe weather-related incidents. Additionally, land degradation and a rise in the number of displaced people are caused by climate change (Allen et al. 2018; Zommers et al. 2016).

Right now, 971 million individuals are estimated to be employed in areas where there is a high or exceptional risk of climate-related hazards. Among those most vulnerable to the effects of climate change are women, children, the elderly, and the impoverished, with many of them residing in least-developed countries (Allen et al. 2018). Pakistan, for instance, ranks ninth among countries that are grappling with climate change. Research has mainly concentrated on evaluating the effects of climate change rather than examining how people perceive and feel it, even though people all over the world have different viewpoints and experiences with its repercussions (Ayeb-Karlsson et al. 2016).

In the past, from 1999 to 2018, Pakistan came in fifth place in the world for the most extreme weather (Eckstein et al. 2019). Rehman et al. (2015) state that Pakistan's economy is heavily dependent on agriculture. Alterations in temperature, water accessibility, or monsoon behaviors can unleash catastrophic impacts on the sustenance of countless individuals. Extreme weather events and climate change only worsen the country's existing issues with food security and poverty. Between 1998 and 2018, Pakistan encountered 152 instances of severe weather phenomena., resulting in 9989 deaths and $3.8 billion in economic losses (Abubakar 2020). Numerous studies have been conducted on how climate change has affected Pakistan's food security (Ahmed and Schmitz 2011). These studies indicate that Pakistan is currently grappling with the repercussions of climate change, evident in the form of escalating temperatures, unpredictable rainfall patterns, glacier retreats, and declining agricultural output. Nevertheless, there remains a dearth of understanding regarding how the population of Pakistan perceives climate change and extreme weather phenomena. Understanding the impact of these phenomena on their daily lives and means of sustenance, as well as the strategies they employ to cope, remains largely unexplored. Additionally, there is scant information available regarding the challenges and hindrances individuals encounter while endeavoring to adjust to both gradual shifts and abrupt alterations in climate conditions.

There is a substantial amount of material on how people are affected by climate change. By using a narrative-based, people-centered approach to illuminate the everyday experiences of Pakistani residents who are directly impacted by climate change, this chapter seeks to close this gap. The goal is to encourage collaboration and coordinated actions, acknowledging that the effects of climate change transcend disciplinary and sectoral boundaries, in addition to language (Moezzi et al. 2017).

Global climate change is predicted to pick up speed in the future due to rising air temperatures and atmospheric CO_2 levels that disrupt rainfall patterns (Yang et al. 2019). A significant body of literature has been written about how humans are affected by climate change. The increased evaporation from the soil, caused by intense heat events and dry winds, leads to a loss of available moisture. This can worsen existing drought conditions, even though the primary factor contributing to the region's heightened drought stress is usually a lack of rainfall (Cohen et al. 2021). Drought stress and climate change are global issues. Furthermore, other stresses like heat stress, salt stress, and combination stress all of which are thought to be important contributors to plant water deficit are made worse by drought stress (Adnan et al. 2020).

Some effects of climate change may be beneficial to agricultural productivity. For instance, higher rates of photosynthesis have been associated with increased CO_2, indicating that high environmental CO_2 concentrations may lead to higher yields in the future (Brown et al. 2018). However, climate variability generally has a negative impact on both agricultural and natural environments (Ishfaq et al. 2022). The melting of glaciers due to the air temperature rising and flooding of agricultural fields that are flat or have little slope (Cook et al. 2014). Moreover, the disappearance of glaciers results in smaller water reservoirs, limiting the amount of available water for crops. The state of affairs is getting worse. Global warming has resulted in a decline in the amount of cumulative yearly precipitation in many rain-fed agricultural regions worldwide (Warner and Afifi 2014). Loss of soil water content and plant water content are further effects of global warming. Global warming raises temperatures, which causes plants to lose a large quantity of their internal water to the atmosphere. This exacerbates water scarcity problems that already exist in many agricultural systems across the world (Sultan et al. 2019). A fifth of the world's population may experience severe water shortages if air temperatures rise by the expected 2 °C (Ray et al. 2019).

Rain-fed regions are more vulnerable to losses in food production due to their reliance on rainfall events, as opposed to those that are irrigated by canals, rivers, or water channels (Konapala et al. 2020). As such, the connectedness of drought occurrences in various regions is closely linked to the amount of rainfall they receive each year. Thus, there is a good possibility that we will encounter water stress at specific times (Konapala et al. 2020). Industrialization, deforestation, and urbanization are further human activities that cause disturbances in rainfall patterns (Fatima et al. 2020).

In Pakistan, the early spring and winter have more erratic rainfall, but the early fall and summer have little to no rainfall, leading to more frequent dry and hot seasons. The detrimental effects of drought stress on plant growth and maturation are intensified by the increased atmospheric water requirement for vegetation, increased rates of transpiration and evaporation, and decreased summer rainfall. However, rainfall distribution and intensity within and between years remain the primary determinants of plant water availability and drought situations (Karandish and Šimůnek 2017).

Globally, the classification of monsoon is known as a significant summer main source of rainfall. It is primarily driven by temperature (Ali and Anjum 2016). If current trends persist, rain-fed regions should anticipate a 70% decrease by the onset of the third century, (Winston et al. 2013). Estimates advise that the concentration of CO_2 rising in the environmental atmosphere will result in rainfall increased, which will have detrimental effects on agricultural output, lead to widespread flooding, and cause substantial economic losses in densely populated agricultural nations (Reddy 2015).

In these conditions, changes in the amount, frequency, and length of monsoon rainfall still affect the moisture levels in the rhizosphere. As a result, plant productivity in specific regions of the world is also affected. It is surprising to note that two-thirds of the global population is currently facing food insecurity, largely due to significant changes in dry and wet seasonal rainfalls resulting from shifts in monsoon patterns (Aryal et al. 2020).

Climate change has a direct impact on precipitation and weather patterns. Surface evaporation increases in tandem with temperature, resulting in elevated levels of atmosphere water vapor and more intense downpours. This can have significant consequences in mountainous areas of rainfall (Trenberth 2011). Conversely, the increased evaporation resulting in the phenomenon of global warming has intensified potential drought conditions in regions of the desert.

2 Combined Drought and Heavy Metal Stress

Human activities like mining, industrialization, automobile emissions, sewage water, and the careless use of chemical fertilizers are the main causes of the rise in heavy metal exposure among agricultural plants (Akhter et al. 2022). Of all the heavy metals, toxicity to cadmium poses the greatest concern worldwide and is producing worrying circumstances (El Rasafi et al. 2022). In normal growing settings, plants are often subjected to a combination of abiotic stressors that adversely affect their growth and cause severe yield losses. One of the main limiting factors that makes the effects of heavy metal contamination worse and puts more stress on plants is drought. Drought exceeds the threshold limits of certain metals (Fig. 1.)

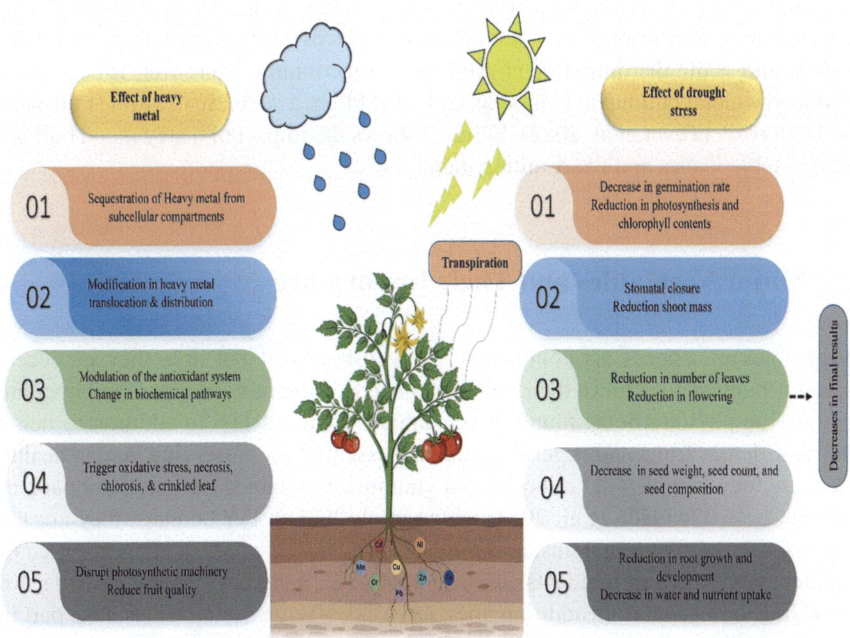

Fig. 1 The combined impact of drought stress and heavy metal contamination on vegetables

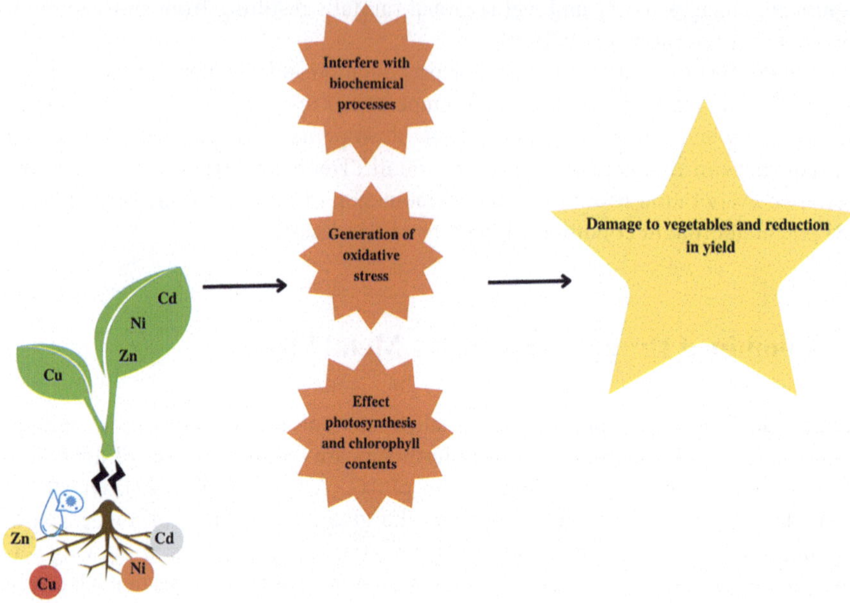

Fig. 2 Impact of drought stress in triggering the heavy metal contamination in vegetables causing changes in functioning and leading to poor yield

required by plants throughout their regular growth cycle, hence impairing their functionality. The absorption of heavy metals induces oxidative bursts, growth inhibition, and acute disruption of physiological functioning. Moreover, it is obvious that heavy metal buildup in plants causes leaf folding, a decrease in nitrate transport, and chlorosis (Tuver et al. 2022). Figure 2 shows the impact of heavy metal buildup in vegetable leaves and the resulting damages.

3 Major Vegetables and Their Importance

Vegetables are a necessary component of a well-balanced diet since they are necessary for humans to achieve their nutritional needs. Minerals, vitamins, and fiber are among the phytonutrients found in vegetables (Ryder 2011). In addition to being nutrient-dense, fruits and vegetables are also essential for preserving human health. Because they include vital minerals and vitamins, vegetables are important parts of human diets (Owolade et al. 2017; Ishfaq et al. 2023a). For humans, they are the main source of fiber, vitamins, and minerals. Vegetables are also a great source of carotene and other antioxidants like vitamins A, C, and E, which are essential for good health in general. In addition, they serve as cofactors for enzymes, take part in a variety of processes, and support oxidation and reduction reactions, among other tasks (Naseer et al. 2019).

Table 1 Contamination of vegetables by heavy metals in different regions of the world

Source	Heavy metals	Vegetable	Region	References
Wastewater	Pb, Cu, Ni, Zn	Pumpkin	Bangladesh	Ratul et al. (2018)
Soil	Hg	Spinach	China	Yu et al. (2018)
Mining	As, Cd, Cu, Pb	Eggplant	China	Yang et al. (2018)
Urbanization	As, Cd, Cr, Hg	Tomato	China	Sawut et al. (2018)
Anthropogenic	Cr, Cu, Ni, Pb	Spinach	Pakistan	Latif et al. (2018)
Glass industry	Cd, Cr, Cu, Fe	Mustard green	India	Kumar and Chopra (2015)
Industry	Pb, Fe, Cr, Cu	Pepper	Bangladesh	Ahmad and Goni (2010)
Organic fertilizer	Cu, Fe, ZN	Garlic	Pakistan	Akhter et al. (2022)

A worldwide vegetable census found that there are 230 species and 69 families represented among the 402 vegetable crops (Kays 2011). About 0.381 million ha, or 1.67 percent of all cultivated land, was devoted to vegetable cultivation in 1994, and Pakistan produced 4.06 million metric tons of vegetables overall (sweet potatoes and potatoes excluded) (Khan et al. 2011).

Leafy vegetables make up the majority of vegetables consumed (53%) and are followed in consumption by vegetable fruits (15%) and plants having edible organs underground (17%). Vegetable harvests of many sorts are eaten. Most vegetables are sold in the market mostly fresh, with minimal processing, due to their perishability. Vegetables should be used promptly after harvest to preserve their quality (Dias 2012). Though new veggies may be added to and introduced to the list, it is not exhaustive. However, it is significant to remember that soil conditions lacking in moisture encourage the absorption of heavy metals. Which can contaminate vegetables and pose a threat to consumers. Moreover, heavy metal contamination in vegetables is exacerbated by various sources in the era of climate change. These sources are summarized in Table 1. It is imperative to prioritize the production of high-quality vegetables with minimal water usage in order to meet current demands. Unfortunately, drought aggravates the effects of heavy metals, causing a disastrous reduction in yield and growth.

4 Essential Nutrients for Vegetable Growth

Growing awareness of the importance of nutrition in our diets has led to a greater awareness of the minerals and vitamins found in both raw and cooked foods. The factors that affect soil and plant chemistry, as well as the effects of fertilizers and management choices, are of primary concern to growers. Three elements are essential to photosynthetic reactions: hydrogen, oxygen, and carbon. The essential macronutrients after these are, in that particular order, nitrogen, phosphorus, potassium, calcium, magnesium, and sulfur (Nonnecke 1989).

5 Drought Impact on Availability of Nutrients for Vegetables

Drought stress significantly impacts plants in diverse physiological and biochemical manners, representing one of the most intricate abiotic stresses in the environment and posing substantial threats to global agricultural output. Among crops, vegetable crops are notably susceptible to drought, with their growth, development, and susceptibility to pests and diseases influenced by carbon dioxide, a pivotal greenhouse gas (Fig. 3). Altering climatic conditions results in hindered vegetable production marked by crop failures, reduced yields, compromised quality, and heightened pest and disease pressures. Both drought and salinity contribute to decreasing soil water potential, primarily affecting growth by inducing water shortage and osmotic stress. Nutrient uptake by roots and their subsequent translocation to shoots diminishes due to reduced transpiration rates, altered membrane permeability, and limited active transport (Ishfaq et al. 2022). Drought stress is indicated by physiological markers such as cell membrane integrity, water utilization efficiency, and maximal quantum yield of PSII (Fv/Fm), relative water content, and water potential. Diminished leaf water potential during drought events leads to reduced cell turgor pressure (Fig. 3), hindering soil nutrient uptake. Stomatal closure, occurring at specific water potential thresholds in plants like tomatoes and peppers, further exacerbates nutrient uptake limitations and reduces photosynthetic rates (Ishfaq et al. 2022). Extended drought stress also disrupts biochemical processes, including

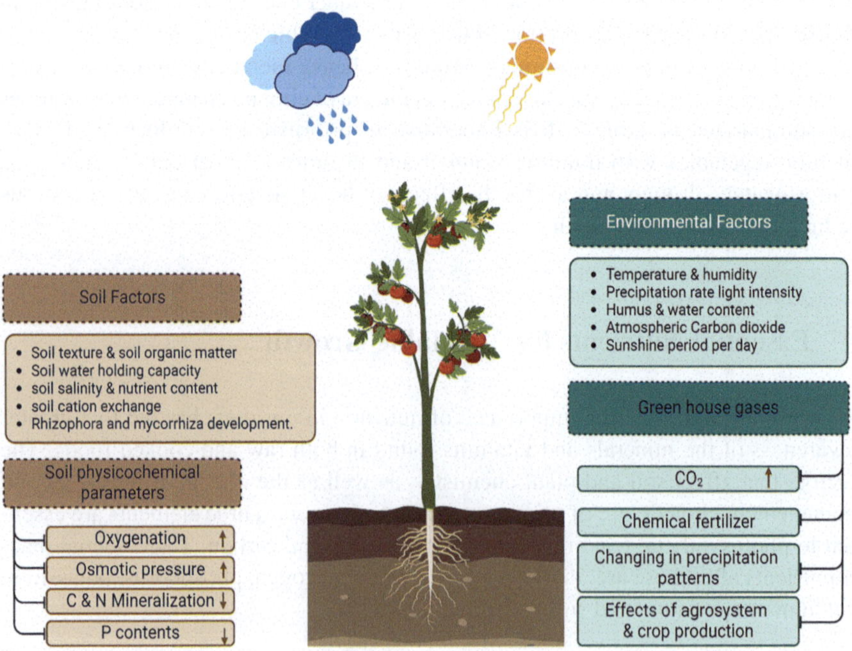

Fig. 3 Impact of drought stress and soil properties on nutrient availability for vegetables

carboxylation efficiency, Rubisco quantity and regeneration, and PSII activity suppression, leading to reduced photosynthetic rates due to photo inhibition. Environmental stresses such as drought can compromise cell membrane integrity, resulting in ion leakage, which can be quantified by measuring electrolyte efflux (Chatterjee and Solankey 2015).

Drought damages plants in both structural and non-structural ways, which lowers the quality of all crops especially vegetables. These changes affect various components such as carbohydrates, proteins, lipids, and carotenoids. Two non-structural carbohydrate components, namely starch and sugar content, are particularly important in determining the final quality of the crops. During vegetative growth, (Fig. 3) drought stress limits the amount of carbohydrates that can be dropped in the leaves and stems, leading to a constraint on seed mass (Marcaida III et al. 2014). Lack of water during the reproductive phases also causes early grain drying, preventing seeds from developing properly. In starchy tuber crops such as potatoes and cassava, which controls starch biosynthesis and sugar levels (up to 8 times) and sucrose synthesis in tubers (two to three times higher), drought has been shown to decrease starch concentration.

Total soluble solids accumulation is a critical adaptation process in many agricultural plants resulting in water stress. Nevertheless, not all crops exhibit this mechanism to the same extent, such as vegetables (Babita et al. 2010). For example, a study detected a normal importance of total soluble solids (TSS), using °Brix, 5.78 for irrigation with little water and 4.30 for irrigation with the most water. Water stress has been shown to increase the TSS levels in tomato fruits (Patanè et al. 2011). In comparison, it performs that water stress does not affect TSS levels in onions (Favati et al. 2009).

Some reports see more sugar in the storage organs of sugar beets, while others less. Beta vulgaris var. conditiva Aleph, the red beet, has demonstrated a decrease in partial sugar application that is less than mild and can result in extremely stressful situations (Stagnari et al. 2014).

Reduced lipid content in harvested crops is a result of drought stress. Apart from crop genotypes, the precise effects on control can vary depending on the species. Reduced lipid content in the membrane due to scarcity of water. This reduction is associated with the suppression of lipid synthesis, which is brought on by degradative mechanisms such as lipolytic activation and peroxidative activity (Labusch et al. 2013).

Drought permanently reduces the amount of water available in the soil, which has an impact on how well plants absorb nutrients. Due to lowered transpiration rates, it may also limit the flow of nutrients to the higher sections of the plants. Additionally, drought can weaken membrane permeability and active transport systems (Misle et al. 2014).

Reduced soil N mineralization during a drought results in lower N availability. Because calcareous soils (such as those in Mediterranean regions) have high pH values, or because semiarid tropical regions have low pH levels due to the presence of iron and aluminum oxide, have low pH values, and arid zones have increased phosphorus fixation. Moreover, there is a clear correlation between soil K+ mobility and soil water content (Wakeel and Ishfaq 2022).

The K^+, and Mg^{2+} concentrations of watermelon are positively impacted by both the frequency of watering and the grafting combination (Rouphael et al. 2008). P and Ca^{2+} absorptions, however, did not significantly change. Important nutrients including Ca^{2+}, Na^+, Zn, and Fe increased in plants in a drought-prone location, but the principal mineral nutrient, K^+, was found to be less concentrated (Martínez-Ballesta et al. 2010; Wakeel and Ishfaq 2022). Water stress appears to be the cause of sunflowers' accumulation of K, since it is essential for osmotic adjustment in plant tissues and organs (Esmaeilian et al. 2012), while depleting K+ concentration in maize and Rhus typhina (Ocimum gratissimum L., Lupinus albus, and Lupinus mutabilis) (Yu et al. 2007). The effects of water stress on Ca2+ and Mg2+ levels are very erratic; some studies indicate a general increase while others indicate a decrease (Yu et al. 2007).

6 Drought-Resistance Management Techniques

By making use of the best plant genome now available, the effects of drought stress can be lessened. Additionally, modifying rhizosphere management and implementing effective agronomic strategies, such as soil management and plant density, can also help alleviate the impact of drought. It is important to note that each of these strategies comes with its own set of advantages and risks. Therefore, we propose a comprehensive approach to tackle the problem of water scarcity in farming, in order to yield the desired outcomes.

6.1 Rhizosphere Enrichment

Numerous natural processes that impact the availability and uptake of nutrients and water are influenced by soil properties (Ishfaq et al. 2023a). Plants, soil organisms, climate, and type of soil all have an impact on the organic matter pools in the soil. However, in naturally humid and sub-humid environments, human activities have upset the delicate balance between living and non-living materials. An increased evapotranspiration, good soil porosity, favorable rooting conditions, and improved soil quality all contribute to these conditions, which make the environment ideal for plant growth and closed-cycle nutrient exchanges between the vegetation that has adapted to these conditions (Ishfaq et al. 2023a). This, in turn, reduces erosion and runoff while enhancing water permeability.

Anthropogenic activities, land usage systems, plant species, and management techniques also impact soil biological activities. The rate at which organic matter breaks down is influenced by temperature and plant residue breakdown. Tropical ecosystems face challenges in maintaining high organic matter concentration as a result of the heat and reduced water availability, resulting in faster decomposition. High annual rates of organic inputs are necessary for agricultural soils in these conditions to maintain an ideal variable soil organic matter pool (Ishfaq et al. 2023a). Organic waste, such as carbon-rich materials or animal feces, can contribute to increasing soil organic matter. Biochar is one method of enhancing soil nutrient status by enhancing the quality of the soil water as well as nutrient holding capacity. Through its high porosity, biochar derived from biomass provides natural nutrients and supports microbial growth. It also enhances plant growth, antioxidant activity, and resilience against drought and osmotic stress. Biochar can act as a carrier for bacterial inoculants and stimulate microbial activity, thereby improving soil geochemistry. Composting is another technique for improving soil fertility and quality, especially in sand areas that have a low capacity to retain water and nutrients. Composts enhance topsoil structure, hydraulic conductivity, porosity, water retention, and aggregation. Manure inputs also contribute to soil fertility, but the kind of soil and how often it is applied to determine how compost affects the availability of water.

6.2 Phytohormones

The mechanism by which plants tolerate drought stress is complex. Plant hormones are essential for communicating with and directing a variety of internal processes, especially during periods of dietary stress and drought (Zhou et al. 2013; Ishfaq et al. 2021). Salicylic acid (SA), a phenolic molecule with several functions in plant development, plays a critical role in mediating systemic acquired resistance during pathogen invasion. According to recent studies, SA is also necessary for plants' responses to external stresses and symbiotic connections. Plant tolerance to drought is increased by increasing photosynthetic rate, leaf area, relative water content, membrane stability, and enzyme activity (Hussain et al. 2008). SA pretreatment increases proline content in response to drought stress by downregulating the proline dehydrogenase gene (PDH), which suppresses cell death, and upregulating the transcription of pyrroline-5-carboxylate synthase, which synthesizes proline (Hayat and Ahmad 2007). Treatment with 500 M of SA increases stomatal conductance and increases CO_2 absorption in drought-stressed *H. vulgare* plants, both of which improve plant dry weight. SA may benefit plants experiencing oxidative stress by regulating the enzymatic and non-enzymatic elements of the glyoxalase system and the ascorbic acid-glutathione pathway (Alam et al. 2013). *Z. mays* leaves treated with SA exhibit a notably better antioxidant defense system than leaves of the control plant (Saruhan et al. 2012). According to Kang et al. (2014), wheat plants treated with 0.5 mm SA exhibit increased height and weight, reduced leaf wilting, and membrane lipid peroxidation. SA levels have been connected to the amounts of ortho-hydroxy-cinnamic acid that *O. sativa* requires to acquire drought tolerance (Pál et al. 2014). Jasmonic acid (JA) increases antioxidant activity, which promotes plant development and tolerance to drought stress (Bao et al. 2009). JA, along with its creator 12-OPDA, regulates stomata closure to minimize water loss in plants. The regulation of 12-OPDA production by 13-lipoxygenase LOX6 enables plants to withstand drought conditions by closing their stomata without the aid of ABA. In response to drought stress, cytokinin is generated, which inhibits leaf senescence (Pospíšilová et al. 2000). Furthermore, ethylene causes an increase in the expression of the drought-tolerant gene SodERF3, which causes plants' stomata to close (Desikan et al. 2006; Trujillo et al. 2008).

6.3 Nanoparticles and Liquid Nano Clay

Highly reactive materials with surfaces the size of nanometers, pore widths that may be adjusted, and the ability to target specific plant cell organelles are known as nanoparticles (NPs). These NPs boost antioxidant activity in plants, helping them withstand the damaging impacts of the drought. Nevertheless, there is still little proof of sustainable nanotechnology application in agriculture (An et al. 2022; Yousaf et al. 2023; Ahmad et al. 2023). The use of NPs enhances nitrogen metabolism, boosts physiological functions, enhances seedling germination, and boosts antioxidant activity in plants to fortify them against drought stress. Increased antioxidant levels, co-precipitation, aiding in the immobilization of heavy metals, improved nutrient absorption, and nutrient ion absorption are some of the ways that NPs reduce stress. Because NPs are so mobile, nutrients reach the highest levels of the plant very quickly. Metal oxides, such as ZnO NPs, which improve seedling development and drought-sustained plant nitrogen use efficiency, and TiO_2 nanoparticles, which boost SOD activity, have been used in many studies (Laware and Raskar 2014). Furthermore, it has been shown that using silicon nanoparticles reduces wheat's stress from water (Baybordi 2005). A recent study indicated that by reducing the detrimental impacts of coupled drought and cadmium stressors, the application of Fe-NPs to wheat increased photosynthetic rates and decreased oxidative stress in plants (Adrees et al. 2020). Sorghum still contains a comparable beneficial portion of ZnO-NPs, which improves nitrogen uptake in soil with low moisture levels (Dimkpa et al. 2019). When chitin is N-deacetylated, a non-toxic, biodegradable, and ecologically friendly polymer called chitosan is created. Wheat treated with 90 parts per million of chitosan nanoparticles had encouraging outcomes in reducing drought stress (Behboudi et al. 2019). Soil reclamation with liquid Nano-clay (LNC) is a novel technology. The process of treating desert soil by combining sand particles, water, and clay NPs has been patented by a scientist from Norway by the name of Kristian Morten Olesen. Since 2005, he has been engaged in LNC activities. The LNC technique significantly alters the physical properties of the sand particles by coating them with nanostructured clay, which allows the sand particles to hold onto water. As a result, poor-quality sandy soil can be transformed into highly productive soil. When planting, it is important to use appropriate fertilizers. When compared to conventional approaches, the LNC treatment dramatically reduces crop water requirements by 12–34% while increasing the yield of all soil-grown plants. Deserts can be turned into green spaces, surface temperatures can be lowered by about 15°, and CO_2 emissions can be decreased by 15 to 17 t ha^{-1} with the application of LNC technology. If farmers can afford it, this innovative technology can convert deserts into profitable fields.

6.4 Balanced Nutrition

Plants thrive when nutrients are abundant in the soil. The presence of soil organic matter increases the usefulness of fertilizers. Fertilizers not only increase crop yields but also aid in keeping the soil's nutrient pool full (Ishfaq et al. 2023a). Maintaining a balanced fertilizer schedule improves soil health, on the other hand, illness and soil mining might result from uneven use. It is well known that uneven nitrogen availability causes nutritional scarcity by depleting the soil's nutrient storage in addition to producing production that is lower and of inferior quality. Under drought conditions, the application of P-fertilizer has been found to have a considerable improvement in root biomass and water-extracting capacity, both of which enhance physiological processes (Tariq et al. 2018). Nitrogen availability can disturb deficiency via controlling the absorption of root water in crops. Compared to nitrate feeding, ammonium nutrition has been demonstrated to improve plant functioning during drought (Guo et al. 2007), which is connected to changes in the expression of the gene for aquaporin. Root water intake is also influenced by the type and quantity of nitrogen that is available; high nitrate levels greatly enhance nitrate uptake and root water absorption. Nitrate has been found to affect growth inhibition under drought stress through ABA reorganization (Ding et al. 2018). Depending on the species, different nutrient resorption processes and drought and nitrogen enrichment reactions occur (Zhao et al. 2020; Ullah et al. 2023; Ullah et al. 2024).

A variety of inorganic nutrients, including selenium (Se), potassium, zinc, nitrogen, and phosphorus have been demonstrated to lessen stress (Nawaz et al. 2020). By enhancing antioxidant scavenging capacity and preserving membrane integrity, silicone plant seed priming protects crops from the detrimental impacts of drought stress. Under drought conditions, silicon has a well-established impact that increases leaf hydration and surface area. Which will eventually result in increased plant biomass (Gong et al. 2003). Dehydration stress in plants leads to a shortage of potassium ions. Foliar potassium treatment applied during various stages of drought-stressed wheat, especially during the reproductive stage, enhances nutrient absorption while reducing sodium acceptance (Raza et al. 2013; Ishfaq et al. 2023b). Similar to this, using borate increases a plant's ability to withstand stress and encourages growth metrics (Abdel-Motagally and El-Zohri 2018). Cakmak (2000) has also documented the benefits of zinc in decreasing stress; in drought-stricken situations, foliar spraying zinc boosted yield by 16%. In dry conditions, applying zinc, manganese, and boron topically along with inorganic fertilizers boosts grain yield and micronutrient concentration. These micronutrient therapies applied topically improve photosynthetic rate, pollen and spike viability, grain content per spike, and water usage efficiency. According to certain theories, foliar treatment throughout the booting to anthesis stages may lessen the detrimental effects of drought, which often occur in the last stages of grain growth (Karim and Rahman 2015).

6.5 Management of Water

The world is presently facing several issues relating to water, such as scarcity and contamination, which render water bodies unfit for farming. Water is becoming more and more significant since it is essential to creating and implementing solutions to the main water problems facing the globe. The issue is especially worse in emerging nations because of human activity, increasing resource demand, and population growth is all interconnected. Therefore, rather than depending entirely on already-depleted freshwater supplies, it is imperative to use other water sources for agriculture.

6.6 Rainwater Harvesting

Drylands have low agricultural outputs due to inconsistent or inadequate rainfall and a significant portion up to 40% of precipitation being lost as runoff right away. To lessen the consequences of drought, the soil needs to absorb and store as much precipitation as it can for later use. Depending on its properties and biological activity, soil can have varying capacities for storing and releasing water. Enhancing water infiltration, decreasing soil evaporation, and boosting the soil's moisture-holding capacity are the three primary methods for preserving soil moisture. These properties are intimately related to soil organic matter, which controls the physical properties of the soil. Because of its capacity for bonding and adhesion, organic matter both directly and indirectly influences soil aggregation and soil porosity. Organic matter considerably increases the soil's ability to hold onto moisture when paired with mineral soil content. Recharging groundwater aquifers and gathering rainwater for storage are the two aspects of rainfall harvesting. Groundwater reserves are refilled more quickly through managed aquifer recharge than through natural recharge. Man-made structures and other artificial recharge systems serve to elevate groundwater supplies, Increase the amount of water in wells and tube wells during the dry season, and through dilution, increase the productivity of groundwater that is already present.

6.7 Rhizosphere Engineering with Plant Growth-Promoting Rhizobacteria

The essential microorganisms known as plant growth-promoting bacteria (PGPR) live in the rhizosphere of the soil or the region around plant roots. Among their many functions is the solubilization of fixed phosphorus, which is inaccessible to plants. According to Bhattacharyya and Jha (2012), the bacteria that produce ACC deaminases have the ability to colonize roots and form biofilms, which enables them to directly and indirectly influence plant development. Additionally, PGPR aids in plant growth under water stress and aids in plant adaptation to stressful environments. Plants influence and select their microbial communities, with selection being strongest around the roots (Berendsen et al. 2012). They operate as the main home for the microflora and are susceptible to a variety of root actions. Root exudates are the primary producers of the microbiome in the rhizosphere. Root exudates and microbiota can interact to alter the microflora to meet plant requirements, creating a new area of study to guarantee food security (Williams and de Vries 2020). PGPR has been shown to assist plants in withstand harsh conditions, successfully withstand a variety of pressures, and improve drought stress tolerance by boosting plant development in desert regions or high-stress environments (Zia et al. 2021). Furthermore, the strain of Phyllobacterium brassicacearum has been reported to improve water status and mitigate drought (Bresson et al. 2014). Bacillus cereus, Bacillus subtilis, and Serratia sp. have also been demonstrated to increase drought tolerance in cucumbers (Wang et al. 2012).

6.8 PGPR-Mediated Amelioration of Drought Stress

Numerous documented routes for plant stress tolerance are mediated by PGPR. These mechanisms include the creation of phytohormones, the activation of antioxidant machinery to reduce oxidative stress, and the increased manufacture of osmoprotectants to buffer cellular membranes, according to Chakraborty et al. (2013). Stress causes almost all plants to release more of the plant hormone ethylene. Conversely, when stress-induced ethylene synthesis takes place, plant growth and development are significantly inhibited (Iqbal et al. 2017). One of the primary ways PGP bacteria lessen stress is through the hydrolysis of 1-aminocyclopropane-1-carboxylic acid (ACC) by the enzyme ACC deaminase. In plants, the hormone ethylene is immediately preceded by ACC. According to several reports, some PGPRs include the enzyme ACC deaminase, which reduces the amount of ethylene in plants by converting ACC to ammonia (Raghuwanshi and Prasad 2018).

7 Conclusion

The issue of water shortage will persist as a major obstacle to agricultural productivity in the era of climate change. To increase the quantity of food for the growing population, the main approach is to adopt climate-smart practices that improve agricultural production. This chapter elucidates the mechanisms by which water stress induces nutritional and hormonal imbalances, hence hindering plant growth and reducing yield. Mineral nutrients are essential for the synthesis of crucial chemical components such as amino acids and proteins, and disturbances in nutrition can significantly affect biological processes. Drought disrupts the relationship between mineral nutrients, impeding plant growth and development and reducing agricultural output. Mass flow-dependent absorption and mineral nutrients moving from the roots to the shoot are typically affected by drought, influencing all metabolic processes in plant physiology. Adequate supplies of macro and micronutrients, however, may lessen the impact of drought on the development and growth of plants. Most of the studies have focused on the impact of drought on macronutrient needs rather than micronutrient needs. A plausible reason for this is that plants necessitate lesser amounts of micronutrients to maintain regular metabolic functions. However, the application of micronutrient fertilization, specifically selenium and silicon, has proven to be a successful approach in reducing the negative effects of drought on crops. The heightened occurrence and lengthened duration of droughts in arid and semi-arid areas due to global climate change may lead to food insecurity for both farmers and governments.

Introducing drought-resistant vegetation with high nutrient-use efficiency could potentially serve as a technical solution to cope with, or at least reduce this problem. Hence, it is imperative to endeavor to identify strategies that mitigate the impact of drought on plants. Plants can now withstand drought better thanks to a variety of methods, but the molecular underpinnings of plant-microbiome interactions are still mostly unknown. An essential component of plant health is the microbiome, sometimes called the second genome of the plant. Plants can alter the rhizosphere microbiome, leading to a variety of microbial populations supported by various plant species cultivated in the same soil, according to recent studies on plant-microbe interactions. Gaining further insight into the parameters of a favorable relationship between plants and plant growth-promoting rhizobacteria (PGPR) would help manage microorganisms and harness their potential for plant benefit. In this chapter, an innovative method for controlling plant drought stress is presented as a component of an integrated management plan. y modifying the microbiome, several strategies can be used, depending on the kind of soil, the length of the drought, and changes in the climate. Additional research on the physiological and biochemical elements of stress will provide new approaches to problems associated with food insecurity. Subsequent investigations ought to concentrate on determining the right bacteria to withstand drought stress, in addition to addressing the problems of delivery methods and field assessment of relevant species.

References

Abbasi T, Abbasi SA (2010) Biomass energy and the environmental impacts associated with its production and utilization. Renew Sust Energ Rev 14(3):919–937

Abdel-Motagally FMF, El-Zohri M (2018) Improvement of wheat yield grown under drought stress by boron foliar application at different growth stages. J Saudi Soc Agric Sci 17(2):178–185

Abubakar SM (2020) Pakistan 5th most vulnerable country to climate change, reveals Germanwatch report. Dawn, January, 16

Adnan M, Fahad S, Zamin M, Shah S, Mian IA, Danish S et al (2020) Coupling phosphate-solubilizing bacteria with phosphorus supplements improve maize phosphorus acquisition and growth under lime induced salinity stress. Plan Theory 9(7):900

Adrees M, Khan ZS, Ali S, Hafeez M, Khalid S, ur Rehman MZ et al (2020) Simultaneous mitigation of cadmium and drought stress in wheat by soil application of iron nanoparticles. Chemosphere 238:124681

Ahmad JU, Goni M (2010) Heavy metal contamination in water, soil, and vegetables of the industrial areas in Dhaka, Bangladesh. Environ Monit Assess 166(1):347–357

Ahmad MA, Adeel M, Shakoor N, Ali I, Ishfaq M, Haider FU, Deng X (2023) Unraveling the roles of modified nanomaterials in nano enabled agriculture. Plant Physiol Biochem 202:107944

Ahmed M, Schmitz M (2011) Economic assessment of the impact of climate change on the agriculture of Pakistan. Bus Economic Horiz (BEH) 4(1232–2016-101145):1–12

Akhter P, Khan ZI, Hussain MI, Ahmad K, Farooq Awan MU, Ashfaq A et al (2022) Assessment of heavy metal accumulation in soil and garlic influenced by waste-derived organic amendments. Biology 11(6):850

Alam MM, Hasanuzzaman M, Nahar K, Fujita M (2013) Exogenous salicylic acid ameliorates short-term drought stress in mustard (*Brassica juncea* L.) seedlings by up-regulating the antioxidant defense and glyoxalase system. Aust J Crop Sci 7(7):1053

Ali N, Anjum MM (2016) Drought stress: major cause of low yield and productivity. Austin Environ Sci 1:10–12

Allen MR, Coninck HD, Dube OP, Hoegh-Guldberg O, Jacob D (2018) Technical summary. In: global warming of 1.5° C. An IPCC special report on the impacts of global warming of 1.5° C above pre-industrial levels and related global greenhouse gas emission pathways

An C, Sun C, Li N, Huang B, Jiang J, Shen Y et al (2022) Nanomaterials and nanotechnology for the delivery of agrochemicals: strategies towards sustainable agriculture. J Nanobiotechnol 20(1):1–19

Aryal JP, Sapkota TB, Khurana R, Khatri-Chhetri A, Rahut DB, Jat ML (2020) Climate change and agriculture in South Asia: adaptation options in small holder production systems. Environ Dev Sustain 22(6):5045–5075

Ayeb-Karlsson S, Van der Geest K, Ahmed I, Huq S, Warner K (2016) A people-centred perspective on climate change, environmental stress, and livelihood resilience in Bangladesh. Sustain Sci 11(4):679–694

Babita M, Maheswari M, Rao LM, Shanker AK, Rao DG (2010) Osmotic adjustment, drought tolerance and yield in castor (*Ricinus communis* L.) hybrids. Environ Exp Bot 69(3):243–249

Bao AK, Wang SM, Wu GQ, Xi JJ, Zhang JL, Wang CM (2009) Overexpression of the Arabidopsis H+-PPase enhanced resistance to salt and drought stress in transgenic alfalfa (*Medicago sativa* L.). Plant Sci 176(2):232–240

Baybordi A (2005) Effect of zinc, iron, manganese and copper on wheat quality under salt stress conditions. J Soil Water Conserv 140:150–170

Behboudi F, Tahmasebi-Sarvestani Z, Kassaee MZ, Modarres-Sanavy SAM, Sorooshzadeh A, Mokhtassi-Bidgoli A (2019) Evaluation of chitosan nanoparticles effects with two application methods on wheat under drought stress. J Plant Nutr 42(13):1439–1451

Berendsen RL, Pieterse CM, Bakker PA (2012) The rhizosphere microbiome and plant health. Trends Plant Sci 17(8):478–486

Bhattacharyya PN, Jha DK (2012) Plant growth-promoting rhizobacteria (PGPR): emergence in agriculture. World J Microbiol Biotechnol 28(4):1327–1350

Bresson J, Vasseur F, Dauzat M, Labadie M, Varoquaux F, Touraine B, Vile D (2014) Interact to survive: Phyllobacterium brassicacearum improves Arabidopsis tolerance to severe water deficit and growth recovery. PLoS One 9(9):e107607

Brown S, Nicholls RJ, Lázár AN, Hornby DD, Hill C, Hazra S et al (2018) What are the implications of sea-level rise for a 1.5, 2 and 3° C rise in global mean temperatures in the Ganges-Brahmaputra-Meghna and other vulnerable deltas? Reg Environ Chang 18(6):1829–1842

Cakmak I (2000) Tansley review no. 111 possible roles of zinc in protecting plant cells from damage by reactive oxygen species. New Phytol 146(2):185–205

Chakraborty U, Chakraborty BN, Chakraborty AP, Dey PL (2013) Water stress amelioration and plant growth promotion in wheat plants by osmotic stress tolerant bacteria. World J Microbiol Biotechnol 29(5):789–803

Chatterjee A, Solankey SS (2015) Functional physiology in drought tolerance of vegetable crops: an approach to mitigate climate change impact. Clim Dyn Hortic Sci 1:149–171

Cohen I, Zandalinas SI, Huck C, Fritschi FB, Mittler R (2021) Meta-analysis of drought and heat stress combination impact on crop yield and yield components. Physiol Plant 171(1):66–76

Cook BI, Smerdon JE, Seager R, Coats S (2014) Global warming and 21st century drying. Clim Dyn 43(9):2607–2627

Daryanto S, Wang L, Jacinthe PA (2017) Global synthesis of drought effects on cereal, legume, tuber and root crops production: a review. Agric Water Manag 179:18–33

Desikan R, Last K, Harrett-Williams R, Tagliavia C, Harter K, Hooley R et al (2006) Ethylene-induced stomatal closure in Arabidopsis occurs via AtrbohF-mediated hydrogen peroxide synthesis. Plant J 47(6):907–916

Dias JS (2012) Major classes of phytonutriceuticals in vegetables and health benefits: a review. J Nutr Ther 1(1):31–62

Diatta AA, Fike JH, Battaglia ML, Galbraith JM, Baig MB (2020) Effects of biochar on soil fertility and crop productivity in arid regions: a review. Arab J Geosci 13(14):1–17

Dimkpa CO, Singh U, Bindraban PS, Elmer WH, Gardea-Torresdey JL, White JC (2019) Zinc oxide nanoparticles alleviate drought-induced alterations in sorghum performance, nutrient acquisition, and grain fortification. Sci Total Environ 688:926–934

Ding L, Lu Z, Gao L, Guo S, Shen Q (2018) Is nitrogen a key determinant of water transport and photosynthesis in higher plants upon drought stress? Front Plant Sci 9:1143

Dong Z, Chen M, Srivastava AK, Mahmood UH, Ishfaq M, Shi X, Zhang Y, Moussa MG, Li X, Hu C, Zhang F (2024) Climate changes altered the citrus fruit quality: a 9-year case study in China. Sci Total Environ 923:171406

Eckstein D, Künzel V, Schäfer L, Winges M (2019) Global climate risk index 2020. Germanwatch

El Rasafi T, Oukarroum A, Haddioui A, Song H, Kwon EE, Bolan N, Tack FM, Sebastian A, Prasad MNV, Rinklebe J (2022) Cadmium stress in plants: a critical review of the effects, mechanisms, and tolerance strategies. Crit Rev Environ Sci Technol 52(5):675–726

Esmaeilian Y, Sirousmehr AR, Asghripour MR, Amiri E (2012) Comparison of sole and combined nutrient application on yield and biochemical composition of sunflower under water stress. Int J Appl Sci Technol 2:214–220

Fatima A, Farid M, Safdar K, Fayyaz A, Ali SM, Adnan S et al (2020) Loss of agro-biodiversity and productivity due to climate change in continent Asia: a review. In: Plant Ecophysiology and Adaptation under Climate Change: Mechanisms and Perspectives I. Springer, pp 51–71

Favati F, Lovelli S, Galgano F, Miccolis V, Di Tommaso T, Candido V (2009) Processing tomato quality as affected by irrigation scheduling. Sci Hortic 122(4):562–571

Gong HJ, Chen KM, Chen GC, Wang SM, Zhang CL (2003) Effects of silicon on growth of wheat under drought. J Plant Nutr 26(5):1055–1063

Guo S, Kaldenhoff R, Uehlein N, Sattelmacher B, Brueck H (2007) Relationship between water and nitrogen uptake in nitrate-and ammonium-supplied *Phaseolus vulgaris* L. plants. J Plant Nutr Soil Sci 170(1):73–80

Hayat S, Ahmad A (eds) (2007) Salicylic acid-a plant hormone. Springer Science & Business Media, pp 1–14

Hussain M, Malik MA, Farooq M, Ashraf MY, Cheema MA (2008) Improving drought tolerance by exogenous application of glycinebetaine and salicylic acid in sunflower. J Agron Crop Sci 194(3):193–199

Iqbal MS, Jabbar B, Sharif MN, Ali Q, Husnain T, Nasir IA (2017) In silico MCMV silencing concludes potential host-derived miRNAs in maize. Front Plant Sci 8:372

Ishfaq M, Zhong Y, Wang Y, Li X (2021) Magnesium limitation leads to transcriptional down-tuning of auxin synthesis, transport, and signaling in the tomato root. Front Plant Sci 12:802399

Ishfaq M, Kiran A, Ur Rehman H, Farooq M, Ijaz NH, Nadeem F, Azeem I, Li X, Wakeel A (2022) Foliar nutrition: potential and challenges under multifaceted agriculture. Environ Exp Bot 200:104909

Ishfaq M, Wang Y, Xu J et al (2023a) Improvement of nutritional quality of food crops with fertilizer: a global meta-analysis. Agron Sustain Dev 43:74

Ishfaq M, Kiran A, Wakeel A, Tayyab M, Li X (2023b) Foliar-applied potassium triggers soil potassium uptake by improving growth and photosynthetic activity of wheat and maize. J Plant Nutr 46:2691–2706

Kang G, Li G, Guo T (2014) Molecular mechanism of salicylic acid-induced abiotic stress tolerance in higher plants. Acta Physiol Plant 36(9):2287–2297

Karandish F, Šimůnek J (2017) Two-dimensional modeling of nitrogen and water dynamics for various N-managed water-saving irrigation strategies using HYDRUS. Agric Water Manag 193:174–190

Karim MR, Rahman MA (2015) Drought risk management for increased cereal production in Asian least developed countries. Weather Clim Extrem 7:24–35

Kays SJ (2011) Cultivated vegetables of the world: a multilingual onomasticon: a multilingual Onomasticon. Wageningen Academic Publishers

Khan A, Islam M, Ahmad S, Abbas G, Athar M (2011) Technology transfer for cucumber (*Cucumis sativus* L.) production under protected agriculture in uplands Balochistan, Pakistan. Afr J Biotechnol 10(69):15538–15544

Konapala G, Mishra AK, Wada Y, Mann ME (2020) Climate change will affect global water availability through compounding changes in seasonal precipitation and evaporation. Nat Commun 11(1):1–10

Kumar V, Chopra AK (2015) Heavy metals accumulation in soil and agricultural crops grown in the province of Asahi India glass ltd., Haridwar (Uttarakhand), India. Adv Crop Sci Technol 2015:1–6

Labusch C, Shishova M, Effendi Y, Li M, Wang X, Scherer GF (2013) Patterns and timing in expression of early auxin-induced genes imply involvement of phospholipases a (pPLAs) in the regulation of auxin responses. Mol Plant 6(5):1473–1486

Latif A, Bilal M, Asghar W, Azeem M, Ahmad MI, Abbas A et al (2018) Heavy metal accumulation in vegetables and assessment of their potential health risk. J Environ Anal Chem 5(234):2380–2391

Laware SL, Raskar S (2014) Effect of titanium dioxide nanoparticles on hydrolytic and antioxidant enzymes during seed germination in onion. Int J Curr Microbiol App Sci 3(7):749–760

Marcaida M III, Li T, Angeles O, Evangelista GK, Fontanilla MA, Xu J et al (2014) Biomass accumulation and partitioning of newly developed green super Rice (GSR) cultivars under drought stress during the reproductive stage. Field Crop Res 162:30–38

Martínez-Ballesta MC, Dominguez-Perles R, Moreno DA, Muries B, Alcaraz-López C, Bastías E et al (2010) Minerals in plant food: effect of agricultural practices and role in human health. A Rev Agron Sustain Dev 30(2):295–309

Misle E, Kahlaoui B, Garrido E, Hachicha M (2014) Using an allometric model for the accumulation of mineral nutrients in crops under saline and water stress: a field experience in fertigation. In: Improvement of crops in the era of climatic changes. Springer, New York, NY, pp 81–106

Moezzi M, Janda KB, Rotmann S (2017) Using stories, narratives, and storytelling in energy and climate change research. Energy Res Soc Sci 31:1–10

Naseer S, Hussain S, Zahid Z (2019) Nutritional and antioxidant potential of common vegetables in Pakistan. RADS J Biol Res Appl Sci 10(1):36–40

Nawaz A, Shahbaz M, Imran A, Marghoob MU, Imtiaz M, Mubeen F (2020) Potential of salt tolerant PGPR in growth and yield augmentation of wheat (*Triticum aestivum* L.) under saline conditions. Front Microbiol 11:2019

Nonnecke IL (1989) Vegetable production. Springer Science & Business Media

O'Connell E (2017) Towards adaptation of water resource systems to climatic and socio-economic change. Water Resour Manag 31(10):2965–2984

Okorie VO, Mphambukeli TN, Amusan SO (2019) Exploring the political economy of water and food security nexus in BRICS. Afr Insight 48(4):21–38

Owolade SO, Akinrinola AO, Popoola FO, Aderibigbe OR, Ademoyegun OT, Olabode IA (2017) Study on physico-chemical properties, antioxidant activity and shelf stability of carrot (*Daucus carota*) and pineapple (*Ananas comosus*) juice blend. Int Food Res J 24(2):534–540

Pál M, Kovács V, Szalai G, Soós V, Ma X, Liu H et al (2014) Salicylic acid and abiotic stress responses in rice. J Agron Crop Sci 200(1):1–11

Passioura JB, Angus JF (2010) Improving productivity of crops in water-limited environments. Adv Agron 106:37–75

Patanè C, Tringali S, Sortino O (2011) Effects of deficit irrigation on biomass, yield, water productivity and fruit quality of processing tomato under semi-arid Mediterranean climate conditions. Sci Hortic 129(4):590–596

Pospíšilová J, Synková H, Rulcová J (2000) Cytokinins and water stress. Biol Plant 43(3):321–328

Raghuwanshi R, Prasad JK (2018) Perspectives of rhizobacteria with ACC deaminase activity in plant growth under abiotic stress. In: Giri B, Prasad R, Varma A (eds) Root biology. Soil biology, vol 52. Springer, Cham, pp 303–321

Ratul AK, Hassan M, Uddin MK, Sultana MS, Akbor MA, Ahsan MA (2018) Potential health risk of heavy metals accumulation in vegetables irrigated with polluted river water. Int Food Res J 25(1):329–338

Ray DK, West PC, Clark M, Gerber JS, Prishchepov AV, Chatterjee S (2019) Climate change has likely already affected global food production. PLoS One 14(5):e0217148

Raza S, Farrukh Saleem M, Mustafa Shah G, Jamil M, Haider Khan I (2013) Potassium applied under drought improves physiological and nutrient uptake performances of wheat (*Triticum aestivun* L.). J Soil Sci Plant Nutr 13(1):175–185

Reddy PP (2015) Impacts of climate change on agriculture. In: Climate resilient agriculture for ensuring food security. Springer, New Delhi, pp 43–90

Rehman A, Jingdong L, Shahzad B, Chandio AA, Hussain I, Nabi G, Iqbal MS (2015) Economic perspectives of major field crops of Pakistan: An empirical study. Pac Sci Rev B 1(3):145–158

Rouphael Y, Cardarelli M, Colla G, Rea E (2008) Yield, mineral composition, water relations, and water use efficiency of grafted mini-watermelon plants under deficit irrigation. HortScience 43(3):730–736

Ryder E (2011) World vegetable industry: production, breeding, trends. Hortic Rev 38:299

Saruhan N, Saglam A, Kadioglu A (2012) Salicylic acid pretreatment induces drought tolerance and delays leaf rolling by inducing antioxidant systems in maize genotypes. Acta Physiol Plant 34(1):97–106

Sawut R, Kasim N, Maihemuti B, Hu L, Abliz A, Abdujappar A, Kurban M (2018) Pollution characteristics and health risk assessment of heavy metals in the vegetable bases of Northwest China. Sci Total Environ 642:864–878

Stagnari F, Galieni A, Speca S, Cafiero G, Pisante M (2014) Effect of light and water supply on morphological and physiological leaf traits of red beet. Agron J 106(2):459–468

Sultan B, Defrance D, Iizumi T (2019) Evidence of crop production losses in West Africa due to historical global warming in two crop models. Sci Rep 9(1):1–15

Tariq A, Pan K, Olatunji OA, Graciano C, Li Z, Sun F et al (2018) Phosphorous fertilization alleviates drought effects on *Alnus cremastogyne* by regulating its antioxidant and osmotic potential. Sci Rep 8(1):1–11

Teuling AJ, Van Loon AF, Seneviratne SI, Lehner I, Aubinet M, Heinesch B et al (2013) Evapotranspiration amplifies European summer drought. Geophys Res Lett 40(10):2071–2075

Trenberth KE (2011) Changes in precipitation with climate change. Clim Res 47(1–2):123–138

Trujillo LE, Sotolongo M, Menendez C, Ochogavia ME, Coll Y, Hernandez I et al (2008) SodERF3, a novel sugarcane ethylene responsive factor (ERF), enhances salt and drought tolerance when overexpressed in tobacco plants. Plant Cell Physiol 49(4):512–525

Tuver GY, Ekinci M, Yildirim E (2022) Morphological, physiological, and biochemical responses to combined cadmium and drought stress in radish (*Raphanus sativus* L.). Rend Lincei Sci Fis e Nat 33(2):419–429

Ullah F, Ullah H, Ishfaq M, Gul SL, Kumar T, Li Z (2023) Improvement of nutritional quality of tomato fruit with Funneliformis mosseae inoculation under greenhouse conditions. Horticulturae 9(4):448

Ullah F, Ullah H, Ishfaq M et al (2024) Genotypic variation of tomato to AMF inoculation in improving growth, nutrient uptake, yield, and photosynthetic activity. Symbiosis 92:111–124

Wakeel A, Ishfaq M (2022) Potash use and dynamics in agriculture, 1st edn. Springer, Singapore

Wang RK, Li LL, Cao ZH, Zhao Q, Li M, Zhang LY, Hao YJ (2012) Molecular cloning and functional characterization of a novel apple MdCIPK6L gene reveals its involvement in multiple abiotic stress tolerance in transgenic plants. Plant Mol Biol 79(1):123–135

Warner K, Afifi T (2014) Where the rain falls: evidence from 8 countries on how vulnerable households use migration to manage the risk of rainfall variability and food insecurity. Clim Dev 6(1):1–17

Williams A, de Vries FT (2020) Plant root exudation under drought: implications for ecosystem functioning. New Phytol 225(5):1899–1905

Winston HY, Yang YC, Savitsky A, Alford D, Brown C (2013) The Indus basin of Pakistan: the impacts of climate risks on water and agriculture. World bank publications

Yang J, Ma S, Zhou J, Song Y, Li F (2018) Heavy metal contamination in soils and vegetables and health risk assessment of inhabitants in Daye. China J Int Med Res 46(8):3374–3387

Yang H, Huntingford C, Wiltshire A, Sitch S, Mercado L (2019) Compensatory climate effects link trends in global runoff to rising atmospheric CO_2 concentration. Environ Res Lett 14(12):124075

Yousaf N, Ishfaq M, Qureshi HA, Saleem A, Yang H, Sardar MF, Zou C (2023) Characterization of root and foliar-applied iron oxide nanoparticles (α-Fe_2O_3, γ-Fe_2O_3, Fe_3O_4, and bulk Fe_3O_4) in improving maize (*Zea mays* L.) performance. Nano 13(23):3036

Yu X, Du X, Song L (2007) Effects of water stress on the growth and ecophysiology of seedlings of the *Rhus typhina*. Sci Silvae Sin 43:57–61

Yu H, Li J, Luan Y (2018) Meta-analysis of soil mercury accumulation by vegetables. Sci Rep 8(1):1–10

Zhao Q, Guo J, Shu M, Wang P, Hu S (2020) Impacts of drought and nitrogen enrichment on leaf nutrient resorption and root nutrient allocation in four Tibetan plant species. Sci Total Environ 723:138106

Zhou M, Li D, Li Z, Hu Q, Yang C, Zhu L, Luo H (2013) Constitutive expression of a miR319 gene alters plant development and enhances salt and drought tolerance in transgenic creeping bentgrass. Plant Physiol 161(3):1375–1391

Zia R, Nawaz MS, Yousaf S, Amin I, Hakim S, Mirza MS, Imran A (2021) Seed inoculation of desert-PGPR induce biochemical alterations and develop resistance against water stress in wheat. Physiol Plant 172(2):990–1006

Zommers Z, van der Geest K, De Sherbinin A, Kienberger S, Roberts E, Harootunian G, James R (2016) Loss and damage: the role of ecosystem services. United Nations Environment Programme, Nairobi

The Influence of Drought Stress on Tomatoes: A Brief Understanding

Muhammad Tahir Shehzad, Huda Ghazanfar, Musawar Hussain, Zia-ur-Rehman Farooqi, Muhammad Mahroz Hussain, and Hamna Bashir

Abstract Globally, tomatoes hold great economic importance, particularly among temperate zones of southeastern countries of Asia. Notable production losses could result from drought stress conditions. Significant work is being done to enhance tolerance to these agriculturally important varieties of tomato including traditional breeding to novel methods. However, only some genes responsible for drought tolerance processes have been recognized until now. Thus, more studies are needed for contributing the separation and functional characterization of tolerance against abiotic stress for improvement of salinity and drought stress tolerance by genetic modification. Recently, techniques such as grafting, use of plant biostimulants and microbiome engineering are introduced to ameliorate drought tolerance in tomato, which are also safe toward the environment. This chapter elucidates the economic significance and responses of tomato against drought stress. The crucial role of gene modification and genetic engineering, especially microbiome engineering, to fight with drought stress will also be discussed. A brief conception of physiological and productive losses due to drought stress and plant's natural defense mechanism in such stress will be investigated.

Keywords Drought · Abiotic stress · Physiological · Genetic engineering · Growth reduction

M. T. Shehzad (✉) · H. Ghazanfar · M. Hussain · Z.-u.-R. Farooqi · M. M. Hussain
H. Bashir
Institute of Soil and Environmental Sciences, University of Agriculture Faisalabad, Faisalabad, Pakistan
e-mail: muhammad.shehzad@uaf.edu.pk

U. K. Chaudhry et al. (eds.), *Drought Stress*,
https://doi.org/10.1007/978-3-031-80610-0_8

1 Introduction

Tomato with scientific name Lycopersicon esculentum, is from family Solanaceae (night shade). Tropical America is its native land where it is commonly known as tomati. Asia brought tomatoes from Europe and Europe brought them from Mexico. As a valuable crop, it is cultivated in home and market-based gardens and by truck farmers. Greenhouses can also be used for its production. We use it in various products either in processed form or in fresh form. It has an important role in maintaining health and stamina. Ripen tomatoes comprises of antimicrobial properties that can cure wounds and overcome the vitamin A, B and C deficiencies in humans. It is commonly utilized in cooking and in raw form as salad (Oranges 2018).

Consumption and popularity of tomato is spreading widely in short span of time. Tomato has wide consumption and is becoming increasingly popular in a short span. It is highly vulnerable to high and low temperature but still it can grow in multiple types of climates. Pakistan cultivated 55,258 hectares area with annual production of 561,293 tons of tomato (FAOSTAT 2020). Khyber Pakhtunkhwa (KPK) is the largest producer with the cultivation area of 4230 hectares and an annual production of 51,063 tons. Peshawar also contributes to national production with 994 acres cultivation area and 2.978 thousand tons yield (Tahir et al. 2012).

Water act as a limiting factor for growth of plants. Drought stress affects the plants more than any other environmental factor and results in lack of irrigation and rainfalls. Plants response to drought stress through various processes, like lowering the rate of growth and photosynthesis. When some species are subjected to drought stress, they usually change the color of their leaves into blue green. Water scarcity results in wilting and defoliation that eventually cause death of plants. Water shortage appears as salt stress in plants because the osmotic effect of roots cause water loss in higher concentrations of salts (Udmale et al. 2014). Keen observation of cultural and environmental circumstances can indicate key problems. Water loss of plants not only depend upon the soil moisture but also on other environmental factors including wind speed, relative humidity, temperature, and light intensity. Another factor in drought stress development is the previous environmental conditions of plants. Plants that have experienced drought before it, and that were well irrigated before the onset of drought, are more tolerant of drought. It is estimated that till 2050 half of the fertile land would get salt affected due to drought and salinity (Khan et al. 2015). Increased world population results in increased food production and thereby, increased environmental stresses. Such circumstances can be coped through the advanced plant biotechnology and farming. Water scarcity is an essential environmental factor in agricultural field. Many improvements have been made for yield enhancement of crops under drought conditions. Natural selection ensures species survival in harsh conditions, while breeding promotes the financial benefits of species. Historical work of more than 80 years on breeding resulted in improvement of yield of some crops even in drought conditions (Bruinsma 2017). Basic research helps to understand molecular and physiological reactions of plants under water scarce conditions, but still yield under stressed conditions differ from that under optimal ones. Food security is needed to reduce these production gaps and increase sustainability under drought conditions.

2 Natural Defense Mechanism of Tomato in Drought Stress

We know that water scarce conditions limit growth, production, and development of plants. Major losses attributed to drought include disrupted photosynthetic activities, reduced uptake of essential elements, increased oxidative stress due to increased reactive oxygen species (ROS), reduced water potential causing plant wilting, and disrupted stomatal functions. Plants have a wide range of self-protective mechanisms to deal with such situations. They try to escape from drought, avoid the drought, or create tolerance against drought, where with these are three common responses, they adapt to survive in drought conditions (Ibrahim et al. 2020; Shohat et al. 2021). Plants reduce their overall lifecycle in drought to escape it, slow down all the processes and water usage to make osmotic managements, ROS reduction and oxidant production to tolerate it (Shohat et al. 2021).

The first and most important mechanism in tomato to cope with water deficit conditions is the alteration of the levels of growth regulators and stress hormones, especially abscisic acid and gibberellins (Fig. 1; Ibrahim et al. 2020; Shohat et al. 2021). These hormones are involved in basic growth processes such as germination, turgor management, fruiting and flowering processes, water potential maintenance, osmotic regulations, stomatal conductance, signaling and expression of genes, and production of soluble sugars, osmolytes, oxidants and other organic compounds under abiotic stresses (Shohat et al. 2021; Ueguchi-Tanaka et al. 2005; Živanović et al. 2020). Gibberellins are important phytohormones, and many studies have

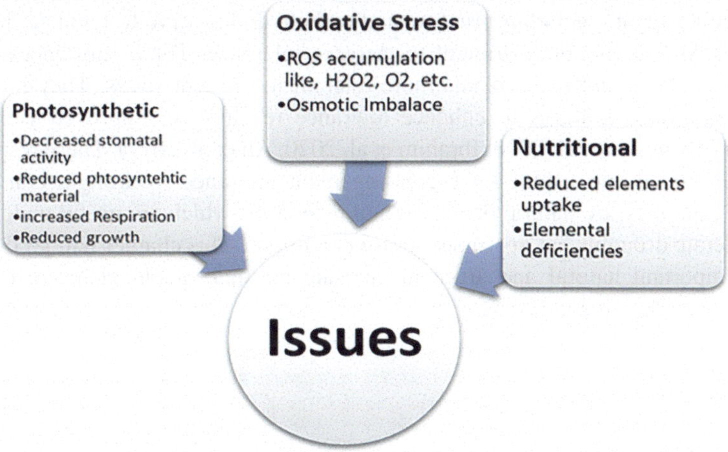

Fig. 1 Issues faced under drought stress by a plant

shown that their levels decrease under drought or other environmental stresses to enhance tolerance. Low levels of gibberellin activity enhance the ability of plants to cope with abiotic stresses, particularly water deficit and salt stress. Overexpression of dehydration-related transcription factors in some tomato families leads to reduced levels of gibberellins and enhanced tolerance, possibly involving the Dehydration-Responsive Element Binding (DREB) class of transcription factors. Under drought conditions low level of gibberellins act to increase the number of osmolytes, antioxidants and decrease the level of oxidative stress caused by ROS. Experiments shown that drought suppress the genes that are responsible for gibberellins production and activate the genes that reduce gibberellins' functioning to increase tolerance in tomato (Shohat et al. 2021; Ueguchi-Tanaka et al. 2005). Another most important phytohormone is abscisic acid that is vital in plant growth regulation and abatement of environmental stresses. It indicates stress in plants. The primary function of abscisic acid is to maintain water potential through the regulation of stomatal function. Stomatal closure, turgor pressure, and transpiration are important for increasing water use efficiency under drought conditions, which is controlled by osmoregulation, reduction of ROS toxicity, and management of osmolytes that directly linked to level of abscisic acid. Abscisic acid act to close the stomata in stress conditions and decreases the photosynthesis. Moreover, it produces antioxidants, osmolytes and some other organic amino acids that limit ROS activity and toxicity. Another important role of abscisic acid is the regulation of soluble sugars, all of which act as osmoprotectants and enhance drought tolerance.

Water deficit conditions lead to an increase in ROS that react with the chemical and biochemical structures of plants and cause damage. They are useful in some processes under drought such as stomatal closure and lack of photosynthetic pigments to limit water use but cause more harm than benefit. Plants counteract this by accumulating osmolytes, antioxidant organic compounds and enzymes, and soluble sugars. Osmolytes are nontoxic proteins that useful to reduce the ROS, such as proline, antioxidants contribute to reduce the oxidative stress such as oxidative enzymes, and soluble sugars including fructose and carbohydrates serve to regulate the biochemical processes under drought to increase tolerance. These substances act as osmoprotectants and make plants more tolerant to drought stress. They maintain cell hydration conditions to enhance tolerance resulting in improved yield and growth (Živanović et al. 2020; Ibrahim et al. 2020; Ali et al. 2017). These responses are interrelated and useful for increasing plant tolerance to drought conditions (Fig. 2), but they are natural processes and responses, which are useful under mild to moderate droughts but not under severe conditions. This chapter will go through other important natural and artificial mechanisms that enable plants to tolerate severe water deficit conditions.

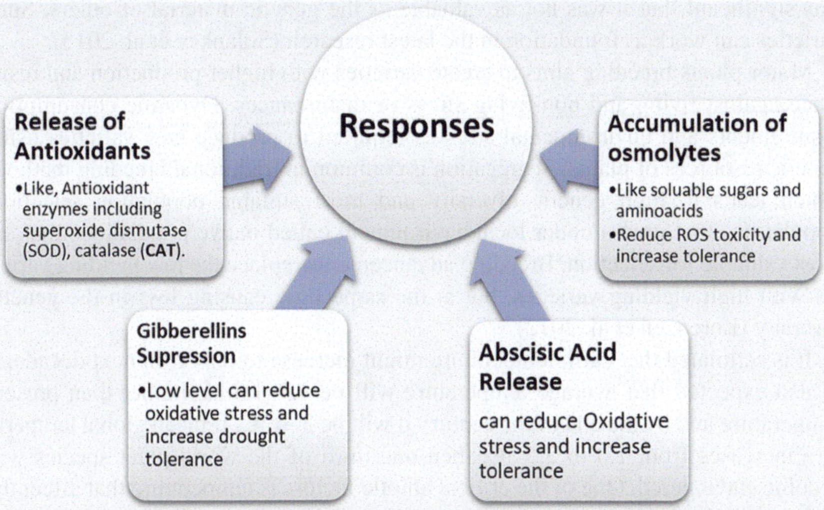

Fig. 2 Natural defense mechanisms in response to drought stress by plants

3 High Temperature Tolerance in Tomato

Tomato production, development, and reproduction can be controlled by temperature in tropical regions. Most suitable temperature for tomato production is 25 to 30 °C and 20 °C during day and night respectively (Carvalho et al. 2011). Minor increase from optimum temperature affects the reproduction stages, seed and fruit formation that ultimately reduces yields. Temperature above 35 °C badly affects the seedling, germination, vegetative stage, flowering, and fruits (Hasanuzzaman et al. 2013). The temperature stress significantly threats the setting of fruit in different varieties of tomato (Alsamir et al. 2021) and thereby, selection of heat tolerant varieties for breeding is essential for developing temperature resistant cultivars (Gray and Brady 2016). The progress of tomato lines which are heat tolerant and consequent release of well-adapted, tropical cultivars has been significantly contributed. Cross among heat tolerant tropical varieties produce genetically diverse varieties which shows higher production and high temperature tolerance. Heat tolerant lines of tomato were grown with the help of breeding lines and genetically heterogeneous varieties which showed tolerance against heat stress such as VC-11-2-5, VC-11-3-1-8, and Divisoria-2 from the Philippines while PI289309 and Tamu Chico-III from the United States. But in the heat tolerant lines, lower productivity remains a major problem (Solankey et al. 2015).

Tomato breeding lines are affected at temperature 35–40 °C particularly due to pollen disabilities that leads to reduced weight of fruit and yield. However, genotypes VRT-101 A, EC-538180, EC-381263, EC-620421 and EC-620419 show more production than other genotypes in heat stress. Although the mutation in EC-620421 was significant, but it was not as valuable as the genetic material of others. Such varieties can work as foundation in the latest research (Solankey et al. 2015).

Major plants breeding aims to create varieties with higher production and resistance against living and non-living stress or disturbances. Dynamic consumption requirements and environmental stresses required to produce new varieties using genetic resources of plants. Segregation is common in traditional breeding methods which leads to high genetic diversity and most suitable population selection. Populations that grown under local environment called native populations that are most valuable for selection. Breeding advancements replace the low yielding varieties with high yielding varieties, but at the same time causing loss in the genetic diversity (Lorenzen et al. 2012).

It is estimated that earth temperature might increase to 0.3 °C in next decade. It is also expected that average temperature will be 1.3–1.8 °C higher than present temperature and by the end of the century it will be 3–6 °C. If mean global temperature increases from 1.0 to 2.5 °C, then one third of the world plant species will become endangered. One of the critical abiotic factors is temperature that affect the biological, physical, chemical, and production characteristics of plants (Nahar and Ullah 2012). Development of cultivars using the variety of genes is very essential for genetic advancements that can mitigate the harmful impacts of heat stress. Classification, conservation, and utilization of genetics pools is very crucial. Selection on the basis of phenotypic features, which are effected by the ecological conditions is entirely wrong description of breeding because stress reactions in plants are very complex. Experiments show that overall, genetically modified organisms are more temperature resistant in moist and high temperature conditions with appropriate production and fruit set. Extreme level tolerance shown by just less than 1% of cultivars or breeds (George et al. 2011). No doubt field screening is best but molecular markers are also useful for enhancement of efficiency to achieve accuracy. Molecular markers can be the best strategy for tomato to develop tolerance. In breeding generally preferred varieties are those which are tolerant to heat stress screened by molecular techniques to select the best parents for breeding (Terzopoulos and Bebeli 2008; Paupière et al. 2014). Many efficient, fast, and reliable breeding methods such as tissue culturing, technology of recombinant DNA and molecular genetics are used for the conservation of genetic resources.

Non-tolerant varieties may harm environment with increasing yield, but temperature tolerant varieties can be useful to cope climate change. We need to utilize all genotypic varieties to discover more about the heat tolerance varieties. Pacific and Southeast Asian regions contain a tomato line CL5915, which has the high heat tolerance with fruit set 15–30% at temperature 35 °C even in heat vulnerable varieties. Characterization of temperature resistance with respect to fruit yield and fruit set in CL5915 depends on dominant and additive effects (Paupière et al. 2014). It is indicated by the patterns of fruit set and breed lines retrieval that temperature resistance can be control by some basic modifiers and genes. High temperature resistant varieties particularly CL5915 have been contributed to new breeding lines development. Temperature resistance and sensitivity are best found in the CL5915 cultivars groupings for overall yield and fruit set (Hasanuzzaman et al. 2013; Khan et al. 2015).

4 Water-Use Efficiency of Tomato and Drought Tolerance

Inadequate precipitation and altered rainfall patterns result in drought, which is often followed by comparatively higher temperatures and enhanced evapotranspiration that affects photosynthetic kinetics and production of crop. Plants might escape drought stress by reducing their life cycle, when under water deficit condition. Nonetheless, the photosynthetic processes, energy dissipation capacity and protection of metabolism against ROS are damaged by the oxidative stress due to rapid dehydration; and these are crucial for plants to survive in conditions of drought (Nahar and Ullah 2012). In crops, tolerance for severe dehydration is unusual but is present in native species of dry environments such as xerophytes (Yang et al. 2021). Drought tolerance genetic variability is limited and insufficient in S. lycopersicum.

Examination of drought stress susceptibility through pot experimentation showed that the tomato genotypes EC-538408, EC-538455, EC-625644, EC-625660 and Pant T-5 have potential of good yield. The EC-538408 and EC-625660 tomato genotypes had maximum yield, under normal conditions. In future research, these genotypic varieties can be utilized as a mean for drought tolerance.

At the resource center of tomato genetics, University of California, Davis collected a group of tomato germplasm reportedly known to be stress tolerant including S. cheesmanii, S. chilense, S. habrochaites, S. lycopersicum, S. lycopersicum var. cerasiforme, S. pennellii, S. peruvianum, and S. pimpinellifolium accessions, native to arid/semi-arid regions of South America. These species have undetermined growth pattern and produce fruit which are small and green. S. chilense is found in places with no other plantation and is acclimatized to desert regions of North Chile. This specie is distinguished in its characteristics of well-developed and expanded roots, leaves, and length from the other grown tomato (Ghebremariam et al. 2013). The experiments state that S. chilense is five-fold more wilt tolerant than other grown tomato. Although, S. pennellii have innate ability to enhance water usage efficiency in dry environments, it possesses round, thick, waxy leaves and its trichomes contain acyl-sugars, unlike the cultivated S. lycopersicum (Nahar and Ullah 2012; Oranges 2018). From these drought-resistant species, introgressive hybridization and then gene utilization would increase tolerance for drought among tomato cultivars. To achieve this goal, fertile progenies were produced by extensive crosses with S. pennellii. Moreover, S. chilense has cross incompatibility with S. lycopersicum and in order to produce offspring plants from these crosses, embryo rescue by tissue culture is needed. The work is in progress to determine genetic factors fundamental to drought tolerance among wild varieties and to transmit these characteristics into the cultivated tomatoes (Verma et al. 2020).

5 Grafting for Drought Tolerance in Tomato

A fundamental approach in the development and promotion of horticulture is grafting. It involves the growth of new plants of superior characteristics through the combination of scions and rootstocks of different plants. It is a useful tool in vegetative propagation. In the past few years, it is not only effective in biotic stress but also used to cope with abiotic stress such as water scarcity, salt stress, metal contamination, heat stress, pH control and nutrient deficiency (Li et al. 2014; Zhang et al. 2019). The topic of plant adaptations to water scarce conditions has been widely studied and reviewed comprehensively. Grafting is a method of improving plant growth and yield under drought conditions. Improved adaptations have observed in the grafted tomatoes, grapevines, apples, and cucumber. Grafting can substitute breeding and improvement of horticultural crops. Grafting of tomato cultivars with drought tolerant cultivars under drought stress is likely to increase tomato production (Zhang et al. 2020).

Root-shoot links are important for drought stress tolerance in tomato, particularly when we graft tomato cuttings onto tolerant rootstocks (Zhang et al. 2019). Drought adversely affects the uptake of micro and macro nutrients which ultimately affects the growth and vigor of roots and shoots. Water deficit conditions reduces nutrient uptake in leaves and increases organic acids and carbohydrates in both leaves and roots. When the vulnerable varieties grafted onto drought-tolerant varieties it was shown to enhance the uptake and storage of essential elements (Cu, Fe, K, P, and N) (Sánchez-Rodríguez et al. 2014). NO^{3-} and nitrate reductase activity can also be improved by grafting (Sánchez-Rodríguez et al. 2013). Grafting with drought tolerant varieties of tomato can increase resistance to drought conditions, improve growth and increase yield compared to grafting with drought susceptible varieties. Grafting of tomato plants provide with increased organic and inorganic osmotic substances that improve the water uptake and promote plants growth under drought conditions (Zhang et al. 2020).

Moreover, drought conditions reduce rate of photosynthesis through various phenomenon, including disturbance in photosynthetic pigments and light absorption. Fruit weight and growth severely affected by the drought conditions by restricting the root growth that ultimately reduce plant growth. Grafting increases the photosynthesis and reduces ROS toxicity that improves the growth and production of tomato. Drought causes oxidative stress in plants that cause reduction of photosynthetic contents and it can be due to degradation of chlorophyll pigments and photo oxidation. Plants usually reduce the chlorophyll contents in response of the drought to resist against wilting and minimize the activity of chloroplasts. Nevertheless, it is observed that the grafted plants reduce significantly less amount of chlorophyll content compared to non-grafted cultivars. The activity of photosynthesis is also affected by the disturbance in light absorption and utilization under water scarce conditions. Grafted plants maintain the concentration of photosynthetic pigments, reduce the production ROS, and achieve more production of antioxidant enzymes such as superoxide dismutase that reduce the oxidative stress and maintain the tolerance in drought conditions (Sánchez-Rodríguez et al. 2012; Zhang et al. 2019). Overall grafting is a useful tool for drought tolerance in tomato.

6 Genotypic Screening of Drought Tolerant Tomato

The presence of genotypic variations among species might serve as a possible source to screen contrary to drought and heat stress. The study of biochemical, morphological, and physiological characteristics and their interrelations is crucial to effectively select and design appropriate architecture for plants against abiotic stresses. Heat and drought tolerance through screening methods has been stated. The breeding suitability of the favorable genotypic varieties against abiotic stress can be determined by providing stress conditions and the stability in such conditions can be judged through many biometrical approaches (Amrutha et al. 2019).

7 Plant Breeding Technique for Drought Tolerance

For enhancing the crop production under conditions of stress, techniques are required in addition to conventional methods, which are incapable of avoiding production losses. Whole Genome Sequencing (WGS) is an application of genomics to determine innovative and high-throughput technologies for both genetic and molecular levels (Paulsen et al. 2005). Genes are detected and their stress tolerance functions are well understood, with the help of WGS. This allows the genetic modification related to environmental stress tolerance. These tools assure quick returns but are requisite of huge expenses. These genetic and molecular tools and techniques are successfully used to perform several tasks. Research on molecular genetics of plants is used to increase conventional breeding and acquire benefits from genetic engineering to maintenance of crop yield in response to abiotic stresses (Araus and Cairns 2014).

As a crop, tomato acts as an ideal model with its entirely sequenced genome and mapped quantitative trait loci (QTLs) for many of its characteristics. Markers based on polymerase chain reaction (PCR) and molecular maps have been designed on the basis of distant hybridization. Construction of molecular maps of high density is done by the utilization of distant cross of L. pennellii × L. esculentum (Wu and Tanksley 2010). It uses more than 2000 markers of 750 $kbp \cdot cM^{-1}$ on average and the distance of marker on average is less than <1 cM (Foolad 2007). In tomato, the use of breeding assisted by markers is largely restricted because of limited polymorphism of markers inside the domesticated species. However, 12 chromosomes of tomato were used for the mapping of 400 physiological, morphological, isoenzyme, genetic markers which were disease resistant (Foolad 2007). Due to absence of reliable markers based on PCR and issues with drag linkage, these physiological and biochemical gene markers will be an excellent supplement and alternative. Work is being done to significantly contribute for the development of high-throughput genetic markers with high resolution, encompassing single nucleotide polymorphisms (SNPs).

Plant breeding has been revolutionized through genetic modification with the help of technology of molecular marker. Genetic and genomic advancements enhanced the structural and functional understanding of plant genomes. Molecular markers used as a tool for selection allows for incresing breeding efficiency potential by reduction of variations in environment, promotion of prior selection, and decrease of successive sizes of population for field tests (Mir et al. 2012). The effective superior alleles introgression from wild species into reproduction is facilitated and the pyramiding of quantitative traits of gene control are enabled and high production, stress tolerant tomato varieties are enhanced by the molecular markers. Analysis of molecular marker for vegetable stress tolerance is restricted, but work is in progress to determine QTLs fundamental to abiotic stress tolerance (Araus and Cairns 2014; Foolad 2007; Mir et al. 2012).

8 Genetic Engineering for Drought Tolerance in Tomato

The very first marketed genetically modified food crop was tomato in the United States. Many biotechnical methods are primarily concerned about tomato as the individuals of genus Solanum are altered easily and bioengineered tomatoes often produced are co-cultivated with Agrobacterium radiobacter. The genetic sequencing of euchromatin area of all the 12 chromosomes of tomato is done, which makes tomato a model crop for studying functional genomes in plant systems. The Cornell University, North Carolina State University (R. G. Gardner, fresh market), Ohio State University (D. M. Francis, processing), Pennsylvania State University (M. R. Foolad, processing and fresh market), and the University of Florida has initiated public programs on tomato breeding (Foolad 2007).

In tomato, the tolerance against environmental stress is a complicated trait encompassing several genes (Mir et al. 2012). The profiles of protein and RNA expression alter to respond the stress conditions. By utilizing microarrays, about 130 genes of drought response were identified including genes controlling transpiration, metabolism of carbohydrate, transcriptional gene modulation, and ion transport. Transcription factors such as CBF, DREBIA and HSF genes were also involved for response against heat and drought in tomato (Bartels and Sunkar 2005). Sucrose synthase (SUSY) and cell wall invertase (INV) contributes to partition of carbohydrates in plants and regulates the metabolism of carbohydrate in leaves that might indicate part of the overall cellular response to adaptation and under stress condition, plays role in osmotic adjustment. The ERECTA gene is responsible for the regulation of transpiration efficiency of plant. Transport of ion involves the AVPI and NHX genes (Masle et al. 2005). Many other genes are associated with stress response and the identification of those genes which presents a tolerant phenotype in the tomato is a challenge (Table 1).

Table 1 Biostimulants and their mode of action under drought stress in tomato

Biostimulants	Mechanism	References
Non-microbial biostimulants: Calcium based biostimulant	Improved dry weight and quality of fruit, reduction in the amount of cracked fruits	Della Lucia et al. (2022)
Hydrosylate based biostimulant	Improved water status and viability of pollens and enhanced yeild.	Francesca et al. (2021)
Pepton based biostimulant	Enhanced antioxidant protection and promoted hormonal balance.	Casadesús et al. (2019)
Microbial biostimulants: *Glomus species*	Higher ROS scavenging action in leaves and roots, lowering of lipid peroxidation.	Fadiji et al. (2021)
Rhizophagus intaradices, Funneliformis mosseae	Raised efficiency of water usage, fruit production, biomass and potential of leaf-water.	Malhi et al. (2020)

Although, the role of these genes is explained, the overexpression of them results in a few changes contributing for a tolerant phenotype. Increased activity in soil water deficient conditions is a result of expression of a vacuolar H+ pyrophosphatase (AVPI) in tomato. The larger and stronger system of roots in a bioengineered tomato provide better uptake by roots in restricted water conditions. After 5 days of no water, the control plants undergo irreparable damage against bioengineered tomatoes, in which damage due to water stress starts to appear after 13 days; but as soon as the water availability is ensured these completely recover. In tomato, genetic engineering of drought tolerance is successfully done using the CBF/DREB1 genes (Chen and Murata 2011). Many crops possess orthologous genes of CBF and conservation of pathway in these species of plants is indicated by functional tests. In some tomato species, the cold, heat, salt and water stress tolerance are resulted from overexpression of NHX1, CBF genes. Additionally, the bioengineered plants were stunted and dark green in color, with higher soluble sugar and proline concentrations in stress tolerance. Enhanced stress tolerance can be attained without retarding the growth of plant by the utilization of stress inducible promoters as these under normal growth conditions expresses the CBF gene with low background (Wani et al. 2016). Under stress conditions, when the genetically engineered tomato were grown in the presence of NaCl about 200 mM, they flowered as well as set fruits and high concentrations of sodium were accumulated in their leaves, but the tomatoes grown had low levels of sodium. Tolerance against a particular stress, demonstrated by many constituent traits and managed by related genes, can give crucial information for the identification of major genes for tolerance against stress in tomato (Deinlein et al. 2014).

9 Biostimulants for Drought Tolerance in Tomato

There are several kinds of biostimulants for plants, but one can classify them in two broad groups i.e., non-microbial biostimulants encompassing hydrosylate proteins, humic acids, and extract of seaweed, and microbial biostimulants, including microorganisms (Rouphael et al. 2019). Studies show that various biostimulants can be utilized to cope with the detrimental physiological impacts of drought stress among different crop plants. An experiment was conducted by using biostimulant Megafol on tomato plants resulted that treatment of the tomato plants with Megafol under drought stress conditions enhanced drought stress tolerance through increased productivity with respect to both chlorophyll florescence and biomass. Molecular level studies illustrate that Megafol can effectively serve in reducing the physiological damage caused by drought stress in tomato (Petrozza et al. 2014). Other biostimulants for plants such as Boosten and Isabion were also seen to contribute to growth promotion among tomato seedlings under temperature stress, but Boosten was found to be more efficient in this regard (Niu et al. 2022).

Furthermore, it was observed that by treating the genotype E42 of tomato with a hydrosylate protein based biostimulant in conditions of drought resulted in enhancement in mean weight of each fruit of treated tomato in contrast to the untreated ones, which indicates an improved fruit set and also mitigates the shrinkage of fruit induced by drought stress (Francesca et al. 2021). The application of biostimulants leads to significant increase in average production of fruits of tomato, irrespective of the soil-water circumstances. It is suggested that it might be due to influence of the biostimulants on the rate of electron transport, photosystem II quantum yield (Fv/Fm), translocation of metabolic substances responsible for flowering and antioxidant enzymes, which enhances their activity under its presence (Peripolli et al. 2021).

Another physiological analysis was performed on the tomato plants that involved the use of a biostimulant based on calcium (SOB01), which revealed that an enhanced response of stomata at development and maturation stage of fruit, an increment in the chlorophyll content and thereby in the rate of photosynthesis, and a reduction in the amount of cracked fruits was observed in water deficit conditions due to foliar application of this biostimulant (Della Lucia et al. 2022). Hence, it is evident that such biostimulants can serve plants to deal with drought stress.

Similarly, it is also known that Arbuscular Mychorrhizal Fungi (AMF) based microbial biostimulants increase tolerance of tomato in response to drought. AMF strain, such as Rhizophagus intaradices, presented great physiological response in extreme drought stressed conditions by promoting efficiency of water consumption in tomato plant (Rouphael and Colla 2020). It was reported that tomato plants which were drought stressed, when treated with a biostimulant constituting AMF resulted in 18% higher shoot and 14% higher root on dry weight basis in comparison with the untreated tomato (Lau et al. 2022). Biostimulants attenuate the defense system of plants to abiotic stressors, by providing defensive properties in response (De Vasconcelos and Garófalo Chaves 2019).

At present, the use of biostimulants as well as microbiome engineering for plants are in great demand for being environmentally safe methods. Both these techniques can be applied to ameliorate crop yield and plant's tolerance in response to environmental stresses (Lau et al. 2022).

10 Microbiome Engineering for Drought Tolerance in Tomato

Methods directly involved in microbiome engineering comprise introduction of synthetic microbial consortia, particular probiotic microbial inoculation, and transplanting and breeding of microbiome. Indirect methods, on the other hand, include utilization of selective substrates or amendments of soil (Arif et al. 2020).

The rhizospheric microbiome holds great potential such that it comprises growth promoting endophytes, mycorrhizas and rhizobacteria, which form association with roots. Thereby, it can significantly transform the approaches for induction of drought resilience among plants (Mathur and Roy 2021). The Microbacterium sp. is reported to significantly increase tolerance against drought in tomato (Aslam et al. 2022). Similarly, inoculating Achromobacter piechaudi showing activity of ACC deaminase into drought stressed tomato reduced the effects of stress, and also resulted in higher dry and fresh weight of plant than the normal plant under stress (Huang et al. 2017). Recently, incubating seedlings of tomato with (E25) Pseudomonas stutzeri and (CR71) Stenotrophomonas maltophilia was found to be growth promoting with improving both biomass of root and capacity of photosynthesis (Zhang et al. 2021).

In plants, microbiomes have great importance in attenuation of abiotic stress. By lowering oxidative stress and toxicity of metals, improving nutrient bioavailability and intake of water, forming plant hormones, and managing numerous signaling pathways such as degeneration of 1-Aminocyclopropane 1-carboxylic acid (ACC), which acts as a stress hormone precursor (Arif et al. 2020).

The method of rhizospheric engineering based on microbiome can bring about sustainable agriculture through guaranteeing management of biotic and abiotic stress and assisting in improvement of crop (Tyagi et al. 2022). Rhizo-bacteriome presents abundance of Plant Growth-Promoting Rhizobacteria (PGPR) in facilitating drought stress resistance for plants (Yadav et al. 2020). Endophytic microbiome are also found to enhance drought resilience among plants (Liu et al. 2017).

11 Conclusion

Tomato is a sensitive crop towards drought stress and severely decreases the yield under drought stress. Numerous experiments have been done to enhance the drought tolerance in tomato crop including grafting, biotechnological approaches (including use of biostimulants), traditional breeding, genetic studies, and more recently on microbiome engineering. Thus, it is concluded that the genetic modification, more particularly microbiome engineering, can be a feasible approach to enhance the drought tolerance in this crop. It is dire need of time to incorporate some drought tolerant genes in such a valuable and important crop to cope with the drought stress and the food security issues in the world, where availability of good quality water is decreasing day by day and global temperature is rising.

References

Ali F, Bano A, Fazal A (2017) Recent methods of drought stress tolerance in plants. Plant Growth Regul 82(3):363–375

Alsamir M, Mahmood T, Trethowan R, Ahmad N (2021) An overview of heat stress in tomato (*Solanum lycopersicum* L.). Saudi J Biol Sci 28(3):1654–1663

Amrutha S, Muneera Parveen AB, Muthupandi M, Sivakumar V, Nautiyal R, Dasgupta MG (2019) Variation in morpho-physiological, biochemical and molecular responses of two eucalyptus species under short-term water stress. Acta Bot Croat 78(2):125–134

Araus JL, Cairns JE (2014) Field high-throughput phenotyping: the new crop breeding frontier. Trends Plant Sci 19(1):52–61

Arif I, Batool M, Schenk PM (2020) Plant microbiome engineering: expected benefits for improved crop growth and resilience. Trends Biotechnol 38(12):1385–1396

Aslam MM, Okal EJ, Idris AL, Qian Z, Xu W, Karanja JK, Yuan W (2022) Rhizosphere microbiomes can regulate plant drought tolerance. Pedosphere 32(1):61–74

Bartels D, Sunkar R (2005) Drought and salt tolerance in plants. Crit Rev Plant Sci 24(1):23–58

Bruinsma J (2017) World agriculture: towards 2015/2030: an FAO perspective. Routledge

Carvalho RF, Takaki M, Azevedo RA (2011) Plant pigments: the many faces of light perception. Acta Physiol Plant 33(2):241–248

Casadesús A, Polo J, Munné-Bosch S (2019) Hormonal effects of an enzymatically hydrolyzed animal protein-based biostimulant (pepton) in water-stressed tomato plants. Front Plant Sci 10:758

Chen TH, Murata N (2011) Glycinebetaine protects plants against abiotic stress: mechanisms and biotechnological applications. Plant Cell Environ 34(1):1–20

De Vasconcelos ACF, Garófalo Chaves HL (2019) Biostimulants and their role in improving plant growth under abiotic stresses. In: Mirmajlessi SM, Radhakrishnan R (eds) Biostimulants in plant science. IntechOpen, London, pp 1–14

Deinlein U, Stephan AB, Horie T, Luo W, Xu G, Schroeder JI (2014) Plant salt-tolerance mechanisms. Trends Plant Sci 19(6):371–379

Della Lucia MC, Baghdadi A, Mangione F, Borella M, Zegada-Lizarazu W, Ravi S, Nardi S (2022) Transcriptional and physiological analyses to assess the effects of a novel biostimulant in tomato. Front Plant Sci 12:781993

Fadiji AE, Babalola OO, Santoyo G, Perazzolli M (2021) The potential role of microbial biostimulants in the amelioration of climate change-associated abiotic stresses on crops. Front Microbiol 12:829099

FAOSTAT (2020) Statistical division of the UN Food and Agriculture Organization of the United Nations. http://faostat.fao.org/. Accessed 14 Sept 2022

Foolad MR (2007) Genome mapping and molecular breeding of tomato. Int J Plant Genomics 2007:64358

Francesca S, Cirillo V, Raimondi G, Maggio A, Barone A, Rigano MM (2021) A novel protein hydrolysate-based biostimulant improves tomato performances under drought stress. Plan Theory 10(4):783

George W, Scott J, Splittstoesser W (2011) Parthenocarpy in tomato. Hortic Rev 6:65–84

Ghebremariam KM, Liang Y, Li C, Li Y, Qin L (2013) Screening of tomato inbred-lines for drought tolerance at germination and seedling stage. J Agric Sci 5(11):93

Gray SB, Brady SM (2016) Plant developmental responses to climate change. Dev Biol 419(1):64–77

Hasanuzzaman M, Nahar K, Alam MM, Roychowdhury R, Fujita M (2013) Physiological, biochemical, and molecular mechanisms of heat stress tolerance in plants. Int J Mol Sci 14(5):9643–9684

Huang X-F, Zhou D, Lapsansky ER, Reardon KF, Guo J, Andales MJ, Manter DK (2017) *Mitsuaria* sp. and *Burkholderia* sp. from Arabidopsis rhizosphere enhance drought tolerance in *Arabidopsis thaliana* and maize (*Zea mays* L.). Plant Soil 419(1):523–539

Ibrahim MF, Elbar OHA, Farag R, Hikal M, El-Kelish A, El-Yazied AA et al (2020) Melatonin counteracts drought induced oxidative damage and stimulates growth, productivity and fruit quality properties of tomato plants. Plan Theory 9(10):1276

Khan SH, Arsalan K, Uzma L, Shah AS, Khan MA, Muhammad B, Ali MU (2015) Effect of drought stress on tomato cv. Bombino. J Food Process Technol 6(7):1–6

Lau SE, Teo WFA, Teoh EY, Tan BC (2022) Microbiome engineering and plant biostimulants for sustainable crop improvement and mitigation of biotic and abiotic stresses. Discov Food 2(1):1–23

Li H, Wang F, Chen XJ, Shi K, Xia XJ, Considine MJ, Zhou YH (2014) The sub/supra-optimal temperature-induced inhibition of photosynthesis and oxidative damage in cucumber leaves are alleviated by grafting onto figleaf gourd/luffa rootstocks. Physiol Plant 152(3):571–584

Liu H, Carvalhais LC, Crawford M, Singh E, Dennis PG, Pieterse CM, Schenk PM (2017) Inner plant values: diversity, colonization and benefits from endophytic bacteria. Front Microbiol 8:2552

Lorenzen K, Beveridge MC, Mangel M (2012) Cultured fish: integrative biology and management of domestication and interactions with wild fish. Biol Rev 87(3):639–660

Malhi G, Kaur M, Kaushik P, Alyemeni M, Alsahli A, Ahmad P (2020) Arbuscular mycorrhiza in combating abiotic stresses in vegetables: an eco-friendly approach. Saudi J Biol Sci 28(2):1465–1476

Masle J, Gilmore SR, Farquhar GD (2005) The ERECTA gene regulates plant transpiration efficiency in Arabidopsis. Nature 436(7052):866–870

Mathur P, Roy S (2021) Insights into the plant responses to drought and decoding the potential of root associated microbiome for inducing drought tolerance. Physiol Plant 172(2):1016–1029

Mir RR, Zaman-Allah M, Sreenivasulu N, Trethowan R, Varshney RK (2012) Integrated genomics, physiology and breeding approaches for improving drought tolerance in crops. Theor Appl Genet 125(4):625–645

Nahar K, Ullah S (2012) Morphological and physiological characters of tomato (*Lycopersicon esculentum* Mill) cultivars under water stress. Bangladesh J Agric Res 37(2):355–360

Niu C, Wang G, Sui J, Liu G, Ma F, Bao Z (2022) Biostimulants alleviate temperature stress in tomato seedlings. Sci Hortic 293:110712

Oranges WN (2018) World journal of biology and biotechnology. WORLD 209:213

Paulsen IT, Press CM, Ravel J, Kobayashi DY, Myers GS, Mavrodi DV, Madupu R (2005) Complete genome sequence of the plant commensal *Pseudomonas fluorescens* Pf-5. Nat Biotechnol 23(7):873–878

Paupière MJ, Van Heusden AW, Bovy AG (2014) The metabolic basis of pollen thermo-tolerance: perspectives for breeding. Meta 4(4):889–920

Peripolli M, Dornelles SH, Lopes SJ, Tabaldi LA, Trivisiol VS, Rubert J (2021) Application of biostimulants in tomato subjected to water deficit: physiological, enzymatic and production responses. Rev Bras de Eng Agríc e Ambiental 25:90–95

Petrozza A, Santaniello A, Summerer S, Di Tommaso G, Di Tommaso D, Paparelli E, Cellini F (2014) Physiological responses to Megafol® treatments in tomato plants under drought stress: a phenomic and molecular approach. Sci Hortic 174:185–192

Rouphael Y, Colla G (2020) Editorial: biostimulants in agriculture. Front Plant Sci 11:40

Rouphael Y, Cardarelli M, Bonini P, De Pascale S, Colla G (2019, January) Implications of microbial and non-microbial biostimulatory action on the quality of leafy and fruit vegetables. In: XI international symposium on protected cultivation in mild winter climates and I international symposium on nettings and 1268, pp 13–18

Sánchez-Rodríguez E, del Mar Rubio-Wilhelmi M, Blasco B, Leyva R, Romero L, Ruiz JM (2012) Antioxidant response resides in the shoot in reciprocal grafts of drought-tolerant and drought-sensitive cultivars in tomato under water stress. Plant Sci 188:89–96

Sánchez-Rodríguez E, Romero L, Ruiz JM (2013) Role of grafting in resistance to water stress in tomato plants: ammonia production and assimilation. J Plant Growth Regul 32(4):831–842

Sánchez-Rodríguez E, Leyva R, Constán-Aguilar C, Romero L, Ruiz JM (2014) How does grafting affect the ionome of cherry tomato plants under water stress? Soil Sci Plant Nutr 60(2):145–155

Shohat H, Cheriker H, Kilambi HV, Illouz Eliaz N, Blum S, Amsellem Z, Weiss D (2021) Inhibition of gibberellin accumulation by water deficiency promotes fast and long-term 'drought avoidance' responses in tomato. New Phytol 232(5):1985–1998

Solankey S, Singh R, Baranwal D, Singh D (2015) Genetic expression of tomato for heat and drought stress tolerance: an overview. Int J Veg Sci 21(5):496–515

Tahir A, Shah H, Sharif M, Akhtar W, Akmal N (2012) An overview of tomato economy of Pakistan: comparative analysis. Pak J Agric Res 25(4):288–294

Terzopoulos P, Bebeli P (2008) DNA and morphological diversity of selected Greek tomato (*Solanum lycopersicum* L.) landraces. Sci Hortic 116(4):354–361

Tyagi R, Pradhan S, Bhattacharjee A, Dubey S, Sharma S (2022) Management of abiotic stresses by microbiome-based engineering of the rhizosphere. J Appl Microbiol 133:254–272

Udmale P, Ichikawa Y, Manandhar S, Ishidaira H, Kiem AS (2014) Farmers' perception of drought impacts, local adaptation and administrative mitigation measures in Maharashtra state, India. Int J Disaster Risk Reduc 10:250–269

Ueguchi-Tanaka M, Ashikari M, Nakajima M, Itoh H, Katoh E, Kobayashi M, Matsuoka M (2005) Gibberellin insensitive dwarf1 encodes a soluble receptor for gibberellin. Nature 437(7059):693–698

Verma A, Niranjana M, Jha S, Mallick N, Agarwal P (2020) QTL detection and putative candidate gene prediction for leaf rolling under moisture stress condition in wheat. Sci Rep 10(1):1–13

Wani SH, Kumar V, Shriram V, Sah SK (2016) Phytohormones and their metabolic engineering for abiotic stress tolerance in crop plants. Crop J 4(3):162–176

Wu F, Tanksley SD (2010) Chromosomal evolution in the plant family Solanaceae. BMC Genomics 11(1):1–11

Yadav VK, Raghav M, Sharma SK, Bhagat N (2020) Rhizobacteriome: promising candidate for conferring drought tolerance in crops. J Pure Appl Microbiol 14(1):73–92

Yang X, Lu M, Wang Y, Wang Y, Liu Z, Chen S (2021) Response mechanism of plants to drought stress. Horticulturae 7(3):50

Zhang Z, Cao B, Gao S, Xu K (2019) Grafting improves tomato drought tolerance through enhancing photosynthetic capacity and reducing ROS accumulation. Protoplasma 256(4):1013–1024
Zhang Z, Liu Y, Cao B, Chen Z, Xu K (2020) The effectiveness of grafting to improve drought tolerance in tomato. Plant Growth Regul 91(1):157–167
Zhang C, Wang MY, Khan N, Tan LL, Yang S (2021) Potentials, utilization, and bioengineering of plant growth-promoting methylobacterium for sustainable agriculture. Sustain For 13(7):3941
Živanović B, Milić Komić S, Tosti T, Vidović M, Prokić L, Veljović Jovanović S (2020) Leaf soluble sugars and free amino acids as important components of abscisic acid-mediated drought response in tomato. Plan Theory 9(9):1147

An Overview of the Impact of Drought Stress on Potatoes in the Era of Climate Change

M. Lateef, E. K. Naawe, Z. Hasan, and M. E. Çalışkan

Abstract Abiotic stress has always been a limiting factor for horticulture crops worldwide but now changing climate has made it more difficult. Among all abiotic stresses (cold, heat, drought, flood, salinity, nutrition deficiency, high or low intensity of light including ultraviolet radiation), drought stress has put a huge impact on agricultural crop yield and production. Plants generate tolerance to combat these stresses by adapting different mechanisms. Abiotic stress affects several mechanisms in plants like photosynthesis, biochemical pathways, and stress tolerance. Potato is one of the third most important crops worldwide and it consists of a lot of dietary energy that helps in reducing hunger and enhancing nutrition. Unfortunately, potatoes are also very sensitive to drought stress which effect on quality and production. Climate change and increasing abiotic stress, especially drought stress, that affects potato has been discussed in this chapter. Furthermore, response towards drought stress and phytohormone, and genetic engineering advanced methodology have been elaborated. Genetic editing method as CRISPR/Cas9 and other methods like QTL and GWAS are explained.

Keywords Sustainability climate change · Abiotic stress · Drought stress · Phytohormone · GWAS · QTL

1 Introduction

Any kind of external or environmental barrier that hinders plant productivity or plant growth is known as abiotic stress. In the past half-century, plant researchers are focused on abiotic stress that is created because of the boom in the human

M. Lateef (✉) · E. K. Naawe · M. E. Çalışkan
Department of Agricultural Genetic Engineering, Niğde Ömer Halisdemir University, Niğde, Türkiye

Z. Hasan
Department of Biosciences, Jamia Millia Islamia, New Delhi, India

U. K. Chaudhry et al. (eds.), *Drought Stress*,
https://doi.org/10.1007/978-3-031-80610-0_9

population, which is also intimidating global food security. Increasing food demand necessitated an investigation into the frequently ignored causes of decreased crop production. Approximately at this time, scientists began estimating and investigating the causes of crop loss due to environmental constraints that could bring down productivity up to 70% (Boyer 1982). A lot of research regarding crop reduction because of abiotic stress has started to be published since then (Wu et al. 2007). The production yields and quality of crops get affected by unrepairable injury due to stress. An estimation has been made by Cramer et al. (2011) that 96.5% of rural land worldwide took hold due to abiotic stress on agricultural land, which is based on FAO reports. An adverse change in climate put a barrier on plant performance that led to a remarkable drop in the sustainability of crop production (Jaleel et al. 2009; Farooq et al. 2009). The severity and effects of drought stress are a serious threat to agriculture, and it distorts major physiological changes in plants. It is reported that more than 25% of field deduction in yield is just because of drought stress (Vanani et al. 2020). Long-term drought causes a reduction in relative water amounts as well as a number of metabolic and photosynthetic abnormalities in growing crops (Yao et al. 2021).

More than one hundred countries produce potato which is consumed by globally a billion people. They are enriched with essential nutrition and full of carbohydrates and affluent source of antioxidants and protein and minerals (Bach et al. 2012). Due to lower carbon absorption rates and lowered assimilate partitioning to tubers, distortion in tuberization, bulking, and tuber production gets affected by early stress (Obidiegwu et al. 2015). It is expected that potato production would decrease significantly because of rising global atmospheric temperature and water deficit by 2055. Global potato production will be lowered by 18–32% between 2040 and 2069 stated by Hijmans (2003). To improve potato yield, we must thus discover the most efficient production methods and must look forward to producing new varieties that suit changing climate change very well. Among all abiotic stresses, drought stress has become a significant environmental factor that affects the growth of potatoes and most prominently yield (Fahad et al. 2017). The influence of global climate change on the variability of precipitation, in conjunction with increased water demand and dwindling water availability, may aggravate the conditions of water shortage (Lobell et al. 2011). This will influence all organisms, but it will have a bigger effect on plants lacking mobile features that allow them to migrate during shortages of water and mineral nutrients, consequently decreasing crop productivity. Major effects of drought include poor germination (Farooq et al. 2009), reduction in the growth of seedlings, reduced availability of nutrients which affect leaf number and leaf sizes of plants and thus affecting plants' photosynthetic rates (Manickavelu et al. 2006), fresh and dry weight of plants. Drought is a complicated stress that frequently increases a plant's susceptibility to other abiotic stress factors (Whitmore and Whalley 2009). Climate change exacerbates the unpredictability of the timing, length, and severity of drought in addition to the difficulty of comprehending drought stress (Jenkins and Warren 2015). However, the genetic content of the plant has been connected to the highest degree of drought stress complexity (Blum 2002). This includes the tetraploid genetics seen in farmed potatoes, which makes it difficult to breed a drought tolerance crop because of the complex transmission of features following genetic crosses. Moreover, Anithakumari et al. (2011) showed in a diploid potato population that drought is controlled by many loci on the genome, each of which contributes negligibly to drought responses. Consequently, finding drought-resistance quantitative trait loci (QTLs) in the genome of cultivated potatoes may be difficult and insufficient. Approaches from molecular research have been implemented to comprehend the basis of the drought response systems in potatoes and other plants.

2 Climatic Changes and Abiotic Stress

Climate change is commonly identified as among the future difficulties facing the agricultural sector. It is believed that a rise in temperature will have the greatest impact on the amount and quality of water. This natural resource's accessibility affects both human existence and crop yields. Water is a vital resource for both ecosystems and humans. In several places of the world, especially in the agriculture sector, which accounts for nearly 70% of worldwide water consumption, there is a growing demand for water (FAO 2012). The fundamental drivers of climate change are post-industrialization fossil fuel consumption and the increase in the dense number of hazardous gases of greenhouses. It is presumed that uprising development of industrialization are the main reason behind the environmental crisis (Dutta et al. 2020). It is a hazard that has perturbed the paradigm of growth of a variety of creatures. The natural greenhouse effect, in which the atmosphere of the earth absorbs the emission of the sun and heats the surface, is essential to life on the planet. The troposphere, where life exists, is affected by this greenhouse effect in the atmosphere. Without the greenhouse effect, the average surface temperature of the Earth would be roughly 19 °C, as opposed to the current usual of 14 °C (Cassia et al. 2018). But the constantly changing situation of environmental circumstances resulting from increasing anthropogenic actions like urbanization, fast industrialization, production of non-biodegradable production, blazing agricultural forest, and deforestation represents a crucial serious risk to the climate. In addition, hazardous gas emissions from companies, factories, equipment, and cars contribute significantly to climate change (Chaudhry and Sidhu 2021).

Climate change has caused changes in temperature, precipitation, and air conditions, displaying the plants to harsh and drastic environmental circumstances that negatively strike those on a different levels like morphological, developmental, cellular, and molecular mechanisms. The growth and yield of the plants were significantly affected by unfavorable climatic circumstances, such as high temperature and elevated CO_2 levels (Raza et al. 2019). Their long-life period and harsh environmental conditions do not let them adjust, so their exposure to drought stress goes very long (Lindner et al. 2010). Plant growth and reproduction can be altered with the varying temperature in regular situations also, and severe environmental changes can impair growth-related molecular pathways, consequently influencing plant development (Bita and Gerats 2013). According to Reddy et al. (2010) changed mechanisms of photosynthetic carbon absorption account for the effect of climate change on plant development. Climatic change-induced drought stress drastically impacts plants' stomatal conductance, water utilization of the plant, CO_2 absorption, and photosynthetic pigments, resulting in lower plant productivity (Anjum et al. 2011).

The distribution, outgrowth, and progress of plants are significantly impacted by temperature. Every plant has an optimum temperature in which it grows, and its morphological and physiological processes and characteristics depend on it (Hatfield and Prueger 2015). It is stated in the IPCC report that climate change is taking place and average temperature rise in the coming 30–50 years will be 1.2–3.7 °C. The rise in temperature is way more than the favorable range and causes heat injury in plants. For the usual plant fruiting and developments, mandatory interaction between intermolecular can be destructed by rising in temperature in summertime (Bita and Gerats 2013). Extended exposure on rising temperature on plants may cause severe metabolic alterations and cellular disorganization, resulting in stunted outgrowth, and progress (Rai et al. 2021).

In terms of meteorology, a time span without enough precipitation is known as drought (Jaleel et al. 2009). Whenever water accessibility becomes low through evaporation and transpiration in plants due to any climate change, then drought stress is initiated (Mishra and Cherkauer 2010). It is said that negligence in mandatory moisture for plant development and outgrowth can be considered drought conditions in agriculture (Farooq et al. 2012). Drought is the most devastating environmental stress, drastically affecting crop yields, rate of growth, and biomass (Anjum et al. 2011). Over the past few decades, its ferocity has increased, posing a threat to world food security. More than 50% of agricultural land could be damaged by irregular plant development habits as a result of stress by 2050 under this paradigm (Vurukonda et al. 2016). Due to an extreme scarcity of water, the dry and semi-dry zones of growing nations are especially vulnerable to drought stress (Nawaz et al. 2015). During drought, turgor and potential of water are influenced and then drought stress modifies the morphological, physiological, and molecular properties of plants (Nawaz et al. 2015; Vurukonda et al. 2016).

The world population is anticipated to get 9.1 billion by the year 2050, which poses significant difficulties to agriculture around the world (Iqbal et al. 2014). Drought stress is the most important variable that inhibits plant development and production. Salt or salinity stress points out the detrimental influence of abundant mineral salts, such as Na + and Cl on plants (Parihar et al. 2015). Restricted precipitation, elevated evaporation, transpiration rates, increasing warmth, and inadequate water approachability increase salt stress spread in dry and semiarid locations more widely (Nawaz et al. 2020). An overabundance of salts like sodium chloride and sodium carbonate play part in salt stress (Nawaz et al. 2020). If a soil's electric conductivity (EC) is 4 dS m-1 and its osmotic pressure is 0.2 MPa, it is said to be saline (Munns and Tester 2008). Based on adaptation mechanism of plants that thrive in salt are divided into two parts. One is that plants are categorized as halophytes that thrive in salt and they are very much developed in saline condition, but,

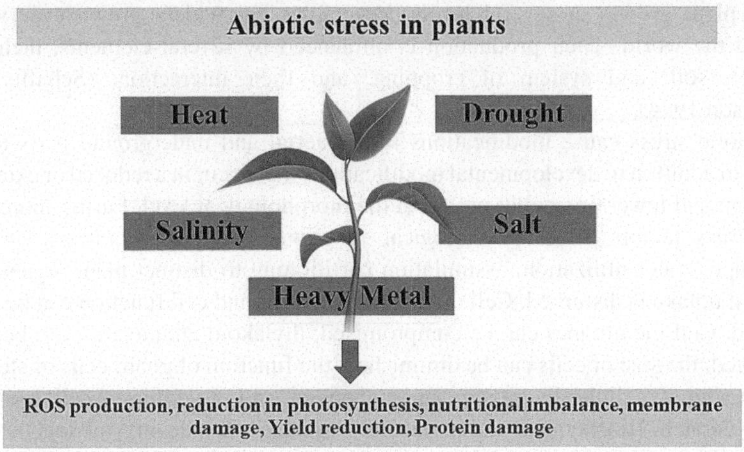

Fig. 1 A General view of the types of abiotic stress in plants

on the contrary, glycophytes are those which highly susceptible to salt and saline environment (Acosta-Motos et al. 2017). Under about 100–200 mM NaCl, the growth of the bulk of crop plants, which are glycophytes, slows or ceases. In contrast, halophytes can tolerate concentrations of 300–500 mM NaCl, which may be the result of plants' increased salt tolerance (Acosta-Motos et al. 2017).

The roots of a plant are the first to recognize saline stress, which both temporarily and permanently stunts plant development. Reduced water supplies produce osmotic stress in plants, resulting in a short-term effect (Parihar et al. 2015). On the contrary, long-term consequences owing to an overabundance of salt cause salt-generated ion toxicity in plants (Parihar et al. 2015). Variation in climatic circumstances resulting from climate change significantly affects plant productivity and output. Agricultural yield is affected by two variables (heat and drought) that are getting worse on daily basis due to the global environmental crisis. Crop output is diminished by the detrimental impact of water deficit on plant growth, physiology, and their developmental pattern (Daryanto et al. 2016) (Fig. 1).

3 Impact of Abiotic Stresses on Vegetable Production

Heat, salt, and drought are the main abiotic stressors that result in substantial yield loss in agricultural and horticulture crops. Horticultural crops include an extensive variety of incredibly important goods, such as fruits and vegetables (Schaffer and Anderson 1994). The major source of crop losses on a global scale is environmental stress, which reduces average crop yields by over 50% for the most essential crops. The morphological, physiological, and molecular function of plants get modified by the effects of these abiotic stresses, that make crop producibility deteriorate and affect plant growth at several levels. Vegetables are widely grown everywhere around the world. Their production is influenced by several elements, including climate, soil, and system of cropping, and their interaction. (Schaffer and Anderson 1994).

Abiotic stress cause modifications in the aerial and underground parts of the crops, in addition to developmental modifications that occur in a reduced or extended life span and fewer or smaller organs at the morphological level. Furthermore, abiotic stress factors affect physiological functions like photosynthesis, gaseous exchange, water utilization, assimilation partitioning to distinct plant organs, and mineral uptake is disturbed. Cell organ arrangements and cell function can be jeopardized. Cell membranes can be compromised, thylakoid architecture can become disrupted, the size of cells can be diminished, the function of guard cells of stomata can be altered, cellular hydration can be changed, and programmed cell death can be accelerated. Biochemical and molecular responses include enzyme inactivation, reactive oxygen species production, osmotic damage, and modifications in metabolite profiles apart from water and ion uptake movements that are changed at the hormonal level. Predictive alterations in plant development and physio-biochemical mechanisms, such as differences in plant structure, growth rate, tissue osmotic potential, and antioxidant defenses, occur during the acclimatization of plants to water constraints (Duan et al. 2007).

To keep things in normal functioning in physiology and membrane transport in plant, water is essential, therefore, supplying adequate water is essential for boosting the yield of horticultural plants. Drought stress occurs when there is minimal moisture content for optimum plant development or water supply restoration (Lombardini and Rossi 2019). Stomatal closure is triggered by low relative water content (RWC) which causes a reduction CO_2 availability in leaf mesophyll and the rate of photosynthesis. Similarly, water deficit affects cell cytokinesis and enlargement, abscisic acid (ABA), and soluble sugar accumulation, resulting in a decrease in crop yield (Lombardini and Rossi 2019). Changes in the environment will have huge impact on environmental stresses that will affect production of vegetable crops. Vegetables, as succulents, often contain more than 90% water. Thus, water has a significant impact on the quantity and quality of vegetables, and drought circumstances dramatically diminish vegetable yield. Due to the increased solute concentration in the environment produced by drought stress, plant cells lose water via osmosis. Solute concentration gets elevated in plant cells, reducing their water potential and interfering with photosynthetic and cellular membrane processes (Rao et al. 2016).

Water stress itself a huge disaster whether it is a drought or flooding, in present time floods due to climate change has turned out to be a biggest problem worldwide. Untimely rain and melting glaciers rising water level in rivers can also cause flood condition. More than a third of agricultural lands are affected by flooding, regardless of how often it occurs. It may be caused by high precipitation, incorrect irrigation, uneven topography, inadequate drainage, or dense soil. The root system experiences multiple morphological, physiological, biochemical, and anatomical alterations during floods (Rao et al. 2016). In addition, increased temperatures, decreased irrigation water supply, flooding, and salinity are the most significant problems preventing the maintenance and expansion of vegetable produce.

As heat waves produce major yield losses, the growing risk of global environmental change is already having a huge influence on crop yield around the world, posing grave threats to the planet's future food security. Numerous crops are restricted in their distribution and yield by temperature.

4 Impact of Drought Stress on Potato

Global climate change affects sustainable crop yield performance and production in the form of intense heat and drought (Hatfield and Prueger (2015). Over a hundred countries cultivate potatoes, which are consumed globally by a billion people. They are an abundant reservoir of carbohydrates and other vital nutrients, that include antioxidants, protein, vitamins, and fiber (Bach et al. 2012). Early stress causes a deleterious effect on tuberization, bulking, and tuber yield because of diminished rates of carbon assimilation (Obidiegwu et al. 2015). Therefore, with global warming and drought, potato yield will be reduced by 2055 as forecasted by scientists. Potatoes require a cool growing season with an average daily temperature of 15–18 °C (Kabira et al. 2006). Temperatures above 21 °C inhibit growth. When potato plants are cultivated at typical daytime temperatures between 14 and 22 °C, they provide the highest yields of tubers for most commercial potato types (van Dam et al. 1999). Due to the crop's weak root system and incapacity to recuperate from water stress, potatoes are also prone to drought (Iwama and Yamaguchi 2006). Potatoes have a shallow root system with a depth ranging from 0.5 to 1 m (Zarzyńska et al. 2017). Approximately 85% of the total root length is concentrated in the first 0.3 meters of soil. Therefore, potatoes extract less available soil water than other crops (Weisz et al. 1994). Drought affects crop yield by directly or indirectly reducing the processes of photosynthesis due to restrictions of photosynthate transport to sink organs (Jones and Corlett 1992).

Nevertheless, the precise relationship is based on additional constructs, including genotype, climate condition, and the tale of drought. Tuber production in potatoes is related to the harvest index (HI) and dry matter concentration; consequently, understanding the impact of genotypic variation on assimilation allocation and dry matter accumulation is critical for breeding (Jefferies 1993). Drought stress in potatoe reduces productivity since the crop may consume less accessible soil water (Weisz et al. 1994), and negatively impact the quality of production (Mane et al. 2008). Plants have a proclivity to detect and respond to drought stress at the molecular and biochemical levels by constantly adjusting regulatory signals via transcription and protein expression, thereby impacting various biochemical reactions and, as a result, physiological and growth processes (Mane et al. 2008; Vasquez-Robinet et al. 2008).

Drought-induced photosynthesis decreases are caused by both stomatal and nonstomatal impacts, with the latter including reduced electron transport due to photo-inhibitory degradation and chloroplast regulatory control of energy dissipation (Baker 2008). During drought stress, the closure of stomatal limits CO_2 supply in the leaf mesophyll and causes a change in electron transfer and metabolic activities which result in a reduction in photosynthesis. During the biochemical changes, ribulose bisphosphate (RuBP) production is reduced, Rubisco activity is reduced, and carboxylation efficiency is reduced (Faver et al. 1996). When the quantity of reactive oxygen species (ROS) exceeds the scavenger potential, irreversible harm to lipids, proteins, and deoxyribonucleic acid (DNA) results in the death of plant cells (Obidiegwu et al. 2015).

5 Morphological and Physio-Biochemical Changes in Potato Against Drought Stress

The sessile nature of plants exposed them to constant abiotic and biotic stress conditions in nature. Recently, human activities have caused an environmental disturbance that exacerbated the frequency of abiotic stresses, causing increased difficulty in securing an adequate food supply. Abiotic stress factors such as water stress cause several changes in the plants ranging from morphological and physio-biochemical changes (Zhang et al. 2022). In this section, the morphological and physio-biochemical changes in potatoes that occur due to drought stress condition will be summarized.

In plants, the roots in the soil are the first point to sense water stress and communicate this information to the shoot (Janiak et al. 2016). Plants response of to drought is a survival mechanism to alleviate the detrimental impact of the stress on their growth and productivity (Yang et al. 2021). The plant has an inbuilt mechanism that enables it to sense and transmit the stress signal to cause a biochemical, molecular, morphological, and physiological response and regulate its gene expression. Potato plants are known to be abiotic stress sensitive. This nature of the potato crop is mainly because of the shallow rooting system. Drought stress inhibits important physiological and biochemical mechanisms resulting in low tuber yield and deficient performance of the crop (Gervais et al. 2021; Nasir and Toth 2022). The extent of damage to the crop is dependent on the growth stage of the plant, cultivar, severity, and the exposure time duration of the plant to the drought episodes. The early stages of potato growth are the most sensitive period to drought stress. It is at this point that stolons are initiated, and the formation of tubers start (Monneveux et al. 2013). Water deficit during this period cause reduction in photosynthetic rates, leaf area, and disruption of the assimilates partitioning to tubers leading to poor tuber initiation, bulking, and tuber yield (Aliche et al. 2020; Gervais et al. 2021). During the tuberization, drought causes fewer stolons per stem, and at the tuber bulking stage, fewer and smaller-sized tubers are formed resulting in lower tuber number and yield. Drought caused a reduction in the shoot height and increase root depth and root architecture.

Vascular tissues of plants play vital roles in water and photo-assimilates transport, and metabolic activities. The plant stem links the underground and above-ground parts of the plants and serves as a column for water movement to sustain the functionality of the water potential (Aliche 2018; Nawaz et al. 2020). The xylem consists of tracheids and vessel members, and transports nutrients and water to all parts of the plant. The transpiration pull drives the xylem transport that operates under negative pressure. Photo-assimilates are the main transport components of the phloem from source tissues to sink organs of the plant for the purposes of storage, growth, and respiration. The phloem substance is transported by the mechanism of mass flow which is facilitated by an osmotic pressure gradient between the source and sinks tissues (Aliche et al. 2020; Aluko et al. 2021). It has sieve elements (SE) that are specialized living cells that connect the phloem companion cells (CC) to the CC-SE complexes (Aliche et al. 2020) that are driven by hydrostatic pressure.

During drought conditions, there is cavitation in potatoes due to the interactions between the xylem and water molecules (Aliche 2018; Aliche et al. 2020; Singh et al. 2013). This causes breakage in the transpiration stream and impairs the hydraulic conductance of xylem vessels which lowers the xylem pressure. The low turgor pressure in the phloem also makes the potato plant vulnerable to water limitation stress. In response, sucrose is retrieved during the drought from lateral sinks into the sieve tube (Aliche 2018). This generates negative water potential in the SEs and causes water from the neighboring xylem to transport (Aliche et al. 2020) and therefore breaks the cavitation in the phloem vessels. These xylem-phloem transport interactions are very important to maintain the above-ground and underground tissue competition for photo-assimilates in potatoes (Aliche et al. 2020; van Bel 2021).

The stem of the plant aside from the communication is also the channel of transport between the terminal sink tissues and assimilate-exporting source leaves (Aliche 2018; Aliche et al. 2020). Morpho-physiologically, a higher narrow diameter-vessel number of xylems per unit area occurs in potato stem under water stress (Aliche 2018; Aliche et al. 2020). Also, a diurnal rhythm change occurs in potatoes during drought stress, which reduces the day and night ratio of xylem flux, while a high ratio between the xylem flux and phloem flux occurs that linked to the dehydration sensitivity in potatoes (Aliche et al. 2020), indicating the xylem water fraction used for phloem transport (Sevanto 2018). Characteristically potato has a higher diurnal xylem flow pattern during the day than at night, also higher values in normal growth conditions than when the plants are experiencing water stress (Aliche et al. 2020). An adaptive mechanism of the water-conducting vessels (smaller-sized and increased xylem density per stem area) of potatoes occurs to facilitate water and assimilate transport during water deficit conditions (Aliche et al. 2020). These interactions between xylem-phloem in potato serves to prevent the negative effects of drought on the vascular tissues of the stem survival purposes.

Water stress causes a reduction (Bhattacharya 2021) in potato plant growth and by effectively decreasing the new leaves proliferation and stem elongation which infringes on the plants' photosynthetic capacity (Aliche et al. 2020). These growth reductions in potato crop water stress or dehydration are adaptive mechanisms in potatoes and other plants purposely for the plants to reduce transport distance to aid an efficient transport of nutrients and water in the plants. Water stress results in reduced height in potato plants. Water stress causes detrimental effects on sugars which affect growth-regulatory factors such as physiological changes, hormonal signals, and physical properties of the plant. These regulatory factors under drought stress control plant height to aid the distribution of water, nutrients, and assimilates.

Potato is characterized by shallow root architecture which is the main feature making the plant susceptible to water deficit (Joshi et al. 2016). The crop has evolved architecturally over time in response to drought stress (Bhattacharya 2021) in the best possible way to absorb water for survival. Potato crops when exposed to water develop small canopies to reduce evapotranspiration and defused root systems to enable them to absorb water. Water uptake and utilization by crops have positive correlations with the root architecture of the crops (Tron et al. 2015). Drought stress in potatoes leads to a reduction in leaf number, leaf area, stomatal conductance, and development of hairy and narrower leaves size to lessen the light absorbance and prevent photooxidative damage (Wasson et al. 2012; Lambers and Oliveira 2019).

Drought adversely affects photosynthesis, nutrient acquisition, and cellular metabolism due to chlorophyll concentration reduction (Maghsoudi et al. 2016), as a result, decreases leaf water content, stability in the cell membrane, and chlorophyll concentration, thus resulting in an increase in oxidation-induced damage. Physiologically drought inhibits photosynthesis. Drought induces the stomata to close which is a regulatory mechanism to reduce the rate of water loss through transpiration and conserve the plant water status (Agurla et al. 2018). The closure of the stomata restricts the rates of CO_2 diffusion in the leaf. As such, CO_2 which is a substrate of the Calvin cycle is limited (Xu et al. 2016; Harrison et al. 2020). Thus, the rates of generation of ATP and NADPH in the photosynthetic electron transport chain exceed their utilization by the Calvin cycle resulting in the accumulation of ATP and NADPH (Gervais et al. 2021). This creates an energy imbalance in the chloroplast level and the generation (Gervais et al. 2021) of ROS that causes oxidative stress and damage to cell components. Potato plants improve instantaneous water use efficiency (WUE) at the whole plant and physiological levels by reducing water loss by transpiration and consequently conserving leaf water status through the decrease in stomatal conductance, leaf number, and leaf area (Gervais et al. 2021; Liu et al. 2022). In this case, the improvement of WUE decreases the leaf photosynthetic surface area and has a reductive impact on carbohydrate synthesis. Thus, resulting in the reduction of stomata conductance of potatoes under drought stress.

Tuber quality traits are adversely impacted by drought stress. Potato tubers defects including hollow heart, secondary growth, tuber cracking, internal brown spot, sugar ends, and malformations are triggered under drought stress (Thornton et al. 2020), also, a-chaconine and a-solanine glycoalkaloid contents are increased (Aziz et al. 2012). In potatoes, chlorophyll a-b binding protein coding genes which are important for photosynthetic enzymes and sucrose biosynthetic (Li et al. 2017b) enzyme regulators are downregulated during drought stress (Schlicke et al. 2014), while there is an upregulation of sucrose breaking enzymes and transcription

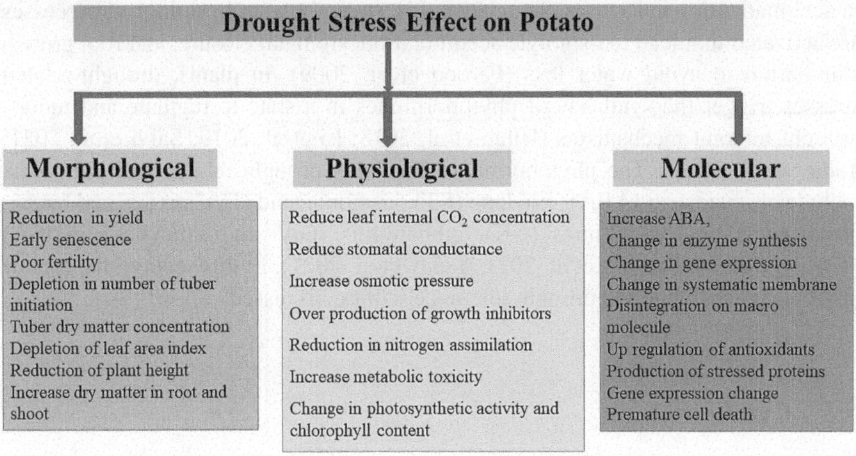

Fig. 2 Response of Potato plant to drought stress at different levels

factors. Among these factors are Dehydrin responsive element binding protein (DREB1A), deeper rooting (DRO), the ethylene response factor (ERF), the LRR receptor-like serine/threonine-protein kinase ERECTA (ERECTA), and MYB (Uga et al. 2013; Joshi et al. 2016).

At the biochemical level, like other crops, accumulation of compatible solutes and osmolytes are also increase in potato under drought stress. These solutes and osmolytes include sugar alcohol, proline, and glycine betaine (Gervais et al. 2021). They mainly reduce the leaf water potential while keeping the turgor pressure, ensuring the capability of leaf cells to imbibe more soil water for leaf water status maintenance and the enablement of plant survival in drought episodes (Monneveux et al. 2013) (Fig. 2).

6 Role of Phytohormones for Stress Tolerance

The sessile nature of plants exposed them to a continuous threat of environmental stress episodes under changing climate dynamics. Under abiotic stress, such as drought conditions, plants are exposed to significant constraints on plant growth, development, productivity, and survival, leading to an alteration in the quality and quantity of crop yield (Waqas et al. 2019; Zhu 2016; Zhang et al. 2022). Drought stress trigger a reduction in the transpiration, turgor pressure, stomatal conductance, and leaf water potential because of the rise in leaf osmotic pressure and the reduction in the osmotic hydraulic conductance and sap flow rate in the roots (Reddy et al. 2021).

Phytohormones are said to be naturally occurring small organic molecules that coordinate cellular activities and influence plant growth and physiological activities in very low concentrations (Chokshi et al. 2017; Peres et al. 2019). This help plants in acclimatization to stresses. To achieve this, several plant physiological processes are activated that lead to osmolyte accumulation, stomatal closure, and root growth stimulation to avoid water loss (Farooq et al. 2009). In plants, drought-related stresses trigger the synthesis of phytohormones in a state to regulate and induce drought tolerant mechanisms (Ullah et al. 2018; Li et al. 2019; Salvi et al. 2021; Yadav et al. 2021). The phytohormones that play drought-related roles in plants include abscisic acid (ABA), ethylene (ET), jasmonic acid (JA), auxins, and brassinosteroids (BRs), cytokinins (CK), gibberellins (GA) and salicylic acid (SA) (Tiwari et al. 2017; Salvi et al. 2021; Yadav et al. 2021). In this section, the role of phytohormones in potato drought tolerance will be discussed.

6.1 Cytokinins

The phytohormone cytokinin (CK) is a group of essential hormones in plants necessary for growth and developmental processes (Spaepen 2015; Ryu and Cho 2015). They were first identified in maize. Studies of their metabolism, signal transduction pathway, and functions in plant growth and development revealed that CKs play several roles in preventing flower abortion, improving fruit development, vascular differentiation, root architecture, and nitrogen partitioning (Albacete et al. 2014).

In potato plants, CKs reduce lateral root cell division and differentiation of phloem and metaxylem in roots (Sharma and Zheng 2019). This mechanism enables the epical roots' cell division and differentiation so that roots can grow deeper to absorb water. CK also inhibits the differentiation of cells in leaves and shoots and reduces the rate of senescence in the leaves (Aloni et al. 2006). This regulatory mechanism results in a reduction in plant height and photosynthesis protection during drought (Rivero et al. 2009; Skalák et al. 2019).

6.2 Abscisic Acid

Abscisic acid (ABA) is a weak acid and one of the most important phytohormones that were first isolated from cotton fruits (Ullah et al. 2018). It participates in various important physiological processes in plants such as stress responses, development, and reproduction. In potato plants, drought stress episodes cause osmotic stress to occur (Hasanuzzaman et al. 2014). When osmotic stress occurs, signal perception is initiated through the plasma membrane leading to the activation of ABA synthesis in the plastids in the roots (Xu et al. 2013; Takahashi et al. 2020). The ABA synthesis activates gene expression which confers adaptive physiological changes in the plant (Asad et al. 2019; Li et al. 2021). As the drought stress condition persists, the level of ABA levels increases which is then transported via vascular tissues to all parts of the plants (Takahashi et al. 2020). As this happens, ABA moves into the cells, and the cell-to-cell transport of the ABA is enhanced by transporters. In this mechanism, the transporters found in the plasma membrane where ATP-binding cassette (ABC) hydrolyze ATP to transport it, and the low-affinity nitrate transporter family. This ABA perception and signal transduction are mediated by two pathways, which are ABA-dependent and ABA-independent (Bhattacharya 2021; Ullah et al. 2018; Takahashi et al. 2020).

ABA significantly regulates stomatal closure in potato plants and other crops when they are exposed to drought stress and triggers the senescence of leaves and synthesis of storage lipids (Ahmadi et al. 2010; Farooq et al. 2012). These physiological actions such as the stomata closure and leaf senescence are adaptive mechanisms to reduce the rate of water loss through the leaves. This action regulates stomatal conductance to control water loss during stress conditions, regulates osmosis and turgidity, and plays a significant role in gas exchange between tissues and the atmosphere (de Campos et al. 2011). The regulation of the gaseous exchange is due to the guard cells' response to ABA, which causes water to exit the guard cells, making them flaccid, and thus reducing the loss of water by the regulation of the stomata. The movements of the stomata are mediated by cell turgor. This is controlled by the transport of ions across membranes (Andrés et al. 2014; Ullah et al. 2018). ABA triggers stomatal pores via efflux of potassium (K+) and anions from guard cells as well as the removal of organic osmolytes (Ullah et al. 2018). The closure of the stomata is the first step to reducing water loss during drought conditions (Ullah et al. 2018). The storage lipids have the capacity to conserve and maintain the water potential of the plant giving it tolerant behavior to drought.

As the ABA is affecting the above-ground parts of the plant to reduce water loss, it promotes root elongation to reach deeper for soil water, causing a more profuse higher root length density and deeper root systems in the soil as drought stress (Uga et al. 2013) adaptative mechanisms. In effect, ABA is a stress-responsive phytohormone that regulates the correlation between stomatal conductance and photosynthesis (Bhattacharya 2021) reduction during drought stress.

6.3 *Jasmonic Acid*

One other important phytohormone in plants' adaptive response to environmental stress conditions is jasmonic acid (JA). JA is obtained from α-linolenic acid and its active derivatives significantly regulate plants' responses to abiotic stresses when exposed. In plants, JA functions in the growth and development of the root, shoot, and involved in fruit ripening, tendril coiling, and viable pollen production (Santino et al. 2013; Ullah et al. 2017, 2018).

In potatoes and other Solanaceae families, JA induces stomata closure and regulation during drought stress conditions and enhances the antioxidant activities in the plant to counteract the effect of ROS (Kar 2011; Ullah et al. 2017). Alongside with other hormones, JA improves the water conductivity of potato crops. This action is in relation to the fact that during drought stress, the vascular tissues decrease in size in the shoot system and thus ensure the water conductivity activity. Moreover, the presence of JA is required to increase ABA levels which function in the closure of the stomata, and reduction in plant shoot density among others. It has been found in other crops that JA prevents the accumulation of malondialdehyde (MDA), cadmium (Cd), and H_2O_2 and thus mitigates the deleterious effects of Cd stress (Faizan et al. 2021). As in other crops, in potatoes, these mechanisms occur alongside increasing the concentrations and activities of glycine betaine, proline, catalase (CAT), superoxide dismutase (SOD), gibberellin (GR), and ascorbate peroxidase (APX) (Hoque et al. 2007). These actions reduce the impact of oxidative stress on the plant. There is also the upregulation and downregulation of some transcription factors with differential gene expression in potatoes. These regulatory mechanisms of jasmonate signaling help in different processes to initiate appropriate responses of the potato to drought stress through various signal transductions. Thus, JA confers tolerance to the plants when they are exposed to drought.

6.4 *Auxin*

Auxins are important phytohormones that can initiate their effect at a long-distance transport system to affect all the life aspects of a plant such as during the development of the embryo, division, and differentiation of the cells, the architectonics of the plant, and its orientation. There are many auxins in plants but phenylacetic acid (PAA), indole-3-acetic acid (IAA), and 4-chloroindole-3-acetic acid (4-Cl-IAA) are the common natural auxins, while 1-naphthaleneacetic acid (NAA), and 2,4-dichlorophenoxyacetic acid (2,4-D) is the synthetic forms (Raggi et al. 2020).

During drought stress conditions in potatoe, auxin triggers a tolerant response by stimulating the growth of roots and tubers through stress-induced changes in the root and the stolons (Joshi and Ginzberg 2021; Choudhary et al. 2021). This triggers epical cell elongation through cell division and differentiation and arrests the initiation of the lateral cell in the root. This growth mechanism response is an adaptive mechanism to acquire water from deep down the soil, and so confers drought tolerance to the potato crop. There are, however, differences between different genotypes of potatoes in this adaptive behavior, which is an area of focus in potato breeding programs to identify drought tolerant breeding lines in the wake of climate change. The auxin accumulation in the root system reduces the potato plant water usage during the day and night. The plant's hydraulic properties are also modulated by the accumulation of auxins. Thus, the expression of water-saving traits associated with enhanced yields during drought is enabled (Sadok and Schoppach 2019).

6.5 *Salicylic Acid*

Salicylic acid (SA) is a phytohormone that have a significant role to ameliorate abiotic (drought) stress and regulating plant defense against microbial pathogens. Also, SA is known to enhance photosynthesis protective reactions, and promote stomatal conductance, and chlorophyll and carotenoid content. During normal plant growth conditions, SA helps plant cell elongation, nutrient uptake, photosynthetic activity, photosynthetic pigment levels, source-to-sink regulation, and regulating plant growth and development (Farooq et al. 2009).

In potato plants under water deficit, the application of SA reduces the plant cell membrane damage in leaves. This is done by enhancing the antioxidant enzymes ascorbate peroxidase (APOD), SOD, ascorbate (AsA), CAT, glutathione (GSH), and guaiacol peroxidase (GPOD) activities which reduce the cellular lipid peroxidation and reactive oxygen species accumulation. Also, when potatoes are exposed to drought, the non-enzymatic antioxidants GSH and AsA directly detoxify H_2O_2 and mitigate their high-efficiency oxidative stress induced by abiotic stress. These actions in the plant lead to the ultimate conversion of ROS to water. Studies show that SA is involved in plant tolerance to drought stress by increasing CO_2, leaf gas exchange, SOD and APX activities, and dry weight accumulation and growth in drought-stressed plants (González-Villagra et al. 2022). These regulatory mechanisms within the potato plant confer drought tolerance properties on the plant though the rate of response is variety specific. Thus, SA enables the potato plant to perform well under drought stress.

7 Modern Genetic Approaches for Trait Improvement in Potato

Genome-wide, the cultivated potato is a highly heterozygous autotetraploid with high allelic diversity and complex segregation with reduced sexual fertility (Ghislain and Douches 2020; Tang et al. 2022) Potato crops faced several environmental impediments in their vicinity including drought stress in recent years due to the changing climatic conditions (Nadakuduti et al. 2019). The classical plant breeding approach has failed to meet the complexities of the genetic basis of abiotic stresses. In potato breeding, the advancement in technology has proved potent in addressing the failed aspects of the classical plant breeding strategies in the recent era. In this section, the role of modern genetic approaches such as QTL mapping and GWAS, and genome editing CRISPR/Cas9 technology in drought stress improvement will be discussed.

7.1 QTL Mapping and GWAS

The advent of genome-wide association studies (GWAS) and quantitative trait loci (QTL) along with next-generation sequencing (NGS) technologies (Li et al. 2017a; Wang et al. 2020) facilitated the breeding of potatoes for drought stress conditions. GWAS are a powerful tool for complex trait analysis in crop plants (Sharma et al. 2018; Uffelmann et al. 2021). GWAS and QTL data revealed gene expression information that is powerful for biological function.

GWAS and QTL applications including NGS, such as RNA sequencing, and whole genome re-sequencing (WGRS) led to the identification of drought-tolerant markers in potato cultivars (Saidi and Hajibarat 2020). These NGS tools essentially help in the identification of marker genes responding to drought in the gene networks (Varshney et al. 2021). QTL and GWAS have widely been used in gene detection associated with growth and developmental processes (Le et al. 2021). In potato breeding, the use of GWAS and QTL in plant breeding is advantageous for potato breeders to identify molecular markers that relate to drought stress in potato plants (Saidi and Hajibarat 2020). Obidiegwu et al. (2015) state that drought stress accounts for about 26% effect on plants and thus making it among the most yield-limiting abiotic stresses plants are exposed to. Thus, knowledge of the genetic basis of drought tolerance and its nucleotide variation (Saidi and Hajibarat 2020) in the various loci will efficiently and effectively help potato breeders to improve crops.

Intricately, water deficit in potatoes affects several morphological parameters (root architecture, branching, leaf size and number, and stems), and yield and yield quality (Begum et al. 2018; Bhattacharya 2021). The most yield effect of drought in potatoes occurs at the tuber initiation and bulking stages of the crop. Due to the complex genetic composition and heterozygote cells of potatoes, simple sequence repeat (SSR) markers and single nucleotide polymorphism (SNPs) are used to detect the existence or absence of an allele on chromosomes (Saidi and Hajibarat 2020). GWAS has emerged by combining genetic information on SNPs in a gene that increases the chances for novel discovery of genes and the generation of more effective results (Saidi and Hajibarat 2020) on plant drought response and how they relate to the genes' influence on complex traits (Tak and Farnham 2015). Thus, GWAS and QTLs are important for the dissection of gene functionality and connections between genes in traits of interest (Liu et al. 2021). These are hence useful in genomic studies and genomic selection for potato cultivar improvement to drought stress. Through GWAS and QTL analysis, genes related to drought stress tolerance have been identified at the genomic and molecular levels, conferring drought tolerance mechanisms on the crop (Abdelraheem et al. 2021; Sallam et al. 2022). The products of drought-related genes play a key role in stimulating initial stress response and in inducing stress tolerance at the cellular level (Gervais et al. 2021).

7.2 *Genome Editing CRISPR/Cas9*

The advent of genome editing tools is a step forward in plant breeding for the improvement of traits of interest. It addresses the limitations of molecular and traditional (conventional) breeding approaches along with the techniques of genetic engineering in plants (Ahmar et al. 2020; Gao 2021; Rahman et al. 2022). Compared to genome editing, the molecular breeding approaches, the traditional (conventional) breeding and genetic engineering techniques consume a lot of time and are cumbersome, accompanied by the resistive force of nonacceptance of the genetically modified varieties because of regulatory concerns. The genome editing tool has come to address these setbacks and facilitate the improvement of agriculturally important traits (Ammann 2019; Backiyarani et al. 2022). The CRISPR/Cas (clustered regularly interspaced short palindromic repeat) system is a tool that is simple, adaptable, flexible, and has a wide rate of applicability, with multiple sequence-specific nucleases that facilitate a precise gene modification whose resultant effect is the development of climate resilient crops (Ali et al. 2022; Naeem et al. 2020; Joshi et al. 2020).

The CRISPR/Cas system operates on a single guide RNA in combination with the complex Cas endonuclease. The Cas endonuclease cuts open the double-stranded DNA at the target site and later the endogenic cell mending mechanisms repair the cut section resulting in the incorporation of the trait of interest. The engineered CRISPR/Cas9 contains a precursor crRNA (precrRNA), a trans-activating crRNA (tracrRNA), and the main domains (CRISPR-associated protein 9, Cas9) (Jiang and Doudna 2017; Salsman and Dellaire 2017). Cas9 is an endonuclease (DNA endonuclease) containing an HNH nuclease domain and takes part in the crRNA maturation process and the crRNA-guided DNA cleavage which is engineered based on the type II CRISPR (Anders et al. 2016; Swartjes et al. 2020). By this, the CRISPR Cas9 system works by targeting a specific sequence in the genome with sgRNA (Jiang and Doudna 2017) that requires an introduction of a 20 bp spacer sequence and thus making CRISPR Cas9 much simpler and easier for manipulation (Bao et al. 2019; Khadempar et al. 2019).

During drought stress in potatoes, the genes associated with drought stress and the stress signaling transcription factors trigger the signal molecules and metabolites accumulation that enhances the plant drought tolerance by the overexpression of the tolerant genes or transcription factors (Bhargava and Sawant 2013). Drought-sensitive gene expression induced ROS production, hormonal imbalance, and reduced antioxidant activity that facilitate drought conditions in plants (Sofo et al. 2015; Laxa et al. 2019). Genome editing by CRISPR/Cas9 technology can improve drought tolerance in sensitive potato cultivars by silencing the drought-sensitive genes that will naturally demonstrate drought tolerance and be used to introduce novel drought tolerant alleles in the genome to target the plants molecular, morphological, and physiological responses during the episode of water deficit that makes the plant susceptible to drought (Rosero et al. 2020). Inserting genes responsible for deep and dense rooting, stomatal opening, photosynthetic rates, and other factors into the genome of the potato makes them tolerant to drought. Gene knockout approach has been used to improve potato water absorption capacity, water use efficiency, photosynthetic rate, the antioxidant activities among others that improve potato drought stress tolerance capacity (Chilcoat et al. 2017; Rosero et al. 2020). Thus, CRISPR/Cas9 technology has two main functions with regards to improving potato tolerance to drought stress through gene silencing or the gene knockout approach.

8 Conclusion and Future Perspectives

In this era, changes in climatic conditions globally have an unfathomable impact on plant developmental and yield processes. Water deficit as an abiotic stress factor affects every growth process of plants causing a reduction in the plant's growth and economic yield. Potato crops are naturally sensitive to drought. The episodes of drought stress in potatoe trigger the stomata to close, cause impairment in photosynthetic and transpiration rates, and result in earlier crop maturity and low production tuber yield and quality. The effect of drought stress results in transpiration and turgor pressure reduction, low stomatal conductance, and leaf water potential because of the rise in leaf osmotic pressure and the reduction in the osmotic hydraulic conductance. In response to these negative impacts of drought, potato crops among others have device adaptive biochemical, morphological, and physiological mechanisms to counteract the effects of drought. Such mechanisms include reduced shoot growth, dense and defused root systems, smaller leaves, closure of stomata, and synthesis of antioxidants to scavenge the ROS. These responses to drought are regulated by factors such as the hydraulic status of the plant, phytohormones, ROS signaling, and osmotic adjustment.

Phytohormones, such as ABA, JA, auxins, CK, GA and SA, play a significant role in potato response and acclimatization to drought stress. Phytohormones activate osmolyte accumulation, closure of stomata, and root growth stimulation. These physiological processes regulate water loss and induce drought tolerant mechanisms.

In the recent era, the advent of modern genetic approaches is a step forward in plant breeding and addresses the limitations of molecular and classical plant breeding approaches for the improvement of traits of interest in plants. In potato breeding, the use of GWAS and QTL techniques are powerful tools aiding the identification of various novel genetic markers that relate to drought stress tolerance. The GWAS and QTL tools have paved the way for genome editing by CRISPR/Cas9 technology. By CRISPR/Cas9 technology, traits of interest can be inserted into the genome of the crop plant, or undesired traits can be silenced or knocked out. Prospectively the negative impact of drought stress on potatoes can be improved by incorporating modern genetic approaches such as GWAS and QTL studies and the recently clustered CRISPR/Cas9 technology. These technologies can be employed in the improvement of other abiotic stress in potatoes and other crops for the advancement of plant breeding.

References

Abdelraheem A, Thyssen GN, Fang DD, Jenkins JN, McCarty JC, Wedegaertner T, Zhang J (2021) GWAS reveals consistent QTL for drought and salt tolerance in a MAGIC population of 550 lines derived from intermating of 11 Upland cotton (*Gossypium hirsutum*) parents. Mol Gen Genomics 296(1):119–129

Acosta-Motos JR, Ortuño MF, Bernal-Vicente A, Diaz-Vivancos P, Sanchez-Blanco MJ, Hernandez JA (2017) Plant responses to salt stress: adaptive mechanisms. Agronomy 7(1):18

Agurla S, Gahir S, Munemasa S, Murata Y, Raghavendra AS (2018) Mechanism of stomatal closure in plants exposed to drought and cold stress. Surviv Strateg Extreme Cold Desiccation 1081:215–232

Ahmadi SH, Andersen MN, Plauborg F, Poulsen RT, Jensen CR, Sepaskhah AR, Hansen S (2010) Effects of irrigation strategies and soils on field-grown potatoes: gas exchange and xylem [ABA]. Agric Water Manag 97(10):1486–1494

Ahmar S, Gill RA, Jung KH, Faheem A, Qasim MU, Mubeen M, Zhou W (2020) Conventional and molecular techniques from simple breeding to speed breeding in crop plants: recent advances and future outlook. Int J Mol Sci 21(7):2590

Albacete AA, Martínez-Andújar C, Pérez-Alfocea F (2014) Hormonal and metabolic regulation of source–sink relations under salinity and drought: from plant survival to crop yield stability. Biotechnol Adv 32(1):12–30

Ali Z, Rai SK, Jan S, Raina K (2022) An assessment on CRISPR Cas as a novel asset in mitigating drought stress. Genet Resour Crop Evol 69:1–17

Aliche EB (2018) Water-saving potatoes: exploring and characterizing drought tolerance mechanisms. Doctoral dissertation, Wageningen University

Aliche EB, Prusova-Bourke A, Ruiz-Sanchez M, Oortwijn M, Gerkema E, Van As H et al (2020) Morphological and physiological responses of the potato stem transport tissues to dehydration stress. Planta 251(2):1–15

Aloni R, Aloni E, Langhans M, Ullrich CI (2006) Role of cytokinin and auxin in shaping root architecture: regulating vascular differentiation, lateral root initiation, root apical dominance and root gravitropism. Ann Bot 97(5):883–893

Aluko OO, Li C, Wang Q, Liu H (2021) Sucrose utilization for improved crop yields: a review article. Int J Mol Sci 22(9):4704

Ammann K (2019) Selected innovative solutions for the regulation of GM crops in times of gene editing. In: Plant biotechnology: Progress in genomic era. Springer, Singapore, pp 3–41

Anders C, Bargsten K, Jinek M (2016) Structural plasticity of PAM recognition by engineered variants of the RNA-guided endonuclease Cas9. Mol Cell 61(6):895–902

Andrés Z, Pérez-Hormaeche J, Leidi EO, Schlücking K, Steinhorst L, McLachlan DH et al (2014) Control of vacuolar dynamics and regulation of stomatal aperture by tonoplast potassium uptake. Proc Natl Acad Sci 111(17):E1806–E1814

Anithakumari AM, Dolstra O, Vosman B, Visser RG, van der Linden CG (2011) In vitro screening and QTL analysis for drought tolerance in diploid potato. Euphytica 181(3):357–369

Anjum SA, Wang LC, Farooq M, Hussain M, Xue LL, Zou CM (2011) Brassinolide application improves the drought tolerance in maize through modulation of enzymatic antioxidants and leaf gas exchange. J Agron Crop Sci 197(3):177–185

Asad MAU, Zakari SA, Zhao Q, Zhou L, Ye Y, Cheng F (2019) Abiotic stresses intervene with ABA signaling to induce destructive metabolic pathways leading to death: premature leaf senescence in plants. Int J Mol Sci 20(2):256

Aziz A, Randhawa MA, Butt MS, Asghar A, Yasin M, Shibamoto T (2012) Glycoalkaloids (α-chaconine and α-solanine) contents of selected Pakistani potato cultivars and their dietary intake assessment. J Food Sci 77(3):T58–T61

Bach S, Yada RY, Bizimungu B, Sullivan JA (2012) Genotype by environment interaction effects on fibre components in potato (Solanum tuberosum L.). Euphytica 187(1):77–86

Backiyarani S, Anuradha C, Uma S (2022) Genomic designing for biotic stress resistant banana. In: Genomic designing for biotic stress resistant fruit crops. Springer, Cham, pp 25–74

Baker NR (2008) Chlorophyll fluorescence: a probe of photosynthesis in vivo. Annu Rev Plant Biol 59:89–113

Bao A, Burritt DJ, Chen H, Zhou X, Cao D, Tran LSP (2019) The CRISPR/Cas9 system and its applications in crop genome editing. Crit Rev Biotechnol 39(3):321–336

Begum M, Saikia M, Sarmah A, Ojah NJ, Deka P, Dutta PK, Ojah I (2018) Water management for higher potato production: a review. Int J Curr Microbiol App Sci 7(5):24–33

Bhargava S, Sawant K (2013) Drought stress adaptation: metabolic adjustment and regulation of gene expression. Plant Breed 132(1):21–32

Bhattacharya A (2021) Effect of soil water deficit on growth and development of plants: a review. In: Soil water deficit and physiological issues in plants. Springer, Singapore, pp 393–488

Bita CE, Gerats T (2013) Plant tolerance to high temperature in a changing environment: scientific fundamentals and production of heat stress-tolerant crops. Front Plant Sci 4:273

Blum A (2002) Drought tolerance is a complex trait. In: Saxena NP, O'Toole JC (eds) Field screening for drought tolerance incrop plants with emphasis of rice. ICRISAT, Patancheru, pp 17–22

Boyer JS (1982) Plant productivity and environment. Science 218(4571):443–448

Cassia R, Nocioni M, Correa-Aragunde N, Lamattina L (2018) Climate change and the impact of greenhouse gasses: CO_2 and NO, friends and foes of plant oxidative stress. Front Plant Sci 9:273

Chaudhry S, Sidhu GPS (2021) Climate change regulated abiotic stress mechanisms in plants: a comprehensive review. Plant Cell Rep 41:1–31

Chilcoat D, Liu ZB, Sander J (2017) Use of CRISPR/Cas9 for crop improvement in maize and soybean. Prog Mol Biol Transl Sci 149:27–46

Chokshi K, Pancha I, Ghosh A, Mishra S (2017) Nitrogen starvation-induced cellular crosstalk of ROS-scavenging antioxidants and phytohormone enhanced the biofuel potential of green microalga *Acutodesmus dimorphus*. Biotechnol Biofuels 10(1):1–12

Choudhary S, Zehra A, Mukarram M, Naeem M, Khan M, Hakeem KR, Aftab T (2021) An insight into the role of plant growth regulators in stimulating abiotic stress tolerance in some medicinally important plants. In: Plant growth regulators. Springer, Cham, pp 75–100

Cramer GR, Urano K, Delrot S, Pezzotti M, Shinozaki K (2011) Effects of abiotic stress on plants: a systems biology perspective. BMC Plant Biol 11(1):1–14

Daryanto S, Wang L, Jacinthe PA (2016) Global synthesis of drought effects on maize and wheat production. PLoS One 11(5):e0156362

de Campos MKF, de Carvalho K, de Souza FS, Marur CJ, Pereira LFP, Bespalhok Filho JC, Vieira LGE (2011) Drought tolerance and antioxidant enzymatic activity in transgenic 'Swingle' citrumelo plants over-accumulating proline. Environ Exp Bot 72(2):242–250

Duan B, Yang Y, Lu Y, Korpelainen H, Berninger F, Li C (2007) Interactions between water deficit, ABA, and provenances in Picea asperata. J Exp Bot 58(11):3025–3036

Dutta P, Chakraborti S, Chaudhuri KM, Mondal S (2020) Physiological responses and resilience of plants to climate change. In: New Frontiers in stress Management for Durable Agriculture. Springer, Singapore, pp 3–20

Fahad S, Bajwa AA, Nazir U, Anjum SA, Farooq A, Zohaib A et al (2017) Crop production under drought and heat stress: plant responses and management options. Front Plant Sci 8:1147

Faizan M, Bhat JA, Hessini K, Yu F, Ahmad P (2021) Zinc oxide nanoparticles alleviates the adverse effects of cadmium stress on *Oryza sativa* via modulation of the photosynthesis and antioxidant defense system. Ecotoxicol Environ Saf 220:112401

FAO (2012) Coping with water scarcity: an action framework for agriculture and food security. Food and Agriculture Organization of the United Nations, Rome

Farooq M, Wahid A, Kobayashi NSMA, Fujita DBSMA, Basra SMA (2009) Plant drought stress: effects, mechanisms and management. In: Sustainable agriculture. Springer, Dordrecht, pp 153–188

Farooq M, Hussain M, Wahid A, Siddique KHM (2012) Drought stress in plants: an overview. In: Plant responses to drought stress. Springer, Berlin/Heidelberg, pp 1–33

Faver KL, Gerik TJ, Thaxton PM, El-Zik KM (1996) Late season water stress in cotton: II. Leaf gas exchange and assimilation capacity. Crop Sci 36(4):922–928

Gao C (2021) Genome engineering for crop improvement and future agriculture. Cell 184(6):1621–1635

Gervais T, Creelman A, Li XQ, Bizimungu B, De Koeyer D, Dahal K (2021) Potato response to drought stress: physiological and growth basis. Front Plant Sci 12:1630

Ghislain M, Douches DS (2020) The genes and genomes of the potato. In: The potato crop. Springer, Cham, pp 139–162

González-Villagra J, Reyes-Díaz MM, Tighe-Neira R, Inostroza-Blancheteau C, Escobar AL, Bravo LA (2022) Salicylic acid improves antioxidant defense system and photosynthetic performance in Aristotelia chilensis plants subjected to moderate drought stress. Plan Theory 11(5):639

Harrison EL, Arce Cubas L, Gray JE, Hepworth C (2020) The influence of stomatal morphology and distribution on photosynthetic gas exchange. Plant J 101(4):768–779

Hasanuzzaman M, Nahar K, Gill SS, Fujita M (2014) Drought stress responses in plants, oxidative stress, and antioxidant defense. In: Tuteja N, Gill SS (eds) Climate change and plant abiotic stress tolerance. Wiley-Blackwell, Hoboken, pp 209–237

Hatfield JL, Prueger JH (2015) Temperature extremes: effect on plant growth and development. Weather Clim Extremes 10:4–10

Hijmans RJ (2003) The effect of climate change on global potato production. Am J Potato Res 80(4):271–279

Hoque MA, Banu MNA, Okuma E, Amako K, Nakamura Y, Shimoishi Y, Murata Y (2007) Exogenous proline and glycinebetaine increase NaCl-induced ascorbate–glutathione cycle enzyme activities, and proline improves salt tolerance more than glycinebetaine in tobacco Bright Yellow-2 suspension-cultured cells. J Plant Physiol 164(11):1457–1468

Iqbal N, Umar S, Khan NA, Khan MIR (2014) A new perspective of phytohormones in salinity tolerance: regulation of proline metabolism. Environ Exp Bot 100:34–42

Iwama K, Yamaguchi J (2006) Abiotic stresses. In: Handbook of potato production, improvement, and postharvest management. CRC Press, pp 231–278

Jaleel CA, Manivannan P, Wahid A, Farooq M, Al-Juburi HJ, Somasundaram R, Panneerselvam R (2009) Drought stress in plants: a review on morphological characteristics and pigments composition. Int J Agric Biol 11:100–105

Janiak A, Kwaśniewski M, Szarejko I (2016) Gene expression regulation in roots under drought. J Exp Bot 67(4):1003–1014

Jefferies RA (1993) Cultivar responses to water stress in potato: effects of shoot and roots. New Phytol 123(3):491–498

Jenkins K, Warren R (2015) Quantifying the impact of climate change on drought regimes using the Standardised Precipitation Index. Theor Appl Climatol 120(1):41–54

Jiang F, Doudna JA (2017) CRISPR-Cas9 structures and mechanisms. Annu Rev Biophys 46(1):505–529

Jones HG, Corlett JE (1992) Current topics in drought physiology. J Agric Sci 119(3):291–296

Joshi M, Ginzberg I (2021) Adventitious root formation in crops-potato as an example. Physiol Plant 172(1):124–133

Joshi M, Fogelman E, Belausov E, Ginzberg I (2016) Potato root system development and factors that determine its architecture. J Plant Physiol 205:113–123

Joshi RK, Bharat SS, Mishra R (2020) Engineering drought tolerance in plants through CRISPR/Cas genome editing. 3 Biotech 10(9):1–14

Kabira JN, Macharia M, Karanja MW, Muriithi LM (2006) Potato seed: how to to grow and market healthy planting material, KARI Technical Note, vol 20. Agricultural Research Institute, Nairobi

Kar RK (2011) Plant responses to water stress: role of reactive oxygen species. Plant Signal Behav 6(11):1741–1745

Khadempar S, Familghadakchi S, Motlagh RA, Farahani N, Dashtiahangar M, Rezaei H, Gheibi Hayat SM (2019) CRISPR-Cas9 in genome editing: its function and medical applications. J Cell Physiol 234(5):5751–5761

Lambers H, Oliveira RS (2019) Photosynthesis, respiration, and long-distance transport: photosynthesis. In: Plant physiological ecology. Springer, Cham, pp 11–114

Laxa M, Liebthal M, Telman W, Chibani K, Dietz KJ (2019) The role of the plant antioxidant system in drought tolerance. Antioxidants 8(4):94

Le TD, Gathignol F, Vu HT, Nguyen KL, Tran LH, Vu HTT et al (2021) Genome-wide association mapping of salinity tolerance at the seedling stage in a panel of Vietnamese landraces reveals new valuable QTLs for salinity stress tolerance breeding in rice. Plan Theory 10(6):1088

Li H, Su G, Jiang L, Bao Z (2017a) An efficient unified model for genome-wide association studies and genomic selection. Genet Sel Evol 49(1):1–8

Li Y, Li H, Li Y, Zhang S (2017b) Improving water-use efficiency by decreasing stomatal conductance and transpiration rate to maintain higher ear photosynthetic rate in drought-resistant wheat. Crop J 5(3):231–239

Li P, Yang H, Wang L, Liu H, Huo H, Zhang C et al (2019) Physiological and transcriptome analyses reveal short-term responses and formation of memory under drought stress in rice. Front Genet 10:55

Li Q, Xu F, Chen Z, Teng Z, Sun K, Li X et al (2021) Synergistic interplay of ABA and BR signal in regulating plant growth and adaptation. Nat Plants 7(8):1108–1118

Lindner M, Maroschek M, Netherer S, Kremer A, Barbati A, Garcia-Gonzalo J et al (2010) Climate change impacts, adaptive capacity, and vulnerability of European forest ecosystems. For Ecol Manag 259(4):698–709

Liu X, Zhang J, Xiong X, Chen C, Xing Y, Duan Y et al (2021) An integrative analysis of transcriptome and GWAS data to identify potential candidate genes influencing meat quality traits in pigs. Front Genet 12:1984

Liu J, Zhang J, Zhu M, Wan H, Chen Z, Yang N et al (2022) Effects of plant growth promoting rhizobacteria (PGPR) strain *Bacillus licheniformis* with biochar amendment on potato growth and water use efficiency under reduced irrigation regime. Agronomy 12(5):1031

Lobell DB, Schlenker W, Costa-Roberts J (2011) Climate trends and global crop production since 1980. Science 333(6042):616–620

Lombardini L, Rossi L (2019) Ecophysiology of plants in dry environments. In: Dryland ecohydrology. Springer, Cambridge, pp 71–100

Maghsoudi K, Emam Y, Pessarakli M (2016) Effect of silicon on photosynthetic gas exchange, photosynthetic pigments, cell membrane stability and relative water content of different wheat cultivars under drought stress conditions. J Plant Nutr 39(7):1001–1015

Mane SP, Robinet CV, Ulanov A, Schafleitner R, Tincopa L, Gaudin A et al (2008) Molecular and physiological adaptation to prolonged drought stress in the leaves of two Andean potato genotypes. Funct Plant Biol 35(8):669–688

Manickavelu A, Nadarajan N, Ganesh SK, Gnanamalar RP, Chandra Babu R (2006) Drought tolerance in rice: morphological and molecular genetic consideration. Plant Growth Regul 50(2):121–138

Mishra V, Cherkauer KA (2010) Retrospective droughts in the crop growing season: implications to corn and soybean yield in the Midwestern United States. Agric For Meteorol 150(7–8):1030–1045

Monneveux P, Ramírez DA, Pino MT (2013) Drought tolerance in potato (*S. tuberosum* L.): can we learn from drought tolerance research in cereals? Plant Sci 205:76–86

Munns R, Tester M (2008) Mechanisms of salinity tolerance. Annu Rev Plant Biol 59:651–681

Nadakuduti SS, Starker CG, Voytas DF, Buell CR, Douches DS (2019) Genome editing in potato with CRISPR/Cas9. In: Plant genome editing with CRISPR systems. Humana Press, New York, pp 183–201

Naeem M, Majeed S, Hoque MZ, Ahmad I (2020) Latest developed strategies to minimize the off-target effects in CRISPR-Cas-mediated genome editing. Cells 9(7):1608

Nasir MW, Toth Z (2022) Effect of drought stress on potato production: a review. Agronomy 12(3):635

Nawaz F, Ahmad R, Ashraf MY, Waraich EA, Khan SZ (2015) Effect of selenium foliar spray on physiological and biochemical processes and chemical constituents of wheat under drought stress. Ecotoxicol Environ Saf 113:191–200

Nawaz F, Shehzad MA, Majeed S, Ahmad KS, Aqib M, Usmani MM, Shabbir RN (2020) Role of mineral nutrition in improving drought and salinity tolerance in field crops. In: Agronomic crops. Springer, Singapore, pp 129–147

Obidiegwu JE, Bryan GJ, Jones HG, Prashar A (2015) Coping with drought: stress and adaptive responses in potato and perspectives for improvement. Front Plant Sci 6:542

Parihar P, Singh S, Singh R, Singh VP, Prasad SM (2015) Effect of salinity stress on plants and its tolerance strategies: a review. Environ Sci Pollut Res 22(6):4056–4075

Peres ALG, Soares JS, Tavares RG, Righetto G, Zullo MA, Mandava NB, Menossi M (2019) Brassinosteroids, the sixth class of phytohormones: a molecular view from the discovery to hormonal interactions in plant development and stress adaptation. Int J Mol Sci 20(2):331

Raggi S, Doyle SM, Robert S (2020) Auxin: at the crossroads between chemistry and biology. In: Geelen D, Xu L (eds) The chemical biology of plant biostimulants. Wiley, New York, pp 123–153

Rahman SU, McCoy E, Raza G, Ali Z, Mansoor S, Amin I (2022) Improvement of soybean; a way forward transition from genetic engineering to new plant breeding technologies. Mol Biotechnol 65:1–19

Rai A, Rai GK, Dubey RS (2021) Heat stress and its effects on plant growth and metabolism. In: Abiotic stress tolerance mechanisms in plants. CRC Press, pp 203–265

Rao NK, Laxman RH, Shivashankara KS (2016) Physiological and morphological responses of horticultural crops to abiotic stresses. In: Abiotic stress physiology of horticultural crops. Springer, New Delhi, pp 3–17

Raza A, Razzaq A, Mehmood SS, Zou X, Zhang X, Lv Y, Xu J (2019) Impact of climate change on crops adaptation and strategies to tackle its outcome: a review. Plan Theory 8(2):34

Reddy AR, Rasineni GK, Raghavendra AS (2010) The impact of global elevated CO_2 concentration on photosynthesis and plant productivity. Curr Sci 99:46–57

Reddy YN, Reddy YP, Ramya V, Suma LS, Reddy AN, Krishna SS (2021) Drought adaptation: approaches for crop improvement. In: Millets and pseudo cereals. Woodhead Publishing, Duxford, pp 143–158

Rivero RM, Shulaev V, Blumwald E (2009) Cytokinin-dependent photorespiration and the protection of photosynthesis during water deficit. Plant Physiol 150(3):1530–1540

Rosero A, Berdugo-Cely JA, Šamajová O, Šamaj J, Cerkal R (2020) A dual strategy of breeding for drought tolerance and introducing drought-tolerant, underutilized crops into production systems to enhance their resilience to water deficiency. Plan Theory 9(10):1263

Ryu H, Cho YG (2015) Plant hormones in salt stress tolerance. J Plant Biol 58(3):147–155

Sadok W, Schoppach R (2019) Potential involvement of root auxins in drought tolerance by modulating nocturnal and daytime water use in wheat. Ann Bot 124(6):969–978

Saidi A, Hajibarat Z (2020) Application of next generation sequencing, GWAS, RNA seq, WGRS, for genetic improvement of potato (*Solanum tuberosum* L.) under drought stress. Biocatal Agri Biotechnol 29:101801

Sallam A, Eltaher S, Alqudah AM, Belamkar V, Baenziger PS (2022) Combined GWAS and QTL mapping revealed candidate genes and SNP network controlling recovery and tolerance traits associated with drought tolerance in seedling winter wheat. Genomics 114(3):110358

Salsman J, Dellaire G (2017) Precision genome editing in the CRISPR era. Biochem Cell Biol 95(2):187–201

Salvi P, Manna M, Kaur H, Thakur T, Gandass N, Bhatt D, Muthamilarasan M (2021) Phytohormone signaling and crosstalk in regulating drought stress response in plants. Plant Cell Rep 40(8):1305–1329

Santino A, Taurino M, De Domenico S, Bonsegna S, Poltronieri P, Pastor V, Flors V (2013) Jasmonate signaling in plant development and defense response to multiple (a) biotic stresses. Plant Cell Rep 32(7):1085–1098

Schaffer B, Anderson PC (1994) Handbook of environmental physiology of fruit crops, 2 vols. CRC Press, Boca Raton

Schlicke H, Hartwig AS, Firtzlaff V, Richter AS, Glässer C, Maier K et al (2014) Induced deactivation of genes encoding chlorophyll biosynthesis enzymes disentangles tetrapyrrole-mediated retrograde signaling. Mol Plant 7(7):1211–1227

Sevanto S (2018) Drought impacts on phloem transport. Curr Opin Plant Biol 43:76–81

Sharma A, Zheng B (2019) Molecular responses during plant grafting and its regulation by auxins, cytokinins, and gibberellins. Biomol Ther 9(9):397

Sharma SK, MacKenzie K, McLean K, Dale F, Daniels S, Bryan GJ (2018) Linkage disequilibrium and evaluation of genome-wide association mapping models in tetraploid potato. G3 Genes Genomes Genet 8(10):3185–3202

Singh A, Shamim MD, Singh KN (2013) Genotypic variation in root anatomy, starch accumulation, and protein induction in upland rice (*Oryza sativa*) varieties under water stress. Agric Res 2(1):24–30

Skalák J, Vercruyssen L, Claeys H, Hradilová J, Černý M, Novák O et al (2019) Multifaceted activity of cytokinin in leaf development shapes its size and structure in Arabidopsis. Plant J 97(5):805–824

Sofo A, Scopa A, Nuzzaci M, Vitti A (2015) Ascorbate peroxidase and catalase activities and their genetic regulation in plants subjected to drought and salinity stresses. Int J Mol Sci 16(6):13561–13578

Spaepen S (2015) Plant hormones produced by microbes. In: Principles of plant-microbe interactions. Springer, Cham, pp 247–256

Swartjes T, Staals RH, van der Oost J (2020) Editor's cut: DNA cleavage by CRISPR RNA-guided nucleases Cas9 and Cas12a. Biochem Soc Trans 48(1):207–219

Tak YG, Farnham PJ (2015) Making sense of GWAS: using epigenomics and genome engineering to understand the functional relevance of SNPs in non-coding regions of the human genome. Epigenetics Chromatin 8(1):1–18

Takahashi F, Kuromori T, Urano K, Yamaguchi-Shinozaki K, Shinozaki K (2020) Drought stress responses and resistance in plants: from cellular responses to long-distance intercellular communication. Front Plant Sci 11:556972

Tang D, Jia Y, Zhang J, Li H, Cheng L, Wang P et al (2022) Genome evolution and diversity of wild and cultivated potatoes. Nature 606:1–7

Thornton M, Olsen N, Liang X (2020) Physiological disorders. In: Potato production systems. Springer, Cham, pp 447–478

Tiwari S, Lata C, Singh Chauhan P, Prasad V, Prasad M (2017) A functional genomic perspective on drought signalling and its crosstalk with phytohormone-mediated signalling pathways in plants. Curr Genomics 18(6):469–482

Tron S, Bodner G, Laio F, Ridolfi L, Leitner D (2015) Can diversity in root architecture explain plant water use efficiency? A modeling study. Ecol Model 312:200–210

Uffelmann E, Huang QQ, Munung NS, De Vries J, Okada Y, Martin AR et al (2021) Genome-wide association studies. Nat Rev Method Primer 1:59

Uga Y, Sugimoto K, Ogawa S, Rane J, Ishitani M, Hara N et al (2013) Control of root system architecture by DEEPER ROOTING 1 increases rice yield under drought conditions. Nat Genet 45(9):1097–1102

Ullah A, Sun H, Yang X, Zhang X (2017) Drought coping strategies in cotton: increased crop per drop. Plant Biotechnol J 15(3):271–284

Ullah A, Manghwar H, Shaban M, Khan AH, Akbar A, Ali U et al (2018) Phytohormones enhanced drought tolerance in plants: a coping strategy. Environ Sci Pollut Res 25(33):33103–33118

van Bel AJ (2021) The plant axis as the command centre for (re) distribution of sucrose and amino acids. J Plant Physiol 265:153488

van Dam J, Levin I, Struik PC, Levy D (1999) Genetic characterisation of tetraploid potato (*Solanum tuberosum* L.) emphasising genetic control of total glycoalkaloid content in the tubers. Euphytica 110(1):67–75

Vanani FR, Shabani L, Sabzalian MR, Dehghanian F, Winner L (2020) Comparative physiological and proteomic analysis indicates lower shock response to drought stress conditions in a self-pollinating perennial ryegrass. PLoS One 15(6):e0234317

Varshney RK, Bohra A, Yu J, Graner A, Zhang Q, Sorrells ME (2021) Designing future crops: genomics-assisted breeding comes of age. Trends Plant Sci 26(6):631–649

Vasquez-Robinet C, Mane SP, Ulanov AV, Watkinson JI, Stromberg VK, De Koeyer D et al (2008) Physiological and molecular adaptations to drought in Andean potato genotypes. J Exp Bot 59(8):2109–2123

Vurukonda SSKP, Vardharajula S, Shrivastava M, SkZ A (2016) Enhancement of drought stress tolerance in crops by plant growth promoting rhizobacteria. Microbiol Res 184:13–24

Wang Q, Tang J, Han B, Huang X (2020) Advances in genome-wide association studies of complex traits in rice. Theor Appl Genet 133(5):1415–1425

Waqas MA, Kaya C, Riaz A, Farooq M, Nawaz I, Wilkes A, Li Y (2019) Potential mechanisms of abiotic stress tolerance in crop plants induced by thiourea. Front Plant Sci 10:1336

Wasson AP, Richards RA, Chatrath R, Misra SC, Prasad SS, Rebetzke GJ et al (2012) Traits and selection strategies to improve root systems and water uptake in water-limited wheat crops. J Exp Bot 63(9):3485–3498

Weisz R, Kaminski J, Smilowitz Z (1994) Water deficit effects on potato leaf growth and transpiration: utilizing fraction extractable soil water for comparison with other crops. American Potato Journal 71(12):829–840

Whitmore AP, Whalley WR (2009) Physical effects of soil drying on roots and crop growth. J Exp Bot 60(10):2845–2857

Wu G, Zhang C, Chu LY, Shao HB (2007) Responses of higher plants to abiotic stresses and agricultural sustainable development. J Plant Interact 2(3):135–147

Xu ZY, Kim DH, Hwang I (2013) ABA homeostasis and signaling involving multiple subcellular compartments and multiple receptors. Plant Cell Rep 32(6):807–813

Xu Z, Jiang Y, Jia B, Zhou G (2016) Elevated-CO_2 response of stomata and its dependence on environmental factors. Front Plant Sci 7:657

Yadav B, Jogawat A, Gnanasekaran P, Kumari P, Lakra N, Lal SK et al (2021) An overview of recent advancement in phytohormones-mediated stress management and drought tolerance in crop plants. Plant Gene 25:100264

Yang X, Lu M, Wang Y, Wang Y, Liu Z, Chen S (2021) Response mechanism of plants to drought stress. Horticulturae 7(3):50

Yao GQ, Li FP, Nie ZF, Bi MH, Jiang H, Liu XD et al (2021) Ethylene, not ABA, is closely linked to the recovery of gas exchange after drought in four Caragana species. Plant Cell Environ 44(2):399–411

Zarzyńska K, Boguszewska-Mańkowska D, Nosalewicz A (2017) Differences in size and architecture of the potato cultivars root system and their tolerance to drought stress. Plant Soil Environ 63(4):159–164

Zhang H, Zhu J, Gong Z, Zhu JK (2022) Abiotic stress responses in plants. Nat Rev Genet 23(2):104–119

Zhu JK (2016) Abiotic stress signaling and responses in plants. Cell 167(2):313–324

microRNA-Mediated Regulation of Drought Stress Responses in Potatoes: Implications for Climate Resilience

Arslan Asim and Fariya Baig

Abstract Drought stress is a significant constraint on crop productivity, particularly for potatoes, which are highly susceptible to drought-induced yield losses due to their shallow root systems. Climate change impacts crop productivity through both abiotic and biotic stresses, with less water stress posing a major challenge, particularly for drought-sensitive crops like potatoes and these stress factors are expected to intensify by 2050. Drought stress affects potatoes physiologically by reducing photosynthetic efficiency, impairing water-use efficiency, and hindering tuber development. This study examines the role of microRNAs (miRNAs) in regulating plant responses to drought stress at the transcriptional and post-transcriptional levels. miRNAs influence various developmental processes and stress responses, including auxin signaling, ABA-mediated regulation, and antioxidant scavenging. Numerous miRNAs have been identified as key players in these regulatory pathways, highlighting their importance in drought tolerance. Understanding the functional mechanisms of these miRNAs and their target genes is essential, especially under combined stress conditions due to climate change. This knowledge is pivotal for developing drought-tolerant crop varieties, enhancing agricultural resilience in a changing climate.

Keywords Drought stress · Potato yield · MicroRNAs (miRNAs) · Climate change

A. Asim (✉)
Department of Agricultural Genetic Engineering, Ayhan Şahenk Faculty of Agricultural Sciences and Technologies, Niğde Ömer Halisdemir University, Niğde, Turkiye

F. Baig
Department of Earth and Environmental Science, Faculty of Arts and Science, University of Central Asia, Khorog, Tajikistan

U. K. Chaudhry et al. (eds.), *Drought Stress*,
https://doi.org/10.1007/978-3-031-80610-0_10

1 Introduction

The impact of climate change on crop productivity is multifaceted, with both abiotic and biotic stresses being affected. While changes in biotic stresses may vary, abiotic stress factors such as heat, salinity, and drought are anticipated to increase in severity. Among these, drought stress emerges as a significant constraint, particularly for crops like potatoes that are highly susceptible to drought-induced yield losses. Traditionally, the vulnerability of potatoes to drought has been linked to their shallow root systems. However, this chapter underscores the importance of other factors such as canopy development and cultivar selection in determining drought tolerance and its effect at transcriptional and post-transcriptional level. microRNAs (miRNAs) primarily regulate genes at the post-transcriptional level (Asim et al. 2021). The production of miRNA involves several key steps: transcription, precursor processing, methylation, and the assembly of the miRNA-induced silencing complex (miRISC). miRNA-mediated regulatory mechanisms are essential in plants, influencing various developmental processes like leaf formation, flowering timing, and organ polarity, as well as responses to abiotic and biotic stress factors (Chen et al. 2005; Jeong and Green 2013). Some studies reported the transmissible mRNAs and proteins directly related to phytochrome regulation, plant development, and stress response (Hao et al. 2020). Numerous miRNAs have been identified as key players in plant responses to drought stress, acting through signaling pathways such as auxin signaling, ABA-mediated regulation, osmoprotectant biosynthesis, and antioxidant scavenging (Candar-Cakir and Cakir 2016; Ding et al. 2013). Several studies have identified the Endogenous short RNAs; miRNAs adaptations and functions in enhancing the drought-resistance (Islam et al. 2022). Drought-associated miRNAs and their respective target genes have been discovered in various plants, including tomato, Arabidopsis and cotton (Candar-Cakir and Cakir 2016; Liu et al. 2008; Xie et al. 2015). For instance, a study using degradome deep sequencing and small-RNA in tomatoes identified 699 conserved miRNAs, some of which were linked to the response of drought through plant hormone signaling and the breeding of drought-tolerant tomato varieties (Candar-Cakir and Cakir 2016). Additionally, a citation Rank-based literature mining analysis in cotton identified 337 miRNAs, revealing those linked with the expression of key genes related to the drought conditions (Xie et al. 2015). Recently, several key drought-responsive miRNAs have been identified in plants. A genome-wide analysis in six Oryza sativa genotypes highlighted miR169, miR398, and miR5505 as potential candidates for developing drought tolerance (Xia et al. 2020). miR169 showed varied responses to drought stress across species: in *A. thaliana* (Li et al. 2016a, b) and *N. tabacum* (Yin et al. 2014), expression was downregulated, whereas in *O. sativa*, it was upregulated (Xia et al. 2020). Specifically, in *N. tabacum*, miR160 regulates the expression of auxin response factor genes, which play a vital role in plant growth, development, and drought tolerance (Li et al. 2016a, b). Additionally, miR156 and miR172 are commonly involved in responsive mechanisms of abiotic stresses and are known for their roles in regulating transcription factors (Pagano et al. 2021). Furthermore, conserved miRNAs such as miR159, miR156, and miR319 were identified as significant players in drought tolerance in seedlings of N. tabacum and Z. mays (Li et al. 2013). However, with climate change leading to not just single stressors but a combination of stress conditions throughout plant development, there is a need for deeper comprehension of the roles and functions of miRNAs in plants when facing multiple stressors simultaneously (Öztürk Gökçe et al. 2021). For this purpose identifying stress specific miRNAs and targets of miRNAs is crucial for understanding their functional mechanisms (Devi et al. 2018). Overall, miRNAs hold significant promise for the development of new drought-tolerant crop varieties in the future (Shi and Hussain 2020).

2 Historical Perspective of Potato (*Solanum tuberosum*)

Potato cultivation has its origins in modern world, while its wild relatives continue to thrive (Spooner et al. 2004). The inception of potato farming in South America dates back approximately 8000 years (Lutaladio and Castaldi 2009), with Spanish conquistadors introducing the crop to Europe during the Columbian exchange (Lutaladio and Castaldi 2009). By the end of the sixteenth century, tubers, specifically potatoes, had been assimilated into the agricultural methodologies prevalent in the United Kingdom and Ireland (de Haan and Rodriguez 2016). However, the adoption of monocultural practices in Europe resulted in severe losses due to late blight, particularly impacting Ireland. Consequently, breeders shifted their focus towards developing cultivars resistant to diseases and with higher yields (Lutaladio and Castaldi 2009).

The global output of Solanum tuberosum has experienced substantial increase, escalating from 270 MT in 1961 to 375 MT in 2022. This increase can primarily be ascribed to the persistent advancements in the yield capacity of various potato cultivars. Interestingly, despite a decrease in the area dedicated to potato cultivation, the yield potential has surged by 58.7% over the last fifty years (Fig. 1). Major potato-producing nations include China, India, Russia, the USA, and Ukraine, with Netherlands, Poland, Belarus, France and Germany following closely (Fig. 2). Europe emerges as the second-highest potato-producing region globally, yielding 125.43 million tons, only trailing Asia's 140.6 million tons (Fig. 3) (FAOSTAT 2021).

Nevertheless, within the geographical confines of Europe, the production of potatoes has undergone a significant reduction, declining from 137.1 million metric tons in 1994 to a mere 99 million metric tons by 2022. The apex of production was recorded in 1996, subsequent to which there was a consistent downward trend. This reduction is predominantly ascribed to the contraction of the harvested area, which experienced a decrease of 53% from 1994 to 2022. In 1994, an area of 9.7 million hectares across Europe was allocated for potato harvesting, whereas, by 2022, only 4.63 million hectares were employed for potato cultivation (Fig. 4). The potato possesses substantial agricultural importance in both advanced and emerging economies due to its multifunctional applications (Lutaladio and Castaldi 2009). It is

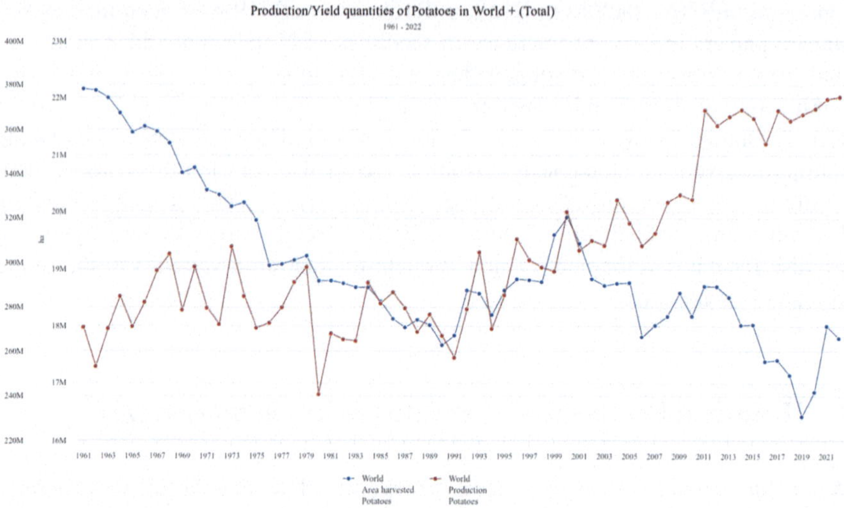

Fig. 1 World potato production and area harvested during 1961–2022. (Source: FAOSTAT (June 10, 2024))

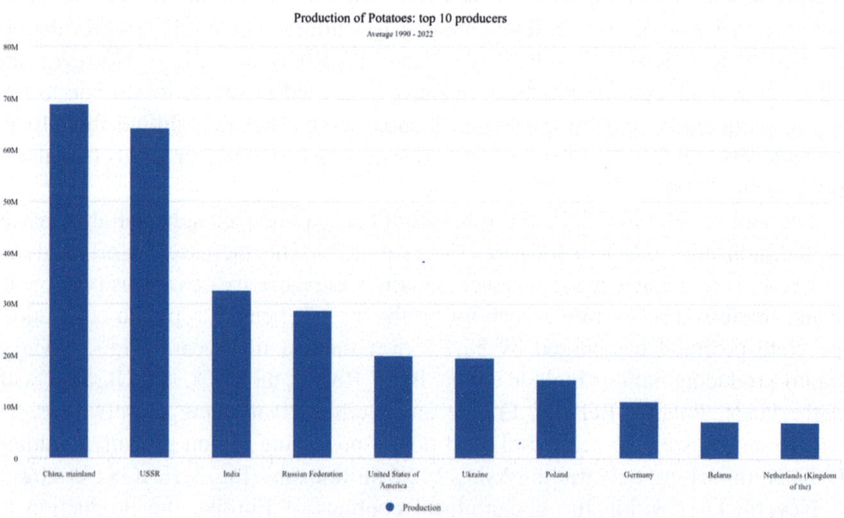

Fig. 2 Top 10 potato producers in the world during 1994–2022. (Source: FAOSTAT (June 7, 2024))

significant to note that 85% of the biomass derived from a potato plant can be consumed, which constitutes a considerably larger fraction in comparison to the 50% identified in cereal crops (Lutaladio and Castaldi 2009). Within the European framework, the potato is ranked as the fourth most crucial agricultural commodity regarding production volume, following wheat, sugar beet, and maize, and is also classified

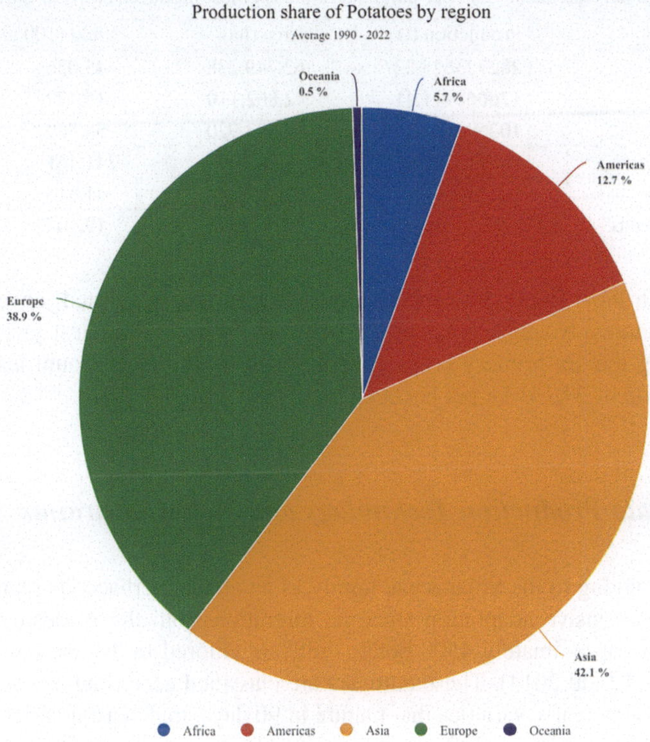

Fig. 3 Regional share in potato production during 1994–2022. (Source: FAOSTAT (June 7, 2024))

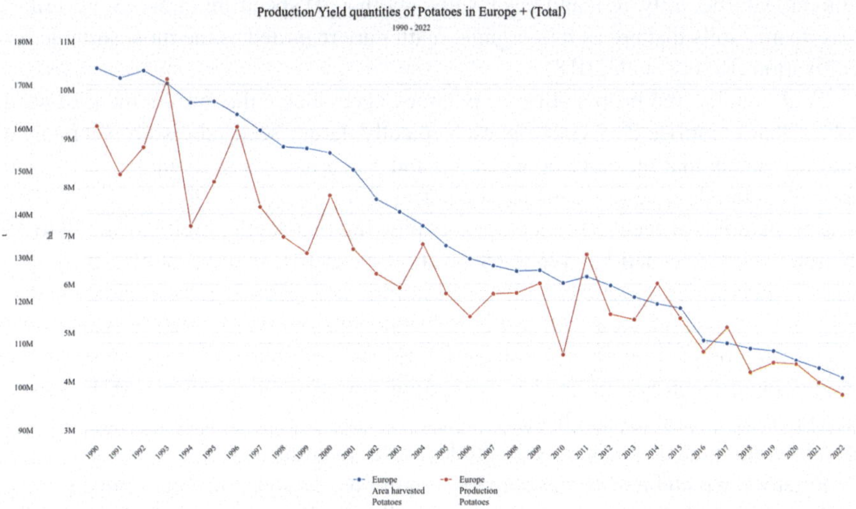

Fig. 4 Potato production and area allocated for potato harvest in Europe during 1994–2022. (Source: FAOSTAT (June 7, 2024))

Table 1 Area harvested and yield per unit land of the six most produced crops in Europe in 2022

Crop	Production (t)	Area (ha)	Yield (100 g/ha)
Wheat	282723625.22	62,749,238	45,056
Sugar Beet	176068911.41	2,862,110	615,172
Maize	102694188.84	17,533,920	58,569
Potatoes	98112778.9	4,063,302	241,461
Barley	91110001.99	21,787,364	41,818
Sunflower seeds	38324650.63	20,060,525	19,105

as the ninth most important crop in terms of land area harvested, covering 4.63 million hectares. Notably, potatoes demonstrate the second-highest yield per unit area among the six primary crops cultivated across European countries, with an average yield of 24,161 kg per hectare (Table 1).

2.1 Potato Production Technology and Yield Constraints

Potato, belonging to the Solanaceae family, is an annual herbaceous plant that has undergone extensive adaptation since its migration from the Andean highlands, resulting in approximately 4500 potato cultivars tailored to diverse environments (Pieterse and Judd 2014). These cultivars are classified according to their maturation periods into early varieties that mature in 90 days, mid-season varieties maturing in 100 days, and late varieties that mature in 110 days. As a crop of short-duration, the potato integrates effectively into diverse intercropping systems across the globe. Potatoes demonstrate adaptability to a variety of edaphic conditions—spanning alluvial, laterite, hilly, red, and black soils—with a pH spectrum of 5–7.5; nevertheless, loamy soils that are rich in organic matter are regarded as the most suitable for cultivation (Reddy et al. 2018).

Traditionally, the propagation of potatoes necessitates the employment of seed tubers that comprise 2–3 viable eyes. Typically, larger seed tubers are segmented into fragments that measure between 3.5 and 5 centimeters in diameter, with each fragment containing a few eyes for the purpose of propagation (Reddy et al. 2018). The preparation of seedbeds encompasses plowing to a depth ranging from 20 to 35 centimeters, succeeded by tillage (Salavati et al. 2018). In areas subjected to elevated temperatures, practices such as soil turning and fallowing are implemented to alleviate issues associated with soil-borne pathogens and weeds (Reddy et al. 2018). The sowing of seed tubers is conducted in ridges, with burrow irrigation being the predominant technique utilized for irrigation purposes (Singh et al. 2021). Given that potatoes can attain heights of up to 3.5 feet depending on the cultivar, earthing up is recommended approximately 30 days after planting.

Potatoes, which favor cooler climatic conditions, demonstrate suboptimal growth performance under elevated temperature conditions (Haverkort and Verhagen 2008; Pulatov et al. 2015). The optimal thermal range for vegetative growth is between 16 and 25 °C, while temperatures spanning from 4 to 18 °C are preferentially observed during the phases of tuber formation and its bulk (Reddy et al. 2018). The process of tuber formation generally initiates 20–25 days subsequent to sowing, with the plants producing flowers that may be white or purple, contingent upon prevailing weather conditions. The water requirements for potatoes throughout their growth cycle vary significantly, ranging from 350 to 550 mm, influenced by climatic variables, soil types, and cultivar (Reddy et al. 2018). Irrigation practices are typically halted 10–15 days preceding the harvest, which is generally performed before ambient temperatures surpass 30 °C.

Although potato cultivation exhibits adaptability to diverse climatic conditions, achieving optimal production levels remains contingent on environmental variables and potato variety, with yields potentially reaching 35 t ha^{-1} (Reddy et al. 2018). However, a multitude of abiotic and biotic determinants impose limitations on the productivity of Solanum tuberosum, resulting in a reduction of yield outcomes. Biotic factors encompass diseases induced by fungi, bacteria, nematodes, and viruses. Notably, prevalent fungal and bacterial diseases in potatoes include late blight, early blight, black scurf, stem canker, powdery scab, common scab, bacterial wilt, blackleg, and soft rot (Abebe and Tadesse 2018; Ding et al. 2013; Gilchrist et al. 2011; Horsfield et al. 2010; Naher et al. 2014; Ngadze 2018; Wang et al. 2019). These diseases can cause substantial yield losses, with late blight alone resulting in losses of up to 71%, underscoring the significant impact of biotic constraints on potato productivity.

Abiotic stresses constitute a substantial risk to the productivity of Solanum tuberosum, with projections indicating that such environmental stressors account for as much as 50% of the average yield reductions observed across a variety of crops globally (Wang et al. 2003). Scholars have thoroughly examined the effects of abiotic determinants on potato yield, encompassing variables such as temperature, solar irradiation, photoperiodicity, soil salinity and drought conditions (Camargo et al. 2015; Battisti and Naylor 2009).

Variations in temperature significantly influence potato production outcomes. Optimal temperature ranges between up to 25 °C during the growth phase and up to 18 °C during the bulk stage of tubers are essential for achieving maximal growth (Reddy et al. 2018). Nonetheless, considerable departures from these specified ranges can exert detrimental impacts on tuber yield. Elevated temperatures during the vegetative growth stage may cause increased respiration rates, physiological wilting, diminished photosynthetic efficiency, and truncated life cycles. In contrast, increased temperatures during the reproductive phase can lead to the formation of smaller tubers, slower rates of tuberization, and shortened reproductive periods, ultimately culminating in diminished tuber yields (Hatfield and Prueger 2015; Hijmans 2003; Levy and Veilleux 2007).

Conversely, reductions in ambient temperature, notably those falling below 0 °C during the initial phases of development, have the potential to inflict harm on seedlings, disrupt the hydrodynamic processes within the plant, and obstruct the absorption of both water and essential nutrients (Hijmans 2003; Liao et al. 2016).

Additional decrease in temperature, specifically those descending below −3 °C, may lead to extensive damage across potato cultivation areas (Pino et al. 2007).

Soil salinity is a major abiotic stressor that negatively impacts potato yield, as potatoes are highly sensitive to saline conditions (Obidiegwu et al. 2015). Factors such as suboptimal water management, inefficient irrigation systems, reduced precipitation, and elevated evaporation rates in arid and semi-arid climates further aggravate soil salinization (Athar and Ashraf 2009).

Although potatoes are regarded as water-efficient crops, their shallow root systems make them particularly vulnerable stress caused by drought (Sun et al. 2015). Drought stress represents a significant abiotic limitation to potato production, disrupting key physiological and biochemical processes essential for optimal yield and productivity (Fahad et al. 2017; Hirut et al. 2017). Thus, ensuring adequate water availability is crucial for maintaining potato productivity under drought conditions.

Water deficit conditions induce morphological alterations in the root system of potatoes (Wishart et al. 2013, 2014). These changes include the expansion of lateral root proliferation, an increase in root thickness, a reduction in root elongation, and decreased hair production, all aimed at enhancing the maintenance of root-soil linkage during dry periods (Haling et al. 2013; Schmidt et al. 2012). Furthermore, plants' drought tolerance is augmented by the hydraulic resistance of roots associated with xylem vessels, alongside a decrease in the rate of water extraction from the soil without rapid depletion of available water resources. Changes in hydraulic conductivity occur not only within the roots but also depend on root type and tissue age, mediated by aquaporin expression and cell membrane permeability (Doussan et al. 1998).

Water scarcity initially leads to an increase in root hydraulic conductivity to sustain water uptake; however, continuous water extraction may subsequently decrease root hydraulic conductivity to conserve soil moisture, with these alterations influenced by aquaporin regulation and activity (Bellon-Maurel et al. 2010). Additionally, an increase in tuberization of the root exodermis/endodermis further reduces root hydraulic conductivity (Vandeleur et al. 2009). Severe and prolonged water scarcity can negatively impact root hydraulic conductivity due to xylem embolism (Cruiziat et al. 2002).

Certain potato genotypes with larger root systems, particularly those bearing stolons, exhibit enhanced resilience to drought stress, thereby ensuring higher potato yields (Puértolas et al. 2014). However, it's noteworthy that genotypes with mega root systems also exhibit rapid canopy closure, which prolongs their duration and photosynthetic activity but increases water loss through soil evaporation rather than plant transpiration. Breeding efforts aimed at enhancing root hydraulic resistance, increasing root hair length, and promoting greater numbers of seminal roots have been recommended (Bengough et al. 2011).

Given the increasing scarcity of water resources globally, plants with enhanced abilities to withstand abiotic stresses are expected to play significant roles in agriculture (Çalışkan 2016). In terms of water availability, potatoes are considered drought-sensitive, especially during tuber formation, emphasizing the plant's need for water. Environmental conditions can significantly affect certain potato cultivars based on their susceptibility to abiotic stress (Çalışkan 2016).

Recent observations have shed light on how water dynamics at the soil-root interface modulate soil physical conditions, soil-root contact, and the uptake of moisture from moist or dry soil (Koebernick et al. 2017).

3 Potato Water Needs and the Impact of Climate Change

Potatoes are recognized for their high water-use efficiency, producing more calories per unit of water compared to many other crops (Sun et al. 2015). As reported by Renault and Wallender (2000), the water requirement for producing 1 kilogram of potatoes is only 105 L, significantly lower than that of other major crops such as maize, rice, and wheat which need 710 L, 1408 L, 1159 L, respectively, for the same yield. Despite this efficiency, potatoes remain highly susceptible to drought stress due to their considerable need for irrigated water (Schafleitner et al. 2009).

The irrigation water requirement for potatoes varies depending on agroclimatic conditions and soil moisture levels, typically reaching up to 313 mm, as noted by Knox et al. (1997). Studies by Li et al. and Vishnoi et al. indicate that optimal potato yields in China and India require between 126–381 mm and 212–226 mm of water, respectively (Vishnoi et al. 2012; Li et al. 2022). In the United Kingdom, during drier years such as 2018, the minimum irrigation need increased to 154 mm (Knox and Hess 2019). According to Byrd et al., in the United States, potatoes consume approximately 10 mm of water every 24–36 h from the post-flowering stage until harvest, resulting in total irrigation needs of up to 610 mm (Byrd et al. 2014). In contrast, many other crops in Europe, such as sugar beet, cereals, carrots, and strawberries, require significantly less irrigation water (Knox et al. 1997).

Cultivars of potato typically have shallow roots, limiting their ability to access water from deeper soil layers during periods of water scarcity (van Loon 1981). However, root length variation among cultivars has been noted, with longer roots being positively associated with increased tuber yield under drought conditions (Lahlou and Ledent 2005). Canopy characteristics also influence transpiration rates, with stem-type canopies showing better performance under stress of drought (Aliche et al. 2018). Nevertheless, under drought conditions, these traits can still lead to significant yield reductions. For instance, the cultivar Désirée experienced up to an 87% yield decline due to reductions in plant height and leaf number (Luitel et al. 2015).

Considering the higher susceptibility of potatoes to drought stress (Schafleitner et al. 2009), change in climate is projected to have a substantial impact on global potato production. As climate change manifests differently across regions, its effects are being examined at global, regional, and local scales (Arnell and Reynard 1996). Temperature increases associated with climate change are predicted to extend the growing season in some regions, potentially boosting potato production (Daccache et al. 2011). However, changes in precipitation patterns, with more rainfall in winter and less in summer, alongside overall water scarcity, are anticipated to have adverse effects on tuber yield (Rial-Lovera et al. 2017). Estimates suggest that a significant portion of the rainfed area suitable for potato production in the United Kingdom may be lost due to water scarcity, necessitating increased irrigation to sustain yields (Daccache et al. 2012). Consequently, the demand for irrigation water is expected to rise, leading to increased production costs for potatoes in rainfed areas (Daccache et al. 2012).

4 Global Impact of Drought

Drought is a complex and multidimensional abiotic stress that profoundly affects plant morphology, physiology, ecology, biochemistry, and molecular traits (Anjum et al. 2011; Salehi-Lisar and Bakhshayeshan-Agdam 2016). The definition of drought varies across disciplines: meteorological drought refers to extended periods of minimal or no rainfall, whereas hydrological drought emphasizes reduced water potential in plant tissues caused by soil moisture deficits (Mitra 2001; Solh and van Ginkel 2014). In agriculture, drought is viewed as a period of water shortage leading to soil moisture deficits that adversely affect plant yields (Demirel et al. 2020).

Hydrological drought can be classified into intermittent and terminal types. Intermittent drought allows for partial recovery of soil moisture, whereas terminal drought leads to prolonged soil moisture deficits, significantly inhibiting development in plants and, in severe cases, causing premature wilting of plant (Neumann 2008). While water and rain deficiency is the primary cause of drought stress, it is also influenced by different environmental factors such as temperature, along with non-environmental factors including land cover, type of soil and human activities (Cook et al. 2018).

Throughout history, humanity has experienced devastating drought events, including the European drought of 1976, the Dust Bowl in the United States during the 1930s, and food crises in Russia and China in the 1920s, which collectively resulted in significant human suffering and loss of life (CRED n.d.; Schubert et al. 2004). The twenty-first century has witnessed numerous droughts globally, including prolonged dry spells in Australia, Europe, the Amazon rainforest, the Iberian Peninsula, Russia, China, Scandinavia, and Africa, each causing substantial ecological, economic, and social impacts (Andreu and Pérez-Martín 2009; Cattiaux et al. 2010; Grumm 2011; Lewis et al. 2011; Lu et al. 2011; McGrath et al. 2012; Rebetez et al. 2009; Robine et al. 2008; Viste et al. 2013).

Drought events in regions such as the central and southern USA, Russia, Brazil, New Zealand, Central Europe, Namibia, and Scandinavia in recent years have intensified agricultural, economic, and environmental challenges (Dettinger and Cayan 2014; Aghakouchak et al. 2014). The long duration of these droughts is regarded as one of the most destructive natural threats, affecting billions of people globally and underscoring the need for urgent measures to mitigate its impacts (Kazmierczak and Georgi 2012; Wilhite 2016).

Agricultural practices, including extensive irrigation, are employed to sustain crop yields throughout the growing season, contributing to significant freshwater usage, despite current freshwater shortages in many regions (Nouri et al. 2019). Arid and semi-arid areas, covering 40% of the available agricultural land of the world, face particularly severe water scarcity challenges (Ehsanpour and Razavizadeh 2005; Salvin et al. 2000).

The water crisis is anticipated to intensify as a result of climate change, with climate models predicting reduced rainfall, increased temperatures, and more frequent droughts, adversely affecting global production of crops (Monneveux et al. 2013; Thiele et al. 2010). Climate projections indicate a rise in drought severity and frequency over the next several decades, highlighting the urgent need for effective drought mitigation strategies (Dai 2013; Pachauri et al. 2014).

In response to the escalating threat of drought, it is imperative to study the drought tolerance mechanisms of key crops to develop strategies for maintaining agricultural productivity in the face of changing climatic conditions (Hijmans 2003; Monneveux et al. 2013).

5 Influence of Drought Stress on Potato

Drought stress induces morphological changes in the root system of potatoes, including increased lateral root proliferation, root thickness, and decreased root elongation and hair production. These alterations aid in maintaining root-soil linkage during dry periods (Chang et al. 2018; Kesiime et al. 2016; Shao et al. 2009). Drought stress causes a significant decline in the quality and quantity of potato yield, it causes physiological and biological changes, making maintenance of osmotic adjustment highly important (Ozturk et al. 2021). Additionally, plants enhance drought tolerance through hydraulic resistance of roots, mediated by changes in hydraulic conductivity alongside aquaporin regulation and activity (Jefferies and MacKerron 1989; Szabolcs and Pessarakli 1999; Weisz et al. 1994).

The response to drought varies among potato genotypes, with those possessing larger root systems, particularly stolon-bearing roots, exhibiting improved drought resilience and higher yields (Romero et al. 2017). However, genotypes with larger root systems may experience quick canopy closure, prolonging photosynthetic activity but also leading to increased water loss through evaporation from the soil (Lahlou et al. 2003).

As water resources decline globally, crops capable of coping with abiotic stresses, such as drought, are becoming increasingly important for agricultural sustainability (Najm et al. 2013). Despite being declared drought-sensitive, especially during tuber formation, potatoes show variability in response to environmental conditions based on cultivar susceptibility (Michel et al. 2019).

Drought affects potato growth and development depending on its severity, duration, and the plant's developmental stage. The phenomenon of stomatal closure, elevated concentrations of sugars within leaf tissues, and the buildup of reactive oxygen species are prevalent physiological responses observed in plants subjected to drought stress. (Ávila-Valdés et al. 2020; Gaur et al. 2017; Pieters and El Souki 2005; Pourasadollahi et al. 2019). These physiological changes can lead to reduced CO_2 uptake, inhibition of photosynthesis, and oxidative damage at the cellular level (Flexas et al. 2006; Lawlor 2002; Tezara et al. 1999).

Visible signs of drought stress in potatoes include reduced vegetative growth, manifested as decreased plant height and leaf size and number (Vu et al. 1987). Drought can impair leaf photosynthesis, affecting chlorophyll content, leaf area index, and duration, consequently impacting yield (Bota et al. 2004; Cornic 2000; Stalham et al. 2007). Furthermore, drought stress can shorten the growth cycle, reduce tuber size and number, and affect tuber quality, ultimately influencing overall potato yield and quality (Albiski et al. 2012; Joshi et al. 2016; Steckel and Gray 1979).

6 Drought Stress Effects on Above Ground Potato Parts

The development of the canopy in plant species, encompassing the generation of foliar structures, stems (specifically stolons), as well as an augmentation in the dimensions of individual leaves and overall plant stature, exhibits a marked sensitivity to hydric stress (Shao et al. 2009). It has been documented that drought-induced stress impedes stem elongation (Chang et al. 2018), the development of young leaves (Aliche et al. 2018), the quantity of stems produced (Chang et al. 2018), and the area of individual leaves (Kesiime et al. 2016) in *Solanum tuberosum*. This inhibition is attributed to the reduction in water availability, which is essential for maintaining high turgor pressure necessary for cell expansion and growth (Kesiime et al. 2016).

The impact of drought on canopy growth is considerable, as the growth of foliage in Solanum tuberosum ceases when the availability of soil moisture falls beneath specific thresholds (Szabolcs and Pessarakli 1999; Weisz et al. 1994). Generally, the diminution of stem and leaf development represents initial discernible outcomes of water scarcity in potato cultivars (Jefferies and MacKerron 1989). Both early and late manifestations of drought stress have been shown to impede canopy growth, with early drought conditions postponing canopy maturation and late drought precipitating the abscission of mature leaves while simultaneously suppressing the development of young leaves (Aliche et al. 2018). For example, Chang et al. (2018) reported a marked decrease in the potato plants stem length subjected to early conditions of drought.

The impact of drought stress on canopy development varies among potato cultivars, with late-maturing cultivars generally exhibiting lower sensitivity to late-season drought compared to early-maturing cultivars (Aliche et al. 2018; Romero et al. 2017). This variation is attributed to the longer vegetative growth period of late-maturing cultivars, which enables them to postpone full canopy development until later in the drought period, thereby mitigating the stress effects (Romero et al.

2017). Furthermore, certain traits, such as stem number, may be less influenced by late-season drought, as plants often achieve optimal numbers of stem before onset of late drought stress (Lahlou et al. 2003).

The Leaf Area Duration (LAD) and Leaf Area Index (LAI) are critical factors influencing tuber yield, with drought stress causing significant reductions in both metrics in potato crops (Najm et al. 2013). Research consistently indicates that drought stress decreases LAI across various potato cultivars (Michel et al. 2019; Pourasadollahi et al. 2019). This decrease in Leaf Area Index (LAI) is primarily due to reduced cell expansion under drought conditions, which tends to affect late-maturing cultivars more severely than ones that mature early (Gaur et al. 2017; Pourasadollahi et al. 2019; Salavati et al. 2018). Additionally, the variability in Leaf Area Index (LAI) among cultivars of potato in drought conditions is influenced by differences in architecture of the canopy (Michel et al. 2019).

Moreover, a decrease in LAI impacts the duration of radiation interception, which in turn affects production of biomass (Chang et al. 2018). Jefferies and MacKerron (1989) emphasized that Leaf Area Duration (LAD) plays a more critical role than LAI in determining biomass production. This was further supported by a recent study showing a significant reduction in tuber yield due to the effect of drought stress on LAD in potato cultivars (Ávila-Valdés et al. 2020).

Drought stress adversely affects photosynthesis in potato plants by decreasing both the number of leaves and the area of individual leaves, which directly diminishes photosynthetic capacity (Gaur et al. 2017). Shortages of water and carbon dioxide reduce the photosynthetic rate by impairing the production of ribulose bisphosphate (RuBP), a critical carbon dioxide acceptor in the photosynthetic process (Flexas et al. 2006; Tezara et al. 1999). Additionally, stomatal closure under conditions of water scarcity further limits carbon dioxide uptake, leading to reduced availability of substrates crucial for respiration and plant development (Cornic 2000; Flexas et al. 2006). The magnitude of these effects can vary depending on the cultivar, as well as duration and time of the stress caused by drought (Bota et al. 2004).

6.1 *Effect of Drought Stress at Underground Part and Tuber Mass of Potato*

Potatoes possess underground components including stolons, roots and tubers, and tubers are economically valuable. The shallow root system of potatoes makes them particularly vulnerable to drought stress (Joshi et al. 2016; Stalham et al. 2007). Although extensive research has been conducted on root length and mass within potato root system architecture, a conclusive drought stress effect on underground development of plant has yet to be established (Albiski et al. 2012).

Research findings regarding the impact of drought stress on potato roots have been inconsistent. Some studies report reductions in root length and dry mass under drought conditions, while others observe increases or no changes (Boguszewska-Mańkowska et al. 2020; Lahlou and Ledent 2005; Steckel and Gray 1979). Similar discrepancies are noted for stolon numbers in response to stress caused by drought (Lahlou and Ledent 2005; Haverkort et al. 1990). These variations are attributed to factors such as genotype, environmental interactions, and the timing of drought stress (Boguszewska-Mańkowska et al. 2020; Joshi et al. 2016; Rudack et al. 2017; Rykaczewska 2015; Gluska 2004; Iwama 2008; Zarzyńska et al. 2017). Different potato cultivars exhibit diverse responses to drought intensity and duration, with late-maturing cultivars often demonstrating deeper and more extensive root systems compared to cultivars that mature early under comparable stress conditions (Kashiwagi et al. 2000).

Experimental variables and errors, location, tuber physiological age, soil type, and root handling techniques further complicate the study of underground parts' response to drought stress (Obidiegwu et al. 2015; Stalham et al. 2007; Steckel and Gray 1979). Despite these challenges, studying the effects of drought stress on underground potato components is essential for cultivar development.

Tuber yield is a principal focus in potato cultivation and has been extensively studied in relation to drought stress over the past few decades. Drought stress significantly impacts various aspects of tuber yield, including fresh tuber mass, tuber number, tuber dry matter, and marketable tuber yield (Carli et al. 2014; Deblonde et al. 1999; Khosravifar et al. 2008; Lahlou et al. 2003).

Fresh tuber yield is highly sensitive to water shortage, as water content contributes significantly to fresh tuber mass (Jefferies and Mackerron 1993). Long-term water stress can drastically reduce the relative water content of tubers, leading to decreased fresh tuber yield (Jefferies and MacKerron 1989). However, the response to water shortage varies among potato cultivars, with some exhibiting more resilience than others (Lahlou et al. 2003). Different cultivars may exhibit varying levels of reduction in weight of tuber under drought stress, with some cultivars experiencing substantial decreases (Aliche et al. 2018; Boguszewska-Mańkowska et al. 2020).

The timing of drought stress also plays an essential role in fresh tuber yield, with both early and late drought affecting potato yield. Early stress tends to reduce fresh tuber weight by delaying tuber growth, especially in early maturing cultivars. Late drought, on the other hand, may impact early maturing cultivars more severely due to their shorter growth periods (Chang et al. 2018).

Besides fresh tuber weight, drought stress affects tuber number, with prolonged or severe stress leading to a reduction in the number of tubers produced (Eiasu et al. 2007; Haverkort et al. 1990). The timing of drought stress influences the extent of this reduction, with tuber initiation being particularly sensitive to water shortage (Chang et al. 2018; King and Stark 1997; Minhas and Bansal 1991). Certain experimental studies have observed tuber number increase in response to stress caused by drought, possibly as an adaptive response to sustain yield or due to other environmental factors such as heat stress (Lahlou et al. 2003; Rykaczewska 2017).

Total tuber dry mass, which is crucial for yield and quality, is influenced by drought stress through its impact on canopy growth and photosynthetic activities (Jefferies and MacKerron 1989). Reductions in net photosynthesis, chlorophyll content, and area of leaf under water stress directly affect tuber dry matter (Ahmadi et al. 2010; Li et al. 2016a, b; Ohashi et al. 2006; Sadeghipour and Aghaei 2012). The severity of stress and cultivar characteristics determines the extent of reduction in tuber dry mass, with some cultivars exhibiting higher resilience than others (Ruttanaprasert et al. 2016).

In conclusion, understanding the complex interactions between drought stress and various components of potato yield is essential for developing drought-tolerant cultivars and optimizing potato production in water-limited environments.

7 Drought Mitigation Strategies

Drought poses significant challenges in potato cultivation, and while progress has been made in identifying genes related to drought stress, developing truly drought-tolerant potato genotypes remains a formidable task (Ferrero et al. 2005). Current strategies for mitigating the effects of drought stress in potatoes involve selecting appropriate genotypes for specific climates and improving agronomic practices (Sekhon et al. 2010).

Effective soil management is vital for alleviating drought stress. Techniques such as tillage, mulching, residue management, and the addition of organic matter can improve infiltration of water and mitigate evapotranspiration (Ferrero et al. 2005). However, traditional hill-based potato cultivation limits the application of certain tillage practices. Organic mulches, including animal manure and carbon-rich wastes, can help control evaporation and improve soil nutrient and water-holding capacity (Fazal and Bano 2016; Lipiec et al. 2013). Additionally, biochar and compost have shown promise in alleviating drought stress and enhancing soil structure (Sodhi et al. 2009; Tejada et al. 2009; Thomas et al. 2013).

Soil health is affected by nutrient management and retention of water. Inorganic nutrients such as zinc, nitrogen, phosphorus, potassium, and selenium have been reported in literature to mitigate drought stress symptoms in crops like wheat (Gevrek and Atasoy 2012; Nawaz et al. 2013; Aown et al. 2012). Application of silicon, whether via foliar treatment or soil incorporation, has been demonstrated to enhance tolerance of drought in potatoes (Pilon et al. 2014). Additionally, foliar application of specific micronutrients such as zinc, boron, and manganese, combined with soil application of NPK fertilizers, can improve grain yield and micronutrient concentrations (Pilon et al. 2014).

Plant growth regulators, both natural and synthetic, have the potential to mitigate drought stress. Substances like glycine betaine, 1-aminocyclopropane-1-carboxylic acid, ABA, and gibberellic acid have shown promising results (Beaudette et al. 2007; Farooq et al. 2009; Parent et al. 2009). However, more research is required to fully comprehend their effectiveness in potato cultivation under drought conditions.

Efficient water management practices are essential for drought mitigation. Modern techniques, such as targeted irrigation, can significantly reduce water usage compared to flood irrigation (Lipiec et al. 2013). Timing irrigation to align with critical growth stages such as tuber initiation and bulking can enhance tuber yield. Increasing rainwater harvesting and exploring treated wastewater use for irrigation are also viable options to alleviate water scarcity (Ricart and Rico 2019; Suhani et al. 2020).

While agronomic practices can provide some relief from drought stress, the development of site-specific production technologies and drought-resistant genotypes is crucial. Breeding approaches targeting drought resistance require a deep understanding of molecular responses to drought stress in potatoes. Significant research efforts are underway to unravel the molecular mechanisms underlying potato genotype responses to varying degrees of drought stress (Golan 2016).

8 Plant Responses to Stress Through Transcriptional and Post-transcriptional Regulation

The literature indicates that plants respond to stress not only at the protein level but also at the transcriptional and post-transcriptional levels. Transcriptional gene regulatory mechanisms can be suppressed through RNA silencing, mediated by post-transcriptional gene silencing (PTGS) or RNA interference (RNAi) (Gupta et al. 2014). The first report of RNAi was in petunia plants in the early 1990s (Napoli et al. 1990). Transgenic plants designed to produce dark purple flowers unexpectedly displayed white and chimeric flowers due to homologous gene silencing, a phenomenon termed 'co-suppression' (Cogoni and Macino 2000). Small RNAs play a crucial role as bio-regulators in this silencing process, with various types such as siRNAs, miRNAs, smRNAs, tncRNAs, piRNAs, and stRNAs categorized based on biogenesis and genomic loci structure.

Among these small RNAs, miRNAs have been identified as key players in stress responses (Bej and Basak 2014; Cooper et al. 2019). Movahedi et al. (2018) outlined pathways of gene silencing and gene knockdown in eukaryotic organisms involving small RNAs, typically 20–25 nucleotides in length, including siRNAs and miRNAs. These small RNAs are utilized to suppress gene expression at the transcriptional level, ultimately leading to mRNA destruction and protein production inhibition through RNAi.

RNA silencing, including RNAi, plays a crucial role in plant stress response pathways. Conversely, proteins that inhibit antiviral RNAi activation in plants, known as Viral Suppressors of RNA silencing (VSRs), suppress plant RNA silencing, thereby preventing the initiation of the plant's immune response against viral RNAi (Gupta et al. 2014). Research of present era primarily focuses on plant small RNAs and their role in combating abiotic stresses. Plant signaling molecules encompass chemical compounds that facilitate intercellular communication within plants, coordinating a myriad of physiological processes and reactions to both endogenous and exogenous cues. Among these signaling molecules, such as jasmonic acid, salicylic acid, calcium ions, and nitric oxide, there are pivotal functions in modulating defense mechanisms against pathogens, herbivores, and environmental stresses in plants. Reactive oxygen species (ROS), comprising superoxide radicals, hydrogen peroxide, and hydroxyl radicals, are recognized as signaling molecules capable of contributing to plant defense mechanisms (Mir et al. 2024). Modulating the levels of phytohormones or their corresponding signaling cascades represents a promising approach for enhancing abiotic stress tolerance in potato plants. The roles and interplay of these molecules can exhibit variability contingent upon the type and intensity of stress, alongside the developmental phase of the potato plant. Mir et al. (2024) emphasizes that a comprehensive study of these signaling pathways is indispensable for devising tactics to fortify stress resilience and bolster potato crop viability amidst demanding environmental circumstances.

9 Unlocking the Potential: miRNAs as Guardians Against Drought Stress in Potato

The yield of *Solanum tuberosum* is significantly impacted by the drought stress. miRNA functions at the post-transcriptional level and modulates the gene expression of the plant. They facilitate adaptations in root architecture promoting deeper root growth, hence increasing water uptake during stress conditions (Jiao et al. 2022). Studies on miRNAs involved in regulating potato plant growth have seen significant progress in recent years. The miRbase database now lists 7025 miRNA entries from 86 plant species (Guo et al. 2020). Efforts to identify miRNAs and their target genes in potato plants began in 2009 and are ongoing. Zhang et al. (2009) employed a bioinformatics approach to identify 48 potential miRNAs in potatoes, which were predicted to regulate 186 potential target genes involved in various developmental processes and stress responses. Subsequent RT-PCR analysis confirmed the expression of some of these miRNAs across different tissues. These observations imply a fundamental involvement of miRNAs in numerous physiological processes in plants, such as genome stability maintenance, growth, development, maturation, and stress response initiation. MicroRNAs play an important role in elevating drought stress tolerance in potato plants. Through different signaling pathways, they enhance the plant's tolerance to survive under abiotic stress conditions (Mir et al. 2024). They modulate stress-responsive genes by targeting mRNA either for degradation or for inhibiting translation. Thereby, directly influencing root development, antioxidant production, and other processes such as stomatal closure (Gervais et al. 2021). miRNA regulation helps in efficiently managing the water use in potato plants, whilst protecting the cells from oxidative damage during drought conditions (Seleiman et al. 2021). Consequently, manipulating miRNAs to develop stress-tolerant potato varieties could potentially enhance yield and nutritional attributes. Within this chapter, we explore the specific roles of potato miRNAs across different stages of the plant's life cycle, elucidating their contributions to stress resilience and gene expression modulation (Kaur et al. 2016).

In another study, (Yang et al. 2010) identified 71 potential miRNAs, of which approximately 65 were newly predicted. Validation through qRT-PCR revealed differential expression levels of these miRNAs across various tissues. The differential expression patterns of miRNAs in different tissues may correlate with their roles in regulating potato vegetative development (Li et al. 2013). Singh et al. (2023) reported that miRNAs play a crucial role in modulating the expression of a diverse set of genes responsive to drought stress, they represent promising candidates for genetic engineering interventions aimed at increasing drought tolerance in agricultural crops, thus contributing to the promotion of sustainable farming practices.

Xie et al. (2011) proposed that the number of identified miRNAs in potatoes was low compared to other plants, suggesting the existence of more miRNAs. Using an improved genome comparison strategy, they identified 202 potential miRNAs, including 54 new families specific to potatoes. Expression analysis of selected miRNAs revealed their involvement in various physiological processes, including stress responses and signal transduction.

Furthermore, Kim et al. (2011) predicted 22 miRNAs and 221 target genes in potatoes belonging to the Solanaceae family using an in-silico approach. Zhang et al. (2013, 2014) employed NGS and potato genome sequencing data to identify 259 miRNAs associated with 159 families in potatoes. They also reported potential target genes involved in various biological processes, including transcription regulation, defense mechanisms, and flowering.

Lakhotia et al. (2014) identified conserved and potato-specific miRNAs, including miR160, miR164, miR172, and miR171, and their target genes associated with tuber growth. Din and Barozai (2014) identified 120 novel miRNAs and their potential target genes using a comparative genomic approach. These studies collectively demonstrate the diverse roles of miRNAs in regulating various aspects of potato plant growth and development, including stress responses, metabolism, and tuber formation.

Research on miRNAs that regulate the abiotic stress response of potato plants has garnered significant attention in recent years. Initially, studies primarily focused on identifying miRNAs involved in regulating drought tolerance. For instance, (Hwang et al. 2011a) identified several miRNAs, including stu-miR156a, stu-miR157a, and stu-miR396, whose expression levels were altered under drought stress conditions in potatoes. They observed dynamic changes in the expression patterns of these miRNAs over time following drought application, shedding light on their potential role in stress response. Later on (Liu et al. 2020) reported some key miRNAs that are involved in regulating drought stress in potatoes are; miR164 regulates leaf senescence. miR396 & miR156 are involved in promoting root growth, enhancing the plant's ability to absorb water from deeper soil layers. miR159, miR160 & miR393 influences the production of hormones such as gibberellin and auxin. miR159 is involved in regulating stomatal opening to decrease water loss through transpiration. miR172 influences the tuberization process ensuring the plant continues to grow during stress conditions. These miRNAs work together to promote tolerance in potato plants by regulating various mechanisms.

Furthermore, Hwang et al. (2011b) investigated the expression alterations of miR171 family members under drought stress conditions and found distinct expression patterns in response to different stress application methods. Their findings highlighted the complexity of miRNA regulation in potato drought response. Yang et al. (2013) identified several miRNAs involved in regulating proline synthesis, a crucial mechanism for plant stress tolerance. They demonstrated that miR172, miR396a, miR396c, and miR4233 may regulate the expression of genes responsible for proline synthesis, thereby enhancing drought tolerance in potatoes.

Moreover, (Zhang et al. 2014) detected numerous novel and known miRNAs in control and drought-treated potato samples. They observed significant changes in the expression levels of these miRNAs under stress conditions, suggesting their involvement in drought response pathways. Specific miRNAs are upregulated or downregulated in response to water scarcity. Öztürk Gökçe et al. (2021) conducted a comprehensive analysis of miRNA profiles in response to drought and heat stress in different potato cultivars. Their results highlighted specific miRNAs, such as Novel_105, miR162a-3p, and miR160a-5p, as potential regulators of stress response pathways. Transgenic approaches have been employed to elucidate the functions of specific miRNAs in stress tolerance. For instance, (Şanlı and Gökçe 2021) overexpressed miR160a-5p in potato cultivars and observed improved tolerance to drought and heat stress. Similarly, Yalçin and Gökçe (2021) reported enhanced stress tolerance in transgenic potato plants overexpressing novel_105 miRNA. Asim et al. (2021) demonstrated the involvement of miR172b-3p in drought and heat stress tolerance mechanisms in potatoes. Their findings underscored the potential of miRNA-based strategies for enhancing stress tolerance in potato cultivars. Shen et al. (2022) reported that knocking out osa-miR1366 leads to improved drought tolerance in potato plants through stem xylem development. Recent studies also highlighted that several processes of ABA and non-ABA mediated ecosystems are regulated by miRNAs in potatoes indicating their role in promoting drought resistance (Muhammad Aslam et al. 2022). Drought tolerance is conferred by ABA in combination with phytochromes i.e., ethylene (ET), jasmonic acid (JA), salicylic acid (SA), brassinosteroids, and auxin (AA). In this way, miRNA represses the abundance of target transcripts (Iqbal et al. 2022). Zhang et al. (2022) reported stu-miR159 targets the *R2R3 MYB* domain transcription factors, Suppression of this domain is reported to enhance drought tolerance in potatoes. sit-miR394 has been studied for the same response, indicating its role in drought-resistant pathways. Adaptive responses to water scarcity are initiated and modulated by the interaction between miRNAs and phytochromes like abscisic acid (ABA).

miRNAs can negatively regulate target genes for drought stress tolerance. In potatoes, stu-miR398b-3p targets *Cu/Zn SOD*. *Cu/Zn SOD* scavenges reactive oxygen species (ROS) that are produced under scarce water conditions. miR398b-3p downregulates the expression of *Cu/Zn SOD* and helps plants dealing with stress conditions. It can do so by mimicking the tertiary structure of *Cu/Zn SOD* having 152 amino acids, with 8 α-helices, 48 extended chains, and 14 β-turns (Zheng et al. 2023).

Furthermore, studies in other plant species have provided valuable insights into the regulatory roles of miRNAs in stress response pathways. For example, in Arabidopsis, miR169 has been implicated in drought tolerance mechanisms through the regulation of target genes like *NFYA5* (Li et al. 2008). Similarly, in tomato, miR169 overexpression has been shown to confer drought tolerance by regulating stomatal behavior and transpiration rate (Zhang et al. 2011). Additionally, studies in rice and sweet potato have revealed the involvement of specific miRNAs in cold and chilling stress tolerance mechanisms (Lv et al. 2010; Xie et al. 2017).

Overall, research on miRNAs in potato and other plant species has provided valuable insights into the molecular mechanisms underlying abiotic stress responses. These findings hold significant promise for the development of stress-tolerant crop varieties through targeted genetic manipulation of miRNA pathways. One way by which miRNAs enhance drought stress tolerance is by regulating Auxin Response Factors (ARFs). ARFs are implicated in development, growth, and stress adaptation. During drought conditions miRNAs are upregulated to modulate ARF gene expression, which promotes root development, hence enhancing the uptake of water. This regulation helps the potato plant to adapt the water-scarce conditions, leading to the development of drought-resistant potato varieties. Understanding these interactions can lead to the development of drought-resistant varieties of potatoes, highlighting the important role of miRNAs in crop improvement strategies (Peng et al. 2020).

It is evident from literature that the choice of potato cultivar and its canopy characteristics significantly influence drought resilience. A deeper understanding of the tuberization process in different cultivars, including the gene-environment (GxE) interactions is significant for the development of optimal and functional varieties (Kaur et al. 2024). Late-maturing cultivars show promise in regions experiencing late-season droughts, as they can sustain yield despite water scarcity during critical growth stages. Conversely, under prolonged drought conditions, mid-maturing cultivars with fewer but larger and thicker leaves may offer better resilience. Understanding the diverse responses of different potato genotypes to varying drought intensities is essential for breeders aiming to develop drought-tolerant potato cultivars.

By considering the interactions between cultivar characteristics and drought stress responses, breeders can identify promising genotypes for targeted breeding programs. By integrating molecular insights with agronomic practices, the development of drought-tolerant potato cultivars and sustainable production systems can be accelerated, thereby mitigating the adverse effects of climate change on potato productivity.

References

Abebe S, Tadesse T (2018) Effect of varieties and fungicides on Potato Late Blight (*Phytophthora infestans*, (Mont) de Bary). Manag Agrotechnol 07

Aghakouchak A, Feldman D, Stewardson MJ, Saphores J-D, Grant S, Sanders B (2014) Australia's drought: lessons for California. Science 343:1430–1431

Ahmadi SH, Andersen MN, Plauborg F, Poulsen RT, Jensen CR, Sepaskhah AR, Hansen S (2010) Effects of irrigation strategies and soils on Field grown potatoes: yield and water productivity. Agric Water Manag 97:1923–1930

Albiski F, Najla S, Sanoubar R, Alkabani N, Murshed R (2012) In vitro screening of potato lines for drought tolerance. Physiol Mol Biol Plants 18:315–321

Aliche EB, Oortwijn M, Theeuwen TPJM, Bachem CWB, Visser RGF, van der Linden CG (2018) Drought response in field grown potatoes and the interactions between canopy growth and yield. Agric Water Manag 206:20–30

Andreu J, Pérez-Martín M (2009) Decision support system for drought planning and Management in the Jucar River Basin, Spain

Anjum SA, Wang LC, Farooq M, Hussain M, Xue LL, Zou CM (2011) Brassinolide application improves the drought tolerance in maize through modulation of enzymatic antioxidants and Leaf gas exchange. J Agron Crop Sci 197:177–185

Aown M, Raza S, Saleem MF, Anjum SA, Khaliq T, Wahid MA (2012) Foliar application of potassium under water deficit conditions improved the growth and yield of wheat (*Triticum aestivum* L.). J Anim Plant Sci 22:431–437

Arnell NW, Reynard NS (1996) The effects of climate change due to global warming on river flows in Great Britain. J Hydrol 183:397–424

Asim A, Gökçe ZNÖ, Bakhsh A, Çaylı İT, Aksoy E, Çalışkan S, Çalışkan ME, Demirel U (2021) Individual and combined effect of drought and heat stresses in contrasting potato cultivars overexpressing miR172b-3p. Turk J Agric For 45:651–668

Athar HR, Ashraf M (2009) Strategies for crop improvement against salinity and drought stress: An overview. In: Ashraf M, Ozturk M, Athar HR (eds) Salinity and water stress: improving crop efficiency. Springer, Dordrecht, pp 1–16

Ávila-Valdés A, Quinet M, Lutts S, Martínez JP, Lizana XC (2020) Tuber yield and quality responses of potato to moderate temperature increase during tuber bulking under two water availability scenarios. Field Crop Res 251:107786

Battisti DS, Naylor RL (2009) Historical warnings of future food insecurity with unprecedented seasonal heat. Science 323:240–244

Beaudette PC, Chlup M, Yee J, Emery RN (2007) Relationships of root conductivity and aquaporin gene expression in *Pisum sativum*: diurnal patterns and the response to HgCl2 and ABA. J Exp Bot 58:1291–1300

Bej S, Basak J (2014) MicroRNAs: the potential biomarkers in plant stress response. Am J Plant Sci 5:748–759

Bellon-Maurel V, Fernandez-Ahumada E, Palagos B, Roger J-M, McBratney A (2010) Critical review of chemometric indicators commonly used for assessing the quality of the prediction of soil attributes by NIR spectroscopy. TrAC Trends Anal Chem 29:1073–1081

Bengough AG, McKenzie BM, Hallett PD, Valentine TA (2011) Root elongation, water stress, and mechanical impedance: a review of limiting stresses and beneficial root tip traits. J Exp Bot 62:59–68

Boguszewska-Mańkowska D, Zarzyńska K, Nosalewicz A (2020) Drought differentially affects root system size and architecture of potato cultivars with differing drought tolerance. Am J Potato Res 97:54–62

Bota J, Medrano H, Flexas J (2004) Is photosynthesis limited by decreased rubisco activity and RuBP content under progressive water stress? New Phytol 162:671–681

Byrd SA, Rowland DL, Bennett J, Zotarelli L, Wright D, Alva A, Nordgaard J (2014) Reductions in a commercial potato irrigation schedule during tuber bulking in Florida: physiological, yield, and quality effects. J Crop Improv 28:660–679

Çalışkan ME (2016) New challenges in potato breeding to cope with climate change: dual tolerance to heat and drought. Lucr Ştiinţ- Ser Agron 59:151–154

Camargo DC, Montoya F, Córcoles JI, Ortega JF (2015) Modeling the impacts of irrigation treatments on potato growth and development. Agric Water Manag 150:119–128

Candar-Cakir B, Cakir O (2016) miRNA profiling in plants: current identification and expression approaches. In: Plant omics: trends and applications. Springer, Cham, pp 189–215

Carli C, Yuldashev F, Khalikov D, Condori B, Mares V, Monneveux P (2014) Effect of different irrigation regimes on yield, water use efficiency and quality of potato (*Solanum tuberosum* L.) in the lowlands of Tashkent, Uzbekistan: a Field and modeling perspective. Field Crop Res 163:90–99

Cattiaux J, Vautard R, Cassou C, Yiou P, Masson-Delmotte V, Codron F (2010) Winter 2010 in Europe: a cold extreme in a warming climate. Geophys Res Lett 37

Chang DC, Jin YI, Nam JH, Cheon CG, Cho JH, Kim SJ, Yu H-S (2018) Early drought effect on canopy development and tuber growth of potato cultivars with different maturities. Field Crop Res 215:156–162

Chen PY, Manninga H, Slanchev K, Chien M, Russo JJ, Ju J, Sheridan R, John B, Marks DS, Gaidatzis D, Sander C (2005) The developmental miRNA profiles of zebrafish as determined by small RNA cloning. Genes Dev 19(11):1288–1293

Cogoni C, Macino G (2000) Post-transcriptional gene silencing across kingdoms. Curr Opin Genet Dev 10:638–643

Cook BI, Mankin JS, Anchukaitis KJ (2018) Climate change and drought: from past to future. Curr Clim Chang Rep 4:164–179

Cooper AMV, Silver K, Zhang J, Park Y, Zhu KY (2019) Molecular mechanisms influencing efficiency of RNA interference in insects. Pest Manag Sci 75:18–28

Cornic G (2000) Drought stress inhibits photosynthesis by decreasing stomatal aperture–not by affecting ATP synthesis. Trends Plant Sci 5:187–188

CRED (n.d.) EM-DAT – the international disaster database [WWW Document]. https://www.emdat.be/. Accessed 24 Nov 2021

Cruiziat P, Cochard H, Améglio T (2002) Hydraulic architecture of trees: main concepts and results. Ann For Sci 59:723–752

Daccache A, Weatherhead EK, Stalham MA, Knox JW (2011) Impacts of climate change on irrigated potato production in a humid climate. Agric For Meteorol 151:1641–1653

Daccache A, Keay C, Jones RJA, Weatherhead EK, Stalham MA, Knox JW (2012) Climate change and land suitability for potato production in England and Wales: impacts and adaptation. J Agric Sci 150:161–177

Dai A (2013) Increasing drought under global warming in observations and models. Nat Clim Chang 3:52–58

de Haan S, Rodriguez F (2016) Chapter 1 – Potato origin and production. In: Singh J, Kaur L (eds) Advances in potato chemistry and technology, 2nd edn. Academic, San Diego, pp 1–32

Deblonde PMK, Haverkort AJ, Ledent J-F (1999) Responses of early and late potato cultivars to moderate drought conditions: agronomic parameters and carbon isotope discrimination. Eur J Agron 11:91–105

Demirel U, Morris WL, Ducreux LJ, Yavuz C, Asim A, Tindas I, Campbell R, Morris JA, Verrall SR, Hedley PE (2020) Physiological, biochemical, and transcriptional responses to single and combined abiotic stress in stress-tolerant and stress-sensitive potato genotypes. Front Plant Sci 11:169

Dettinger M, Cayan DR (2014) Drought and the California delta—A matter of extremes. San Francisco Estuary and Watershed Science 12(2)

Devi KJ, Saha P, Chakraborty S, Rajwanshi R (2018) Computational identification and functional annotation of microRNAs and their targets in three species of kiwifruit (*Actinidia* spp.). Indian J Plant Physiol 23:179–191

Din M, Barozai MYK (2014) Profiling microRNAs and their targets in an important fleshy fruit: tomato (*Solanum lycopersicum*). Gene 535:198–203

Ding C, Shen Q, Zhang R, Chen W (2013) Evaluation of rhizosphere bacteria and derived bio-organic fertilizers as potential biocontrol agents against bacterial wilt (*Ralstonia solanacearum*) of potato. Plant Soil 366:453–466

Doussan C, Pagès L, Vercambre G (1998) Modelling of the hydraulic architecture of root systems: an integrated approach to water absorption—model description. Ann Bot 81:213–223

Ehsanpour A, Razavizadeh R (2005) Effect of UV-C on drought tolerance of Alfalfa (*Medicago sativa*) Callus. Am J Biochem Biotechnol 1:107–110

Eiasu BK, Soundy P, Hammes PS (2007) Response of potato (*Solarium tuberosum*) tuber yield components to gel-polymer soil amendments and irrigation regimes. N Z J Crop Hortic Sci 35:25–31

Fahad S, Bajwa AA, Nazir U, Anjum SA, Farooq A, Zohaib A, Sadia S, Nasim W, Adkins S, Saud S, Ihsan MZ, Alharby H, Wu C, Wang D, Huang J (2017) Crop production under drought and heat stress: plant responses and management options. Front Plant Sci 8

FAOSTAT, 2021 [WWW Document] (n.d.) Food Agric. Organ. U. N. URL https://www.fao.org/faostat/en/#data/QCL. Accessed 23 Nov 2021

Farooq M, Aziz T, Wahid A, Lee D-J, Siddique KH (2009) Chilling tolerance in maize: agronomic and physiological approaches. Crop Pasture Sci 60:501–516

Fazal A, Bano A (2016) Role of plant growth-promoting Rhizobacteria (PGPR), biochar, and chemical fertilizer under salinity stress. Commun Soil Sci Plant Anal 47:1985–1993

Ferrero A, Usowicz B, Lipiec J (2005) Effects of tractor traffic on spatial variability of soil strength and water content in grass covered and cultivated sloping vineyard. Soil Tillage Res 84:127–138

Flexas J, Bota J, Galmes J, Medrano H, Ribas-Carbó M (2006) Keeping a positive carbon balance under adverse conditions: responses of photosynthesis and respiration to water stress. Physiol Plant 127:343–352

Gaur D, Singh SP, Sharma K, Sharma SK, Dhakad H, Dangi RS, Patidar R, Sharma R, Dixit JP, Rawat GS (2017) Effect of different phosphorus levels on growth attributes physiological parameter and grading of tuber in potato crop (*Solanum tuberosum* L.). Int J Chem Stud 5:215–219

Gervais T, Creelman A, Li XQ, Bizimungu B, De Koeyer D, Dahal K (2021) Potato response to drought stress: physiological and growth basis. Front Plant Sci 12:698060

Gevrek MN, Atasoy GD (2012) Effect of post Anthesis drought on certain agronomical characteristics of wheat under two different nitrogen application conditions. Turk J Field Crops 17:19–23

Gilchrist E, Soler J, Merz U, Reynaldi S (2011) Powdery scab effect on the potato *Solanum tuberosum* ssp. andigena growth and yield. Trop Plant Pathol 36:350–355

Gluska A (2004) Differentiation of root system size in potato cultivars. Biul Inst Hod Aklim Rosl 232:37–46

Golan T (2016) The fall and rise of the Kishon River. Water 8:283

Grumm RH (2011) The central European and Russian heat event of July–August 2010. Bull Am Meteorol Soc 92:1285–1296

Guo Z, Kuang Z, Wang Y, Zhao Y, Tao Y, Cheng C, Yang J, Lu X, Hao C, Wang T, Cao X (2020) PmiREN: a comprehensive encyclopedia of plant miRNAs. Nucleic Acids Res 48:1114–1121

Gupta K, Sengupta A, Saha J, Gupta B (2014) The attributes of RNA interference in relation to plant abiotic stress tolerance. Gene Technol 3:110

Haling RE, Brown LK, Bengough AG, Young IM, Hallett PD, White PJ, George TS (2013) Root hairs improve root penetration, root–soil contact, and phosphorus acquisition in soils of different strength. J Exp Bot 64:3711–3721

Hao L, Zhang Y, Wang S, Zhang W, Wang S, Xu C, Yu Y, Li T, Jiang F, Li W (2020) A constitutive and drought-responsive mRNA undergoes long-distance transport in pear (*Pyrus betulaefolia*) phloem. Plant Sci 293:110419

Hatfield JL, Prueger JH (2015) Temperature extremes: effect on plant growth and development. Weather Clim. Extrem 10(USDA Research and Programs on Extreme Events):4–10

Haverkort AJ, Verhagen A (2008) Climate change and its repercussions for the potato supply chain. Potato Res 51:223–237

Haverkort AJ, Van de Waart M, Bodlaender KBA (1990) The effect of early drought stress on numbers of tubers and stolons of potato in controlled and field conditions. Potato Res 33:89–96

Hijmans RJ (2003) The effect of climate change on global potato production. Am J Potato Res 80:271–279

Hirut B, Shimelis H, Fentahun M, Bonierbale M, Gastelo M, Asfaw A (2017) Combining ability of highland tropic adapted potato for tuber yield and yield components under drought. PLoS One 12:e0181541

Horsfield A, Wicks T, Davies K, Wilson D, Paton S (2010) Effect of fungicide use strategies on the control of early blight (*Alternaria solani*) and potato yield. Australas Plant Pathol 39:368–375

Hwang EW, Shin SJ, Kwon HB (2011a) Identification of microRNAs and their putative targets that respond to drought stress in *Solanum tuberosum*. J Korean Soc Appl Biol Chem 54:317–324

Hwang EW, Shin SJ, Park SC, Jeong MJ, Kwon HB (2011b) Identification of miR172 family members and their putative targets responding to drought stress in *Solanum tuberosum*. Genes Genomics 33:105–110

Iqbal S, Wang X, Mubeen I, Kamran M, Kanwal I, Díaz GA, Abbas A, Parveen A, Atiq MN, Alshaya H, Zin El-Abedin TK (2022) Phytohormones trigger drought tolerance in crop plants: outlook and future perspectives. Front Plant Sci 12

Islam W, Idrees A, Waheed A, Zeng F (2022) Plant responses to drought stress: microRNAs in action. Environ Res 215:114282

Iwama K (2008) Physiology of the potato: new insights into root system and repercussions for crop management. Potato Res 51:333–353

Jefferies RA, MacKerron DKL (1989) Radiation interception and growth of irrigated and droughted potato (*Solanum tuberosum*). Field Crop Res 22:101–112

Jefferies RA, Mackerron DKL (1993) Responses of potato genotypes to drought. II. Leaf area index, growth and yield. Ann Appl Biol 122:105–112

Jeong DH, Green PJ (2013) The role of rice microRNAs in abiotic stress responses. J Plant Biol 56:187–197

Jiao P, Ma R, Wang C, Chen N, Liu S, Qu J, Guan S, Ma Y (2022) Integration of mRNA and microRNA analysis reveals the molecular mechanisms underlying drought stress tolerance in maize (*Zea mays L.*). Front Plant Sci 13:932667

Joshi M, Fogelman E, Belausov E, Ginzberg I (2016) Potato root system development and factors that determine its architecture. J Plant Physiol 205:113–123

Kashiwagi J, Iwama K, Hasegawa T, Nagata T (2000) The relationship between the root mass in seedlings and tuber generations in potato. In: Proceedings of the 12th symposium of ISTRC. Tsukuba, Japan, pp 446–450

Kaur PJ, Satya S, Pant KK, Naik SN (2016) Eco-friendly preservation of bamboo species: traditional to modern techniques. Bioresources 11:10604–10624

Kaur G, Jain S, Bhushan S, Das N, Sharma M, Sharma D (2024) Role of microRNAs and their putative mechanism in regulating potato (*Solanum tuberosum* L.) life cycle and response to various environmental stresses. Plant Physiol Biochem:108334

Kazmierczak A, Georgi B (2012) Urban adaptation to climate change in Europe: challenges and opportunities for cities together with supportive national and European policies. European Environment Agency, Copenhagen

Kesiime VE, Tusiime G, Kashaija IN, Edema R, Gibson P, Namugga P, Kakuhenzire R (2016) Characterization and evaluation of potato genotypes (*Solanum tuberosum* L) for tolerance to drought in Uganda. Am J Potato Res 93:543–551

Khosravifar S, Yarnia M, Benam MB, Moghbeli AH (2008) Effect of potassium on drought tolerance in potato cv. Agri J Food Agri Environ 6:236–241

Kim HJ, Baek KH, Lee BW, Choi D, Hur CG (2011) In silico identification and characterization of microRNAs and their putative target genes in Solanaceae plants. Genome 54:91–98

King BA, Stark JC (1997) Potato irrigation management Bul. 789 cooperative extension system College of Agriculture Idaho. University of Idaho, Moscow

Knox JW, Hess T (2019) High level review of the optimum water use methodology for agriculture following the 2018 drought in England, Technical Report. Cranfield University

Knox JW, Weatherhead EK, Bradley RI (1997) Mapping the total volumetric irrigation water requirements in England and Wales. Agric Water Manag 33:1–18

Koebernick N, Daly KR, Keyes SD, George TS, Brown LK, Raffan A, Cooper LJ, Naveed M, Bengough AG, Sinclair I, Hallett PD (2017) High-resolution synchrotron imaging shows that root hairs influence rhizosphere soil structure formation. New Phytol 216:124–135

Lahlou O, Ledent J-F (2005) Root mass and depth, stolons and roots formed on stolons in four cultivars of potato under water stress. Eur J Agron 22:159–173

Lahlou O, Ouattar S, Ledent J-F (2003) The effect of drought and cultivar on growth parameters, yield and yield components of potato. Agronomie 23:257–268

Lakhotia N, Joshi G, Bhardwaj AR, Katiyar-Agarwal S, Agarwal M, Jagannath A, Goel S, Kumar A (2014) Identification and characterization of mirnaome in root, stem, leaf and tuber developmental stages of potato (*Solanum tuberosum* L.) by high-throughput sequencing. BMC Plant Biol 14:1–16

Lawlor DW (2002) Limitation to photosynthesis in water-stressed leaves: stomata vs. metabolism and the role of ATP. Ann Bot 89:871–885

Levy D, Veilleux RE (2007) Adaptation of potato to high temperatures and salinity-a review. Am J Potato Res 84:487–506

Lewis SL, Brando PM, Phillips OL, van der Heijden GMF, Nepstad D (2011) The 2010 Amazon drought. Science 331:554–554

Li W-X, Oono Y, Zhu J, He X-J, Wu J-M, Iida K, Lu X-Y, Cui X, Jin H, Zhu J-K (2008) The Arabidopsis NFYA5 transcription factor is regulated transcriptionally and posttranscriptionally to promote drought resistance. Plant Cell 20:2238–2251

Li J, Fu F, An M, Zhou S, She Y, Li W (2013) Differential expression of microRNAs in response to drought stress in maize. J Integr Agric 12:1414–1422

Li W, Wang T, Zhang Y, Li Y (2016a) Overexpression of soybean miR172c confers tolerance to water deficit and salt stress, but increases ABA sensitivity in transgenic *Arabidopsis thaliana*. J Exp Bot 67(1):175–194

Li W, Xiong B, Wang S, Deng X, Yin L, Li H (2016b) Regulation effects of water and nitrogen on the source-sink relationship in potato during the tuber bulking stage. PLoS One 11:e0146877

Li Y, Tang J, Wang J, Zhao G, Yu Q, Wang Y, Hu Q, Zhang J, Pan Z, Pan X, Xiao D (2022) Diverging water-saving potential across China's potato planting regions. Eur J Agron 134:126450

Liao X, Su Z, Liu G, Zotarelli L, Cui Y, Snodgrass C (2016) Impact of soil moisture and temperature on potato production using seepage and center pivot irrigation. Agric Water Manag 165:230–236

Lipiec J, Doussan C, Nosalewicz A, Kondracka K (2013) Effect of drought and heat stresses on plant growth and yield: a review. Int Agrophys 27:463–477

Liu HH, Tian X, Li YJ, Wu CA, Zheng CC (2008) Microarray-based analysis of stress-regulated microRNAs in *Arabidopsis thaliana*. RNA 14:836–843

Liu S, Yang C, Wu L, Cai H, Li H, Xu M (2020) The peu-miR160a – PeARF17.1 / PeARF17.2 module participates in the adventitious root development of poplar. Plant Biotechnol J 18(2):457–469

Lu E, Luo Y, Zhang R, Wu Q, Liu L (2011) Regional atmospheric anomalies responsible for the 2009–2010 severe drought in China. J Geophys Res Atmospheres 116

Luitel BP, Khatri BB, Choudhary D, Paudel BP, Jung-Sook S, Hur O-S, Baek HJ, Cheol KH, Yul RK (2015) Growth and yield characters of potato genotypes grown in drought and irrigated conditions of Nepal. Int J Appl Sci Biotechnol 3:513–519

Lutaladio N, Castaldi L (2009) Potato: the hidden treasure. J Food Compos Anal Int Year Potato 22:491–493

Lv DK, Bai X, Li Y, Ding X-D, Ge Y, Cai H, Ji W, Wu N, Zhu YM (2010) Profiling of cold-stress-responsive miRNAs in rice by microarrays. Gene 459:39–47

McGrath GS, Sadler R, Fleming K, Tregoning P, Hinz C, Veneklaas EJ (2012) Tropical cyclones and the ecohydrology of Australia's recent continental-scale drought. Geophys Res Lett 39

Michel AJ, Teixeira EI, Brown HE, Dellow SJ, Maley S, Gillespie RN, Richards KK (2019) Water stress responses of three potato cultivars. Agron NZ 49:25–37

Minhas JS, Bansal KC (1991) Tuber yield in relation to water stress at stages of growth in potato (*Solanum tuberosum* L.). J Indian Potato Assoc 18:1–8

Mir B, John A, Rahayu F, Martasari C, Husni A, Sukmadjaja D, Prahardini P, Kosmiatin M, Supriadi K, Purwati R, Wani A (2024) Potato stress resilience: unraveling the role of signalling molecules and phytohormones. Plant Gene:100456

Mitra J (2001) Genetics and genetic improvement of drought resistance in crop plants. Curr Sci 80:758–763

Monneveux P, Ramírez DA, Pino M-T (2013) Drought tolerance in potato (*S. tuberosum* L.). Plant Sci 205–206:76–86

Movahedi A, Zhang J, Sun W, Kadkhodaei S, Mohammadi K, Almasizadehyaghuti A, Yin T, Zhuge Q (2018) Plant small RNAs: definition, classification and response against stresses. Biologia (Bratisl) 73:285–294

Muhammad Aslam M, Waseem M, Jakada BH, Okal EJ, Lei Z, Saqib HSA, Yuan W, Xu W, Zhang Q (2022) Mechanisms of Abscisic acid-mediated drought stress responses in plants. Int J Mol Sci 23(3):1084

Naher L, Ali MA, Dey TK, Islam MM, Naher L, Ismail A, Rahman M (2014) Evolution of disease and potential biocontrol activity of Trichoderma SP. against Rhizoctonia solani on potato. Biosci J Online 30:1108–1117

Najm A, Haj Seyed Hadi M, Fazeli F, Darzi M, Shamorady R (2013) Effect of utilization of organic and inorganic nitrogen source. Agron J 105:592–600

Napoli C, Lemieux C, Jorgensen R (1990) Introduction of a chimeric chalcone synthase gene into petunia results in reversible co-suppression of homologous genes in trans. Plant Cell 2:279–289

Nawaz F, Ashraf MY, Ahmad R, Waraich EA (2013) Selenium (Se) seed priming induced growth and biochemical changes in wheat under water deficit conditions. Biol Trace Elem Res 151:284–293

Neumann PM (2008) Coping mechanisms for crop plants in drought-prone environments. Ann Bot 101:901–907

Ngadze E (2018) Calcium soil amendment increases resistance of potato to blackleg and soft rot pathogens. Afr J Food Agric Nutr Dev 18:10.18697/ajfand.81.16220–10.18697/ajfand.81.12991

Nouri H, Stokvis B, Galindo A, Blatchford M, Hoekstra AY (2019) Water scarcity alleviation through water footprint reduction in agriculture: the effect of soil mulching and drip irrigation. Sci Total Environ 653:241–252

Obidiegwu JE, Bryan GJ, Jones HG, Prashar A (2015) Coping with drought: stress and adaptive responses in potato and perspectives for improvement. Front Plant Sci 6

Ohashi Y, Nakayama N, Saneoka H, Fujita K (2006) Effects of drought stress on photosynthetic gas exchange, chlorophyll fluorescence and stem diameter of soybean plants. Biol Plant 50:138–141

Öztürk Gökçe ZN, Aksoy E, Bakhsh A, Demirel U, Çalışkan S, Çalışkan ME (2021) Combined drought and heat stresses trigger different sets of miRNAs in contrasting potato cultivars. Funct Integr Genomics 21:489–502

Ozturk M, Turkyilmaz Unal B, García-Caparrós P, Khursheed A, Gul A, Hasanuzzaman M (2021) Osmoregulation and its actions during the drought stress in plants. Physiol Plant 172(2):1321–1335

Pachauri RK, Allen MR, Barros VR, Broome J, Cramer W, Christ R, Church JA, Clarke L, Dahe Q, Dasgupta P, Dubash NK, Edenhofer O, Elgizouli I, Field CB, Forster P, Friedlingstein P, Fuglestvedt J, Gomez-Echeverri L, Hallegatte S, Hegerl G, Howden M, Jiang K, Jimenez Cisneroz B, Kattsov V, Lee H, Mach KJ, Marotzke J, Mastrandrea MD, Meyer L, Minx J, Mulugetta Y, O'Brien K, Oppenheimer M, Pereira JJ, Pichs-Madruga R, Plattner G-K, Pörtner H-O, Power SB, Preston B, Ravindranath NH, Reisinger A, Riahi K, Rusticucci M, Scholes R, Seyboth K, Sokona Y, Stavins R, Stocker TF, Tschakert P, van Vuuren D, van Ypserle J-P (2014) Climate change 2014: synthesis report. Contribution of working groups I, II and III to the fifth assessment report of the intergovernmental panel on climate change, EPIC3Geneva, 151 p., p 151, ISBN: 978-92-9169-143-2. IPCC, Geneva, Switzerland

Pagano L, Rossi R, Paesano L, Marmiroli N, Marmiroli M (2021) miRNA regulation and stress adaptation in plants. Environ Exp Bot 184:104369

Parent B, Hachez C, Redondo E, Simonneau T, Chaumont F, Tardieu F (2009) Drought and Abscisic acid effects on aquaporin content translate into changes in hydraulic conductivity and leaf growth rate: a trans-scale approach. Plant Physiol 149:2000–2012

Peng Y, Fang T, Zhang Y, Zhang M, Zeng L (2020) Genome-wide identification and expression analysis of auxin response factor (ARF) gene family in longan (*Dimocarpus longan L.*). Plan Theory 9(2):221

Pieters AJ, El Souki S (2005) Effects of drought during grain filling on PS II activity in Rice. J Plant Physiol 162:903–911

Pieterse L, Judd J (2014) Catalogue Universel Des Variétés de Pommes de Terre: Potato Library. Agrimedia

Pilon C, Soratto RP, Broetto F, Fernandes AM (2014) Foliar or soil applications of silicon alleviate water-deficit stress of potato plants. Agron J 106:2325–2334

Pino M-T, Skinner JS, Park E-J, Jeknić Z, Hayes PM, Thomashow MF, Chen THH (2007) Use of a stress inducible promoter to drive ectopic AtCBF expression improves potato freezing tolerance while minimizing negative effects on tuber yield. Plant Biotechnol J 5:591–604

Pourasadollahi A, Siosemardeh A, Hosseinpanahi F, Sohrabi Y (2019) Physiological and agro-morphological response of potato to drought stress and hormone application. J Plant Physiol Breed 9:47–61

Puértolas J, Ballester C, Elphinstone JG, Dodd IC (2014) Two potato (*Solanum tuberosum*) varieties differ in drought tolerance due to differences in root growth at depth. Funct Plant Biol 41:1107–1118

Pulatov B, Linderson M-L, Hall K, Jönsson AM (2015) Modeling climate change impact on potato crop phenology, and risk of frost damage and heat stress in northern Europe. Agric For Meteorol 214–215:281–292

Rebetez M, Dupont O, Giroud M (2009) An analysis of the July 2006 heatwave extent in Europe compared to the record year of 2003. Theor Appl Climatol 95:1–7

Reddy B, Mandal R, Chakroborty M, Hijam L, Dutta P (2018) A review on potato (*Solanum tuberosum* L.) and its genetic diversity. Int J Genet 10:360

Renault D, Wallender WW (2000) Nutritional water productivity and diets. Nutr Water Product Diets 45:275–296

Rial-Lovera K, Davies WP, Cannon ND (2017) Implications of climate change predictions for UK cropping and prospects for possible mitigation: a review of challenges and potential responses. J Sci Food Agric 97:17–32

Ricart S, Rico AM (2019) Assessing technical and social driving factors of water reuse in agriculture: a review on risks, regulation and the yuck factor. Agric Water Manag 217:426–439

Robine J-M, Cheung SLK, Roy SL, Oyen HV, Griffiths C, Michel J-P, Herrmann FR (2008) Death toll exceeded 70,000 in Europe during the summer of 2003. C R Biol 331:171–178

Romero AP, Alarcón A, Valbuena RI, Galeano CH (2017) Physiological assessment of water stress in potato using spectral information. Front Plant Sci 8:1608

Rudack K, Seddig S, Sprenger H, Köhl K, Uptmoor R, Ordon F (2017) Drought stress-induced changes in starch yield and physiological traits in potato. J Agron Crop Sci 203:494–505

Ruttanaprasert R, Jogloy S, Vorasoot N, Kesmala T, Kanwar RS, Holbrook CC, Patanothai A (2016) Effects of water stress on total biomass, tuber yield, harvest index and water use efficiency in Jerusalem artichoke. Agric Water Manag 166:130–138

Rykaczewska K (2015) Morphology and anatomy of the root system of new potato cultivars part I. Morphology of the root system. Biul Inst Hod Aklim Roślin 279:31–43

Rykaczewska K (2017) Impact of heat and drought stresses on size and quality of the potato yield. Plant Soil Env 63:40–46

Sadeghipour O, Aghaei P (2012) Response of common bean (*Phaseolus vulgaris* L.) to exogenous application of salicylic acid (SA) under water stress conditions. Adv Env Biol 6:1160–1168

Salavati S, Valadabadi S, Parvizi K, Sayfzadeh S, Masouleh E (2018) The effect of super-absorbent polymer and sowing depth on growth and yield indices of potato (*Solanum tuberosum* L.) in Hamedan Province. Iran Appl Ecol Environ Res 16:7063–7078

Salehi-Lisar SY, Bakhshayeshan-Agdam H (2016) Drought stress in plants: causes, consequences, and tolerance. In: Hossain MA, Wani SH, Bhattacharjee S, Burritt DJ, Tran L-SP (eds) Drought stress tolerance in plants, Physiology and biochemistry, vol 1. Springer International Publishing, Cham, pp 1–16

Salvin S, Baruah K, Bordoloi S (2000) Drip irrigation studies in banana cv. Barjahaji (Musa AAA group, Cavendish sub-group)

Şanlı BA, Gökçe ZNÖ (2021) Investigating effect of miR160 through overexpression in potato cultivars under single or combination of heat and drought stresses. Plant Biotechnol Rep:1–14

Schafleitner R, Gutierrez R, Legay S, Evers D, Bonierbale M (2009) Drought stress tolerance traits of potato

Schmidt S, Bengough AG, Gregory PJ, Grinev DV, Otten W (2012) Estimating root–soil contact from 3D X-ray microtomographs. Eur J Soil Sci 63:776–786

Schubert SD, Suarez MJ, Pegion PJ, Koster RD, Bacmeister JT (2004) On the cause of the 1930s dust bowl. Science 303:1855–1859

Sekhon HS, Singh G, Sharma P, Bains TS (2010) Water use efficiency under stress environments. In: Climate change and management of cool season grain legume crops. Springer, Berlin/Heidelberg, pp 207–227

Seleiman MF, Al-Suhaibani N, Ali N, Akmal M, Alotaibi M, Refay Y, Dindaroglu T, Abdul-Wajid HH, Battaglia ML (2021) Drought stress impacts on plants and different approaches to alleviate its adverse effects. Plan Theory 10(2):259

Shao H-B, Chu L-Y, Jaleel CA, Manivannan P, Panneerselvam R, Shao M-A (2009) Understanding water deficit stress-induced changes in the basic metabolism of higher plants–biotechnologically and sustainably improving agriculture and the ecoenvironment in arid regions of the globe. Crit Rev Biotechnol 29:131–151

Shen X, He J, Ping Y, Guo J, Hou N, Cao F, Li X, Geng D, Wang S, Chen P, Qin G (2022) The positive feedback regulatory loop of miR160-auxin response factor 17-HYPONASTIC LEAVES 1 mediates drought tolerance in apple trees. Plant Physiol 188(3):1686–1708

Shi B, Hussain SS (2020) Chapter 12 - Toward elucidating the functions of miRNAs in drought stress tolerance. In: Guleria P, Kumar V (eds) Plant small RNA. Academic, pp 231–246

Singh R, Singh A, Singh G, Singh A, Kaur G (2021) Influence of bed size and irrigation methods on potato (*Solanum tuberosum*). Indian J Agric Res 55

Singh A, Jain D, Pandey J, Yadav M, Bansal K, Singh I (2023) Deciphering the role of miRNA in reprogramming plant responses to drought stress. Crit Rev Biotechnol 43:613–627

Sodhi GPS, Beri V, Benbi DK (2009) Soil aggregation and distribution of carbon and nitrogen in different fractions under long-term application of compost in rice–wheat system. Soil Tillage Res 103:412–418

Solh M, van Ginkel M (2014) Drought preparedness and drought mitigation in the developing world's drylands. Weather Clim Extrem 3(High Level Meeting on National Drought Policy):62–66

Spooner DM, van den Berg RG, Rodrigues A, Bamberg JB, Hijmans RJ, Lara-Cabrera S (2004) Wild potatoes (Solanum section Petota; Solanaceae) of North and Central America. The American Society of Plant Taxonomists, [S.l.]

Stalham MA, Allen EJ, Rosenfeld AB, Herry FX (2007) Effects of soil compaction in potato (*Solanum tuberosum*) crops. J Agric Sci 145:295–312

Steckel JR, Gray D (1979) Drought tolerance in potatoes. J Agric Sci 92:375–381

Suhani I, Vaish B, Singh P, Singh RP (2020) Restoration, construction, and conservation of degrading wetlands: a step toward sustainable management practices. In: Restoration of wetland ecosystem: a trajectory towards a sustainable environment. Springer, Berlin/Heidelberg, pp 1–16.

Sun Y, Cui X, Liu F (2015) Effect of irrigation regimes and phosphorus rates on water and phosphorus use efficiencies in potato. Sci Hortic 190:64–69

Szabolcs I, Pessarakli M (1999) Handbook of plant and crop stress. CRC Press, Boca Raton

Tejada M, Hernandez MT, Garcia C (2009) Soil restoration using composted plant residues: effects on soil properties. Soil Tillage Res 102:109–117

Tezara W, Mitchell VJ, Driscoll SD, Lawlor DW (1999) Water stress inhibits plant photosynthesis by decreasing coupling factor and ATP. Nature 401:914–917

Thiele G, Theisen K, Bonierbale M, Walker T (2010) Targeting the poor and hungry with potato science. Potato J 37:75–86

Thomas SC, Frye S, Gale N, Garmon M, Launchbury R, Machado N, Melamed S, Murray J, Petroff A, Winsborough C (2013) Biochar mitigates negative effects of salt additions on two herbaceous plant species. J Env Manag 129:62–68

van Loon CD (1981) The effect of water stress on potato growth, development, and yield. Am Potato J 58:51–69

Vandeleur RK, Mayo G, Shelden MC, Gilliham M, Kaiser BN, Tyerman SD (2009) The role of plasma membrane intrinsic protein aquaporins in water transport through roots: diurnal and drought stress responses reveal different strategies between isohydric and anisohydric cultivars of grapevine. Plant Physiol 149:445–460

Vishnoi L, Roy S, Murty N, Nain A (2012) Study on water requirement of potato (*Solanum tuberosum* L.) using CROPWAT MODEL for Tarai region of Uttarakhand. J Agrometeorol 14:180–185

Viste E, Korecha D, Sorteberg A (2013) Recent drought and precipitation tendencies in Ethiopia. Theor Appl Climatol 112:535–551

Vu JCV, Allen LHJ, Bowes G (1987) Drought stress and elevated CO_2 effects on soybean ribulose bisphosphate carboxylase activity and canopy photosynthetic rates. Plant Physiol 83:573–578

Wang W, Vinocur B, Altman A (2003) Plant responses to drought, salinity and extreme temperatures: towards genetic engineering for stress tolerance. Planta 218:1–14

Wang Z, Li Y, Zhuang L, Yu Y, Liu J, Zhang L, Gao Z, Wu Y, Gao W, Ding G, Wang Q (2019) A rhizosphere-derived consortium of *Bacillus subtilis* and *Trichoderma harzianum* suppresses common scab of potato and increases yield. Comput Struct Biotechnol J 17:645–653

Weisz R, Kaminski J, Smilowitz Z (1994) Water deficit effects on potato leaf growth and transpiration: utilizing fraction extractable soil water for comparison with other crops. Am Potato J 71:829–840

Wilhite DA (2016) Drought-management policies and preparedness plans: changing the paradigm from crisis to risk management. In: Land Restoration. Elsevier, pp 443–462

Wishart J, George TS, Brown LK, Ramsay G, Bradshaw JE, White PJ, Gregory PJ (2013) Measuring variation in potato roots in both field and glasshouse: the search for useful yield predictors and a simple screen for root traits. Plant Soil 368:231–249

Wishart J, George TS, Brown LK, White PJ, Ramsay G, Jones H, Gregory PJ (2014) Field phenotyping of potato to assess root and shoot characteristics associated with drought tolerance. Plant Soil 378:351–363

Xia Q, Tao H, Li Y, Pan D, Cao J, Liu L, Zhou X, Barba FJ (2020) Characterizing physicochemical, nutritional and quality attributes of wholegrain *Oryza sativa* L. subjected to high intensity ultrasound-stimulated pre-germination. Food Control 108:106827

Xie F, Frazier TP, Zhang B (2011) Identification, characterization and expression analysis of MicroRNAs and their targets in the potato (*Solanum tuberosum*). Gene 473:8–22. https://doi.org/10.1016/j.gene.2010.09.007

Xie F, Jones DC, Wang Q, Sun R, Zhang B (2015) Small RNA sequencing identifies miRNA roles in ovule and fibre development. Plant Biotechnol J 13(3):355–369

Xie Z, Wang A, Li H, Yu J, Jiang J, Tang Z, Ma D, Zhang B, Han Y, Li Z (2017) High throughput deep sequencing reveals the important roles of microRNAs during sweetpotato storage at chilling temperature. Sci Rep 7:1–12

Yalçin M, Gökçe ZNÖ (2021) Investigation of the effects of overexpression of Novel_105 miRNA in contrasting potato cultivars during separate and combined drought and heat stresses. Turk J Bot 45:397–411

Yang W, Liu X, Zhang J, Feng J, Li C, Chen J (2010) Prediction and validation of conservative microRNAs of *Solanum tuberosum* L. Mol Biol Rep 37:3081–3087

Yang J, Zhang N, Ma C, Qu Y, Si H, Wang D (2013) Prediction and verification of microRNAs related to proline accumulation under drought stress in potato. Comput Biol Chem 46:48–54

Yin F, Gao J, Liu M, Qin C, Zhang W, Yang A, Xia M, Zhang Z, Shen Y, Lin H, Luo C, Pan G (2014) Genomewide analysis of water-stress-responsive microRNA expression profile in tobacco roots. Funct Integr Genomics 14:319–332

Zarzyńska K, Boguszewska-Mańkowska D, Nosalewicz A (2017) Differences in size and architecture of the potato cultivars root system and their tolerance to drought stress. Plant Soil Env 63:159–164

Zhang W, Luo Y, Gong X, Zeng W, Li S (2009) Computational identification of 48 potato microRNAs and their targets. Comput Biol Chem 33:84–93

Zhang X, Zou Z, Gong P, Zhang J, Ziaf K, Li H, Ye Z (2011) Over-expression of microRNA169 confers enhanced drought tolerance to tomato. Biotechnol Lett 33:403–409

Zhang R, Marshall D, Bryan GJ, Hornyik C (2013) Identification and characterization of miRNA transcriptome in potato by high-throughput sequencing. PLoS One 8:e57233

Zhang N, Yang J, Wang Z, Wen Y, Wang J, He W, Liu B, Si H, Wang D (2014) Identification of novel and conserved microRNAs related to drought stress in potato by deep sequencing. PLoS One 9:e95489

Zhang F, Yang J, Zhang N, Wu J, Si H (2022) Roles of microRNAs in abiotic stress response and characteristics regulation of plant. Front Plant Sci 13:919243

Zheng Z, Yang J, Wang X, Zhang N, Si H (2023) Potato Stu-miR398b-3p negatively regulates cu/Zn-SOD response to drought tolerance. Int J Mol Sci 24(3):2525

Drought Stress and the Role of NAC Transcription Factors in Drought Response

Cüneyt Uçarlı

Abstract The increasing susceptibility of plants to various abiotic stresses adversely affect the plant growth and development as well as productivity of major crops. Drought has a considerable threat to sustainable agriculture and the global food security due to global warming and climate change. Drought stress decreases the plant growth by affecting various physiological and biochemical functions including photosynthesis, nutrient metabolism, respiration, reactive oxygen species generation. Several drought stress-responsive genes, playing metabolic or regulatory roles, have been identified and characterized in plants using modern genetics and genomic approaches including transcriptomics and proteomics. Transcription factors (TFs), including NAC, WRKY, AP2/ERF, MYB, are regulatory genes that have important roles in response to stress by regulating the downstream stress-responsive genes. NACs are unique plant TFs and have essential regulatory roles in abiotic and biotic stress responses, and plant development. The purpose of this chapter is to summarize the current knowledge of drought stress and function of NAC transcription factors in response to drought stress in plants. The determination of NAC function and relationship with other stress-responsive genes in drought stress will continue to be a major challenge in the coming years.

Keywords Abiotic stress · ABA · NAC · ROS · Water deficit

1 Introduction

Human civilization is facing a dangerous challenge due to climate change, which is affecting food production and restricting people's access to food. The world population is growing uncontrollably and is estimated to reach 9 billion by 2050 (FAO 2017). The FAO estimates that food production must increase by at least 60% in the

C. Uçarlı (✉)
Depertment of Molecular Biology and Genetics, Istanbul University, İstanbul, Türkiye
e-mail: ucarlicu@istanbul.edu.tr

U. K. Chaudhry et al. (eds.), *Drought Stress*,
https://doi.org/10.1007/978-3-031-80610-0_11

coming decades to meet the increased demand driven by population growth and dietary changes. Plants are immobile organisms and subjected to diverse environmental stresses, such as drought, salinity, extreme temperature, excess water, and pathogens at all developmental stages and during their life span (Seleiman et al. 2021). Plants need sufficient water for growth, development, and reproduction. The arid and semi-arid regions comprise more than 30% of the world lands. In addition, unpredictable droughts frequently occur in most other land regions of the world and the frequency of occurrence is increasing year by year all over the world due to climatic changes (Fang and Xiong 2015). Drought is one the most severe of environmental stresses that limits plant growth and development, and therefore reduces agricultural productivity worldwide, especially in arid and semi-arid regions. Drought causes over 34% of crop and livestock production loss mostly in the least developed and lower-middle-income countries, with a total of 37 billion USD loss in the agricultural sector (FAO 2021). The comparable water reduction (approximately 40%) leads serious yield decrease in world's major crops including rice, maize and wheat by 50, 39 and 21%, respectively (Daryanto et al. 2017; Leng and Hall 2019). Some parameters, including plant species, developmental stage, exposure time and drought level, affect the yield, leading a reduction in productivity under drought stress (Abobatta 2020).

2 Drought Stress

Drought, or water-deficit stress, occurs when available soil water becomes insufficient due to little, or no precipitation, and when atmospheric conditions cause constant water loss through transpiration or evaporation. Drought stress give rise to water loss from cell, resulting with increased cellular dehydration, osmotic stress, and reactive oxygen species (ROS) generation (Singh et al. 2019). In plants, the metabolic, biochemical, and physiological effects of drought lead to cell damage and delay in plant development, resulting decrease in product quality and loss of crop yield (Joshi et al. 2016). Therefore, as immobile organisms, plants have developed some intricate physiological, morphological, and molecular adaptation strategies, including decrease in water loss, improvement in reactive oxygen species (ROS) scavengers, production of osmoprotectants, including polyols, amonoacids and polyamines, and hormonal and transcriptional modulations, to overcome adverse effects of drought stress (Fig. 1).

Drought tolerance can be termed as the ability of a plant to maintain biomass production and to produce its maximum economic yield under drought conditions (Rollins et al. 2013). Plants adapt to drought conditions through three major strategies including drought avoidance, drought tolerance and drought escape without clear boundaries among them (Levitt 1980) (Fig. 2). Drought avoidance (DA) is the maintenance of high-water status during moderate drought conditions by increasing water uptake and/or reducing water loss. DA is a mechanism that retards plant development associated with reduced photosynthesis and low cell metabolism, and is achieved by increasing rooting, hydraulic conductance, and reducing

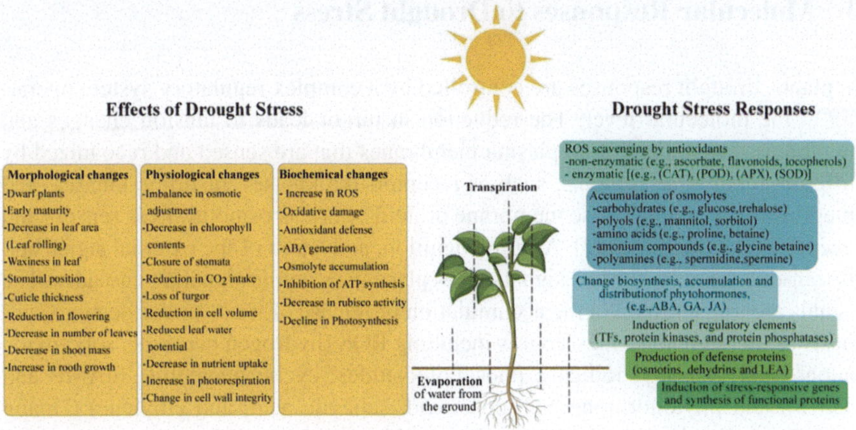

Fig. 1 Effects of drought stress in plants and drought response of the plants. (Modified from Reddy et al. 2004)

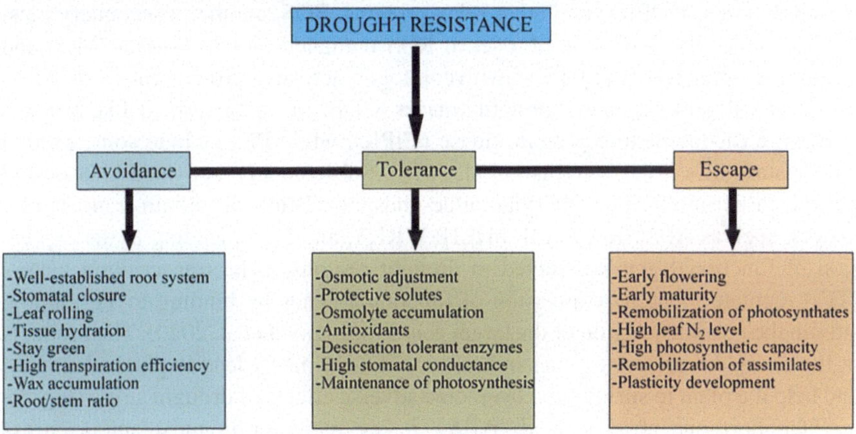

Fig. 2 Mechanisms of drought resistance in plants. (Modified from Panda et al. 2021)

transpiration and transpiration area (Shavrukov et al. 2017). Drought tolerance (DT) is the ability of plants to withstand vigorous dehydration and maintenance of functional growth under low leaf water status through osmotic adjustment and synthesis of osmoprotectants, such as proline, sugar alcohols and polyamines (Luo 2010). The plants maintain cell turgor through not only osmotic adjustment but also by cellular elasticity and by increasing protoplasmic drought resistance (Fang and Xiong 2015). Drought escape (DE), another classical drought adaptive mechanism, involve rapid plant development, such as early flowering time and a shorter vegetative stage, before on set of drought, thus plants do not experience drought stress (Shavrukov et al. 2017). Drought recovery (DR) can also be added to these three basic strategies, as is the ability of plants to recover from a water-deficient state, and maintain growth and yield after exposure to prolonged drought stress (Carraro and Di Iorio 2022).

3 Molecular Responses to Drought Stress

In plants, drought responses are controlled by a complex regulatory system operating at the molecular level. The reduction in turgor leads to tension changes and membrane destabilization in plasma membranes that are sensed and recognized by different membrane proteins, such as receptor-like kinases, histidine kinases, and integrin-like proteins. These membrane proteins work as osmotic stress sensors and receptors (Bandurska 2022). After recognition, activation of the external signal and its transduction into internal signals takes place in systemic signaling cascades. The signals are transmitted via the accumulation or release of Ca^{2+}, and/or de novo synthesis of the secondary messengers including ROS (hydrogen peroxide, superoxide anions, and hydroxyl radical), phosphoinositides, cyclic nucleotide (cAMP and cGMP), and phytohormones (abscisic acid, salicylic acid, and ethylene) (Huang et al. 2012; Joshi et al. 2016; Hrmova and Hussain 2021). These secondary messengers regulate the intracellular Ca^{2+} level. Ca^{2+} sensors, also known as calcium binding proteins, sense the changes in cytosolic Ca^{2+} level. Then, the sensory proteins interact with their respective interacting partners, including calcium-dependent protein kinases (CDPKs) and other relay proteins. ROS, another a secondary messenger, induce the activation of a set of ROS-regulated protein kinases (PKs) and protein phosphatases (PPs). PKs involve mitogen-activated protein kinase (MAPK) cascades, calcium-dependent protein kinases (CDPKs), calcineurin-B-like proteins (CBLs), CBL-interacting protein kinase (CIPK), while PPs include some protein phosphatase 2Cs (PP2Cs) (Khan et al. 2018). Subsequently, activated kinases and phosphatases initiate phosphorylation/de-phosphorylation of certain transcription factors, such as AP2/ERF, MYB, bHLH, WRKY, and NAC, leading to the expression of functional genes involved in drought response. The transcription factors (TFs) regulate the gene expression of functional genes by binding to TF binding sites in the promoter region of the target genes (Mahmood et al. 2020). The products of the drought-responsive genes induced by TFs ultimately lead to plant adaptation and help the plant to survive and overcome adverse effects of drought stress (Fig. 3).

Plant hormone abscisic acid (ABA), a key regulator in most abiotic stress responses, has a crucial role in mediating drought response in coordination with other phytohormones (Abhilasha and Choudhury 2021). ABA's synergistic or antagonistic crosstalks with other plant hormones, including ethylene, auxin, jasmonic acid, salicylic acid, and cytokinins, regulate root growth, induction of stomatal closure and water loss by stomata (Sharp et al. 2004; Ciura and Kruk 2018). Synthesized active ABA is predominantly accumulating in the vascular tissue of leaves and is transported to action sites for stress response (Efetova et al. 2007). Plants respond to drought stress through ABA-dependent or ABA-independent pathways at physiological, molecular, and biochemical level (Shinozaki and Yamaguchi-Shinozaki 2007). The increasing concentration of active ABA can result in both suppression and increased expression of ABA-responsive genes through ABA signaling machinery. In the ABA-dependent pathway, ABA is recognized by ABA receptor Pyrabactin Resistance 1 (PYR1)/PYR-Like (PYL)/Regulatory

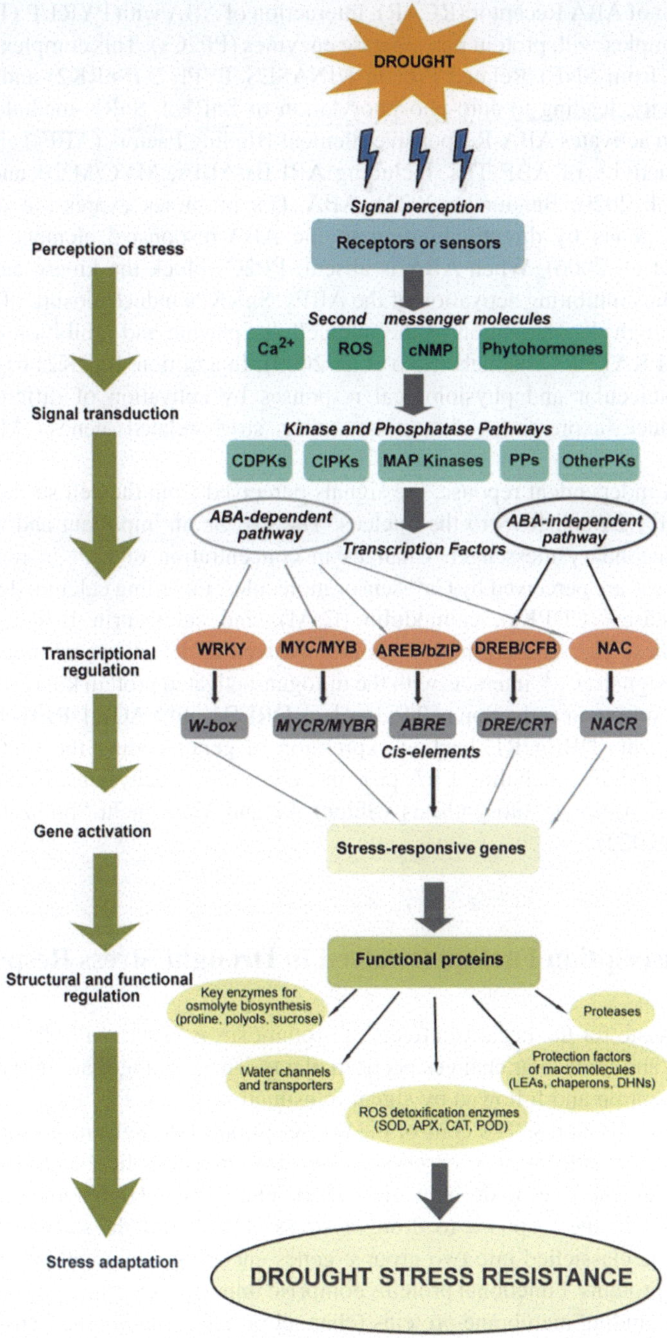

Fig. 3 Molecular responses of drought stress in plants. (Modified from Shinozaki and Yamaguchi-Shinozaki 2007)

Component of ABA Receptor (RCAR). Interaction of ABA with PYR1/PYL/RCAR forms a complex with protein phosphatase enzymes (PP2Cs). This complex dissociates PP2C from SNF1-Related Protein KINASES TYPE 2 (SnRK2) and inhibits PP2C activity, leading to auto-phosphorylation of SnRK2. SnRK-mediated phosphorylation activates ABA-Responsive Element Binding Factors (ABFs) and determine the activity of ABF TFs, including AREBs/ABFs, MYC/MYB and NACs (Yoon et al. 2020; Bandurska 2022). ABA TFs promotes expression of ABA-responsive genes by directly binding to the ABA-responsive element (ABRE) (Furihata et al. 2006). When ABA is absent, PP2Cs block the kinase activity of SnRK2s, thus inhibiting activation of the ABFs. SnRK2s induce closure of stomata mainly through depolarization of the intercellular plasma and inhibition of guard cell-inward KAT1 K^+ channels (Sato et al. 2009). In addition, SnRK2s trigger and regulate molecular and physiological responses by activation of different TFs, which induce expression of various water stress-related genes (Mahmood et al. 2020).

In ABA-independent reponse, the signals perceived from the cell surface mediated by Ca^{2+}, JA and ROS to the nucleus. The Ca^{2+} is an important and the most common secondary messenger. Changes in concentration of Ca^{2+} in reponse to drought stress are perceived by Ca^{2+} sensor molecules, including calcium dependent protein kinases (CDPKs), calmodulin (CaM), and calcineurin B-like proteins (CBLs). These Ca^{2+}-sensing molecules are involved in detection and transmission of cellular signals. Ca^{2+} interacts with the mitogen-activated protein kinase (MAPK) cascade, resulting in activation of TFs, such as DREBs and NACs. DREBs bind cis-acting elements DRE/CRT, leading expression of genes coding the synthesis of functional proteins, including LEA proteins, chaperones, dehydrins, osmotin, and enzymes of osmolyte biosynthesis (Shinozaki and Yamaguchi-Shinozaki 2007; Bandurska 2022).

4 Transcription Factors Involved in Drought Stress Response

To compensate for the negative effects of drought stress, a series of physiological, metabolic, and molecular changes occur in plants. These changes are initiated with signal perception and followed by signal transduction. Then, TFs are activiated by distinct protein kinases. TFs control the transcription of drought-responsive genes by inducing and altering their expression, resulting in metabolic, biochemical, and physiological responses to drought stress (Shao et al. 2015). High number of genes are involved in the response to drought stress. These drought stress-responsive genes can be classiefied into two groups: genes encoding functional proteins, and regulatory proteins. Functional proteins comprise important enzymes and metabolic proteins, including membrane proteins (channel proteins, membrane transporters), enzymes associated with osmolyte (proline, polyamine, sugar, polyol, etc.) biosynthesis, ROS scavenging enzymes (such as SOD, CAT, APX, and GST), proteins involved in lipid biosynthesis and other proteins that play a role in the protection of macromolecules (Shinozaki and Yamaguchi-Shinozaki 2007; Hrmova and Hussain

2021). Regulatory proteins comprise transcription factors, regulating gene expression via DREB, WRKY, MYC/MYB, NAC, and protein kinases (CDPK, MAPK, CIPK, etc.) and proteases are involved in signal transduction (Shinozaki and Yamaguchi-Shinozaki 2007; Shao et al. 2015).

TFs are central regulators of stress-responsive gene expression and regulate gene expression by binding specific DNA sequences, known as cis-elements which are located on the promoter region of target genes, and interacting with different proteins in transcriptional complexes (Baillo et al. 2019). Ultimately, TFs act as molecular switches for gene expression by being involved in multiple signaling pathways and regulating a set of drought stress-responsive genes (Leng and Zhao 2020). Genes encoding TFs constitute 3–10% of transcriptomes in eukaryotic genomes and this ratio is about 7% in plants (Gonzalez 2016; Singh et al. 2021). In plants, TFs can be classified into families according to their DNA-binding domains. Among these TF families, NAC, bZIP, WRKY, MYB, and ERF-DREB have confirmed to be involved in drought tolerance in different plants (Golldack et al. 2011; Leng and Zhao 2020) (Table 1). A typical plant TF consists of four functional domains: a

Table 1 Transcription factor families and genes involved in regulation of drought response

TF family	Gene	Identified in plant	Transgenic plant	Reference
MYB	*FtMYB9*	*Fagopyrum tataricum*	Arabidopsis	Gao et al. (2017)
	1R-MYB	*Cicer arietinum*	Chickpea	Ramalingam et al. (2015)
	GmMYB84	*Glycine max*	Soybean	Wang et al. (2017)
	OsMYB6	*Oryza sativa*	Rice	Tang et al. (2019)
	Mdmyb10	*Malus domestica*	Arabidopsis	Gao et al. (2010)
	HvMYB1	*Hordeum vulgare*	Barley	Alexander et al. (2019)
	GbMYB5	*Gossypium barbadense*	Tobacco	Chen et al. (2015)
	OsMYB55	*Oryza sativa*	Maize	Casaretto et al. (2016)
	GaMYB62L	*Gossypium arboreum*	Arabidopsis	Butt et al. (2017)
bHLH	*AtbHLH68*	*Arabidopsis thaliana*	Arabidopsis	Le Hir et al. (2017)
	PebHLH35	*Populus euphratica*	Arabidopsis	Dong et al. (2014)
	TabHLH39	*Triticum aestivum*	Arabidopsis	Zhai et al. (2016)
	FtbHLH3	*Fagopyrum tataricum*	Arabidopsis	Yao et al. (2017)
WRKY	*OsWRKY47*	*Oryza sativa*	Rice	Raineri et al. (2015)
	AtWRKY53	*Triticum aestivum*	Arabidopsis	Sun and Yu (2015)
	GhWRKY33	*Gossypium hirsutum*	Arabidopsis	Wang et al. (2019a)
	ZmWRKY17	*Zea mays*	Arabidopsis	Cai et al. (2017)
	TaWRKY2	*Triticum aestivum*	Wheat	Gao et al. (2018)
WRKY	*AtWRKY30*	*Arabidopsis thaliana*	Wheat	El-Esawi et al. (2019)
	ZmWRKY40	*Zea mays*	Arabidopsis	Wang et al. (2018)
	GmWRKY27	*Glycine max*	Soybean	Wang et al. (2015)
	MuWRKY3	*Macrotyloma uniflorum*	Groundnut	Kiranmai et al. (2018)
	MtWRKY76	*Medicago truncatula*	Medicago	Liu et al. (2016)

(continued)

Table 1 (continued)

TF family	Gene	Identified in plant	Transgenic plant	Reference
bZIP	*AtABF3*	*Arabidopsis thaliana*	Alfalfa	Wang et al. (2016)
	OsABF1	*Oryza sativa*	Rice	Zhang et al. (2016a)
	IbABF4	*Ipomoea batatas*	Arabidopsis	Wang et al. (2019b)
	AtABF3	*Arabidopsis thaliana*	Soy bean	Kim et al. (2018)
	SlbZIP38	*Solanum lycopersicum*	Tomato	Pan et al. (2017)
	PtrAREB1-2	*Populus trichocarpa*	*Populus trichocarpa*	Li et al. (2019)
	OsbZIP62	*Oryza sativa*	Rice	Yang et al. (2019a)
	CaBZ1	*Capsicum annuum*	Potato	Moon et al. (2015)
	GhABF2	*Gossypium hirsutum*	Arabidopsis	Liang et al. (2016)
NAC	*PbeNAC1*	*Pyrus betulifolia*	Tobacco	Jin et al. (2017)
	ONAC022	*Oryza sativa*	Rice	Hong et al. (2016)
	TtNAC2A	*Triticum turgidum*	Arabidopsis	Mergby et al. (2021)
	TaNAC47	*Triticum aestivum*	Arabidopsis	Zhang et al. (2016b)
	SlNAC4	*Solanum lycopersicum*	Tomato	Zhu et al. (2014)
	SNAC1	*Oryza sativa*	Cotton	Liu et al. (2014)
	ZmNAC55	*Zea mays*	Arabidopsis	Mao et al. (2016)
AP2/ERF	*ZmDREB2A*	*Zea mays*	Arabidopsis	Qin et al. (2007)
	ZmDBF3	*Zea mays*	Arabidopsis	Zhou et al. (2016)
	AhDREB1	*Arachis hypogaea*	Arabidopsis	Zhang et al. (2018)
	AtDREB1A	*Arabidopsis thaliana*	Soybean	de Paiva Rolla et al. (2014)
	SsDREB	*Suaeda salsa*	Tobacco	Zhang et al. (2015)
	VrDREB2A	*Vigna radiata*	Rice	Chen et al. (2016)
	SpERF1	*Stipa purpurea*	Arabidopsis	Yang et al. (2016)
	OsERF71	*Oryza sativa*	Rice	Lee et al. (2016)

highly conserved DNA-binding domain, a nuclear localization signal (NLS), a variable transcriptional regulatory region, and an oligomerization site (Singh et al. 2021). However, some TFs lack either a transcription regulation domain or a specific DNA-binding site (Liu et al. 1999). DNA-binding regions of TFs recognize and bind specific sequences, known as cis-acting element or TF binding site, within the promoter regions of the genes, which they regulate. By binding to these sequences, they either increase or decrease the transcription of their target genes, thus acting as activators or repressors, respectively (Walhout 2006). The ability of a TF to act as an activator or repressor, in most cases, dependent on domains that are located outside, and act independently of the DNA-binding domain (Gonzalez 2016). The recognition of specific DNA sequences by TFs is achieved by interactions established between the side chains of basic amino acids of the DNA binding domain with nucleotides of the target site. The activity of TFs also depends on the presence of the TF in right place that is nucleus. The TFs located in the cytosol need the nuclear localization signals (NLS), positively charged residues, in their structure for translocation to the nucleus, which are recognized by the nuclear import machinery (Stewart 2007).

Plant hormones, such as ABA, jasmonic acid, salicylic acid, and ethylene, regulate signal transduction pathways and interact with TFs during drought stress (Baillo et al. 2019). Mostly the TFs belong to either ABA-dependent or ABA-independent signal transduction pathways. Some genes contain the same cis-elements, and their transcription is controlled by the same TFs. This type of genes constitutes a regulon. Five classes of such extensively studied regulons that are involved in drought response. These are AREB/ABF regulons, DREB/CBF regulons, MYB/MYC regulons, NAC regulons and WRKY regulons (Gahlaut et al. 2016).

5 NAC (NAM, ATAF1/2, CUC2) Transcription Factors

The NAC (NAM, ATAF1/2, CUC2) TFs contain a complex plant-specific super family and identified in many plant species with more than 100 members in most plants; 112 in Arabidopsis, 263 in wheat, and 150 in barley (Shao et al. 2015). The NACs are unique plant TFs and were identified in 149 plant species with 19,997 sequences and served as the basis of the conducted analyses in databases such as Plant TFDB 5.0 (Plant Transcription Factor Database, version 5, http://planttfdb.gao-lab.org/). In higher plants, comparisons of NAC sequences revealed that rapeseed (Brassica napus) has the highest number of NAC sequences with 411 and lowest number (20) of NACs was found in Cowpea (Vigna unguiculata) (Table 2). In addition, it has been reported that filamentous terrestrial algae Klebsormidium flaccidum encodes the lowest number of NAC TFs (Mohanta et al. 2020). The NAC acronym is derived from three earliest identified proteins from petunia NAM (No Apical Meristem), Arabidopsis ATAF1/2 (Transcription Activator Factor 1/2) and CUC2 (Cup-shaped Cotyledon 2) (Puranik et al. 2012). NAC TFs are multifunctional proteins and have crucial roles in various biological processes and pathways, which include growth, development, flower formation, seed development, senescence, formation of secondary wall, and morphogenesis (Nuruzzaman et al. 2010). Besides, NACs are commonly associated wtih signaling pathways in response to various plant hormones (ABA, ethylene and jasmonic acid), biotic stress and abiotic stresses, such as drought, salinity, and cold (Bian et al. 2021).

5.1 *Structure of NAC TFs*

Although different species have different numbers of NAC TF members their structural features are generally similar. A typical NAC TFs contain highly conservative NAC domain at the N- terminal and a highly variable transcriptional regulatory

Table 2 Number of NAC genes in plant species according to Plant TFDB (v5.0)

Species	No of NAC genes	Species	No of NAC genes
Arabidopsis thaliana	112	Soybean	269
Rice (*Oryza Sativa* ssp. indica)	158	Common bean	106
Rice (*Oryza Sativa* ssp. japonica)	170	Mung bean	82
Wheat	263	Chickpea	96
Durum wheat	168	Cowpea	20
Maize	189	Peanut (*Arachis ipaensis*)	83
Barley	150	Peanut (*Arachis hypogaea*)	33
Foxtail millet	165	Poplar (*Populus trichocarpa*)	289
Sorghum	180	Grape (*Vitis vinifera*)	71
Pearl millet	151	Orange (*Citrus sinensis*)	145
Rapeseed	411	Orange (*Citrus clementina*)	129
Brachypodium distachyon	186	Apple (*Malus domestica*)	253
Sunflower	21	Tea (*Camellia sinensis*)	45
Sugar beet	59	Wild strawbeery (*Fragaria vesca*)	127
Tobacco	280	Potato (*Solanum tuberosum*)	131
Tomato	101	Cucumber (*Cucumis sativus*)	102
Camelina sativa	350	Radish (*Raphanus sativus*)	217
Cannabis sativa	75	Sesame (*Sesamum indicum*)	105
Klebsormidium flaccidum	3	Spinach (*Spinacia oleracea*)	45

region (TRR) at the C-terminal. NAC domain, about 150–160 amino acid residues, is divided in to five sub-domains (named A to E) (Fig. 4) (Welner et al. 2016). Subdomain A is likely to participate in the formation of a functional homo- and hetero-dimer with other NAC domain-containing proteins (Olsen et al. 2005). Subdomains C and D contain highly conserved and positively charged amino acid residues that bind to DNA (Chen et al. 2011). Subdomains B and E may be responsible for the various functions of NAC TFs (Puranik et al. 2012). NAC TFs include various nuclear localization signal (NLS). There are mono or bipartite nuclear localization signals with the lysine (K) residues in the subdomain D having crucial roles in nuclear shuttling. In addition, the NAC domain also regulates the binding of protein, which may determine the function and fate of the NAC protein (Olsen et al. 2005). These interactions are important for tolerance of plants to the abiotic and biotic stresses.

The transcriptional regulatory region (TRR) is located at the highly diverged C-terminus of NAC proteins, which can either activate or repress gene transcription under stress conditions. (Puranik et al. 2012) Bioinformatic analysis revealed some common motifs at the C-terminal end of NAC subfamilies (Ooka et al. 2003). However, the TRR provides a diversity for the exceptional functions of NAC proteins. Besides, an α-helical transmembrane motif (TMM) is found at TRRs of many

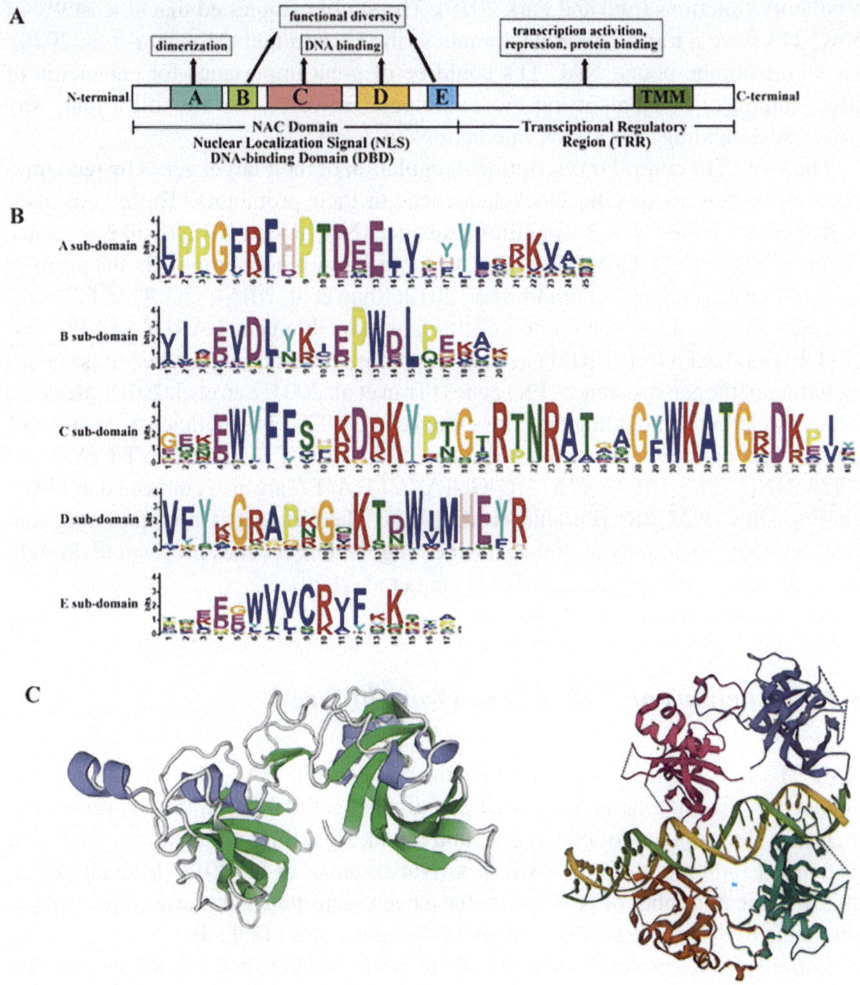

Fig. 4 (**a**) A concenptual diagram of a typical NAC protein showing highly conserved NAC domain which is divided into 5 conserved subdomains (**a–e**) at the N-terminal region. The NAC domain contains nuclear localization signal (NLS) and DNA-binding domain (DBD). A transcriptional regulatory region (TRR) is located on the C-terminal region. Besides, a transmembrane motif (TMM) is located on the C-terminal of some NACs. (**b**) Conserved motifs in subdomains in *Hordeum vulgare* NAC TFs identified using MEME tool. (**c**) 3D structure of *Hordeum vulgare* NAC protein (Left) and structure of NAC-DNA complex (Right)

NAC proteins, which is responsible for the anchoring of newly formed NAC proteins, named NTLs (NTM1-Like), on the membrane of plasma or endoplasmic reticulum. Under suitable conditions, NTLs are usually inactivated. However, NTLs are detached from the membrane by the stimulation of internal and/or environmental signals in case of stress and then enter into the nucleus to perform their regulatory functions (Seo and Park 2010). The studies suggested that at least 9% of NAC TFs have a transmembrane domain at the C- terminal (Mohanta et al. 2020). These membrane-bound NAC TFs could be of great importance for enhancing of stress tolerance of plants using biotechnology. However, we are still a long way from understanding NAC-DNA interactions in detail.

The NAC TFs control transcriptional regulation of their target genes by recognizing and binding to specific DNA sequences in their promoters. These consensus sequences are called NAC recognition sequence (NACRS) or NACRS-like sequence containing the "CGT(G/A)" and "CACG" core-DNA binding motifs in the promoters of downstream genes (Puranik et al. 2012; Shao et al. 2015). The "CACG" motif was also identified in the promoter of drought inducible gene EARLY RESPONSE TO DEHYDRATION1 (ERD1) in Arabidopsis as well as in some other stress genes including pathogen resistance (PR) genes (Tran et al. 2004; Seo et al. 2010). Besides, other cis-elements including the core sequence "GCTT", iron deficiency-responsive IDE2 motif containing the core sequence "CA(A/C)G(T/C)(T/C/A)(T/C/A)" and "(T/A)NN(C/T)(T/C/G)TN7A(A/C)GN(A/C/T) (A/T)" are also confirmed as NAC-binding sites (NACBS) (Puranik et al. 2012). Therefore, it is not surprising that NAC TFs can potentially modulate a wide range of target genes involved in drought stress response (Puranik et al. 2012; Hoang et al. 2020).

5.2 *Functions of NAC TFs in Drought Stress*

NAC TFs have various chimeric domains, as a result, NAC TFs are likely to be involved in the modulation of at least 289 distinct cellular pathways and processes, including development of cell, tissue, and organ, signalling processes, as well as in response to abiotic and biotic stresses (Nuruzzaman et al. 2013; Mohanta et al. 2020). A great number of NAC genes have been identified and transferred to different plants to overcome adverse effects of drought stress (Table 3).

Initially, NAC proteins were shown to be associated with plant growth. But recent functional studies have demonstrated that NAC genes can also be induced by different abiotic stresses including drought, salinity and ABA. For example, Arabidopsis ANAC072 gene is induced by ABA stress, salinity enhances the expression of ANAC019 by salinity (Singh et al. 2021). In Arabidopsis, overexpression of ANAC05, ANAC019 and ANAC072 was reported to improve drought tolerance (Baillo et al. 2019). NAC genes comprise some stress-responsice cis-elements, including ABREs (ABA-responsive elements), DREs (drought responsive elements), LTREs (low-temperature responsive elements), MYB (Myeloblastosis) and MYC (Myelocytomatosis) binding sites on the upstream region of the gene and these cis-elements strictly modulate the transcription of NAC genes and alter their expression (Puranik et al. 2012; Joshi et al. 2016). Post-transcriptional regulation of pre-NAC mRNAs is made by microRNA-mediated cleavage and alternative splicing. Besides, translated NAC proteins instensely subjected to phosphorylation, glycosylation, dimerization, proteolysis and ubiquitinization (Nakashima et al. 2012;

Table 3 Functions of NAC TFs in response to drought by over expression of genes in transgenic plants

Gene	Origin species	Transgenic species	Drought response	Reference
OsSNAC6	*Oryza sativa*	Rice	Increased root number and root diameter; up-regulation in the expression genes involved in membrane modification, nicotianamine (NA) biosynthesis, glutathione relocation, and glycosylation	Lee et al. (2017)
ONAC022	*Oryza sativa*	Rice	Higher survival ratios and better growth; reduction in water loss rate and transpiration, percentage of open stomata as well; increase in proline and soluble sugars accumulation	Hong et al. (2016)
ONAC066	*Oryza sativa*	Rice	Reduction in water loss rate and accumulation of ROS; increase in proline and soluble sugars accumulation; upregulated expression of stress-related genes	Yuan et al. (2019)
OsNAC14	*Oryza sativa*	Rice	Induced drought tolerance in plants both at the vegetative and the reproductive stages of growth, increased survival rate; higher hlorophyll fluorescence (Fv/Fm), enhanced the expression of stress response, DNA damage repair and defense related genes	Shim et al. (2018)
SlJUB1	*Solanum lycopersicum*	Tomato	Higher relative water content; decreased water loss; lower content of malondialdeyhde (MDA); increased expression of *DREB*s and *DELLA*	Thirumalaikumar et al. (2018)
SlNAC11	*Solanum lycopersicum*	Tomato	Longer hypocotyl and root; higher malondialdehyde (MDA) content	Wang et al. (2017)

(continued)

Table 3 (continued)

Gene	Origin species	Transgenic species	Drought response	Reference
SlNAC6	*Solanum lycopersicum*	Tomato	Reduced water loss rate and degree of oxidative damage; higher proline and malondialdehyde content content; enhanced antioxidant enzyme activity such as SOD, POD and CAT activity	Jian et al. (2021)
TtNAC2A	*Triticum turgidum*	Arabidopsis	Higher water retention capacity; attenuated oxidative stress; lower electrolyte leakage; higher proline accumulation	Mergby et al. (2021)
TaNAC48	*Triticum aestivum*	Wheat	Increased proline content; decrease in rates of water loss, malondialdehyde (MDA), H_2O_2 and O_2^- content	Chen et al. (2021)
TaNAC29	*Triticum aestivum*	Arabidopsis	Accumulation of lesser MDA and H_2O_2, higher SOD and CAT activities	Huang et al. (2015)
TaSNAC4-3A	*Triticum aestivum*	Arabidopsis	Enhanced germination rate and root growth; reduced stomatal aperture; lower water loss	Mei et al. (2021)
GmNAC019	*Glycine max*	Arabidopsis	Lower water loss; higher survival rates;, lower cellular hydrogen peroxide level and stronger antioxidant defense with increased SOD, POD, and CAT activities	Hoang et al. (2019)
GmNAC109	*Glycine max*	Arabidopsis	Enhanced lateral root formation and survival rate as well as germination rate; upregulated stress-related genes including *DREB1A, DREB2A, AREB1, AREB2,* and *RD29A*	Yang et al. (2019b)
GmNAC085	*Glycine max*	Arabidopsis	Improved cell membrane integrity, reduction in transpiration rate, water loss, and malondialdehyde content; increased activities of SOD, CAT and APX enzymes; upregulation of drought stress-responsive genes including *NCED3, LEA14, RD20* and *RD29B*	Nguyen et al. (2018)

(continued)

Table 3 (continued)

Gene	Origin species	Transgenic species	Drought response	Reference
VvNAC17	*Vitis Vinifera*	Arabidopsis	Reduced water loss rates, lower MDA and H_2O_2 contents; higher POD, SOD, and CAT activities; higher proline content	Ju et al. (2020)
StNAC053	*Solanum tuberosum*	Arabidopsis	Higher water content; reducued stomatal aperture; higher SOD, CAT, and POD activities; upregulation of stress related genes, including *DREB1A*, *ERD11*, and *ERF5*	Wang et al. (2021)
CaNAC46	*Capsicum annuum*	Arabidopsis	Longer roots and more lateral root; higher MDA content; enhanced the expression of antioxidant genes (*SOD, POD)* and drought stress-related genes, including *RD29B, RD20,* and *P5CS*	Ma et al. (2021)
PwNAC11	*Picea wilsonii*	Arabidopsis	Induced stomatal closure; increased ABA and MDA content; alleviation in efficiency of photosynthesis (Fv/Fm) and quantum yield of PSII electron transport; upregulated *ERD1*	Yu et al. (2021)
GhirNAC2	*Gossypium hirsutum*	Tobacco	Enhanced seed germination and root growth; increased seedling growth	Shang et al. (2020)
GhNAC4	*Gossypium hirsutum*	Tobacco	Enhanced stomatal closure, higher amount of proline content, reduced chlorosis and membrane damage, induction of ROS-scavenging genes, including *NtCAT1, NtAPX,* and *NtMnSOD*	Trishla and Kirti (2021)
GhSNAC3	*Gossypium hirsutum*	Arabidopsis	Increased root length and fresh weight	Liu et al. (2018)
ZmNAC49	*Zea Mays*	Maize	Reduction in the transpiration rate, stomatal conductance, and stomatal density; higher leaf relative water content; lower water loss; upregulation in expression of stomatal development genes, such as *ZmTMM, ZmSDD1* and *ZmMUTE*	Xiang et al. (2021)

(continued)

Table 3 (continued)

Gene	Origin species	Transgenic species	Drought response	Reference
MusaNAC042	*Musa acuminata*	Banana	Higher level of total chlorophyll and lower level of MDA content; higher proline content; better Fv/Fm ratio (photosynthetic efficiency)	Tak et al. (2017)
MdNAC1	*Malus domestica*	Apple	Reduced water loss and electrolyte leakage in leaves; maintenance of photosynthesis and photosynthetic pigments content; Reduction in MDA and ROS contents; increased activites of SOD, POD, and CAT	Jia et al. (2019)
MlNAC10	*Miscanthus lutarioripariu*	Arabidopsis	Higher proline content; higher antioxidant enzyme activity including CAT, POD, and SOD, lower MDA content	He et al. (2019)
MlNAC12	*Miscanthus lutarioripariu*	Arabidopsis	Longer root length; lower water loss; reduced stomatal aperture; decrease in the accumulation of ROS and MDA; improved activites of SOD, POD and CAT	Yang et al. (2018)
SlNAC8	*Suaeda liaotungensis*	Arabidopsis	Increase in seed germination rate, root and shoot growth, proline content and level of chlorophyll fluorescence; enhanced activites of SOD, POD and CAT; lower MDA concentration and electrolyte leakage percentage	Wu et al. (2018)
CmNAC1	*Cucurbita moschata*	Arabidopsis	Lower water loss rate; better water retaining; higher survival rate; high expression of *AtNCED3*	Cao et al. (2017)
SiNAC110	*Setaria italica*	Arabidopsis	Increased seed germination rate, root length, root surface area, fresh weight, and dry weight; upregulation of gene expression involved in proline biosynthesis and drought repsonses	Xie et al. (2017)
CarNAC6	*Cicer arietinum*	Arabidopsis	Increased survival rate; decreased water loss, higher proline content, higher recovery rate after re-watering	Liu et al. (2017)

Shao et al. 2015). The NAC TFs control the the transcription of downstream target genes by binding to NACRS. NACRS are consensus sequences comprising the "CACG" and/or "CGT(G/A)" core-DNA motifs on the promoter of downstream genes (Shao et al. 2015). Not only core motif but flanking sequences also in contact with the recognition motif on NAC proteins and contributes to the recognition of NACRS. Besides, NAC proteins also interact with other type of TFs and NAC TFs to regulate many cell functions, including drought response, in plants (Welner et al. 2016; Mohanta et al. 2020).

NAC TFs rapidly adjust their target gene expression through interactions with other TF types when exposed to stress. Arabidopsis NAC TF ANAC096 directly interacts with bZIP-type TFs ABRE binding factor 2 (ABF2) and ABF4. ANAC096 and ABF2 interaction synergistically activates RD29A (ABA-inducible gene) transcription under dehydration stress and help to plants survive under drought and osmotic stress conditions (Xu et al. 2013). The expression of drought stress-related NAC genes is altered and strictly modulated at the transcriptional level during drought stress. Different number of NAC genes is up-regulated or down-regulated under drought conditions. Ling et al. (2017) have shown that 44 Medicago truncatula MtNAC genes, out of 97 MtNAC putative genes, were regulated under different stress conditions, including drought, ABA, and salt. While expression of 17 MtNAC genes was induced, only MtNAC1 was suppressed under all stress conditions. Recently, Hussain et al. (2017) have shown that 28 GmNAC genes, out of 139 GmNAC genes, were differentially expressed under drought stress in the soybean (Glycine max). In addition, the expression level of eight of 28 GmNAC genes was found to be higher in drought-resistant cultivars than in drought-sensitive cultivars.

In rice, OsNAC5, OsNAC6, OsNAC9, and OsNAC10 TFs dramatically improve drought resistance, resulting in higher yield under drought conditions by changing root architecture (Chung et al. 2018). Moreover, Chung et al. (2018) found that four OsNAC TFs regulated genes associated with biosynthesis of polysaccharide and cell wall. The four drought inducible OsNACs directly target and regulate a different class of TFs, including AP2/EREBP and bHLH. The activation of four OsNAC TFs causes the up-regulation of several ERF (Ethylene responsive factor) domain genes, including OsERF1, OsERF9, OsERF54, OsERF57, OsERF107, and OsERF120. The up-regulation of these genes resulted with enhanced expression of Glucan endo-1,3-beta-glucosidase 6, Laccase-22-like (LAC26), Triacylglycerol lipase 2 (TAG lipase 2), and expansin-A7-like (OsEXP7), leading changes in the root architecture for drought tolerance and increased grain yield in rice (Singh et al. 2021). Consequently, drought inducible OsNAC TFs can bind to promoter of different class of transciription factors, leading up-regulation of them. The results have shown that NACs have also a role in the crosstalk inherent in abiotic stress responses (Chung et al. 2018).

Plant hormones play a role in influencing the signal response by acting with or against each other to maintain cellular homeostasis (Puranik et al. 2012). Among them, ABA plays important roles in regulation of abiotic stress responses, such as drought, salinity, and cold (Bian et al. 2021). ABA-dependent and ABA-independent pathways coordinate NAC genes due to differences in their promoter elements

(Puranik et al. 2012). Multiple stress responses of NAC genes are under the control of the central ABA perception and signaling module, such as ANAC019, ANAC055, and ANAC072/RD26, which play synergistic or antagonistic roles in ABA signaling and osmotic stress responses. (Li et al. 2014). OsNAC5 and OsNAC6 were strongly triggered by ABA and ABREs. Three ABRE sequences (ACGTGG/T C) were identified in the OsNAC5 and OsNAC6 promoters. In rice, OsNAC5 was revealed to interact with stress responsive proteins OsNAC6 and SNAC1. Overexpression of OsNAC5 induced the drought-responsive genes, including OsLEA3, leading to enhanced drought tolerance in rice (Takasaki et al. 2010). Besides, OsNAC6 and SNAC1 were demonstrated to bind to the promoter of OsLEA3. Chen et al. (2021) reported that TaNAC48 is required for crosstalk between the ABA signaling pathway and drought stress responses in wheat. In rice, stress-responsive NAC (SNAC) genes (SNAC1, OsNAC6/SNAC2, OsNAC3, OsNAC4, and OsNAC5) is induced by ABA and drought stress (Hu et al. 2006).

NAC TFs have been confirmed to modulate ROS metabolism and homeostasis during the drought stress response (Bian et al. 2021). In rice, a stress-responsive NAC gene, SNAC3, conferred drought tolerance by regulating ROS homeostasis and dramatically increasing transcript level of a great number of ROS-scavenging genes, such as SOD, APX, and CAT (Fang et al. 2015). A drought-responsive NAC transcription factor NTL4 (NAC WITH TRANS-MEMBRANE MOTIF 1-LIKE 4) can promote ROS generation by directly binding the the promoters of genes encoding ROS biosynthetic enzymes during drought-stress in Arabidopsis. For example, NTL4 has been reported to bind the promoter of AtRBOH (ARABIDOPSIS THALIANA RESPIRATORY BURST OXIDASE HOMOLOG), calcium-dependent NADPH oxidase that generates superoxide, leading ROS generation. The NAC-binding motifs have been found at the 1.5 kb upstream of the transcriptional start site of AtRBOH gene (Lee et al. 2012).

The overexpression of TaNAC47 caused ABA hypersensitivity and increased the drought tolerance in transgenic Arabisopsis. Besides, the transcript level of six drought stress-responsive genes, including AtRD29A, AtRD29B, AtCOR47, AtRD20, AtGSTF6, and AtP5CS1, was enhanced in TaNAC47 transgenic Arabidopsis. The genome wide analysis demonstrated that these up-regulated genes include NAC core motif CGT(G/A) on the promoter their promoter regions (Zhang et al. 2016b). Similarly, over expression of TaNAC67 gene enhanced the expression of five stress-responsive genes, including RD29B, RD29A, RAB18, DREB1A, and ABI5, and improved physiological traits, such as membrane integrity, chlorophyll content, and photosynethetic capability, under drought conditions (Mao et al. 2014). TaNAC47 and TaNA67 show high sequence identiy. However, some intrinsic differences, suh as gene structure difference, were found between these two genes. In addition, their expression level and response time were significantly different under stress conditions due to variation and presence of different cis-elements in their promoter region. Huang et al. (2015) reported that TaNAC29-overexpression lines conferred drought tolerance by enhanced physiological indices, including lower MDA and H2O2 content, and higher acitivity of ROS scavenging enzymes; however, the transcript level of some marker gene involved in ABA response and senescence, such as RD29B, ERD11, ABI5, AIB1, SAG13, and SAG113, was dramatically decreased in the transgenic TaNAC29 plants. In conclusion, wheat NAC TFs can both activate and suprepress downstream genes to regulate the drought response in plants (Gahlaut et al. 2016).

6 Conclusion

Crop plants will face a wider range and number of environmental stresses, including drought, salinity, extreme temperatures, and pathogen attacks, due to recent climate changes, which are probably to occur simultaneously. Among the abiotic stresses, drought stress significantly decreases crop biomass and yield in arid and semi-arid regions of the world, leading global food security. Therefore, it is very urgent to cultivate stress-resistant crop varieties to meet the increasing demand for food productivity due to global population growth.

Drought stress triggers a wide range of plant responses, including alteration of gene expression and cellular metabolism to adapt plant growth and development to these unfavorable conditions. Therefore, understanding the complex mechanism of drought toleance is crucial for crop production. In addition, it is important to develop new drought-tolerant plant genotypes for further agricultural improvement. A large number of drought stress-responsive genes have been identified and transferred to different cultivated plants. Transgenic plants showed enhanced drought tolerance. Among these stess-related genes, transcription factors play crucial role as modern genomic tools to improve tolerance of crops to not only drought stress but also multiple abiotic stresses. The NAC TFs are unique to plants and constitute one of the largest TF families in plants. The NACs regulate a large number of cellular processes including organogenesis, hormone signalling, development of development of root and shoot tissues, flower formation, secondary cell wall development, fruit ripening, during both normal development and under stress conditions. NAC TFs are multifunctional and can be used as promising candidates for generation of multi-stress tolerant crops with high productivity under various adverse conditions. Finally, due to the great number of NAC TFs from different plants, we are still long way from understanding the exact role and the detailed mechanism of NAC TFs in drought stress response.

References

Abhilasha A, Choudhury RS (2021) Molecular and physiological perspectives of abscisic acid mediated drought adjustment strategies. Plan Theory 10:2769

Abobatta WF (2020) Plant responses and tolerance to combined salt and drought stress. In: Hasanuzzaman M, Tanveer M (eds) Salt and drought stress tolerance in plants. Signaling and communication in plants. Springer, Cham

Alexander RD, Wendelboe-Nelson C, Morris PC (2019) The barley transcription factor HvMYB1 is a positive regulator of drought tolerance. Plant Physiol Biochem 142:246–253
Baillo EH, Kimotho RN, Zhang Z, Xu P (2019) Transcription factors associated with abiotic and biotic stress tolerance and their potential for crops improvement. Genes 10(10):771
Bandurska H (2022) Drought stress responses: coping strategy and resistance. Plan Theory 11:922
Bian Z, Gao H, Wang C (2021) NAC transcription factors as positive or negative regulators during ongoing battle between pathogens and our food crops. Int J Mol Sci 22:81
Butt HI, Yang Z, Chen E, Zhao G, Gong Q, Yang Z et al (2017) Functional characterization of cotton GaMYB62L, a novel R2R3 TF in transgenic Arabidopsis. PLoS One 12:e0170578
Cai R, Dai W, Zhang C, Wang Y, Wu M, Zhao Y et al (2017) The maize WRKY transcription factor ZmWRKY17 negatively regulates salt stress tolerance in transgenic Arabidopsis plants. Planta 246:1215–1231
Cao H, Wang L, Nawaz MA, Niu M, Sun J, Xie J et al (2017) Ectopic expression of pumpkin NAC transcription factor CmNAC1 improves multiple abiotic stress tolerance in Arabidopsis. Front Plant Sci 28(8):2052
Carraro E, Di Iorio A (2022) Eligible strategies of drought response to improve drought resistance in woody crops: a mini-review. Plant Biotechnol Rep 16:265–282
Casaretto JA, El-kereamy A, Zeng B, Stiegelmeyer SM, Chen X, Bi YM, Rothstein SJ (2016) Expression of OsMYB55 in maize activates stress-responsive genes and enhances heat and drought tolerance. BMC Genomics 17:312
Chen Q, Wang Q, Xiong L, Lou Z (2011) A structural view of the conserved domain of rice stress-responsive NAC1. Protein Cell Beijing High Educ Press 2:55–63
Chen T, Li W, Hu X, Guo J, Liu A, Zhang B (2015) A cotton MYB transcription factor, GbMYB5, is positively involved in plant adaptive response to drought stress. Plant Cell Physiol 56:917–929
Chen H, Liu L, Wang L, Wang S, Cheng X (2016) VrDREB2A, a DREB-binding transcription factor from *Vigna radiata*, increased drought and high-salt tolerance in transgenic *Arabidopsis thaliana*. J Plant Res 129:263–273
Chen J, Gong Y, Gao Y, Zhou Y, Chen M, Xu Z et al (2021) TaNAC48 positively regulates drought tolerance and ABA responses in wheat (*Triticum aestivum* L.). Crop J 9(4):785–793
Chung PJ, Jung H, Choi YD, Kim JK (2018) Genome-wide analyses of direct target genes of four rice NAC-domain transcription factors involved in drought tolerance. BMC Genomics 19:40
Ciura J, Kruk J (2018) Phytohormones as targets for improving plant productivity and stress tolerance. J Plant Physiol 229:32–40
Daryanto S, Wang L, Jacinthe PA (2017) Global synthesis of drought effects on cereal, legume, tuber and root crops production: a review. Agric Water Manag 179:18–33
de Paiva Rolla AA, Carvalho JDFC, Fuganti-Pagliarini R, Engels C, Do Rio A, Marin SRR et al (2014) Phenotyping soybean plants trans- formed with rd29A: AtDREB1A for drought tolerance in the green- house and field. Transgenic Res 23:75–87
Dong Y, Wang C, Han X, Tang S, Liu S, Xia X, Yin W (2014) A novel bHLH transcription factor PebHLH35 from *Populus euphratica* confers drought tolerance through regulating stomatal development, photosynthesis and growth in Arabidopsis. Biochem Biophys Res Commun 450:453–458
Efetova M, Zeier J, Riederer M, Lee CW, Stingl N, Mueller M et al (2007) A central role of Abscisic acid in drought stress protection of agrobacterium-induced tumors on Arabidopsis. Plant Physiol 145:853–862
El-Esawi MA, Al-Ghamdi AA, Ali HM, Ahmad M (2019) Over-expression of AtWRKY30 transcription factor enhances heat and drought stress tolerance in wheat (*Triticum aestivum* L.). Genes 10:163
Fang Y, Xiong L (2015) General mechanisms of drought response and their application in drought resistance improvement in plants. Cell Mol Life Sci 72:673–689
Fang Y, Liao K, Du H, Xu Y, Song H, Li X, Xiong L (2015) A stress-responsive NAC transcription factor SNAC3 confers heat and drought tolerance through modulation of reactive oxygen species in rice. J Exp Bot 66(21):6803–6817

FAO (2017) Faostat [online]. https://www.faostate.fao.org/
FAO (2021) The impact of disasters and crises on agriculture and food security, Rome. https://doi.org/10.4060/cb3673en
Furihata T, Maruyama K, Fujita Y, Umezawa T, Yoshida R, Shinozaki K, Yamaguchi-Shinozaki K (2006) Abscisic acid-dependent multisite phosphorylation regulates the activity of a transcription activator AREB1. Proc Natl Acad Sci USA 103(6):1988–1993
Gahlaut V, Jaiswal V, Kumar A, Gupta PK (2016) Transcription factors involved in drought tolerance and their possible role in developing drought tolerant cultivars with emphasis on wheat (*Triticum aestivum* L.). Theor Appl Genet 129:2019–2042
Gao JJ, Zhang Z, Peng RH, Xiong AS, Xu J, Zhu B, Yao QH (2010) Forced expression of MdMYB10, a MYB transcription factor gene from apple, enhances tolerance to osmotic stress in transgenic Arabidopsis. Mol Biol Rep 38:205–211
Gao F, Zhou J, Deng RY, Zhao HX, Li CL, Chen H et al (2017) Overexpression of a Tartary Buckwheat R2R3-MYB transcription factor gene, FtMYB9, enhances tolerance to drought and salt stresses in transgenic Arabidopsis. J Plant Physiol 214:81–90
Gao H, Wang Y, Xu P, Zhang Z (2018) Overexpression of a WRKY transcription factor TaWRKY2 enhances drought stress tolerance in transgenic wheat. Front Plant Sci 9:997
Golldack D, Lüking I, Yang O (2011) Plant tolerance to drought and salinity: stress regulating transcription factors and their functional significance in the cellular transcriptional network. Plant Cell Rep 30:1383–1391
Gonzalez DH (2016) Introduction to transcription factor structure and function. In: Gonzalez DH (ed) Plant transcription factors. Academic, London, pp 3–11
He K, Zhao X, Chi X, Wang Y, Jia C, Zhang H et al (2019) A novel Miscanthus NAC transcription factor MlNAC10 enhances drought and salinity tolerance in transgenic Arabidopsis. J Plant Physiol 233:84–93
Hoang X, Nguyen NC, Nguyen YH, Watanabe Y, Tran LP, Thao NP (2019) The soybean GmNAC019 transcription factor mediates drought tolerance in Arabidopsis in an abscisic acid-dependent manner. Int J Mol Sci 21(1):286
Hoang XLT, Nguyen YNH, Thao NP, Tran LSP (2020) NAC transcription factors in drought and salinity tolerance. In: Hasanuzzaman M, Tanveer M (eds) Salt and drought stress tolerance in plants. Signaling and communication in plants. Springer, Cham
Hong Y, Zhang H, Huang L, Li D, Song F (2016) Overexpression of a stress-responsive NAC transcription factor gene ONAC022 improves drought and salt tolerance in rice. Front Plant Sci 7:4
Hrmova M, Hussain SS (2021) Plant transcription factors involved in drought and associated stresses. Int J Mol Sci 22(11):5662
Hu H, Dai M, Yao J, Xiao B, Li X, Zhang Q, Xiong L (2006) Overexpressing a NAM, ATAF, CUC (NAC) transcription factor enhances drought resistance and salt tolerance in rice. Proc Natl Acad Sci 103:12987–11299
Huang GT, Ma SL, Bai LP, Zhang L, Ma H, Jia P et al (2012) Signal transduction during cold, salt, and drought stresses in plants. Mol Biol Rep 39:969–987
Huang Q, Wang Y, Li B, Chang J, Chen M, Li K et al (2015) TaNAC29, a NAC transcription factor from wheat, enhances salt and drought tolerance in transgenic Arabidopsis. BMC Plant Biol 4(15):268
Hussain RM, Ali M, Feng X, Li X (2017) The essence of NAC gene family to the cultivation of drought-resistant soybean (*Glycine max* L. Merr.) cultivars. BMC Plant Biol 17(1):55
Jia D, Jiang Q, van Nocker S, Gong X, Ma F (2019) An apple (*Malus domestica*) NAC transcription factor enhances drought tolerance in transgenic apple plants. Plant Physiol Biochem 139:504–512
Jian W, Zheng Y, Yu T, Cao H, Chen Y, Cui Q et al (2021) SlNAC6, A NAC transcription factor, is involved in drought stress response and reproductive process in tomato. J Plant Physiol 264:153483

Jin C, Li KQ, Xu XY, Zhang HP, Chen HX, Chen YH et al (2017) A novel NAC transcription factor, PbeNAC1, of *Pyrus betulifolia* confers cold and drought tolerance via interacting with PbeDREBs and activating the expression of stress-responsive genes. Front Plant Sci 8:1049

Joshi R, Wani SH, Singh B, Bohra A, Dar ZA, Lone AA et al (2016) Transcription factors and plants response to drought stress: current understanding and future directions. Front Plant Sci 7:1029

Ju YL, Yue XF, Min Z, Wang XH, Fang YL, Zhang JX (2020) VvNAC17, a novel stress-responsive grapevine (*Vitis vinifera* L.) NAC transcription factor, increases sensitivity to abscisic acid and enhances salinity, freezing, and drought tolerance in transgenic Arabidopsis. Plant Physiol Biochem 146:98–111

Khan SA, Li MZ, Wang SM, Yin HJ (2018) Revisiting the role of plant transcription factors in the battle against abiotic stress. Int J Mol Sci 19:1634

Kim HJ, Cho HS, Pak JH, Kwon T, Lee JH, Kim DH et al (2018) Confirmation of drought tolerance of ectopically expressed AtABF3 gene in soybean. Mol Cells 41(5):413

Kiranmai K, Lokanadha Rao G, Pandurangaiah M, Nareshkumar A, Amaranatha Reddy V, Lokesh U et al (2018) A novel WRKY transcription factor, MuWRKY3 (*Macrotyloma uniflorum* Lam. Verdc.) enhances drought stress tolerance in transgenic groundnut (*Arachis hypogaea* L.) plants. Front Plant Sci 9:346

Le Hir R, Castelain M, Chakraborti D, Moritz T, Dinant S, Bellini C (2017) AtbHLH68 transcription factorcontributes to the regulation of ABA homeostasis and drought stress tolerance in *Arabidopsis thaliana*. Physiol Plant 160:312–327

Lee S, Seo PJ, Lee HJ, Park CM (2012) A NAC transcription factor NTL4 promotes reactive oxygen species production during drought-induced leaf senescence in Arabidopsis. Plant J 70(5):831–844

Lee DK, Jung H, Jang G, Jeong JS, Kim YS, Ha SH et al (2016) Overexpression of the OsERF71 transcription factor alters rice root structure and drought resistance. Plant Physiol 172(1):575–588

Lee DK, Chung PJ, Jeong JS, Jang G, Bang SW, Jung H et al (2017) The rice OsNAC6 transcription factor orchestrates multiple molecular mechanisms involving root structural adaptions and nicotianamine biosynthesis for drought tolerance. Plant Biotechnol J 15(6):54–764

Leng G, Hall J (2019) Crop yield sensitivity of global major agricultural countries to droughts and the projected changes in the future. Sci Total Environ 654:811–821

Leng P, Zhao J (2020) Transcription factors as molecular switches to regulate drought adaptation in maize. Theor Appl Genet 133:1455–1465

Levitt J (1980) Responses of plants to environmental stresses, vol II. Water, radiation, salt, and other stresses. Academic, London

Li XY, Li L, Liu X, Zhang B, Zhveng WL, Ma WL (2014) Analysis of physiological characteristics of abscisic acid sensitivity and salt resistance in Arabidopsis ANAC mutants (ANAC019, ANAC072 and ANAC055). Biotechnol Biotechnol Equip 26:2966–2970

Li S, Lin YCJ, Wang P, Zhang B, Li M, Chen S et al (2019) The AREB1 transcription factor influences histone acetylation to regulate drought responses and tolerance in *Populus trichocarpa*. Plant Cell 31:663–686

Liang C, Meng Z, Meng Z, Malik W, Yan R, Lwin KM et al (2016) GhABF2, a bZIP transcription factor, confers drought and salinity tolerance in cotton (*Gossypium hirsutum* L.). Sci Rep 6:35040

Ling L, Song L, Wang Y, Guo C (2017) Genome-wide analysis and expression patterns of the NAC transcription factor family in *Medicago truncatula*. Physiol Mol Biol Plants 23:343–356

Liu L, White MJ, MacRae TH (1999) Transcription factors and their genes in higher plants functional domains, evolution and regulation. Eur J Biochem 262(2):247–257

Liu G, Li X, Jin S, Liu X, Zhu L, Nie Y, Zhang X (2014) Overexpression of rice NAC gene SNAC1 improves drought and salt tolerance by enhancing root development and reducing transpiration rate in transgenic cotton. PLoS One 9:e86895

Liu L, Zhang Z, Dong J, Wang T (2016) Overexpression of MtWRKY76 increases both salt and drought tolerance in *Medicago truncatula*. Environ Exp Bot 123:50–58

Liu Y, Yu X, Liu S, Peng H, Mijiti A, Wang Z, Zhang H (2017) A chickpea NAC-type transcription factor, CarNAC6, confers enhanced dehydration tolerance in Arabidopsis. Plant Mol Biol Report 35:83–96

Liu ZJ, Li F, Wang LG, Liu RZ, Ma JJ, Fu MC (2018) Molecular characterization of a stress-induced NAC gene, GhSNAC3, from *Gossypium hirsutum*. J Genet 97:539–548

Luo LJ (2010) Breeding for water-saving and drought-resistance rice (WDR) in China. J Exp Bot 61(13):3509–3517

Ma J, Wang LY, Dai JX, Wang Y, Lin D (2021) The NAC-type transcription factor CaNAC46 regulates the salt and drought tolerance of transgenic *Arabidopsis thaliana*. BMC Plant Biol 21(1):11

Mahmood T, Khalid S, Abdullah M, Ahmed Z, Shah MKN, Ghafoor A, Du X (2020) Insights into drought stress signaling in plants and the molecular genetic basis of cotton drought tolerance. Cells 9:105

Mao X, Chen S, Li A, Zhai C, Jing R (2014) Novel NAC transcription factor TaNAC67 confers enhanced multi-abiotic stress tolerances in Arabidopsis. PLoS One 9:e84359

Mao H, Yu L, Han R, Li Z, Liu H (2016) ZmNAC55, a maize stress-responsive NAC transcription factor, confers drought resistance in transgenic Arabidopsis. Plant Physiol Biochem 105:55–66

Mei F, Chen B, Li F, Zhang Y, Kang Z, Wang X, Mao H (2021) Overexpression of the wheat NAC transcription factor TaSNAC4-3A gene confers drought tolerance in transgenic Arabidopsis. Plant Physiol Biochem 160:37–50

Mergby D, Moez H, Saidi MN (2021) The durum wheat NAC transcription factor TtNAC2A enhances drought stress tolerance in Arabidopsis. Environ Exp Bot 186:104439

Mohanta TK, Yadav D, Khan A, Hashem A, Tabassum B, Khan AL et al (2020) Genomics, molecular and evolutionary perspective of NAC transcription factors. PLoS One 15(4):e0231425

Moon SJ, Han SY, Kim DY, Yoon IS, Shin D, Byun MO et al (2015) Ectopic expression of a hot pepper bZIP-like transcription factor in potato enhances drought tolerance without decreasing tuber yield. Plant Mol Biol 89:421–431

Nakashima K, Takasaki H, Mizoi J, Shinozaki K, Yamaguchi-Shinozaki K (2012) NAC transcription factors in plant abiotic stress responses. Biochim Biophys Acta 1819:97–103.Xu

Nguyen KH, Mostofa MG, Li W, Van Ha C, Watanabe Y, Le DT et al (2018) The soybean transcription factor GmNAC085 enhances drought tolerance in Arabidopsis. Environ Exp Bot 151:12–20

Nuruzzaman M, Manimekalai R, Sharoni AM, Satoh K, Kondoh H, Ooka H, Kikuchi S (2010) Genome-wide analysis of NAC transcription factor family in rice. Gene 465(1–2):30–44

Nuruzzaman M, Sharoni AM, Kikuchi S (2013) Roles of NAC transcription factors in the regulation of biotic and abiotic stress responses in plants. Front Microbiol 4:248

Olsen AN, Ernst HA, Lo Leggio L, Skriver K (2005) NAC transcription factors: structurally distinct, functionally diverse. Trends Plant Sci 10:79–87

Ooka H, Satoh K, Doi K, Nagata T, Otomo Y, Murakami K et al (2003) Comprehensive analysis of NAC family genes in *Oryza sativa* and *Arabidopsis thaliana*. DNA Res 10:239–247

Pan Y, Hu X, Li C, Xu X, Su C, Li J et al (2017) SlbZIP38, a tomato bZIP family gene down regulated by abscisic acid, is a negative regulator of drought and salt stress tolerance. Genes 8:402

Panda D, Mishra SS, Behera PK (2021) Drought tolerance in Rice: focus on recent mechanisms and approaches. Rice Sci 28(2):119–132

Puranik S, Sahu PP, Srivastava PS, Prasad M (2012) NAC proteins: regulation and role in stress tolerance. Trends Plant Sci 7(6):369–381

Qin F, Kakimoto M, Sakuma Y, Maruyama K, Osakabe Y, Tran LSP et al (2007) Regulation and functional analysis of ZmDREB2A in response to drought and heat stresses in *Zea mays* L. Plant J 50(1):54–69

Raineri J, Wang S, Peleg Z, Blumwald E, Chan RL (2015) The rice transcription factor OsWRKY47 is a positive regulator of the response to water deficit stress. Plant Mol Biol 88:401–413

Ramalingam A, Kudapa H, Pazhamala LT, Garg V, Varshney RK (2015) Gene expression and yeast two-hybrid studies of 1R-MYB transcription factor mediating drought stress response in chickpea (*Cicer arietinum* L.). Front Plant Sci 6:1117

Reddy AR, Chaitanya KV, Vivekanandan M (2004) Drought-induced responses of photosynthesis and antioxidant metabolism in higher plants. J Plant Physiol 161(11):1189–1202

Rollins JA, Habte E, Templer SE, Colby T, Schmidt J, von Korff M (2013) Leaf proteome alterations in the context of physiological and morphological responses to drought and heat stress in barley (*Hordeum vulgare* L.). J Exp Bot 64(11):3201–3212

Sato A, Sato Y, Fukao Y, Fujiwara M, Umezawa T, Shinozaki K et al (2009) Threonine at position 306 of the KAT1 potassium channel is essential for channel activity and is a target site for ABA-activated SnRK2/OST1/SnRK2.6 protein kinase. Biochem J 424:439–448

Seleiman MF, Al-Suhaibani N, Ali N, Akmal M, Alotaibi M, Refay Y et al (2021) Drought stress impacts on plants and different approaches to alleviate its adverse effects. Plan Theory 10:259

Seo PJ, Park CM (2010) A membrane-bound NAC transcription factor as an integrator of biotic and abiotic stress signals. Plant Signal Behav 5:481–483

Seo PJ, Kim MJ, Park JY, Kim SY, Jeon J, Lee YH et al (2010) Cold activation of a plasma membrane-tethered NAC transcription factor induces a pathogen resistance response in Arabidopsis. Plant J 61:661–671

Shang X, Yu Y, Zhu L, Liu H, Chai Q, Guo W (2020) A cotton NAC transcription factor GhirNAC2 plays positive roles in drought tolerance via regulating ABA biosynthesis. Plant Sci 296:110498

Shao H, Wang H, Tang X (2015) NAC transcription factors in plant multiple abiotic stress responses: progress and prospects. Front Plant Sci 6:902

Sharp RE, Poroyko V, Hejlek LG, Spollen WG, Springer GK, Bohnert HJ, Nguyen HT (2004) Root growth maintenance during water deficits: physiology to functional genomics. J Exp Bot 55:2343–2351

Shavrukov Y, Kurishbayev A, Jatayev S, Shvidchenko V, Zotova L, Koekemoer F et al (2017) Early flowering as a drought escape mechanism in plants: how can it aid wheat production? Front Plant Sci 8:195

Shim JS, Oh N, Chung PJ, Kim YS, Choi YD, Kim J (2018) Over- expression of OsNAC14 improves drought tolerance in rice. Front Plant Sci 9:310

Shinozaki K, Yamaguchi-Shinozaki K (2007) Gene networks involved in drought stress response and tolerance. J Exp Bot 58(2):221–227

Singh PK, Srivastava D, Tiwari P, Tiwari M, Verma G, Chakrabarty D (2019) Drought tolerance in plants: molecular mechanism and regulation of signaling molecules. In: Khan MIR, Reddy PS, Ferrante A, Khan NA (eds) Plant signaling molecules. Woodhead Publishing, pp 105–123

Singh S, Koyama H, Bhati KK, Alok A (2021) The biotechnological importance of the plant-specific NAC transcription factor family in crop improvement. J Plant Res 134(3):475–495

Stewart M (2007) Molecular mechanism of the nuclear protein import cycle. Nat Rev Mol Cell Biol 8:195–208

Sun Y, Yu D (2015) Activated expression of AtWRKY53 negatively regulates drought tolerance by mediating stomatal movement. Plant Cell Rep 34:1295–1306

Tak H, Negi S, Ganapathi TR (2017) Banana NAC transcription factor MusaNAC042 is positively associated with drought and salinity tolerance. Protoplasma 254(2):803–816

Takasaki H, Maruyama K, Kidokoro S, Ito Y, Fujita Y, Shinozaki K et al (2010) The abiotic stress-responsive NAC-type transcription factor OsNAC5 regulates stress-inducible genes and stress tolerance in rice. Mol Gen Genomics 284(3):173–183

Tang Y, Bao X, Zhi Y, Wu Q, Guo Y, Yin X et al (2019) Over- expression of a MYB family gene, OsMYB6, increases drought and salinity stress tolerance in transgenic rice. Front Plant Sci 10:168

Thirumalaikumar VP, Devkar V, Mehterov N, Ali S, Ozgur R, Turkan I et al (2018) NAC transcription factor JUNGBRUNNEN 1 enhances drought tolerance in tomato. Plant Biotechnol J 16:354–366

Tran LSP, Nakashima K, Sakuma Y, Simpson SD, Fujita Y, Maruyama K et al (2004) Isolation and functional analysis of Arabidopsis stress-inducible NAC transcription factors that bind to a drought responsive cis-element in the early responsive to dehydration stress 1 promoter. Plant Cell 16:2481–2498

Trishla VS, Kirti PB (2021) Structure-function relationship of *Gossypium hirsutum* NAC transcription factor, GhNAC4 with regard to ABA and abiotic stress responses. Plant Sci 302:110718

Walhout AJM (2006) Unraveling transcription regulatory networks by protein-DNA and protein-protein interaction mapping. Genome Res 16:1445–1454

Wang F, Chen HW, Li QT, Wei W, Li W, Zhang WK et al (2015) GmWRKY27 interacts with GmMYB174 to reduce expression of GmNAC29 for stress tolerance in soybean plants. Plant J 83:224–236

Wang Z, Su G, Li M, Ke Q, Kim SY, Li H et al (2016) Overexpressing Arabidopsis ABF3 increases tolerance to multiple abiotic stresses and reduces leaf size in alfalfa. Plant Physiol Biochem 109:199–208

Wang N, Zhang W, Qin M, Li S, Qiao M, Liu Z, Xiang F (2017) Drought tolerance conferred in soybean (*Glycine max* L.) by GmMYB84, a novel R2R3-MYB transcription factor. Plant Cell Physiol 58:1764–1776

Wang CT, Ru JN, Liu YW, Yang JF, Li M, Xu ZS, Fu JD (2018) The maize WRKY transcription factor ZmWRKY40 confers drought resistance in transgenic Arabidopsis. Int J Mol Sci 19(9):2580

Wang NN, Xu SW, Sun YL, Liu D, Zhou L, Li Y, Li XB (2019a) The cotton WRKY transcription factor (GhWRKY33) reduces transgenic Arabidopsis resistance to drought stress. Sci Rep 9:1–13

Wang W, Qiu X, Yang Y, Kim HS, Jia X, Yu H, Kwak SS (2019b) Sweetpotato bZIP transcription factor IbABF4 confers tolerance to multiple abiotic stresses. Front Plant Sci 10:630

Wang Q, Guo C, Li Z, Sun J, Deng Z, Wen L, Li X, Guo Y (2021) Potato NAC transcription factor StNAC053 enhances salt and drought tolerance in transgenic Arabidopsis. Int J Mol Sci 4, 22(5):2568

Welner DH, Deeba F, Lo Leggio L, Skriver K (2016) NAC transcription factors: from structure to function in stress-associated networks. Elsevier, Plant transcription factors, pp 199–212

Wu D, Sun Y, Wang H, Shi H, Su M, Shan H et al (2018) The SlNAC8 gene of the halophyte *Suaeda liaotungensis* enhances drought and salt stress tolerance in transgenic *Arabidopsis thaliana*. Gene 662:10–20

Xiang Y, Sun X, Bian X, Wei T, Han T, Yan J, Zhang A (2021) The transcription factor ZmNAC49 reduces stomatal density and improves drought tolerance in maize. J Exp Bot 24, 72(4):1399–1410

Xie LN, Ming C, Min DH, Lu FENG, Xu ZS, Zhou YB et al (2017) The NAC-like transcription factor SiNAC110 in foxtail millet (*Setaria italica* L.) confers tolerance to drought and high salt stress through an ABA independent signaling pathway. J Integr Agric 16:559–571

Xu ZY, Kim SY, Hyeon DY, Kim DH, Dong T, Park Y et al (2013) The Arabidopsis NAC transcription factor ANAC096 cooperates with bZIP-type transcription factors in dehydration and osmotic stress responses. Plant Cell 25:4708–4724

Yang Y, Dong C, Li X, Du J, Qian M, Sun X, Yang Y (2016) A novel AP2/ERF transcription factor from Stipa purpurea leads to enhanced drought tolerance in *Arabidopsis thaliana*. Plant Cell Rep 35:2227–2239

Yang X, He K, Chi X, Chai G, Wang Y, Jia C et al (2018) Miscanthus NAC transcription factor MlNAC12 positively mediates abiotic stress tolerance in transgenic Arabidopsis. Plant Sci 277:229–241

Yang S, Xu K, Chen S, Li T, Xia H, Chen L et al (2019a) A stress- responsive bZIP transcription factor OsbZIP62 improves drought and oxidative tolerance in rice. BMC Plant Biol 19:1–15

Yang X, Kim MY, Ha J, Lee SH (2019b) Over-expression of the Soybean NAC gene GmNAC109 increases lateral root formation and abiotic stress tolerance in transgenic Arabidopsis plants. Front Plant Sci 10:1036

Yao PF, Li CL, Zhao XR, Li MF, Zhao HX, Guo JY et al (2017) Overexpression of a tartary buckwheat gene, FtbHLH3, enhances drought/oxidative stress tolerance in transgenic Arabidopsis. Front Plant Sci 8:625

Yoon Y, Seo DH, Shin H, Kim HJ, Kim CM, Jang G (2020) The role of stress-responsive transcription factors in modulating abiotic stress tolerance in plants. Agronomy 10(6):788

Yu M, Liu J, Du B, Zhang M, Wang A, Zhang L (2021) NAC transcription factor PwNAC11 activates ERD1 by interaction with ABF3 and DREB2A to enhance drought tolerance in transgenic Arabidopsis. Int J Mol Sci 28; 22(13):6952

Yuan X, Wang H, Cai J, Bi Y, Li D, Song F (2019) Rice NAC transcription factor ONAC066 functions as a positive regulator of drought and oxidative stress response. BMC Plant Biol 25, 19(1):278

Zhai Y, Zhang L, Xia C, Fu S, Zhao G, Jia J, Kong X (2016) The wheat transcription factor, TabHLH39, improves tolerance to multiple abiotic stressors in transgenic plants. Biochem Biophys Res Commun 473:1321–1327

Zhang X, Liu X, Wu L, Yu G, Wang X, Ma H (2015) The SsDREB transcription factor from the succulent halophyte *Suaeda salsa* enhances abiotic stress tolerance in transgenic tobacco. Int J Genomics 875497

Zhang C, Liu J, Zhao T, Gomez A, Li C, Yu C et al (2016a) A drought-inducible transcription factor delays reproductive timing in rice. Plant Physiol 171:334–343

Zhang L, Zhang L, Xia C, Zhao G, Jia J, Kong X (2016b) The novel wheat transcription factor TaNAC47 enhances multiple abiotic stress tolerances in transgenic plants. Front Plant Sci 6:1174

Zhang B, Su L, Hu B, Li L (2018) Expression of AhDREB1, an AP2/ ERF transcription factor gene from peanut, is affected by histone acetylation and increases abscisic acid sensitivity and tolerance to osmotic stress in Arabidopsis. Int J Mol Sci 19(5):1441

Zhou W, Jia CG, Wu X, Hu RX, Yu G, Zhang XH et al (2016) ZmDBF3, a novel transcription factor from maize (*Zea mays* L.), is involved in multiple abiotic stress tolerance. Plant Mol Biol Rep 34(1):353–364

Zhu M, Chen G, Zhang J, Zhang Y, Xie Q, Zhao Z et al (2014) The abiotic stress-responsive NAC-type transcription factor SlNAC4 regulates salt and drought tolerance and stress-related genes in tomato (*Solanum lycopersicum*). Plant Cell Rep 33:1851–1863

Drought-Responsive miRNAs in Vegetables

Aylin Kobaneri, Deniz Uras, and Bahar Soğutmaz Özdemir

Abstract Plant biotechnology is an emerging area of science as it can generate solutions for food security and sustainable agriculture. The inverse proportion between the increase in human population and negative impacts of climate change and global warming has become a challenge for plant scientists to overcome the limitations in human food demand. Drought stress is one of the most important abiotic stress factors prevalent in many parts of the world and play a major role in limiting plant productivity. Identification of plant stress response mechanisms and pathways is important to better understand how to solve these problems by connecting the plants behavior with biotechnological tools. After their discovery in 1993, non-coding RNAs, mainly microRNAs, have become an important topic of science in terms of research and application. Their role in regulation of gene expression and expressional changes under stress conditions made miRNAs a focus in plant science with a potential solution to tackle the problems in agricultural production. In this chapter, stress responsive miRNAs and their regulatory mechanism in vegetables under drought stress conditions will be discussed to reveal the potential of miRNAs in terms of sustainable agriculture and improved vegetable productivity.

Keywords Abiotic stress · Drought stress · miRNA · Non-coding RNA · Vegetable

1 Introduction

Since their discovery in 1993 in Caenorhabditis elegans, non-coding RNAs (ncRNAs) become a rapidly developing field of research (Lee et al. 1993). In contrast to messenger RNAs (mRNAs), ncRNAs are not translated into proteins (Hüttenhofer et al. 2005). NcRNAs are classified according to their length and

A. Kobaneri · D. Uras · B. Soğutmaz Özdemir (✉)
Department of Genetics and Bioengineering, Faculty of Engineering, Yeditepe University, Istanbul, Türkiye
e-mail: bahar.sogutmaz@yeditepe.edu.tr

U. K. Chaudhry et al. (eds.), *Drought Stress*,
https://doi.org/10.1007/978-3-031-80610-0_12

function. In terms of function, ncRNAs are categorized as housekeeping ncRNAs and regulatory ncRNAs (Fu 2014). On the other hand, in terms of length, they are grouped as small and long ncRNAs (sncRNAs and lncRNAs, respectively). sncRNAs are further grouped into three classes according to their biogenesis, pathways that they regulate, and the way they regulate those pathways. Those classes are microRNAs (miRNAs), small interfering RNAs (siRNAs), and piwi-interacting RNAs (piRNAs) (Fig. 1).

siRNAs are generally 20–24 nucleotides in length and classified as regulatory ncRNAs. They work by means of complementarity with the target molecule and depending on the complementarity level, the target molecule is degraded, or translation is suppressed (Mello and Conte 2004).

piRNAs are 24–31 long regulatory ncRNAs that can be found in mammals. Their function is the suppression of variable elements in the germline (Brennecke et al. 2007; Gunawardane et al. 2007). The last member of sncRNAs is miRNAs which are 20–22 nucleotides in length. They regulate gene expression at the post-transcriptional level (Bartel 2004). The main function of miRNAs is gene silencing which is performed by miRNAs according to their complementarity with the target molecule. If the target mRNA is fully complementary with the miRNA, mRNA is degraded whereas if they are partially complementary, translation is inhibited (Bagga et al. 2005). The number of miRNA genes differs in each plant species, starting from several dozens to hundreds; however, the number of miRNA gene families is much smaller since there are several different miRNAs in the same family (Li and Mao 2007; Nozawa et al. 2012; Zhang et al. 2005a).

The origins of miRNA genes are still unclear; however, several studies illuminate that numerous miRNA families are evolutionarily conserved among all major plant lineages, suggesting that the history of miRNA-mediated gene regulation appears to have been correlated with early stages of plant evolution with tight functional

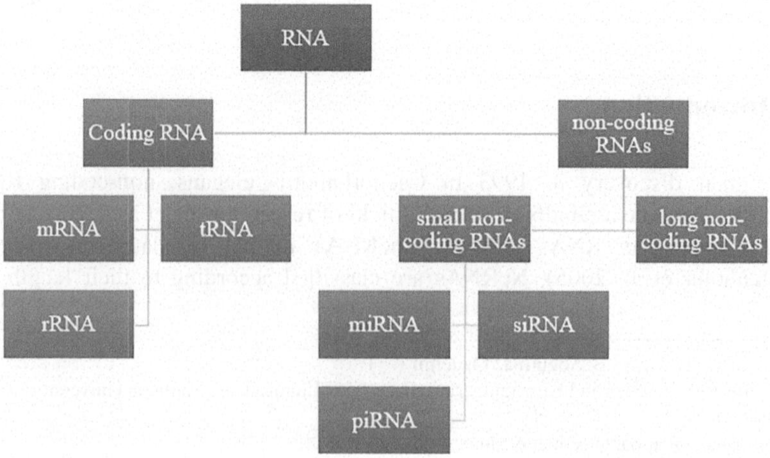

Fig. 1 Classification of RNAs

constraining for more than 425 million years (Zhang et al. 2006a). Several mechanisms were proposed for miRNA origins such as generation from transposable elements, duplication of preexisting miRNA genes or protein-coding genes, and the formation of the hairpin structure during genome evolution. Potentially, the first two explanations are common in plants (Allen et al. 2004; Cuperus et al. 2011; Fahlgren et al. 2007, 2010; Nozawa et al. 2012; Zhou et al. 2013). Plant miRNA repertoires are reported to be dynamically changing since some old miRNAs are lost during evolution while new miRNAs originated. However, new miRNAs have low expression levels, easy to be lost during evolution, and are mostly species-specific (Cuperus et al. 2011; Nozawa et al. 2012).

2 miRNA Biogenesis

miRNAs are transcribed from intergenic MIR genes by RNA polymerase II (RNA Pol II) to the primary transcript, known as primary-miRNA (pri-miRNA) (Lee et al. 2004). Although pri-miRNAs are 100–1000 nucleotides in length as a single-strand and present in the nucleus as a hairpin with a cap and poly-A tail added by RNA Pol II. After this point, miRNA biogenesis differs in mammals and plants (Du and Zamore 2005). In mammals, the cap and poly-A tail were removed by DROSHA (DROSHA ribonuclease III), whereas in plants since there is no DROSHA homologue, the next step of miRNA maturation is regulated by a member of the RNase III family, DICER-like 1 (DCL1), by removing pri-miRNA from the bundle and become a precursor miRNA (pre-miRNA) which is still a hair-pin but shorter in length (~70 nucleotides). During this process, pri-miRNA was stabilized by an RNA-binding protein DAWDLE (DDL). For the completion of this process, interaction of pri-miRNA with C2H2-zinc finger protein SERRATE (SE), double-stranded RNA-binding protein HYPONASTIC LEAVES 1 (HYL1), DCL1, nuclear processing center D-body, and cap-binding complex (CBC) are required (Du and Zamore 2005; Hiraguri et al. 2005; Lu and Fedoroff 2000). In mammals, DROSHA is the responsible enzyme for pre-miRNA formation (Kim 2005). pre-miRNAs are double-stranded as miRNA/miRNA* duplex. The miRNA strand is also known as the guide strand since it is the active form, whereas the miRNA* strand is the passenger strand. Pre-miRNAs needed to be transferred to the cytoplasm to be functional. In mammals, exportin-5 is required for this transfer, whereas in plants an exportin-5 orthologue HASTY transfers the pre-miRNA to the cytoplasm (Bollman et al. 2003; Park et al. 2005). In the cytoplasm, pre-miRNAs are methylated by HEN1 to protect them from degradation by an exonuclease family member Small RNA Degrading Nuclease (SDN). Double-stranded pre-miRNAs are cut by RNA-induced Silencing Complex (RISC) into two single strands. This is followed by the binding of the guide miRNA strand to ARGONAUTE (AGO) to function as silencing of the target mRNA (Baumberger and Baulcombe 2005; Kurihara and Watanabe 2004; Pratt and MacRae 2009) (Fig. 2).

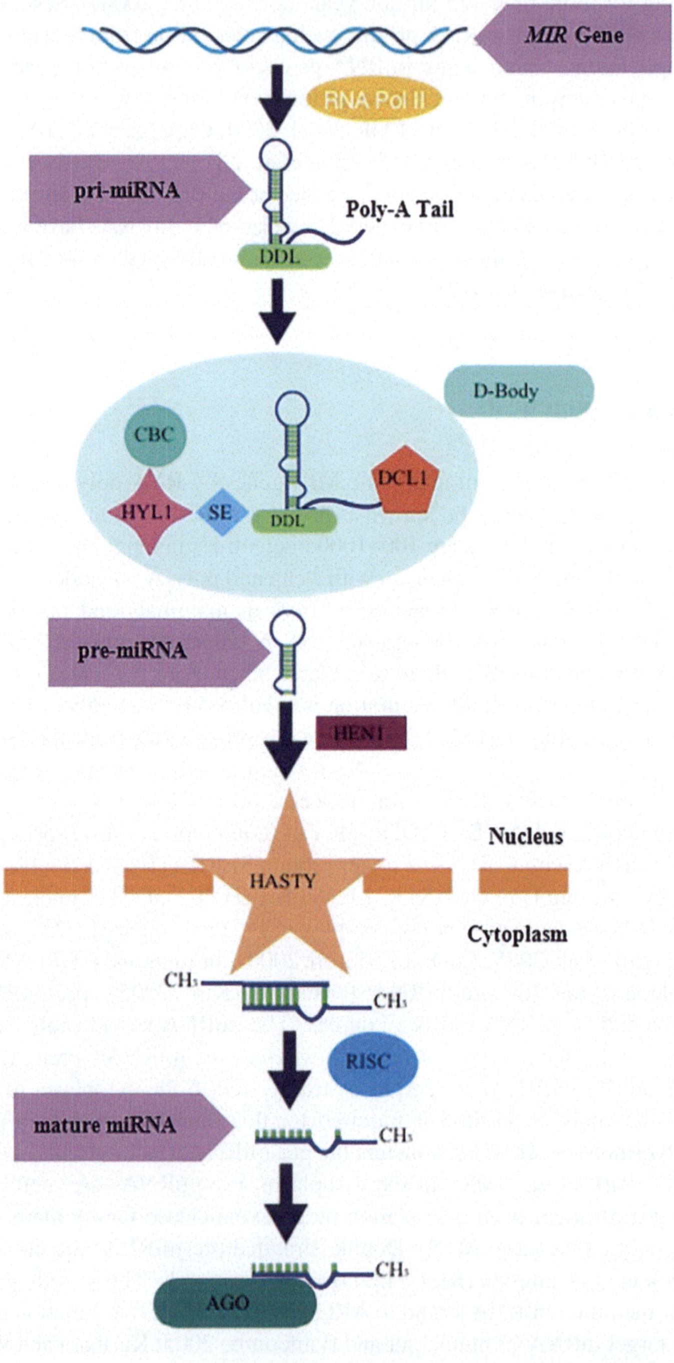

Fig. 2 miRNA biogenesis in plants. (Modified from Eldem et al. 2013). pri-miRNAs that are transcribed from intergenic *MIR* genes, present with a poly-A tail and a cap added by DDL. To become a functioning miRNA, they are processed by DCL1, SE, HYL1, and CBC in D-body to pre-miRNA which is further transferred to the cytoplasm by HEN1. In the cytoplasm, pre-miRNAs are cut into single strands and bind with AGO to regulate gene silencing

After pre-miRNAs are cut into two single strands as miRNA and miRNA strands, miRNA* strands are thought to be degraded since their accumulation is low. However, Meng et al. (2011) reported that miRNA*s are not degraded but instead involved in several biological processes (Meng et al. 2011). The mature miRNAs are involved in gene silencing based on their complementarity with the target mRNA. In a mechanism known as RNA interference (RNAi), mature miRNA forms a complex with RISC and binds to the 3'UTR region of target mRNA (Bartel 2004). After the binding, if miRNA and target mRNA are fully complementary, the target is degraded. On the other hand, if there is a partial complementarity, the translation of the target mRNA is inhibited (Bagga et al. 2005).

3 Approaches for miRNA Identification, Expression, and Functional Analysis in Plants

In order to understand the function of miRNA, its identification is required. miRNAs can be identified by both computational and experimental strategies. Most of the miRNAs are evolutionarily conserved and either identical or have minimal variations in sequences (Axtell 2008; Luo et al. 2013; Weber 2005; Zhang et al. 2006a). Besides, all miRNAs share unique features such as stem-loop hairpin structure, high negative minimal free folding energy (MFE), and high MFE index (MFEI) (Bartel 2004; Zhang et al. 2006a). Several computation-based approaches are designed to identify miRNAs based on these significant features (Zhang et al. 2006d). The most widely used approach is the homologue-based comparative genome approach which provides miRNA identification against all potential nucleotide acid sequences including genome sequence survey (GSS), expressed sequence tags (EST), and genome sequences by using miRNA sequences that are already known (Dehury et al. 2013; Zhang et al. 2005a, 2006d). This approach is mainly used in newly discovered or understudied plant species. Since it requires already known miRNA sequences in publicly available BLASTn analyses, it is easy to use with minimal requirements. Since its development in 2005, the homologue-based comparative genomics approach is adopted to identify numerous miRNAs in a diversity of plant species, including cotton (Qiu et al. 2007; Zhang et al. 2007), corn (Zhang et al. 2006b), switchgrass (Matts et al. 2010; Xie et al. 2010), soybean (Chen et al. 2009.; Zhang et al. 2008a), rapeseed (Xie et al. 2007), wheat (Han et al. 2009; Yin and Shen 2010), potato (Xie et al. 2011; Zhang et al. 2009), strawberry (Dong et al. 2012), Chinese cabbage (Wang et al. 2011a), ginseng (Mathiyalagan et al. 2013), coffee (Rebijith et al. 2013), citrus (Song et al. 2009), peach (Zhang et al. 2012), and apple (Gleave et al. 2008; Ye et al. 2013). It is also a great tool to use identified miRNAs to study the evolution and divergence of plant miRNAs (Axtell and Bartel 2005; Zhang et al. 2006c).

The experimental approaches are used in identifying and defining the functions of plant miRNAs since the discovery of the first miRNA, lin-4, by genetic screening technology (Lee et al. 1993). Also, genetic cloning is used to identify the first plant miRNAs in several studies in the early 2000s and continued to be used since then (Llave et al. 2002; Park et al. 2002; Reinhart et al. 2002). Both approaches are proven to be successful in miRNA identification; however, next-generation deep sequencing technology has become the most commonly used approach for plant miRNA studies due to several reasons. Both direct cloning and genetic screening can work only on a single miRNA per run whereas next-generation deep sequencing generates millions of sequences per run resulting in genome-wide identification of all miRNAs, but also detecting their expression levels based on read number. Thus, next-generation deep sequencing is time- and cost-efficient and powerful not only in the identification of miRNAs but also in defining their expression levels. On the other hand, next-generation deep sequencing requires the support of bioinformatics and computational tools since the huge number of sequences and datasets that are generated at the end of each run. In addition to next-generation deep sequencing, Northern blotting, quantitative real-time polymerase chain reaction (qRT-PCR), and miRNA microarray are other most commonly used methods for the detection of miRNA expression levels.

Since the discovery of the reverse transcriptase enzyme in 1977, qRT-PCR has become the most widely used method for the detection of RNA expression levels due to its ability to distinguish a single nucleotide variation in the sequence and high sensitivity (Chen et al. 2005). Another commonly used method for the detection of expression levels is Northern blotting. It is also useful for miRNA visualization. However, it requires a high amount of quality samples, low throughput, low sensitivity, and is time-consuming; thus, more sensitive and newer methods, such as deep sequencing and microarrays, are often preferred over Northern blotting (Dong et al. 2013). Before deep sequencing became widely used for miRNAs, microarray-based detection is a common method for validating a large number of miRNAs and detecting differential miRNA expression (Li and Ruan 2009). However, microarray-based detection cannot be used to identify unknown miRNAs since it is a probe-based method leading to the detection of only known miRNAs. Thus, other detection techniques such as Northern blotting and/or qRT-PCR were used for validation of most microarray data. As a result of this, high throughput deep sequencing technology currently displaced microarray-based detection technology.

4 Computational Tools and Resources for Identifying Plant miRNAs and Their Targets

In parallel with the increase in miRNA-related studies and identified miRNAs, the number and content of databases and computational tools are increased as well. Several computational tools and databases have been established for miRNA and target identification by using existing databases and datasets generated from sequencing and experimental data. Table 1 represents the major tools, databases, and publicly available websites for plant miRNA research.

Table 1 Major tools, databases, their publicly available websites, and purposes for miRNA research

Database	Website	Purpose	Reference
miRBase	http://www.mirbase.org/	General database	Griffiths-Jones et al. (2008)
PMRD	http://bioinformatics.cau.edu.cn/PMRD/	General database	Zhang et al. (2010)
miRNEST	http://lemur.amu.edu.pl/share/php/mirnest/	General database	Szcześniak et al. (2012)
miRDeepFinder	http://www.leonxie.com/DeepFinder.php	General database	Xie et al. (2012)
miRSpring	http://mirspring.victorchang.edu.au/	General database	Humphreys and Suter (2013)
mirTools	http://122.228.158.106/mr2_dev/	General database	Zhu et al. (2010)
Deepbase	http://deepbase.sysu.edu.cn/	General database	Zheng et al. (2016)
miRExpress	http://mirexpress.mbc.nctu.edu.tw/	General database	Wang et al. (2009)
miRanalyzer	http://web.bioinformatics.cicbiogune.es/microRNA/	General database	Hackenberg et al. (2009)
miRCat2	http://srna-tools.cmp.uea.ac.uk/plant/cgibin/srna-tools.cgi?rm=input_form&tool=mircat	General database	Paicu et al. (2017)
miRDeep*	http://www.australianprostatecentre.org/research/software/mirdeep-star	General database	An et al. (2013)
miRAuto	http://nature.snu.ac.kr/software/miRAuto.htm	General database	Lee et al. (2013)
miRNAkey	http://ibis.tau.ac.il/miRNAkey/	General database	Ronen et al. (2010)
μPC	http://www.biotec.or.th/isl/micropc	General database	Mhuantong and Wichadakul (2009)
psRNATarget	http://plantgrn.noble.org/psRNATarget/	Target prediction	Dai and Zhao (2011)
TAPIR	http://bioinformatics.psb.ugent.be/webtools/tapir/	Target prediction	Bonnet et al. (2010)
miRTour	http://mirnablog.com/mirtour-plant-mirna-and-target-prediction-tool/	Target prediction	Milev et al. (2011)
SoMART	http://bakerlabdev.berkeley.edu/somart-server-plant-mirnatasirna-analysis-resources-and-tools-1	Target prediction	Li et al. (2012)
CleaveLand	http://sites.psu.edu/axtell	Target sequencing	Addo-Quaye et al. (2009)
PMTED	http://pmted.agrinome.org/	Target expression	Sun et al. (2013)

Since its release in 2002, miRBase has been one of the most used databases with published miRNA sequences, references, locations, and annotations. It also includes predicted targets based on the data from target prediction tools. All data found in the database are available for download. Besides plant miRNAs, miRBase also includes animal and virus miRNAs. There are nearly 40,000 miRNA sequences deposited in the database and among them 10,000 miRNA sequences belonging to 82 plant species (Griffiths-Jones 2004; Griffiths-Jones et al. 2008).

PMRD (Plant microRNA Database) is the largest plant-specific database with nearly 10,000 miRNA sequences from 127 plant species. PMRD includes miRNA sequences, target molecules, secondary structures, and expression profiles. In the beginning, miRNA sequences were adopted from miRBase, Rfam, and similar databases, recent ones are adopted from microarray data (Zhang et al. 2010). Since the miRNA databases that were used before PMRD were high in number and not ordered, this database was built for the purpose of gathering the data under a single construct.

miRNEST includes animal, plant, and virus miRNAs collected from 13 different databases and studies. It currently contains 225 animals, 202 plants, and 22 virus miRNA sequences, target molecules, gene structure, mirtrons, and degradome data. The information in the database is adopted from deep sequencing data (Szcześniak et al. 2012).

µPC (MicroPC) is a computational tool that contains miRNA sequences adopted from miRBase and EST (Expressed Sequence Tags) from PlantGDB. It is a useful tool for predicting and comparing miRNAs and their target molecules. Currently, contains 4006 predicted miRNA sequences, 128 miRNA families, and 2995 protein targets from 12 plant species (Mhuantong and Wichadakul 2009).

As next-generation high throughput deep sequencing became a widely used technology for miRNA identification, computational tools and databases have been developed quickly to handle the huge amount of data generated by deep sequencing. Almost all of the databases and tools can be used for the identification of miRNAs based on currently available sequences.

miRDeepFinder was established in 2012 and aimed to close the gaps and missing information in the databases established before. The most important feature of the database is the involvement of pre-miRNAs longer than 200 nucleotides. Former databases do not contain pre-miRNA sequences longer than 200 nucleotides although several pre-miRNAs are reported to be longer. Additionally, the requirement of available genome information to use former databases and tools limits their usage for organisms with limited genome information. miRDeepFinder can be used for both new and conserved miRNAs for non-model plant species as well as producing reliable results for pre-miRNAs longer than 200 nucleotides. The functions of the software package include identification of conserved miRNAs, pre-processing of raw data, miRNA expression profiling, mining, and classifying novel miRNAs, predicting miRNA targets and gene networks, and pathway analysis involving miRNAs (Xie et al. 2012). Also, it can be used together with the CleaveLand package for target molecule identification and degradome sequencing (Addo-Quaye et al. 2009).

psRNATarget is a database containing small plant RNAs, majorly miRNAs, and allows two important analysis tools. It allows reversing complementarity between miRNA and target molecule using a scoring table and, analyze target accessibility by calculating the secondary structure required UPE (unpaired energy) for mRNA-miRNA binding. Additionally, it allows distinguishing translational silencing and mRNA degradation while identifying the target molecule. psRNATarget runs on a Linux cluster and is designed for high-throughput analysis of next-generation deep sequencing data (Dai and Zhao 2011).

PMTED (Plant miRNA Target Expression Database) adopted data from microarrays of plant miRNAs target molecule expressions. Currently, almost 5000 miRNA target molecules from 12 plant species are available and adownloadable (Sun et al. 2013).

In elucidating miRNA function, one of the most critical steps is identifying miRNA targets. In the beginning, computational tools were used by scientists to predict miRNA targets followed by experimental validation. However, the computational prediction was limited because of the low amount of available genome sequences and experimental validation was required. Recently, the most widely used validation method for miRNA targets is the determination of miRNA cleavage site by the rapid amplification of 5’-cDNA ends. Using this method, the 5′-position of the target RNA is shown, which is generally located at the opposite site and between the 10th and 11th position of the corresponding miRNA (Elbashir et al. 2001; German et al. 2009). Nowadays, high-throughput sequencing technologies for both validation and identification of miRNA targets expanded this technique (Addo-Quaye et al. 2009; German et al. 2008; Gregory et al. 2008). Degradome sequencing technology or parallel analysis of RNA ends (PARE) are direct approaches to combine miRNA target identification and confirmation (Addo-Quaye et al. 2009; German et al. 2008, 2009). CleaveLand was developed as a computational tool to efficiently analyze the degradome data and quick identification of miRNA targets (Addo-Quaye et al. 2009).

5 Drought Stress and miRNAs

By 2050, the world population is predicted to increase about 34% and reach 9.1 billion. Most of the increase will be in developing countries and besides urbanization is also predicted to be increased by 20%. Since the increase in the population will lead to an increase in consumption, agricultural production also needs to be increased by 70% which is accepted as the greatest challenge for plant researchers (FAO 2009).

Another parameter that worsens the scenario is that by 2080, the atmospheric temperature is predicted to be 4 °C warmer, leading to a doubling in atmospheric CO^2 (FAO 2009). Only 3.5% of the arable lands are stress-free. In parallel with climate change, the severity and occurrence of drought stress will increase (IPCC 2014). The increase in the atmospheric temperature leads to a reduction in available water in the soil while transpiration and evaporation cause plants to lose water. Since it causes a reduction in plant growth and yield, drought stress or water deficit is one of the biggest problems against sustainable food production (Jaleel et al. 2009). Water content reduction, diminishment of leaf water potential and turgor loss, stomatal closure, cell enlargement, and reduction in growth are characteristics of drought stress (Jaleel et al. 2008).

The sessile structure of plants made them vulnerable to the constantly changing dynamics of their environment. Studies on abiotic stress resistance gain speed and the discovery of miRNAs involvement in abiotic stress response created new approaches.

miRNAs are reported as crucial modulators in drought stress by controlling the expression of drought-responsive genes (Covarrubias and Reyes 2010; Martin et al. 2010; Shuai et al. 2013). The drought-induced miRNAs lead to a downregulation in their target mRNAs, resulting in the downregulation of proteins involved in drought response. On the other hand, miRNAs that are downregulated by drought stress cause an accumulation of their target mRNAs that involves in stress adaptation. Since the majority of miRNA target mRNA encode transcription factors (TFs), miRNAs are at the center of gene regulatory networks (Fig. 3).

5.1 ABA Response

Abscisic acid (ABA) is a key stress hormone that plays a crucial role in the regulation of stress-related gene expression and stomatal closure initiation. Also, it is produced de novo under drought stress (Koornneef et al. 1998; Wilkinson and Davies 2002). The first miRNA involvement in the ABA response pathway is reported by Lu and Fedoroff (2000) by studying the key genes involved in miRNA biogenesis for ABA-hypersensitive mutants. During the germination of Arabidopsis, the HYL1 mutant was reported to be ABA-hypersensitive (Lu and Fedoroff 2000). In another study on ABA sensitivity, key elements of miRNA biogenesis mutants of DCL1 and HEN1 lead to an increase in ABA sensitivity, in addition, mutants of SE and HASTY indicated an increase in ABA sensitivity (Zhang et al. 2008b). These studies reveal the possibility of these mutants of key elements in biogenesis may indicate a particular impaired miRNA involved in the ABA response regulation. Additionally, in Arabidopsis and rice, miR168 whose expression is influenced by water deficit was found to target AGO1 suggesting that feedback regulation might play a role in miRNA activity in plants under drought stress and might induce further changes in numerous miRNA activities (Figs. 3 and 4) (Liu et al. 2008; Zhou et al. 2010).

Both ABA and water deficit induce miR159 in Arabidopsis seeds during germination (Reyes and Chua 2007). miR159a regulates the cleavage of MYB33 and MYB101 in Arabidopsis (Allen et al. 2010; Reyes and Chua 2007). MYB33 and MYB101 mRNA levels are suppressed by the overexpression of miR159a leading to hyposensitivity to ABA, whereas accumulation of cleavage-resistant forms of MYB33 and MYB101 leads to hypersensitivity to ABA (Reyes and Chua 2007).

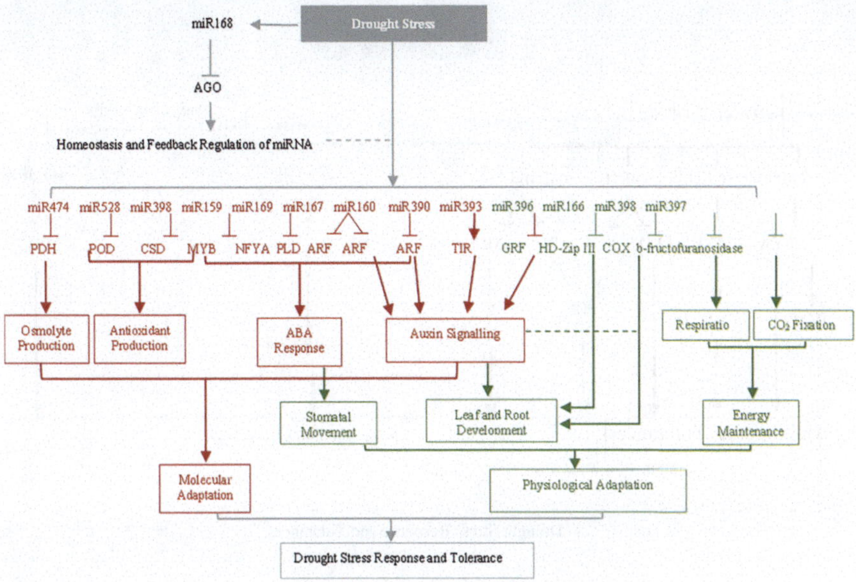

Fig. 3 Regulatory pathways involving miRNAs and their target genes in drought response. (Modified from Ding et al. 2013). Grey indicates the miRNA feedback regulation under drought stress. Black indicates miRNA-associated molecular adaptation of ABA response and auxin signaling. Grey indicates physiological adaptation. Dashed lines indicate potential paths that have not been yet confirmed. *ARF* auxin response factor, *COX* cytochrome C oxidase, *CSD* Cu/Zn-superoxide dismutase, *GRF* growth-regulating factor, *HD-Zip* homeodomain-leucine zipper, *NFY* nuclear factor Y, *PDH* proline dehydrogenase, *PLD* phospholipase D, *POD* peroxidase, *TIR* transport inhibitor response

MYB transcription factors are important elements in drought stress with their role of activation response to dehydration 22 (RD22) by binding cis-elements of its promoter (Abe et al. 2003). These results suggest that miR159 has a crucial role in ABA response by regulating the degradation of MYB mRNAs which are positive regulators of ABA signaling.

miR167d targets phospholipase D (PLD), a positive drought resistance regulator. Its expression is found to be inhibited by drought in maize (Zea mays) leading to an increase in PLD mRNA (Wei et al. 2009). In Arabidopsis, miR167 was reported to be upregulated by drought whereas; it was downregulated by ABA treatments in rice (Liu et al. 2008; Liu et al. 2009b) In another study, ABA response was reported to be directed by PLD and regulate stomatal movement in guard cells (Zhang et al. 2005b). These results revealed that PLD activation by miR167 is crucial in ABA signaling under drought stress.

miR169 is also differentially expressed under drought stress. In Medicago truncatula, miR169 was reported to be downregulated under drought although it is upregulated in rice, especially in roots (Wang et al. 2011c; Zhao et al. 2007). The

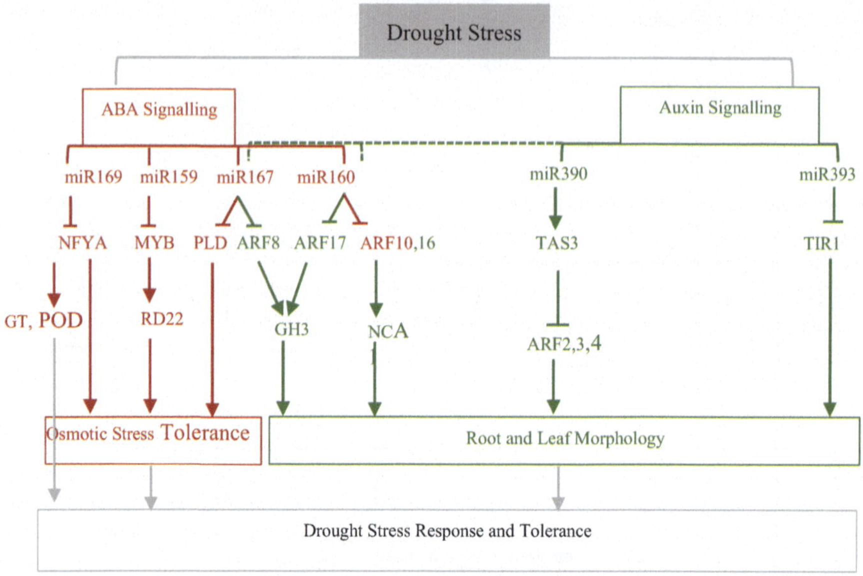

Fig. 4 miRNAs that are involved in modulating ABA and auxin regulatory pathways under drought stress. (Modified from Ding et al. 2013). Black, miR169, and miR159, which guide ABA response; Grey, miR390, and miR393, which are important for determining the cellular auxin level. miR167 and miR160 are involved both in ABA and auxin signaling under drought stress *ARF* auxin response factor, *GT* glutathione transferase, *GH* Gretchen Hagen, *NCA* no catalase activity, *NFY* nuclear factor Y, *PLD* phospholipase D, *POD* peroxidase, *RD* responsive to dehydration, *TAS* tasi-RNA producing locus, *TIR* transport inhibitor response

promoter of MIR169 involved two dehydration responsive elements (DREs) suggesting the involvement of miR169 in drought stress response (Zhao et al. 2007).

5.2 *Auxin Signaling*

Plant growth, including organ patterning, vascular development, and root and leaf architecture, is regulated by a key growth hormone auxin (Millner 1995). Under drought stress, several miRNA families are involved in the regulation of plant growth and development by controlling auxin signaling (Fig. 4). Various studies showed that miR393 is upregulated under drought stress in Arabidopsis, rice, and sugarcane (Ferreira et al. 2012; Sunkar and Zhu 2004; Zhao et al. 2007). TIR1 (transport inhibitor response 1), a positive regulator of auxin signaling, is encoded by the target of miR393 (Dharmasiri and Estelle 2002).

The family of transcription factors involved in auxin signaling is known as auxin response factors (ARFs) which bind to cis-elements in the auxin-inducible genes' cis-elements and several of them are confirmed as targeted by miRNAs. miR160 targets ARF10, ARF16, and ARF17 while ARF6 and ARF8 are regulated by miR167 suggesting they control GH3-like gene expression leading to negative regulation of indole acetic acid (IAA) (Mallory et al. 2005; Teotia et al. 2008). These data reveal the major role of miR160 and miR167 in ABA response under drought stress.

The novel regulatory pathway in auxin signaling miR390-TAS3-ARF2/ARF3/ARF4 was reported to be regulated by miR390 in Arabidopsis (Marin et al. 2010; Yoon et al. 2009). Although miR390 has a protein-coding target, it involves in the expression of tasiRNA (TAS3-derived trans-acting small interfering RNA), which regulates the establishment of organ polarity and emergence of lateral root emergence by targeting transcription factors such as ARF2, ARF3, and ARF4 (Meng et al. 2010). In cowpea (Vigna unguiculata) and Brachypodium distachyon, miR390 was reported to be upregulated suggesting the role of the miR390-dependent trans-acting siRNA pathway in drought response (Barrera-Figueroa et al. 2011; Budak and Akpinar 2011).

5.3 Cell Growth

Drought stress responses are regulated by a complex molecular network causing rapid plant cell growth reprogramming. miRNAs are involved in the growth and development pathways by targeting genes that are involved in those pathways (Carrington and Ambros 2003; Mallory and Vaucheret 2006). miR369 was reported to be differentially expressed under drought stress. In rice and cowpea, miR369 was downregulated while in Arabidopsis and tobacco, it was upregulated under drought (Barrera-Figueroa et al. 2011; Frazier et al. 2011; Liu et al. 2008; Yang and Yu 2009; Zhou et al. 2010). In Arabidopsis, miR369 is found to be target six growth-regulating factor (GRF) transcription factors which involved in the cell division and differentiation during leaf development coordination (Jones-Rhoades and Bartel 2004; Wang et al. 2011b). The overexpression of miR369 resulted in a narrow-leaf phenotype due to the reduction in cell number as a result of the repression of GRF gene expression and cell cycle-related genes including CYCD3;1, histone H4, CYCA2;1, and CYCB1;1 (Liu et al. 2009a; L. Wang et al. 2011b). Additionally, overexpression of miR369 leads to an increase in the tolerance to drought stress in several studies due to the lower stomatal density in the former (Liu et al. 2009a; Yang and Yu 2009).

miR166 is reported to be involved in cell development by regulating class-III homeodomain-leucine zipper (HD-Zip III) at the post-transcriptional level suggesting to be important for lateral root development, axillary meristem initiation, and leaf polarity (Boualem et al. 2008; Chen 2005; Hawker and Bowman 2004). In response to drought stress, miR166 is found to be downregulated in barley and Triticum dicoccoides (Kantar et al. 2010, 2011).

5.4 Osmotic Adjustment

Osmotic adjustment is one of the main mechanisms to preserve cell turgor and stabilize protein structure under drought stress. Production and accumulation of several osmoprotectants and osmolytes including amino acids, sugars, sugar alcohols, and amines are the mechanisms to cope with drought stress for many organisms (Bartels and Sunkar 2005; Seki et al. 2007). Proline is a key amino acid that involves in several defense mechanisms acting as an osmolyte, a free radical scavenger, and a stress-related signal (Nanjo et al. 1999). Proline is accumulated both by the activation of proline biosynthesis and the inactivation of proline degradation, which is catalyzed by proline dehydrogenase (PDH) (Reddy et al. 2004). In maize, Rayapati and Stewart (1991) reported the involvement of miR474 in proline degradation by targeting the PDH gene. Several other studies also revealed that miR474 was upregulated under drought stress leading to downregulation of PDH and thereby accumulation of proline resulting in increased tolerance to drought stress (Trindade et al. 2010; Wei et al. 2009; Zhou et al. 2010).

5.5 Antioxidant Defense

The enhanced reactive oxygen species (ROS) production, especially in chloroplast and mitochondria is one of the results of drought stress (Cruz de Carvalho 2008). In excess levels, ROS are toxic; thus, plants evolved a defense mechanism involving antioxidative enzymes and low-molecular-weight antioxidants to scavenge ROS such as superoxide dismutase (SOD), catalase (CAT), peroxidase (POD), ascorbate peroxidase (APX), and glutathione reductase (GR) (Jiang and Zhang 2001; Loggini et al. 1999; Mittler 2002; Peltzer et al. 2002). The inclusion of miRNAs in the antioxidant regulation is reported in several studies. In maize seedlings, miR528 was downregulated under drought conditions. The downregulation of miR528 leads to an upregulation in the levels of POD which would induce the excessive H2O2 removal and reduce the injury caused by ROS suggesting that the possibility of antioxidant defense is regulated by miR528 under drought stress (Wei et al. 2009).

CSD1 and CSD2, the two closely related Cu/Zn SODs, and cytochrome C oxidase subunit 5 (COX5b) are targeted by miR398. COX5b involves in electron transport in the mitochondrial respiratory pathway while CSDs function in the detoxification of oxidative stress (Jones-Rhoades and Bartel 2004; Mittler 2002; Sunkar and Zhu 2004). In maize, miR398 was reported to be downregulated under drought stress leading to an increased CSD activity and thereby oxidative stress tolerance (Wei et al. 2009).

5.6 *Photosynthesis and Respiration*

Drought stress diminishes photosynthetic activity and electron transport capacity whereas enhanced respiration (Rizhsky et al. 2002; Schulzc 1986; Shinozaki et al. 2003). Besides photosynthetic ability, drought stress also reduces CO2 assimilation which is required for plant growth. The rate of carbon-oxygen compounds should be sustained for the protection from drought stress (Reddy et al. 2004). The target of miR397 is predicted to be β-fructofuranosidase, which involves in starch and sucrose metabolism, therefore, has a role in the reductive carboxylate cycle (CO2 fixation) and energy supply (Zhou et al. 2010).

6 miRNAs in Vegetables

Recent studies revealed that miRNAs are directly involved in developmental processes such as root development, germination, morphogenesis, flowering, and vegetative-reproductive phase changes (Jones-Rhoades et al. 2006; Jung et al. 2009; Nithin et al. 2015; Nodine and Bartel 2010; Wu et al. 2011; Yang et al. 2011). Additionally, at the molecular level, miRNAs are the key regulators in gene regulatory pathways, metabolism, biotic and abiotic stress response, signaling, protein degradation, regulation of the feedback mechanism of siRNA pathways and genome integrity (Bushati and Cohen 2007; Mallory and Vaucheret 2006). These properties made miRNAs an important field of study both in transgenic studies, and miRNA characterization for plant development.

miR169, an evolutionarily conserved miRNA among all plant species, has a crucial role in plant development as well as the abiotic stress response. A study conducted by Zhang et al. (2011) revealed that upregulation of miR169 caused a decrease in the stomatal opening by 35–49%, and stomatal permeability by 33–45% leading to a 38–55% reduction in transpiration in tomato (Zhang et al. 2011). Likewise, miR169 expression was induced under drought stress in cucumber seedlings (Li et al. 2016). These results suggest that the upregulation of miR169 leads to drought resistance.

Potato is one of the most widely used vegetables on the global stage and it is widely cultivated after wheat, maize, and rice (Djami-Tchatchou et al. 2017). Its optimum growing temperatures is between 17 and 21 °C and environmental factors including drought and high temperature highly affected its cultivation (Shriram et al. 2016). Potato productivity, yield, growth, and quality might be decreased by the timing and severity of abiotic stress conditions. Hwang et al. (2011) investigated drought-responsive miRNAs from potato plants (Solanum tuberosum) and stu-miR172c, stu-miR172d and stu-miR172e were revealed with their expression profiles, and target mRNAs under drought stress. In addition to miR169, Li et al. (2016) found out that miR170, miR172, and miR395 expressions are increased under drought stress in cucumber. In another study, the expression of 32 known miRNAs and 25 novel miRNAs were identified as differential in sweet potato under drought stress by small RNA sequencing (Saminathan et al. 2019).

An aquatic perennial herb, Oenanthe javanica, is known as an edible vegetable with medicinal properties, high mineral, and vitamin content. Since genome, transcriptome, and small RNA information of O. javanica is absent, understanding of the biology of this plant is limited. In the study of Jiang et al. (2015), small RNA and transcriptome sequencing of O. javanica were performed for developing SSR markers, potential target genes of miRNAs, and annotating functional genes. They identified 1223 SSR markers and 29 potential target genes. For abiotic stress tolerance (drought, cold, heat, and salinity), ten miRNAs were confirmed to respond to abiotic stresses and a pair of miRNAs were found with their target genes. As a result of the lack of information on the O. javanica genome and transcriptome, this study with the identification of stress-responsive miRNAs is novel research that might promote further studies on gene mapping, SSR development, novel gene discovery, and processes and pathways of miRNAs. This supports the development of the medicinal science of this plant with its useful medicinal properties.

Lettuce (Lactuca sativa L.), a member of the Asteraceae family, is one of the most popular leaf vegetables cultivated and consumed. Acceptance from consumers of this vegetable is increasing because of its basic preparation for consumption and its valuable healthy compounds including polyphenols, vitamin C and vitamin E (Nicolle et al. 2004). In addition, it has a preventive role against many chronic diseases and this improved the effort for biofortification of lettuce. Borowski et al. (2014) found out stably expressed miRNAs under abiotic stresses might be considered as an alternative approach to normalized gene expression data. According to miRNA expression results, miR172 and miR398 were evaluated as important reference miRNAs for abiotic stress. Further analyses on pathways and mechanisms of stress-responsive miRNAs will be helpful to improve stress tolerance of lettuce against abiotic stress and biofortification studies to improve the health-promoting compounds of this product.

Solanaceae family contains a lot of important vegetable crops and most of them tend to salinity stress conditions. Thus, the identification of miRNAs that regulates salinity stress tolerance is important in the Solanaceae family to better understand the mechanism behind the resistance. In wild eggplant (Solanum linnaeanum), high throughput sequencing was performed to create a miRNA library and quantitative real-time PCR observations revealed that 11 miRNAs were down-regulated while 3 miRNAs were up-regulated. This investigation would be beneficial for the development of strategies for genetic improvement of Solanaceae crops with the help of information from functional characterization (Zhuang et al. 2014).

As a transgenic approach, the role of Novel_105 miRNA was investigated in potato cultivars which are tolerant Unica and sensitive Russet Burbank (Yalçın and Gökçe 2021). To achieve this target, transgenic plants were generated for both cultivars with overexpression of Novel_105, and drought stress was applied to both wild type and transgenic T0 lines. Novel_105 miRNA is targeting E3 ubiquitin-protein ligase XBAT35 gene which is responsible for cell death. Thus, an increase in Novel_105 miRNA expression led to a decrease in XBAT35 expression and this inverse proportion provided improved physiobiochemical functions under abiotic stress conditions in potato (Yalçın and Gökçe 2021).

In a comparative study, cultivated but sensitive Jacquemontia pentantha and wild but tolerant Ipomoea campanulata, which are two closely related species, were analyzed for a few biochemical parameters and miRNA expression under water deficiency stress which might naturally occur under field conditions. Biochemical and morpho-physiological traits including higher accumulation of CuZnSOD and lower degree of lipid peroxidation were observed while miRNA profiling revealed differential expression between both species. Even though further studies are required to better understand the adaptive responses of cultivated J. pentantha and wild I. campanulata, these results are supportive of the role of miRNAs under water deficiency conditions (Ghorecha et al. 2014).

Enhanced expression of sly-miR482d-3p, sly-miR172d-3p, sly-miR164b-3p, and sly-miR398b was observed by using Taqman-based qRT-PCR and miRNA expression analysis in individual heat and drought stresses with additional upregulation impacts under combined stresses. On the other hand, the antagonistic response was observed for sly-miR166a expression in single versus combined stresses while expression of slymiR396b-3p and sly-miR397-5p was decreased in combined stresses compared with the single stress factor. Degradome data was used for in silico identification of several high confidence miRNA targets and Gene Ontology enrichment analysis was utilized for functional annotation into various stress regulatory pathways. Inverse expression regulation was revealed for miRNA target pairs including sly-miR172d:Solyc04g049800, sly-miR396b-3p:Solyc01g102810 sly-miR398b:Solyc07g006180, sly-miR164b-3p:Solyc08g061500, and sly-miR396b-3p:Solyc05g017930 under heat, drought stresses and combined stress of both. These miRNAs might be used as candidates for editing and engineering climate-resilient crop plants since miRNAs are highly conserved across diverse plant species. (Bansal et al. 2021).

In the study of Li et al. (2016), changes in the expression of more than 40 miRNAs were observed in the scion of watermelon grafted onto bottle gourd (Lagenaria siceraria) or squash (Cucurbita maxima × Cucurbita moschata) rootstock. In addition, molecular mechanisms of 17 selected miRNAs were investigated under drought stress by grafting cucumber plants onto pumpkin. These studies are important to better understand the miRNA mechanisms behind stress tolerance of grafting plants.

Candar-Cakir et al. (2016) identified 699 known miRNAs in tomato and it has been revealed that most of them were significantly expressed in response to drought stress in a tissue and genotype-specific manner. KEGG and GO analyses have shown that the majority of these miRNAs included mechanisms in response signal transduction, hormone stimulus/reactive oxygen species/salt stress/water deprivation, fruit, seed, shoot and root development, and biosynthesis of plant hormones and hormone signal transduction pathways and plant-pathogen interactions. Drought and development-associated genes like HD-ZIP, DRP, NAC, MYB, and PSII in root and up ground tissues were revealed as regulated by drought-responsive miRNAs including miR398, sly-miR408, sly-miR827, sly-miR9472, sly-miR9476, sly-miR165, sly-miR166, sly-miR160, sly-miR171, sly- and sly-miR9552. Likewise, sly-miR5641, sly-miR5658, sly-miR169, sly-miR172, sly-miR393, and sly-miR7997 have a role in plant hormone signal transduction pathway and related proteins. All these results supported the idea of identification of stress-responsive miRNAs in drought-sensitive and drought-tolerant tomato genotypes in tissue-specific pattern to develop tomato plants that tolerate drought with the help of miRNAs.

7 Conclusion and Future Prospects

Identification of stress-responsive pathways is crucial for understanding the mechanism behind stress tolerance to make use of this information in the development of stress-tolerant cultivars. Numerous studies have revealed that mRNA expression profiling is not the only way to illuminate stress tolerance pathways. miRNAs are also included in plant stress regulatory pathways. Since miRNAs are short RNA sequences, their basic transformation level, the ability of gene silencing, and non-requirement of translation into protein make miRNAs a qualified candidate for studies that target plant stress tolerance. In conventional and modern plant breeding, not only yield but also nutrient quality/value must be considered by plant scientists since a diet without low nutrients leads to a lot of health disorders. Vegetables are important sources of antioxidants, fibers, vitamins, and minerals which are necessary for both the development and adult stage. As a result of this, sustainable production of vegetables is a challenge for plant biotechnology with the drawbacks of increasing human population and global warming. Identification of stress-responsive miRNAs in vegetables is a significant contribution to fight against these conditions. As a conclusion, both transgenic and genome editing studies on miRNAs focus on increasing the yield and creating stress resistance. Most of the miRNAs that are reported to be involved in stress tolerance have not been used in developmental studies, yet which creates a promising field of study. The involvement of CRISPR/Cas9 technology in the miRNA studies will be beneficial for both the identification of miRNA function and plant development according to food security regulations. In the current situation, the need for qualified researchers in the field of miRNA identification and genome editing will increase.

References

Abe H, Urao T, Ito T, Seki M, Shinozaki K, Yamaguchi-Shinozaki K (2003) Arabidopsis AtMYC2 (bHLH) and AtMYB2 (MYB) function as transcriptional activators in abscisic acid signaling. Plant Cell 15(1):63–78

Addo-Quaye C, Miller W, Axtell MJ (2009) CleaveLand: a pipeline for using degradome data to find cleaved small RNA targets. Bioinformatics 25(1):130–131

Allen E, Xie Z, Gustafson AM, Sung GH, Spatafora JW, Carrington JC (2004) Evolution of microRNA genes by inverted duplication of target gene sequences in *Arabidopsis thaliana*. Nat Genet 36(12):1282–1290

Allen RS, Li J, Alonso-Peral MM, White RG, Gubler F, Millar AA (2010) MicroR159 regulation of most conserved targets in Arabidopsis has negligible phenotypic effects. Silence 1(1):1–18

An J, Lai J, Lehman ML, Nelson CC (2013) MiRDeep: an integrated application tool for miRNA identification from RNA sequencing data. Nucleic Acids Res 41(2):727–737

Axtell MJ (2008) Evolution of microRNAs and their targets: are all microRNAs biologically relevant? Biochimica et Biophysica Acta (BBA)-Gene Regulatory Mechanisms 1779(11):725–734

Axtell MJ, Bartel DP (2005) Antiquity of microRNAs and their targets in land plants. Plant Cell 17(6):1658–1673

Bagga S, Bracht J, Hunter S, Massirer K, Holtz J, Eachus R, Pasquinelli AE (2005) Regulation by let-7 and lin-4 miRNAs results in target mRNA degradation. Cell 122(4):553–563

Bansal C, Balyan S, Mathur S (2021) Inferring the regulatory network of the miRNA-mediated response to individual and combined heat and drought stress in tomato. J Plant Biochem Biotechnol 30:862–877

Barrera-Figueroa BE, Gao L, Diop NN, Wu Z, Ehlers JD, Roberts PA et al (2011) Identification and comparative analysis of drought-associated microRNAs in two cowpea genotypes. BMC Plant Biol 11(1):1–11

Bartel DP (2004) MicroRNAs: genomics, biogenesis, mechanism, and function. Cell 116(2):281–297

Bartels D, Sunkar R (2005) Drought and salt tolerance in plants. Crit Rev Plant Sci 24(1):23–58

Baumberger N, Baulcombe DC (2005) Arabidopsis ARGONAUTE1 is an RNA slicer that selectively recruits microRNAs and short interfering RNAs. Proc Natl Acad Sci 102(33):11928–11933

Bollman KM, Aukerman MJ, Park MY, Hunter C, Berardini TZ, Poethig RS (2003) HASTY, the Arabidopsis ortholog of exportin 5/MSN5, regulates phase change and morphogenesis. Development 130(8):1493–1504

Bonnet E, He Y, Billiau K, van de Peer Y (2010) TAPIR, a web server for the prediction of plant microRNA targets, including target mimics. Bioinformatics 26(12):1566–1568

Borowski JM, Galli V, da Silva Messias R, Perin EC, Buss JH, dos Anjos e Silva SD, Rombaldi CV (2014) Selection of candidate reference genes for real-time PCR studies in lettuce under abiotic stresses. Planta 239(6):1187–1200

Boualem A, Laporte P, Jovanovic M, Laffont C, Plet J, Combier JP et al (2008) MicroRNA166 controls root and nodule development in Medicago truncatula. Plant J 54(5):876–887

Brennecke J, Aravin AA, Stark A, Dus M, Kellis M, Sachidanandam R, Hannon GJ (2007) Discrete small RNA-generating loci as master regulators of transposon activity in Drosophila. Cell 128(6):1089–1103

Budak H, Akpinar A (2011) Dehydration stress-responsive miRNA in *Brachypodium distachyon*: evident by genome-wide screening of microRNAs expression. Omics: A J Integr Biol 15(11):791–799

Bushati N, Cohen SM (2007) microRNA functions. Annu Rev Cell Dev Biol 23:175–205

Candar-Cakir B, Arican E, Zhang B (2016) Small RNA and degradome deep sequencing reveals drought-and tissue-specific microRNAs and their important roles in drought-sensitive and drought-tolerant tomato genotypes. Plant Biotechnol J 14(8):1727–1746

Carrington JC, Ambros V (2003) Role of microRNAs in plant and animal development. Science 301(5631):336–338

Chen X (2005) microRNA biogenesis and function in plants. FEBS Lett 579(26):5923–5931

Chen C, Ridzon DA, Broomer AJ, Zhou Z, Lee DH, Nguyen JT et al (2005) Real-time quantification of microRNAs by stem–loop RT–PCR. Nucleic Acids Res 33(20):e179–e179

Chen R, Hu Z, Zhang H (2009) Identification of microRNAs in wild soybean (*Glycine soja*). J Integr Plant Biol 51(12):1071–1079

Covarrubias AA, Reyes JL (2010) Post-transcriptional gene regulation of salinity and drought responses by plant microRNAs. Plant Cell Environ 33(4):481–489

Cruz de Carvalho MH (2008) Drought stress and reactive oxygen species. Plant Signal Behav 3(3):156–165

Cuperus JT, Fahlgren N, Carrington JC (2011) Evolution and functional diversification of MIRNA genes. Plant Cell 23(2):431–442

Dai X, Zhao PX (2011) psRNATarget: a plant small RNA target analysis server. Nucleic Acids Res 39(suppl_2):W155–W159

Dehury B, Panda D, Sahu J, Sahu M, Sarma K, Barooah M, Modi MK (2013) In silico identification and characterization of conserved miRNAs and their target genes in sweet potato (*Ipomoea batatas* L.) expressed sequence tags (ESTs). Plant Signal Behav 8(12):e26543

Dharmasiri S, Estelle M (2002) The role of regulated protein degradation in auxin response. Auxin Mol Biol 49:401–409

Ding Y, Tao Y, Zhu C (2013) Emerging roles of microRNAs in the mediation of drought stress response in plants. J Exp Bot 64(11):3077–3086

Djami-Tchatchou AT, Sanan-Mishra N, Ntushelo K, Dubery IA (2017) Functional roles of microRNAs in agronomically important plants-potential as targets for crop improvement and protection. Front Plant Sci 22:378

Dong QH, Han J, Yu HP, Wang C, Zhao MZ, Liu H et al (2012) Computational identification of microRNAs in strawberry expressed sequence tags and validation of their precise sequences by miR-RACE. J Hered 103(2):268–277

Dong H, Lei J, Ding L, Wen Y, Ju H, Zhang X (2013) MicroRNA: function, detection, and bioanalysis. Chem Rev 113(8):6207–6233

Du T, Zamore PD (2005) microPrimer: the biogenesis and function of microRNA. Development 132(21):4645–4652

Elbashir SM, Lendeckel W, Tuschl T (2001) RNA interference is mediated by 21- and 22-nucleotide RNAs. Genes Dev 15(2):188–200

Eldem V, Okay S, Ünver T (2013) Plant microRNAs: new players in functional genomics. Turk J Agric For 37(1):1–21

Fahlgren N, Howell MD, Kasschau KD, Chapman EJ, Sullivan CM, Cumbie JS et al (2007) High-throughput sequencing of Arabidopsis microRNAs: evidence for frequent birth and death of MIRNA genes. PLoS One 2(2):e219

Fahlgren N, Jogdeo S, Kasschau KD, Sullivan CM, Chapman EJ, Laubinger S et al (2010) MicroRNA gene evolution in *Arabidopsis lyrata* and *Arabidopsis thaliana*. Plant Cell 22(4):1074–1089

FAO (2009) How to feed the world in 2050. Food and Agriculture Organization

Ferreira TH, Gentile A, Vilela RD, Costa GGL, Dias LI, Endres L, Menossi M (2012) microRNAs associated with drought response in the bioenergy crop sugarcane (*Saccharum* spp.). PLoS One 7(10):e46703

Frazier TP, Sun G, Burklew CE, Zhang B (2011) Salt and drought stresses induce the aberrant expression of microRNA genes in tobacco. Mol Biotechnol 49(2):159–165

Fu XD (2014) Non-coding RNA: a new frontier in regulatory biology. Natl Sci Rev 1(2):190–204

German MA, Pillay M, Jeong DH, Hetawal A, Luo S, Janardhanan P et al (2008) Global identification of microRNA-target RNA pairs by parallel analysis of RNA ends. Nat Biotechnol 26(8):941–946

German MA, Luo S, Schroth G, Meyers BC, Green PJ (2009) Construction of parallel analysis of RNA ends (PARE) libraries for the study of cleaved miRNA targets and the RNA degradome. Nat Protoc 4(3):356–362

Ghorecha V, Patel K, Ingle S, Sunkar R, Krishnayya NSR (2014) Analysis of biochemical variations and microRNA expression in wild (*Ipomoea campanulata*) and cultivated (*Jacquemontia pentantha*) species exposed to in vivo water stress. Physiol Mol Biol Plants 20(1):57–67

Gleave AP, Ampomah-Dwamena C, Berthold S, Dejnoprat S, Karunairetnam S, Nain B, MacDiarmid RM (2008) Identification and characterization of primary microRNAs from apple (*Malus domestica* cv. Royal Gala) expressed sequence tags. Tree Genetics Genomes 4(2):343–358

Gregory BD, O'Malley RC, Lister R, Urich MA, Tonti-Filippini J, Chen H et al (2008) A link between RNA metabolism and silencing affecting Arabidopsis development. Dev Cell 14(6):854–866

Griffiths-Jones S (2004) The microRNA registry. Nucleic Acids Res 32(suppl. 1):D109–D111

Griffiths-Jones S, Saini HK, van Dongen S, Enright AJ (2008) miRBase: tools for microRNA genomics. Nucleic Acids Res 36(suppl. 1):D154–D158

Gunawardane LS, Saito K, Nishida KM, Miyoshi K, Kawamura Y, Nagami T et al (2007) A slicer-mediated mechanism for repeat-associated siRNA 5′ end formation in Drosophila. Science 315(5818):1587–1590

Hackenberg M, Sturm M, Langenberger D, Falcón-Pérez JM, Aransay AM (2009) miRanalyzer: a microRNA detection and analysis tool for next-generation sequencing experiments. Nucleic Acids Res 37(suppl. 2):W68–W76

Han Y, Luan F, Zhu H, Shao Y, Chen A, Lu C, Zhu B (2009) Computational identification of microRNAs and their targets in wheat (Triticum aestivum L.). Sci China Ser C Life Sci 52(11):1091–1100

Hawker NP, Bowman JL (2004) Roles for class III HD-zip and Kanadi genes in Arabidopsis root development. Plant Physiol 135(4):2261–2270

Hiraguri A, Itoh R, Kondo N, Nomura Y, Aizawa D, Murai Y, Fukuhara T (2005) Specific interactions between Dicer-like proteins and HYL1/DRB-family dsRNA-binding proteins in *Arabidopsis thaliana*. Plant Mol Biol 57(2):173–188

Humphreys DT, Suter CM (2013) miRspring: a compact standalone research tool for analyzing miRNA-seq data. Nucleic Acids Res 41(15):e147–e147

Hüttenhofer A, Schattner P, Polacek N (2005) Non-coding RNAs: Hope or hype? Trends Genet 21(5):289–297

Hwang EW, Shin SJ, Park SC, Jeong MJ, Kwon HB (2011) Identification of miR172 family members and their putative targets responding to drought stress in *Solanum tuberosum*. Genes and Genomics 33(2):105–110

IPCC (2014) Climate change 2014: synthesis report. Contribution of working groups I, II and III to the fifth assessment report of the intergovernmental panel on climate change. IPCC, Geneva. Accessible through https://www.ipcc.ch/report/ar5/syr/

Jaleel CA, Manivannan P, Lakshmanan GMA, Gomathinayagam M, Panneerselvam R (2008) Alterations in morphological parameters and photosynthetic pigment responses of *Catharanthus roseus* under soil water deficits. Colloids Surf B: Biointerfaces 61(2):298–303

Jaleel CA, Manivannan P, Wahid A, Farooq M, Somasundaram R, Panneerselvam R (2009) Drought stress in plants: a review on morphological. Int J Agric Biol 11(1):100–105

Jiang M, Zhang J (2001) Effect of abscisic acid on active oxygen species, antioxidative defence system and oxidative damage in leaves of maize seedlings. Plant Cell Physiol 42(11):1265–1273

Jiang Q, Wang F, Tan HW, Li MY, Xu ZS, Tan GF, Xiong AS (2015) De novo transcriptome assembly, gene annotation, marker development, and miRNA potential target genes validation under abiotic stresses in *Oenanthe javanica*. Mol Gen Genomics 290(2):671–683

Jones-Rhoades MW, Bartel DP (2004) Computational identification of plant microRNAs and their targets, including a stress-induced miRNA. Mol Cell 14:787–799

Jones-Rhoades MW, Bartel DP, Bartel B (2006) MicroRNAs and their regulatory roles in plants. Annu Rev Plant Biol 57:19–53

Jung JH, Seo PJ, Park CM (2009) MicroRNA biogenesis and function in higher plants. Plant Biotechnol Rep 3(2):111–126

Kantar M, Unver T, Budak H (2010) Regulation of barley miRNAs upon dehydration stress correlated with target gene expression. Funct Integr Genomics 10(4):493–507

Kantar M, Lucas SJ, Budak H (2011) miRNA expression patterns of *Triticum dicoccoides* in response to shock drought stress. Planta 233(3):471–484

Kim VN (2005) MicroRNA biogenesis: coordinated cropping and dicing. Nat Rev Mol Cell Biol 6(5):376–385

Koornneef M, Leon-Kloosterziel KM, Schwartz SH, Zeevaart JA (1998) The genetic and molecular dissection of abscisic acid biosynthesis and signal transduction in Arabidopsis. Plant Physiol Biochem 36(1–2):83–89
Kurihara Y, Watanabe Y (2004) Arabidopsis micro-RNA biogenesis through Dicer-like 1 protein functions. Proc Natl Acad Sci 101(34):12753–12758
Lee RC, Feinbaum RL, Ambros V (1993) The C. Elegans heterochronic gene lin-4 encodes small RNAs with antisense complementarity to lin-14. Cell 75(5):843–854
Lee Y, Kim M, Han J, Yeom KH, Lee S, Baek SH, Kim VN (2004) MicroRNA genes are transcribed by RNA polymerase II. EMBO J 23(20):4051–4060
Lee J, Kim DI, Park JH, Choi IY, Shin C (2013) MiRAuto: An automated user-friendly MicroRNA prediction tool utilizing plant small RNA sequencing data. Mol Cells 35(4):342–347
Li A, Mao L (2007) Evolution of plant microRNA gene families. Cell Res 17(3):212–218
Li W, Ruan K (2009) MicroRNA detection by microarray. Anal Bioanal Chem 394(4):1117–1124
Li F, Orban R, Baker B (2012) SoMART: a web server for plant miRNA, tasiRNA and target gene analysis. Plant J 70(5):891–901
Li C, Li Y, Bai L, He C, Yu X (2016) Dynamic expression of miRNAs and their targets in the response to drought stress of grafted cucumber seedlings. Hortic Plant J 2:41–49
Liu HH, Tian X, Li YJ, Wu CA, Zheng CC (2008) Microarray-based analysis of stress-regulated microRNAs in *Arabidopsis thaliana*. RNA 14(5):836–843
Liu D, Song Y, Chen Z, Yu D (2009a) Ectopic expression of miR396 suppresses GRF target gene expression and alters leaf growth in Arabidopsis. Physiol Plant 136(2):223–236
Liu Q, Zhang YC, Wang CY, Luo YC, Huang QJ, Chen SY et al (2009b) Expression analysis of phytohormone-regulated microRNAs in rice, implying their regulation roles in plant hormone signaling. FEBS Lett 583(4):723–728
Llave C, Kasschau KD, Rector MA, Carrington JC (2002) Endogenous and silencing-associated small RNAs in plants. Plant Cell 14(7):1605–1619
Loggini B, Scartazza A, Brugnoli E, Navari-Izzo F (1999) Antioxidative defense system, pigment composition, and photosynthetic efficiency in two wheat cultivars subjected to drought. Plant Physiol 119(3):1091–1100
Lu C, Fedoroff N (2000) A mutation in the Arabidopsis HYL1 gene encoding a dsRNA binding protein affects responses to abscisic acid, auxin, and cytokinin. Plant Cell 12(12):2351–2365
Luo Y, Guo Z, Li L (2013) Evolutionary conservation of microRNA regulatory programs in plant flower development. Dev Biol 380(2):133–144
Mallory AC, Vaucheret H (2006) Functions of microRNAs and related small a RNAs in plants. Nat Genet 38(6S):S31–S36
Mallory AC, Bartel DP, Bartel B (2005) MicroRNA-directed regulation of Arabidopsis Auxin Response Factor17 is essential for proper development and modulates expression of early auxin response genes. Plant Cell 17(5):1360–1375
Marin E, Jouannet V, Herz A, Lokerse AS, Weijers D, Vaucheret H et al (2010) mir390, Arabidopsis TAS3 tasiRNAs, and their Auxin Response Factor targets define an autoregulatory network quantitatively regulating lateral root growth. Plant Cell 22(4):1104–1117
Martin RC, Liu PP, Goloviznina NA, Nonogaki H (2010) microRNA, seeds, and Darwin? Diverse function of miRNA in seed biology and plant responses to stress. J Exp Bot 61(9):2229–2234
Mathiyalagan R, Subramaniyam S, Natarajan S, Kim YJ, Sun MS, Kim SY et al (2013) In silico profiling of microRNAs in Korean ginseng (*Panax ginseng* Meyer). J Ginseng Res 37(2):227–247
Matts J, Jagadeeswaran G, Roe BA, Sunkar R (2010) Identification of microRNAs and their targets in switchgrass, a model biofuel plant species. J Plant Physiol 167(11):896–904
Mello CC, Conte D (2004) Revealing the world of RNA interference. Nature 431(7006):338–342
Meng Y, Ma X, Chen D, Wu P, Chen M (2010) MicroRNA-mediated signaling involved in plant root development. Biochem Biophys Res Commun 393(3):345–349
Meng Y, Shao C, Gou L, Jin Y, Chen M (2011) Construction of microRNA- and microRNA-mediated regulatory networks in plants. RNA Biol 8(6):1124–1148
Mhuantong W, Wichadakul D (2009) MicroPC (μPC): a comprehensive resource for predicting and comparing plant microRNAs. BMC Genomics 10:366

Milev I, Yahubyan G, Minkov I, Baev V (2011) miRTour: plant miRNA and target prediction tool. Bioinformation 6(6):248–249

Millner PA (1995) The auxin signal. Curr Opin Cell Biol 7(2):224–231

Mittler R (2002) Oxidative stress, antioxidants and stress tolerance. Trend Plants Sci 7(9):405–410

Nanjo T, Kobayashi M, Yoshiba Y, Sanada Y, Wada K, Tsukaya H et al (1999) Biological functions of proline in morphogenesis and osmotolerance revealed in antisense transgenic *Arabidopsis thaliana*. Plant J 18(2):185–193

Nicolle C, Cardinault N, Gueux E, Jaffrelo L, Rock E, Mazur A et al (2004) Health effect of vegetable-based diet: lettuce consumption improves cholesterol metabolism and antioxidant status in the rat. Clin Nutr 23(4):605–614

Nithin C, Patwa N, Thomas A, Bahadur RP, Basak J (2015) Computational prediction of miRNAs and their targets in *Phaseolus vulgaris* using simple sequence repeat signatures. BMC Plant Biol 15(1):1–16

Nodine MD, Bartel DP (2010) MicroRNAs prevent precocious gene expression and enable pattern formation during plant embryogenesis. Genes Dev 24(23):2678–2692

Nozawa M, Miura S, Nei M (2012) Origins and evolution of microRNA genes in plant species. Genome Biol Evol 4(3):230–239

Paicu C, Mohorianu I, Stocks M, Xu P, Coince A, Billmeier M et al (2017) MiRCat2: accurate prediction of plant and animal microRNAs from next-generation sequencing datasets. Bioinformatics 33(16):2446–2454

Park W, Li J, Song R, Messing J, Chen X (2002) CARPEL FACTORY, a Dicer homolog, and HEN1, a novel protein, act in microRNA metabolism in *Arabidopsis thaliana*. Curr Biol 12(17):1484–1495

Park MY, Wu G, Gonzalez-Sulser A, Vaucheret H, Poethig RS (2005) Nuclear processing and export of microRNAs in Arabidopsis. Proc Natl Acad Sci 102(10):3691–3696

Peltzer D, Dreyer E, Polle A (2002) Differential temperature dependencies of antioxidative enzymes in two contrasting species: *Fagus sylvatica* and *Coleus blumei*. Plant Physiol Biochem 40(2):141–150

Pratt AJ, MacRae IJ (2009) The RNA-induced silencing complex: a versatile gene-silencing machine. J Biol Chem 284(27):17897–17901

Qiu CX, Xie FL, Zhu YY, Guo K, Huang SQ, Nie L, Yang ZM (2007) Computational identification of microRNAs and their targets in *Gossypium hirsutum* expressed sequence tags. Gene 395(1–2):49–61

Rayapati PJ, Stewart CR (1991) Solubilization of a proline dehydrogenase from maize (*Zea mays* L.) mitochondria. Plant Physiol 95(3):787–791

Rebijith KB, Asokan R, Ranjitha HH, Krishna V, Nirmalbabu K (2013) In silico mining of novel microRNAs from coffee (*Coffea arabica*) using expressed sequence tags. J Hortic Sci Biotechnol 88(3):325–337

Reddy AR, Chaitanya KV, Vivekanandan M (2004) Drought-induced responses of photosynthesis and antioxidant metabolism in higher plants. J Plant Physiol 161(11):1189–1202

Reinhart BJ, Weinstein EG, Rhoades MW, Bartel B, Bartel DP (2002) MicroRNAs in plants. Genes Dev 16(13):1616–1626

Reyes JL, Chua NH (2007) ABA induction of miR159 controls transcript levels of two MYB factors during Arabidopsis seed germination. Plant J 49(4):592–606

Rizhsky L, Liang H, Mittler R (2002) The combined effect of drought stress and heat shock on gene expression in tobacco. Plant Physiol 130(3):1143–1151

Ronen R, Gan I, Modai S, Sukacheov A, Dror G, Halperin E, Shomron N (2010) miRNAkey: a software for microRNA deep sequencing analysis. Bioinformatics 26(20):2615–2616

Saminathan T, Alvarado A, Lopez C, Shinde S, Gajanayake B, Abburi VL et al (2019) Elevated carbon dioxide and drought modulate physiology and storage-root development in sweet potato by regulating microRNAs. Funct Integr Genomics 19(1):171–190

Schulze ED (1986) Carbon dioxide and water vapor exchange in response to drought in the atmosphere and in the soil. Annu Rev Plant Physiol 37(1):247–274

Seki M, Umezawa T, Urano K, Shinozaki K (2007) Regulatory metabolic networks in drought stress responses. Curr Opin Plant Biol 10(3):296–302

Shinozaki K, Yamaguchi-Shinozaki K, Seki M (2003) Regulatory network of gene expression in the drought and cold stress responses. Curr Opin Plant Biol 6(5):410–417

Shriram V, Kumar V, Devarumath RM, Khare TS, Wani SH (2016) MicroRNAs as potential targets for abiotic stress tolerance in plants. Front Plant Sci 7:817

Shuai P, Liang D, Zhang Z, Yin W, Xia X (2013) Identification of drought-responsive and novel Populus trichocarpa microRNAs by high-throughput sequencing and their targets using degradome analysis. BMC Genomics 14(1):1–14

Song C, Fang J, Li X, Liu H, Thomas Chao C (2009) Identification and characterization of 27 conserved microRNAs in citrus. Planta 230(4):671–685

Sun X, Dong B, Yin L, Zhang R, Du W, Liu D et al (2013) PMTED: a plant microRNA target expression database. BMC Bioinfor 14(1):1–7

Sunkar R, Zhu JK (2004) Novel and stress-regulated microRNAs and other small RNAs from Arabidopsis. Plant Cell 16(8):2001–2019

Szcześniak MW, Deorowicz S, Gapski J, Kaczyński Ł, Makałowska I (2012) miRNEST database: an integrative approach in microRNA search and annotation. Nucleic Acids Res 40(D1):D198–D204

Teotia PS, Mukherjee SK, Mishra NS (2008) Fine tuning of auxin signaling by miRNAs. Physiol Mol Biol Plants 14(1):81–90

Trindade I, Capitão C, Dalmay T, Fevereiro MP, dos Santos DM (2010) miR398 and miR408 are up-regulated in response to water deficit in *Medicago truncatula*. Planta 231(3):705–716

Wang WC, Lin FM, Chang WC, Lin KY, da Huang H, Lin NS (2009) MiRExpress: analyzing high-throughput sequencing data for profiling microRNA expression. BMC Bioinfor 10:328

Wang J, Hou X, Yang X, Gustafson P (2011a) Identification of conserved microRNAs and their targets in Chinese cabbage (*Brassica rapa* subsp. *pekinensis*). Genome 54(12):1029–1040

Wang L, Gu X, Xu D, Wang W, Wang H, Zeng M et al (2011b) MiR396-targeted AtGRF transcription factors are required for coordination of cell division and differentiation during leaf development in Arabidopsis. J Exp Bot 62(2):761–773

Wang T, Chen L, Zhao M, Tian Q, Zhang WH (2011c) Identification of drought-responsive microRNAs in *Medicago truncatula* by genome-wide high-throughput sequencing. BMC Genomics 12(1):1–11

Weber MJ (2005) New human and mouse microRNA genes found by homology search. FEBS J 272(1):59–73

Wei L, Zhang D, Xiang F, Zhang Z (2009) Differentially expressed miRNAs potentially involved in the regulation of defense mechanism to drought stress in maize seedlings. Int J Plant Sci 170(8):979–989

Wilkinson S, Davies WJ (2002) ABA-based chemical signalling: the co-ordination of responses to stress in plants. Plant Cell Environ 25(2):195–210

Wu XM, Liu MY, Ge XX, Xu Q, Guo WW (2011) Stage and tissue-specific modulation of ten conserved miRNAs and their targets during somatic embryogenesis of Valencia sweet orange. Planta 233(3):495–505

Xie FL, Huang SQ, Guo K, Xiang AL, Zhu YY, Nie L, Yang ZM (2007) Computational identification of novel microRNAs and targets in *Brassica napus*. FEBS Lett 581(7):1464–1474

Xie F, Frazier TP, Zhang B (2010) Identification and characterization of microRNAs and their targets in the bioenergy plant switchgrass (*Panicum virgatum*). Planta 232(2):417–434

Xie F, Frazier TP, Zhang B (2011) Identification, characterization and expression analysis of MicroRNAs and their targets in the potato (*Solanum tuberosum*). Gene 473(1):8–22

Xie F, Xiao P, Chen D, Xu L, Zhang B (2012) miRDeepFinder: a miRNA analysis tool for deep sequencing of plant small RNAs. Plant Mol Biol 80(1):75–84

Yalçın M, Gökçe ZNÖ (2021) Investigation of the effects of overexpression of Novel_105 miRNA in contrasting potato cultivars during separate and combined drought and heat stresses. Turk J Bot 45:397–411

Yang F, Yu D (2009) Overexpression of Arabidopsis MiR396 enhances drought tolerance in transgenic tobacco plants. Acta Bot Yunnanica 31(5):421–426

Yang L, Conway SR, Poethig RS (2011) Vegetative phase change is mediated by a leaf-derived signal that represses the transcription of miR156. Development 138(2):245–249

Ye K, Chen Y, Hu X, Guo J (2013) Computational identification of microRNAs and their targets in apple. Genes and Genomics 35(3):377–385

Yin ZJ, Shen FF (2010) Identification and characterization of conserved microRNAs and their target genes in wheat (*Triticum aestivum*). Genet Mol Res 9(2):1186–1196

Yoon EK, Yang JH, Lim J, Kim SH, Kim SK, Lee WS (2009) Auxin regulation of the microRNA390-dependent transacting small interfering RNA pathway in Arabidopsis lateral root development. Nucleic Acids Res 38(4):1382–1391

Zhang BH, Pan XP, Wang QL, Cobb GP, Anderson TA (2005a) Identification and characterization of new plant microRNAs using EST analysis. Cell Res 15(5):336–360

Zhang W, Yu L, Zhang Y, Wang X (2005b) Phospholipase D in the signaling networks of plant response to abscisic acid and reactive oxygen species. Biochimica et Biophysica Acta (BBA)-Mol Cell Biol Lipids 1736(1):1–9

Zhang BH, Pan XP, Cox SB, Cobb GP, Anderson TA (2006a) Evidence that miRNAs are different from other RNAs. Cell Mol Life Sci 63(2):246–254

Zhang B, Pan X, Anderson TA (2006b) Identification of 188 conserved maize microRNAs and their targets. FEBS Lett 580(15):3753–3762

Zhang B, Pan X, Cannon CH, Cobb GP, Anderson TA (2006c) Conservation and divergence of plant microRNA genes. Plant J 46(2):243–259

Zhang B, Pan X, Wang Q, Cobb GP, Anderson TA (2006d) Computational identification of microRNAs and their targets. Comput Biol Chem 30(6):395–407

Zhang B, Wang Q, Wang K, Pan X, Liu F, Guo T et al (2007) Identification of cotton microRNAs and their targets. Gene 397(1–2):26–37

Zhang B, Pan X, Stellwag EJ (2008a) Identification of soybean microRNAs and their targets. Planta 229(1):161–182

Zhang JF, Yuan LJ, Shao Y, Du W, Yan DW, Lu YT (2008b) The disturbance of small RNA pathways enhanced abscisic acid response and multiple stress responses in Arabidopsis. Plant Cell Environ 31(4):562–574

Zhang W, Luo Y, Gong X, Zeng W, Li S (2009) Computational identification of 48 potato microRNAs and their targets. Comput Biol Chem 33(1):84–93

Zhang Z, Yu J, Li D, Zhang Z, Liu F, Zhou X et al (2010) PMRD: plant microRNA database. Nucleic Acids Res 38(suppl_1):D806–D813

Zhang X, Zou Z, Gong P, Zhang J, Ziaf K, Li H et al (2011) Over-expression of microRNA169 confers enhanced drought tolerance to tomato. Biotechnol Lett 33(2):403–409

Zhang Y, Yu M, Yu H, Han J, Song C, Ma R, Fang J (2012) Computational identification of microRNAs in peach expressed sequence tags and validation of their precise sequences by miR-RACE. Mol Biol Rep 39(2):1975–1987

Zhao B, Liang R, Ge L, Li W, Xiao H, Lin H et al (2007) Identification of drought-induced microRNAs in rice. Biochem Biophys Res Commun 354(2):585–590

Zheng LL, Li JH, Wu J, Sun WJ, Liu S, Wang ZL et al (2016) deepBase v2.0: identification, expression, evolution and function of small RNAs, LncRNAs and circular RNAs from deep-sequencing data. Nucleic Acids Res 44(D1):D196–D202

Zhou L, Liu Y, Liu Z, Kong D, Duan M, Luo L (2010) Genome-wide identification and analysis of drought-responsive microRNAs in *Oryza sativa*. J Exp Bot 61(15):4157–4168

Zhou Z, Wang Z, Li W, Fang C, Shen Y, Li C et al (2013) Comprehensive analyses of microRNA gene evolution in paleopolyploid soybean genome. Plant J 76(2):332–344

Zhu E, Zhao F, Xu G, Hou H, Zhou L, Li X et al (2010) mirTools: microRNA profiling and discovery based on high-throughput sequencing. Nucleic Acids Res 38(suppl_2):W392–W397

Zhuang Y, Zhou XH, Liu J (2014) Conserved miRNAs and their response to salt stress in wild eggplant *Solanum linnaeanum* roots. Int J Mol Sci 15(1):839–849

CRISPR/Cas-Mediated Genetic Improvements in Vegetable Crops

Noman Nazik, Amina Amin Ch., Zujun Yin, Ruhangiz Mammadova, Gulnur Suleimanova, Abdul Qayyum Rao, and Allah Bakhsh

Abstract The reliance on vegetables in daily life is significant, primarily attributable to their nutritional value. With the exponential growth of the population, there is a corresponding rise in the demand for food rich in nutritional value. Vegetable crops, characterized by their high nutrient content, serve as significant alternatives for addressing the dietary needs of an expanding population. Traditional approaches have played a crucial role in enhancing the yield of vegetable crops, improving their nutritional quality, and increasing their resilience to biotic and abiotic stresses. However, numerous limitations are associated with these methods. Researchers developed genome editing tools to address these challenges and selected them for the improved and more dependable outcomes. Utilizing these tools has significantly expedited the breeding process, thereby conserving time, attributable to their inherent features, which include user-friendliness, specificity, versatility, and simplicity. Genome editing has made transgenic plant development significantly more precise and accessible than conventional breeding, allowing the targeted introduction of small indels or replacement of mutations, including single-nucleotide changes. Various genome editing tools have been employed to enhance the environmental sustainability, climate resilience, productivity, nutritional value, safety for consumption, and pathogen resistance of essential vegetable crops. This chapter examines the efforts, applications, and future possibilities of CRISPR/Cas-based

N. Nazik · A. A. Ch. · A. Q. Rao · A. Bakhsh (✉)
Centre of Excellence in Molecular Biology, University of the Punjab, Lahore, Pakistan
e-mail: allahbakhsh@cemb.edu.pk

Z. Yin
Research Base, Zhengzhou University, State Key Laboratory of Cotton Biology,
Institute of Cotton Research of Chinese Academy of Agricultural Sciences, Henan, China

R. Mammadova
Genetic Resources Institute, Ministry of Science and Education Azerbaijan Republic,
Baku, Azerbaijan

Western Caspian University, Baku, Azerbaijan

G. Suleimanova
Kazakh National Agrarian Research University, Almaty, Kazakhstan

U. K. Chaudhry et al. (eds.), *Drought Stress*,
https://doi.org/10.1007/978-3-031-80610-0_13

methodologies in vegetable cultivation, highlighting their potential to enhance sustainable crop productivity.

Keywords Genome editing · CRISPR-Cas9 · Improvements · Vegetables · Challenges

1 Importance of Vegetables

Vegetables are an integral food source for humans as they provide us with numerous nutrients found only in traces in staple foods like cereals, etc. Along with providing us with various nutrients, vegetables also contain compounds that reduce the risk of several diseases (Aune et al. 2017). Some vegetables contain many times the minerals and nutrients of fruits.

2 History of Vegetable Improvement

Like agronomic crops, vegetables have garnered the attention of researchers aiming to enhance various traits, including resistance to biotic and abiotic stresses, nutritional value, flavor, shelf life, and overall yield. To tackle these challenges, researchers have employed various methodologies to achieve a preferred variety (Kim et al. 2021).

Plant scientists primarily concentrate on breeding techniques to enhance the traits of crops and vegetables. Classic breeding involves crossing two different varieties with different characteristics to develop a hybrid variety that may have the desired characteristics of both parents. Outcrossing is based on the principles of random introgression and recombination of alleles, which can result in the desired combination of genetic traits in hybrid plants. However, this approach is often lengthy, labor-intensive, and associated with a low success rate. This approach also results in undesired and unknown changes in the genome of a hybrid variety. Further, some essential or desirable traits are controlled by many genes or a complex network of genes. In these cases, the crossing is not a better option. Molecular breeding emerged in the 1980s. Modern plant breeding has advanced by narrowing the genomic scope to specific genes that control desired traits. Different breeding techniques have been deployed to attain these features. These include recombinant DNA technology, tissue culturing, QTL mapping, genome editing, etc.

Alongside widely used crossing techniques, chemical and physical mutagenesis offers innovative methods for inducing genetic mutations. This strategy provides a cost-effective and accessible means to introduce genetic variations. Physical mutagenesis uses radiation, such as X-rays and gamma rays, to alter plant DNA, successfully inducing mutations. This approach allows for the targeted alteration of genetic material, paving the way for new plant varieties with enhanced traits. Subsequently, mutagenic chemicals became the preferred choice for inducing mutagenesis due to their ease of handling and the higher frequency of mutations observed compared to radiological methods. (Ganapathy et al. 2008). Currently, ethyl methanesulfonate stands as the predominant chemical mutagen employed in research. (Türkoğlu et al. 2022). In contrast, other chemical mutagens, including sodium azide and methyl nitrosourea (Az-MNU), are also commonly used. Chemical mutagens were employed as an alternative strategy to mitigate the off-target effects of radiation on non-target areas of plants, thereby decreasing the associated risks to human health and the environment from such radiation exposure. Mutagenesis breeding has proven effective in inducing point mutations and deletions within the plant genome. It has demonstrated greater efficacy in ornamental plants than in vegetables.

Tissue culture has been a prevalent method for enhancing vegetables and other crop species. Somaclonal variations arising from cell passaging in tissue culture could be beneficial (Barden et al. 1986). *De novo* plant regeneration from callus tissue introduces random genetic variations, sometimes enhancing tolerance to stress or improving traits like yield. Occasionally, these mutations can confer tolerance to specific abiotic or biotic stressors and may also enhance various traits, including yield. Additionally, methods such as marker-assisted selection, RNA interference (RNAi), and double-stranded RNA (dsRNA) have been effectively utilized to improve plant characteristics.

Individuals have expressed concern over consuming genetically modified organisms (GMOs) for various reasons. One is the unidentified genetic composition of GMOs created through traditional methods. Opponents of GMOs believe this artificial tinkering may adversely affect human health and DNA (Hundleby and Harwood 2019). Genome editing effectively mitigates this problem. Genome editing has emerged as a safer, precise, straightforward, specific, resilient, and effective method for enhancing plant characteristics. This technology enables precise modifications of genes, from single-base alterations to entire gene edits, rather than producing random mutations in the genome. Nucleases target a specific allele or genomic sequence found to be linked to a specific physiological trait. Nevertheless, despite several studies endorsing their safety, individuals remain apprehensive about consuming genome-edited plants.

Since its inception, the genome editing technique has been utilized in numerous living species for various research objectives. Numerous genome editing tools are currently employed in, and their number is on the rise. Nevertheless, some of the tools have been used more often owing to their superior results. Mega nucleases, Zinc Finger Nucleases (ZFNs) (Kim et al. 1996), and Transcription Activator-Like Effector Nucleases (TALENs) (Bogdanove and Voytas 2011) are protein-based genome editing tools. In contrast, CRISPR-Cas nucleases, which were adapted for genome editing in 2012, function through RNA-guided mechanisms, setting them apart from earlier technologies. Despite being introduced later, CRISPR-Cas has enabled more genome editing studies than all previous methods combined (Jinek et al. 2012).

Crops having altered genomes are classified into three categories according to the type of editing performed. The classifications are determined by the mutation scale and the subsequent DNA repair process following the double-strand genomic DNA break (Entine et al. 2021). Site-directed nucleases-1 (SDN1) facilitate the knockdown or knockout of a gene by inducing random mutations through insertion or deletion. The NHEJ (Non-homologous end joining) repair pathway succeeds in this. SDN2 is linked to a gain-of-function mutation resulting from inserting a short homologous (template) DNA sequence. SDN2 is associated with an HDR (Homology-directed repair) mechanism. SDN3 employs a comparable methodology to SDN2; however, a significant distinction lies in the DNA sequence length intended for insertion. Moreover, SDN3 employs a novel gene not previously present in the target genome. The HDR route subsequently follows it (Pixley et al. 2022).

This book chapter will highlight the role of the technologies mentioned above in the genome editing of vegetable species and will provide examples of genome editing experiments for economically important species.

3 Genome Editing Technologies

Various tools have been developed for genome editing in living organisms. Each of these tools offers unique advantages and applications for precise genomic modifications, facilitating advances in biotechnology and genetic research.

4 Meganucleases

Meganulceases, despite the lack of versatility, have been used for genome editing. They introduce double-stranded breaks in a genome, which are repaired by either Homologous recombination (HR) or NHEJ. The latter is an error-prone phenomenon and occurs more frequently than HR. Nonetheless, both phenomena are equally preferred due to their different outputs. The success rate and efficiency of a nuclease depend upon its ability to induce HDR-mediated recombination after cutting/digesting a double-stranded DNA molecule. One of the limitations in the application of mega nuclease is their recalcitrance to redesigning because their DNA-binding and catalytic domains cannot be separated from each other. Redesigning other genome editing tools is not as challenging (Kathiria and Eudes 2014).

5 ZFNs

Zinc finger nucleases were the first artificially synthesized nuclease to target a specific genome sequence. They can cleave any double-stranded DNA part. As the name suggests, they comprise a zinc finger domain (Cys2-Hys2) combined with a cleavage domain of a Fok1 endonuclease DNA restriction enzyme. Zinc fingers provide the binding function when the monomers recognize and bind their respective specific DNA sequence. Since they are small domains, they only flank the 5–6 bp sequence, allowing the Fok1 enzyme to start its endonuclease activity in the vicinity. Fok1 introduces DSBs (double-stranded breaks) that are repaired by NHEJ pathway that, in almost every case, results in the introduction of frameshift mutation and, thus, inactivation of the target gene (Wood et al. 2011).

6 TALENs

TALENs are transcription factors that are more efficient at genome editing than ZFNs. In conjunction with a nuclease, these transcription factors can bind DNA sequences at any location, causing DNA breaks. These transcription factors-nucleases complexes are transported into plant cells by *Xanthomonas* bacteria via their type III secretion system. Due to their higher efficiency, they are still used more frequently than ZFNs or meganucleases (Butler et al. 2016).

7 CRISPR Cas

The CRISPR-Cas system is the latest advancement in genome editing techniques. It is inherently present in a wide range of bacteria and archaea species as a component of their acquired immune system to protect them from bacteriophages (Jansen et al. 2002). The CRISPR (clustered regularly interspaced short palindromic repeats) locus in the bacterial genome encodes specialized RNAs that guide CRISPR-associated (Cas) enzymes to the to guide it toward the target foreign genomic sequence. Binding triggers Cas endonuclease activity, consequently cleaving the foreign genome and resulting in pathogen-free bacteria. The CRISPR-Cas system remained unidentified until Ishino and colleagues detected it in *E. coli* in 1987 (Ishino et al. 1987). Its function was not elucidated at that time. Years later, its function was elucidated in 2005 when it was discovered to be a component of the adaptive immune system of bacteria and archaea (Bolotin et al. 2005). Until 2012, ZFNs and TALENs dominated the field of genome editing due to uncertainty about the efficacy of the CRISPR-Cas system as a viable alternative. However, this paradigm shifted when Jennifer Doudna and Emmanuelle Charpentier introduced the CRISPR-Cas system, demonstrating it as a revolutionary and user-friendly genome editing system (Jinek et al. 2012).

CRISPR-Cas systems are categorized into two primary classes: Class I and Class II, with each class subdivided into multiple types and subtypes based on evolutionary relationships and structural diversity. The distinction lies in the number of effector proteins required for interference. Class I systems, which are more prevalent, utilize multi-protein complexes for target recognition and cleavage, whereas Class II systems, such as the widely used Cas9 and Cas12, rely on a single, multi-functional effector protein to perform these tasks. Class II systems have garnered significant attention in genome editing due to their simplicity and adaptability.

Since its discovery and swift integration as a genome editing tool, the CRISPR-Cas system has largely overshadowed earlier genome editing technologies, owing to its numerous advantages. It has emerged as the leading technology for genome manipulation, celebrated for its simplicity, versatility, ease of use, robustness, precision, and efficiency (Pickar-Oliver and Gersbach 2019). Moreover, CRISPR-Cas has expanded its reach beyond genome editing, finding extensive applications in gene regulation, nucleic acid targeting, diagnostics, and functional genomics, (Bock et al. 2022; Goell and Hilton 2021; Kampmann 2020; Nakamura et al. 2021; Zhan et al. 2019; Zhang et al. 2021). As ongoing innovations continue to emerge, the CRISPR-Cas toolkit is rapidly evolving, presenting new capabilities and applications that hold great promise for advancing research and biotechnology (Sun and Wang 2022).

8 Functional Studies

In addition to editing genes to create new and enhanced varieties, gene editing technology can manipulate the genome, thereby facilitating a deeper understanding of the influence of one or more genes on the phenotypic characteristics of plants. Genome editing tools facilitate the identification of genes and their associated physiological functions. In this context, numerous investigations have been undertaken. The initial documentation of gene editing in a horticultural crop occurred in 2013, wherein TALENs were employed to induce double-strand breaks in Cabbage (*Brassica oleracea* L. var. capitata L.) FRI gene that has been reported to play a vital role in the vernalization of Brasicca family (Sun et al. 2013). A plathora of functional studies were performed afterwards in different plant species. CRISPR-Cas9-mediated loss-of-function mutations in the *SlAGL7* gene of tomato (*Solanum lycopersicum*) have resulted in the development of *agl7* mutant plants, which display deficiencies in lamina and petiole formation. The *AUXIN RESPONSE FACTOR* (*ARF*) gene functions as a positive regulator in lamina and petioles development; however, its activity is negatively modulated by the *SlAGL7* RNAs (Bouzroud et al. 2020).

Further advancing the field, Danilo et al. (2022) successfully removed a substantial segment of approximately 3 kb from the genomes of both tomato and oilseed rape using I-SceI meganuclease (Danilo et al. 2022). Moreover, TALENs were successfully utilized to induce mutagenesis in the PROCERA (PRO) gene of tomatoes, underscoring its essential function in the regulation of gibberellin signaling pathways (Lor et al. 2014).

Pioneering studies by Pan et al. have shed light on the inheritance and stability of mutations introduced via CRISPR-Cas9 in tomato plants. By focusing on the SlPDS and SlPIF4 genes, the researchers monitored mutation frequencies from the initial mutant generation (T0) through subsequent generations (T1, T2), finding a reliable transmission of mutations from T0 to both T1 and T2 (Pan et al. 2016). Additionally, genome editing techniques have facilitated important research into various facets of tomato biology, encompassing meristem development (Xu et al. 2015), plant architecture (Xu et al. 2016), fruit size (Soyk et al. 2016), and the phenomenon of parthenocarpy (Ueta et al. 2017).

Albino cassava was developed by gene editing of a staple cassava variety (*Manihot esculenta*); *Phytoene desaturase* (*PE*) was targeted by delivering a CRISPR Cas gRNA cassette through *Agrobacterium*. gRNA against two different exonic regions on *PE* resulted in the same phenotype (Odipio et al. 2017). A study reported the generation of the early flowering mutant plant of cabbage (*B. rapa*) through CRISPR Cas 9 mediated knockout of the genes homologous to the *A. thaliana FLOWERING LOCUS C (FLC)* (Jeong et al. 2019). *BrFLC2* and *BrFLC* were successfully knocked out by indel mutation with an efficiency of 97.7% and 100%, respectively. Another study showed that a frameshift mutation by inserting 1 bp in a negative regulator gene (*BrVRN1*) of a floral repressor gene (*BrFLC1*) resulted in a delayed flowering phenotype in Chinese cabbage (Hong et al. 2021).

DNA-free genome editing has recently been introduced to induce transient gene expression in vegetables. In this technique, the ribonucleoprotein (RNP) complex of CRISPR Cas protein and RNAs (mRNA, gRNA) are inserted into the cells. In 2015, Woo et al. edited the lettuce genome with an RNP complex comprised of CRISPR Cas9 (Woo et al. 2015). The complex was delivered in the cells through PEG-mediated transfection. A recent study showed that introducing a point mutation in the *SOC1* gene in lettuce results in delayed flowering or bolting compared to wild-type lettuce (Choi et al. 2022). Point mutation was introduced through Cas 9-RNP complex delivery. Floral regulatory genes were expressed lower in mutant plants than in wild plants. RNP is also delivered into the cells through the electro-transfection method. Lee and the team did this successfully while editing the cabbage genome, which was more successful than the former delivery method (Lee et al. 2020). Meyer et al. (2022) generated transgene-free carrot plants after introducing point mutations through CRISPR Cas9 fused with the cytidine deaminase.

Other studies reporting the targeted mutagenesis of phytoene desaturase *PDS* gene were also reported in vegetables like Chicory (*Cichorium intybus* L.)(Bernard et al. 2019), Carrot (Xu et al. 2019) and cabbage (*Brassica oleracea*) (Lee et al. 2020; Liu, et al. 2019; Murovec et al. 2018). However, the transfection method varied in these studies. In an earlier study, *BolC.GA4.a* gene in the *B. oleracea* was knocked out through Cas9, which resulted in the dwarf phenotype in T_0 plants (Lawrenson et al. 2015).

Genome editing technology has predominantly been used to knockout a specific genomic sequence from the plant genome. Zhang et al. (2018) targeted the upstream open reading frame of the *LsGGP2* gene in lettuce to increase its expression. *LsGGP2* is associated with the ascorbic acid synthesis in lettuce. It was found that the ascorbic acid content was increased by about 150%. Along with that, the resistance in lettuce against oxidative stress was also increased (Zhang et al. 2018).

Multiplex genome editing has been successfully applied across a variety of vegetable crops. This approach involves the inclusion of multiple single-guide RNAs (sgRNAs), each designed to target distinct genes, within a single construct. These sgRNAs are co-expressed with essential components, such as the Cas nuclease, and delivered into plant cells to enable simultaneous editing of multiple genomic loci. This strategy is hindered by the delivery mechanisms and vector capacity tailored for different plant species (Cong et al. 2013). Ma et al. (2019a, b) used a more efficient technique for multiplex genome editing in the cabbage (*B. oleracea*). They developed a synthetic gene comprising tandemly arranged tRNA-sgRNA oligonucleotides against *BoPDS, BoSRK,* and *BoMS1* genes. Both independent and simultaneous/collective knockout of the selected genes was observed. It indicated the successful utilization of plants' endogenous tRNA processing system for genome editing (Zhu et al. 2020).

A study on Chinese Kale (*Brassica oleracea*) reported the gene that regulates chlorophyll and carotenoid content in the leaves. Targeted mutagenesis of the carotenoid isomerase gene (*BoaCRTISO*) resulted by the CRISPR Cas9 system, resulted in the declination in both the number of leaves and chlorophyll and carotenoid content (Sun et al. 2020). A recent study by Cao et al. (2021) revealed the role of the *BoCER1* gene, associated with the glossy phenotype of cabbage, on the wax production in the leaves. On transforming the *BoCER1* gene into the wild-type cabbage, it was found that the wax content was reduced in the transformed plants. The results indicated that the wax is negatively regulated by the *BoCER1* (Cao et al. 2021).

Some vegetables, like cucumber, are cumbersome to transform. Therefore, a few studies have successfully edited such plants genome. Hu et al., used optimized CRISPR Cas 9 protein to boost the expression of gRNAs inside the target cell. Targeted mutagenesis of *CsWIP1* gene generated gynoecium cucumber line free of any transgene. Gynoecium inbred lines remove the labor needed for crossing and have higher production, making them highly desired (Hu et al. 2017; Robinson 2008). Transgene-free genome-edited plants are chosen because of the drawbacks of integrating Genome editing machinery anywhere inside the plant genome that may lead to unwanted disruption of genes and the off-target effects of genome editing enzymes expressing inside the cells of T1 or T2 plants. To address this concern in carrots, Meyer developed an efficient technology to edit the *CENH3* gene by inserting CRISPR Cas, cytosine base editor, alongside a guide RNA, into the protoplast through PEG mediated delivery (Meyer et al. 2022).

A rather recent advancement in the CRISPR Cas toolkit is the Prime Editing (PE) (Anzalone et al. 2019). Despite lower editing efficiencies compared to traditional CRISPR-Cas9 systems, PE achieved precise nucleotide transitions and insertions, showcasing high targeting fidelity and minimal off-target activity. A study demonstrates the functionality of Prime Editing in the model organism *Physcomitrium patens* and tetraploid potato (*Solanum tuberosum*). While the potential for precision crop breeding is evident, further optimization of editing efficiency and pegRNA design is crucial for practical applications in polyploid species, establishing *P. patens* as a suitable model for refining PE (Perroud et al. 2022). Additionally, optimized PE approaches for dicots, including an RNA chaperone and engineered guide RNAs, achieved up to 9.7% editing efficiency and 38.2% heritable desired edits in *tomato* and *Arabidopsis*. These innovations highlight PE potential for transformative applications in plant biotechnology (Vu et al. 2024).

The continuous advancements in CRISPR/Cas9 genome editing have revolutionized the field, making the technology more precise and adaptable for diverse plant species. A significant breakthrough is the optimization of protoplast-based transient gene expression (PTGE) systems, particularly in *Brassica* crops. (Yu et al. 2024). These systems enhance the validation and screening of CRISPR vectors, allowing researchers to fine-tune genome editing protocols with higher efficiency and reduced time. The integration of improved protoplast isolation and transfection techniques further strengthens CRISPR/Cas9 applications in agriculture and plant functional genomics.

In other less-explored vegetables, genome editing technology has done wonders. There are very few studies of genome editing in vegetables by Geminivirus technology. A study reported the efficient and targeted insertion of a promoter in the tomato genome to overexpress a gene. That gene regulates anthocyanin biosynthesis. Successful genome editing was confirmed by the accumulation of pigments in the tissue of the tomato. CRISPR Cas-based carrot genome editing showed that the *flavanone-3-hydroxylase (F3H)* gene regulates the production of anthocyanins in plants. This was confirmed when the knockout of the F3H gene resulted in the discoloration of the calli (Klimek-Chodacka et al. 2018).

9 Biotic Factors

Diseases significantly hinder vegetable productivity, with pathogens such as viruses, fungi, and insect pests responsible for considerable yield losses. These biotic challenges necessitate effective management strategies to mitigate their impact and enhance crop resilience. Table 1 highlights key efforts in utilizing CRISPR-Cas technology for disease resistance and other agronomic improvements in vegetable crops.

Cucumber (*Cucumis sativus L*) genome editing through CRISPR Cas9 was done for the first time to make mutant plants having broad resistance to viruses (Chandrasekaran et al. 2016). The researchers knocked down the *eukaryotic translation initiation factor 4E (eIF4E)* gene to achieve this resistance against potyviruses and *ipomovirus* (Bastet et al. 2019). Similarly, targeting the eIF4E protein-encoding genes (*ncbp-1/ncbp-2*) in Cassava through CRISPR Cas9 produced a double mutant that showed resistance against Cassava brown streak virus (CBSV) (Gomez et al. 2019). A study utilizing Zinc Finger Nucleases (ZFNs) demonstrated broad-spectrum resistance against begomoviruses by targeting the DNA-binding domain within a conserved sequence of the begomovirus genome (Chen et al. 2014). The predominant category of plant viruses consists of RNA viruses. CRISPR-Cas13 has emerged as a transformative tool for plant viral resistance. For example, Zhan et al. (2019) demonstrated its application by engineering Potato virus Y (PVY) resistance in potato plants using the Cas13a/sgRNA system. This

Table 1 Salient examples of CRISPR/Cas-mediated Genetic Improvements in Vegetable Crops

Catagory	Plant species	Genome editing technology	Gene(s)	Mutation	Phenotype	Repair pathway	Delivery method	Genome	Reference
Functional studies	Tomato	CRISPR Cas 9	*AUXIN RESPONSE FACTOR (ARF)*	Loss of function	Impaired, incomplete leaves	NHEJ	*Agrobacterium. tumefaciens*-mediated	Tomato	Bouzroud et al. (2020)
	Tomato	CRISPR Cas 9	*RIN*	Loss of function	Incompletely ripening fruit	NHEJ	*Agrobacterium. tumefaciens*-mediated	Tomato	Jung et al. (2018)
	Tomato	Meganuclease	*GFFP*	Gene insertion	Normal	HDR	*Agrobacterium. tumefaciens*-mediated	Tomato	Danilo et al. (2022)
	Tomato	TALENs	*PROCERA (PRO)*	Loss of function	Taller plants with lighter green leaves	NHEJ	*Agrobacterium*-mediated	Tomato	Lor et al. (2014)
	Tomato	CRISPR Cas9	*TMF* and *SlBOP*	Knock out	Defected inflorescences	NHEJ	*Agrobacterium. tumefaciens*-mediated	Tomato	Xu et al. (2016)
	Tomato	CRISPR Cas9	*SlIAA9*	Knock out	Parthenocarpic tomato	NHEJ	*Agrobacterium*-mediated	Tomato	Ueta et al. (2017)
	Tomato	CRISPR Cas9	*SELF-PRUNING 5G (SP5G)*	Loss of function	Rapid and increased flowering	NHEJ	*Agrobacterium*-mediated	Tomato	Soyk et al. (2016)
	Chinese Cabbage (*Brassica rapa* L. *ssp. pekinensis*)	CRISPR Cas9	*BrVRN1*	Loss of function	Delayed flowering	NHEJ	*Agrobacterium*-mediated	Chinese Cabbage	Hong et al. (2021)

	Chinese Cabbage (*Brassica rapa* L. *ssp. pekinensis*)	CRISPR Cas9	*BrFLC2, BrFLC3*	Loss of function	Early flowering	NHEJ	*Agrobacterium*-mediated	Chinese Cabbage	Jeong et al. (2019)
	Tomato	CRISPR Cas coupled with cytidine deaminase	*DELLA and ETR1*	Base editing	Normal	HDR	*Agrobacterium*-mediated	Tomato	Shimatani et al. (2017)
	Tomato	CRISPR Cas9	*SlPDS* and *SlPIF4*	Loss of function	Albino	NHEJ	*Agrobacterium*-mediated	Tomato	Pan et al. (2016)
	Lettuce (*Lactuca sativa*)	CRISPR Cas9	*SOC1*	Loss of function	Delayed flowering	NHEJ	RNP transfection in protoplast	Lettuce	Choi et al. (2022)
	Lettuce	CRISPR Cas 9	*BIN2*	Loss of function	Normal	NHEJ	RGEN RNP transfection	Lettuce	Woo et al. (2015)
	Lettuce	CRISPR Cas 9	*LsGGP2*	Increased expression	Enhanced ascorbic acid production and oxidative stress resistance	N/A	*Agrobacterium*-mediated	Lettuce	Zhang et al. (2018)
	Cabbage	CRISPR Cas9	*PDS1*	Loss of function	Normal	NHEJ	RNP PEG-mediated transfection and electro-transfection	Cabbage	Lee et al. (2020)
	Cabbage	CRISPR Cas9	*BoPDS, BoSRK,* and *BoMS1*	Loss of function	Male sterlity	NHEJ	*Agrobacterium*-mediated	Cabbage	Ma et al. (2019a, b)
	Cabbage (*Brassica oleracea* L. *var. capitata* L.)	TALENs	*FRIGIDA*	Loss of function	Early flowering	NHEJ	*Agrobacterium*-mediated	Cabbage	Sun et al. (2013)

(continued)

Table 1 (continued)

Catagory	Plant species	Genome editing technology	Gene(s)	Mutation	Phenotype	Repair pathway	Delivery method	Genome	Reference
	Cabbage	CRISPR Cas9	*BolC.GA4.a*	Loss of function	Dwarf	NHEJ	*Agrobacterium*-mediated	Cabbage	Lawrenson et al. (2015)
	Chicory (*Cichorium intybus* L.)	CRISPR Cas9	*Phytoene desaturase (CiPDS)*	Loss of function	Albino	NHEJ	*Agrobacterium*-mediated	Chicory	Bernard et al. (2019)
	Carrot (*Daucus carota*)	CRISPR Cas 9 coupled with Single Transcription Unit-Cytosine Base Editor 1 (STU-CBE1)	*CENH3*	Loss of function		HDR	PEG-mediated	Carrot	Meyer et al. (2022)
	Carrot	CRISPR Cas9	*DcPDS* and *DcMYB113-like*	Loss of function	Albino and deep purple, respectively	NHEJ	*Agrobacterium*-mediated	Carrot	Xu et al. (2019)

approach effectively reduced viral accumulation within plant cells, providing a novel method for improving disease resistance in crops. The Cas13a protein acts by inhibiting the replication of PVY particles, thereby significantly reducing their overall levels (Zhan et al. 2019). The technique represents a significant advancement in protecting plants from viral pathogens and highlights the potential of CRISPR-Cas13 in enhancing crop resilience. However, knocking a single gene out to create resistance to one virus may unexpectedly make plants more vulnerable to other viral infections (Zafirov et al. 2021). To address this limitation, multiplex editing (Cong et al. 2013) provides a more robust and comprehensive approach for combating viral infections, allowing for the simultaneous targeting of multiple genes.

Tomato brown rugose fruit virus (ToBRFV), a contemporary pandemic, poses a significant threat to global tomato production, rapidly spreading and overcoming existing resistance mechanisms(van de Vossenberg et al. 2021). Recent research has demonstrated that quadruple knockouts of the TOM1 gene family in tomatoes, achieved through CRISPR/Cas9 technology, confer strong resistance to ToBRFV. These mutant plants show no accumulation of viral proteins or adverse effects on growth (Ishikawa et al. 2022).

In a related advancement, Nekrasov et al. (2017) utilized CRISPR-Cas9 technology to engineer tomato plants with enhanced resistance to powdery mildew. By creating a loss-of-function mutation in the *Mlo1* gene, which is normally linked to susceptibility to *Oidium neolycopersici*, they effectively boosted the plant's defense against the disease (Nekrasov et al. 2017). In addition, 15 more Mlo genes may serve as possible targets. Another study reported a mutation in powdery mildew resistance 4 (*SlPMR4*) gene, which encodes callose synthase enzyme, to produce a powdery mildew-resistant mutant. A frameshift mutation in the *SlDMR6-1* gene, the tomato orthologue of *DMR6* (Downy Mildew Resistance 6), resulted in a non-functional protein and led to elevated salicylic acid production, conferring broad resistance to multiple bacterial and fungal diseases (Santillán Martínez et al. 2020). This study, alongside prior investigations on *DMR6*, highlights the efficacy of gene editing for achieving broad-spectrum disease resistance in plants. However, a key challenge lies in the antagonistic interaction between salicylic acid (SA) and jasmonic acid (JA)-two critical hormones that regulate plant defense responses, including stomatal dynamics. Attempts to modulate both pathways simultaneously have had limited success, as suppressing one hormone to enhance the other often increases susceptibility to certain diseases. Ortigosa et al. (2019) employed CRISPR-Cas9 to engineer a tomato line resistant to bacterial speck disease caused by *Pseudomonas syringae* pv. *tomato* (Pto) DC3000. This pathogen employs coronatine, a molecular mimic of JA, to manipulate stomatal opening, thereby facilitating pathogen entry into leaf tissues. The researchers utilized precise gene editing to remove the C-terminal jasmonate domain from a target protein, effectively neutralizing the pathogen's ability to breach plant defenses and thereby conferring enhanced resistance to infection. This modification effectively enhanced resistance against Pto DC3000 without adversely affecting resistance to other diseases. A study on tomato *fusarium wilt* was conducted, in which the Solyc08g075770 gene was disrupted using CRISPR-Cas9 technology.

10 Abiotic Stresses

Abiotic stresses present a significant challenge by reducing yield, quality, and resistance to biotic threats in vegetables. These stress impacts are being exacerbated by climate change, as plants face increasingly unfavourable environmental conditions that limit their growth, development, and overall survival. In response, genome editing technologies have emerged as essential tools for improving plant resilience against various stressors.

In a study aimed at improving drought resistance in tomato plants, Li et al. (2019) modified the non-expressor of pathogenesis-related gene 1 (*NPR1*), previously recognized for its role in plant defense mechanisms against pathogens. This research demonstrated that *NPR1* also plays a critical role in modulating drought stress in tomatoes. Using CRISPR-Cas9 technology, *npr1* mutants were generated, revealing the genetically modified plants' ability to mitigate the effects of drought stress. Another study reported loss-of-function mutations in tomatoes to mitigate the harmful effects of reactive oxygen species (ROS), which accumulate during stress events like heat and light exposure. The researchers utilized CRISPR-Cas technology to target and disrupt the *SlMAPK3* gene, one of 16 Mitogen-Activated Protein Kinases (MAPKs) identified in tomatoes. The resulting mutants showed reduced structural and physiological damage caused by ROS during thermal stress, a benefit linked to the increased activity of antioxidant enzymes that neutralize ROS. Additionally, studies have shown that *SlMAPK3* functions as a negative regulator of drought-responsive genes (Ding et al. 2018). Knocking out *SlMAPK3* led to the upregulation of genes such as *SlDREB*, and *SlGST,* and *SlLOX* that regulate cellular defense mechanisms that protect against oxidative damage (Wang et al. 2017).

11 Ripening and Shelf Life

A significant challenge for food security is the loss of vegetables due to their short shelf life and inadequate storage methods. Recent advancements in genetic engineering have been instrumental in extending the shelf life of various vegetables. Tomatoes, an economically important crop worldwide, face the issue of rapid ripening and short shelf life, prompting extensive research aimed at delaying the ripening process. Recent studies have identified several transcription factors encoded by various genes that regulate ripening (Wang et al. 2020).

Ethylene biosynthesis and DNA methylation are key regulators in the fruit maturation process. In tomatoes, the *RIN* gene is integral to controlling ripening, encoding a transcription factor that governs this developmental phase. CRISPR-Cas9 technology has been used to edit the *RIN* gene, aiming to delay ripening (Ito et al. 2015; Li et al. 2018). By disrupting *RIN*, mutations occur that slow the ripening process and extend shelf life, though this comes at the cost of decreased lycopene content (Ito et al. 2015). In addition to *RIN*, other ripening-related genes have also been targeted to achieve delayed maturation. Gao et al. (2020) explored both the downregulation and overexpression of the *NAC-NOR* gene, a member of the NAC transcription factor family. The mutant variant of this gene inhibits the expression of critical genes associated with carotenoid biosynthesis (such as *geranylgeranyl diphosphate synthase* 2), ethylene production (including *1-aminocyclopropane-1-carboxylic acid synthase* 2), and the softening of fruit tissue (e.g., *pectate lyase*). Similar to *RIN*, *NAC-NOR* functions as a positive regulator of fruit development in tomatoes. Comparisons between wild-type and mutant plants revealed either delayed or incomplete ripening in the modified varieties. Further research revealed that mutations in an alternative regulatory gene, SBP-CNR (colorless non-ripening), also contributed to delayed ripening traits (Gao et al. 2019). In an earlier study, Yu et al. (2017) used CRISPR-Cas9 to target the *ALC* gene in tomatoes, employing HDR to achieve precise genetic modifications.

This gene editing led to an extension in the shelf life. Additionally, Wang et al. (2019) conducted a study to assess the effects of gene deletion on multiple genes that inhibit tomato ripening, including *AP2a, FUL1, FUL2,* and *NAC-NOR*. Similarly, The RNA editing factor SlORRM4 plays a critical role in tomato fruit ripening by regulating mitochondrial function and RNA editing processes. Knockout of *SlORRM4* delays ripening, reduces ethylene production, and alters the expression of key ripening-related genes (Yang et al. 2017). A rather recent study demonstrates that ethylene suppression via CRISPR/Cas9 targeting the *CmACO1* gene extends the shelf-life of Japanese luxury melons (*Cucumis melo* var. reticulatus, 'Harukei-3'). By editing *CmACO1*, responsible for ethylene synthesis, the research showed a significant reduction in ethylene production and increased firmness, preserving fruit quality postharvest (Nonaka et al. 2023).

12 Enhancement of the Quality

The quality of vegetables is as important as their disease resistance, as superior quality not only broadens market potential but also increases their economic value. Taste and nutritional value are among the traits that have not been as much exploited through conventional breeding techniques. Genome editing technologies have been instrumental in enhancing various quality traits in vegetables. A notable achievement is the commercialization of the Super Tomato in Japan, the first genome-edited tomato released worldwide. This variety features increased levels of gamma-aminobutyric acid (GABA), a compound linked to improved neurological function in living organisms (Nonaka et al. 2017; Waltz 2022). Prior to this success, several experiments targeted nutritional improvements in tomatoes and other vegetables.

For instance researchers utilized Zinc Finger Nuclease (ZFN) technology to generate *L1L4* mutants in tomatoes by inactivating the *LEAFY-COTYLEDON1-LIKE4* (*L1L4*) gene, a key nuclear transcription factor (Gago et al. 2017). Although most mutants showed no significant changes, a small subset exhibited alterations in metabolic components such as phenols, oxalic acid, β-carotene, and antioxidants, when compared to wild-type tomatoes. The *L1L4* gene is believed to play a role in regulating the synthesis of seed storage proteins and fatty acids in both tomato fruits and seeds, highlighting areas for future research despite the lack of immediate practical applications. Tomato colour is a yet another trait that has been modified through multiplex gene editing to rapidly develop tomato lines with diverse fruit colors by targeting *PSY1*, *MYB12*, and *SGR1*. Through precise knockouts, researchers generated new phenotypes (yellow, brown, pink, etc.) without affecting key agronomic traits like yield and fruit quality (Yang et al. 2023). In another study, the genome of eggplant (*Solanum melongena* L.) was edited to knock out the genes responsible for browning in the flesh after cutting. Maioli et al. successfully induced knockout mutagenesis of three Polyphenol Oxidase (PPO) genes, with the mutations being stably inherited in progeny (Maioli et al. 2020).

De novo domestication is the innovative approach of using genome editing techniques, such as CRISPR-Cas9, to directly alter the genetic makeup of wild plant species, enabling the re-establishment of lost agronomic traits and the creation of new crop varieties with enhanced yield, resilience, and nutritional profiles (Zsögön et al. 2017). The study on *de novo* domestication of *Solanum pimpinellifolium* using CRISPR-Cas9 exemplifies a transformative approach to crop enhancement. By targeting six critical loci-SELF-PRUNING, OVATE, FRUIT WEIGHT 2.2, MULTIFLORA, and LYCOPENE BETA CYCLASE-the researchers achieved a threefold increase in fruit size and a tenfold increase in fruit number. Remarkably, lycopene content was boosted by 500%, underscoring the potential to restore lost traits and improve nutritional quality. This precision editing strategy not only enhances yield but also reintegrates valuable traits from wild relatives, paving the way for innovative molecular breeding (Zsögön et al. 2018).

Recent advancements in CRISPR/Cas9 technology have enabled significant modifications in the starch composition of potatoes, crucial for enhancing their nutritional benefits and industrial applications. Andersson et al. (2017) successfully employed polyethylene glycol (PEG)-mediated transformation to knockout all four *GBSS* alleles in a tetraploid potato, achieving complete amylopectin-rich tubers in 2% of regenerated lines. In another study, Toinga-Villafuerte et al. (2022) used *Agrobacterium*-mediated transformation to create mutations in the *GBSSI* gene of Yukon Gold, resulting in a total knockout of amylose in one event. Furthermore, Tuncel et al. (2019) investigated the roles of starch branching enzymes (SBEs) by targeting *SBE1* and *SBE2* genes, leading to significant structural changes in amylopectin. Takeuchi et al. (2021) also generated mutations in four alleles of the *SBE3* gene, with an 8% efficiency in inducing mutations across 126 transformed plants. These studies highlight the potential for developing potato varieties with optimized starch compositions to improve health benefits and industrial applicability.

Recently, CRISPR-Cas technology has been utilized to alter cis-regulatory elements (CREs) and cis-regulatory modules (CRMs) in plant genomes, expanding its applications beyond just gene coding regions. This approach allows for more nuanced modifications that can influence gene expression without changing the coding sequence itself, potentially leading to improved traits in plants. These regions, which include enhancers, promoters, insulators, and silencers, play essential roles in controlling gene expression by serving as binding sites for transcription factors. The ability to use CRISPR-Cas to edit or create these regulatory elements has provided valuable insights into the control of both the qualitative and quantitative plant traits (Lee et al. 2019). This emerging approach is gaining traction for its potential to accelerate crop breeding improvements (Jogam et al. 2022).

Another notable study explored CRISPR-Cas9-mediated modifications to various promoters, focusing on their effects on quantitative traits in tomatoes, including branching, fruit size, inflorescence, and plant architecture (Xu et al. 2016). The study demonstrated that mutations in the CLAVATA3 gene contributed to variability in tomato fruit traits, highlighting the potential of CRE mutagenesis for enhancing the agronomic properties of vegetable crops (Rodríguez-Leal et al. 2017).

References

Andersson M, Turesson H, Nicolia A, Fält A-S, Samuelsson M, Hofvander P (2017) Efficient targeted multiallelic mutagenesis in tetraploid potato (*Solanum tuberosum*) by transient CRISPR-Cas9 expression in protoplasts. Plant Cell Rep 36(1):117–128

Anzalone AV, Randolph PB, Davis JR, Sousa AA, Koblan LW, Levy JM, Chen PJ, Wilson C, Newby GA, Raguram A, Liu DR (2019) Search-and-replace genome editing without double-strand breaks or donor DNA. Nature 576(7785):149–157

Aune D, Giovannucci E, Boffetta P, Fadnes LT, Keum NN, Norat T, Greenwood DC, Riboli E, Vatten LJ, Tonstad S (2017) Fruit and vegetable intake and the risk of cardiovascular disease, total cancer and all-cause mortality—a systematic review and dose-response meta-analysis of prospective studies. Int J Epidemiol 46(3):1029–1056

Barden KA, Smith SS, Murakishi HH (1986) Regeneration and screening of tomato somaclones for resistance to tobacco mosaic virus. Plant Sci 45(3):209–213

Bastet A, Zafirov D, Giovinazzo N, Guyon-Debast A, Nogué F, Robaglia C, Gallois JL (2019) Mimicking natural polymorphism in eIF4E by CRISPR-Cas9 base editing is associated with resistance to potyviruses. Plant Biotechnol J 17(9):1736–1750

Bernard G, Gagneul D, dos Santos HA, Etienne A, Hilbert JL, Rambaud C (2019) Efficient genome editing using CRISPR/Cas9 technology in Chicory. Int J Mol Sci 20(5):1155

Bock C, Datlinger P, Chardon F, Coelho MA, Dong MB, Lawson KA, Lu T, Maroc L, Norman TM, Song B, Stanley G, Chen S, Garnett M, Li W, Moffat J, Qi LS, Shapiro RS, Shendure J, Weissman JS, Zhuang X (2022) High-content CRISPR screening. Nat Rev Method Primers 2(1):1–23

Bogdanove AJ, Voytas DF (2011) TAL effectors: customizable proteins for DNA targeting. Science 333(6051):1843–1846

Bolotin A, Quinquis B, Sorokin A, Dusko Ehrlich S (2005) Clustered regularly interspaced short palindrome repeats (CRISPRs) have spacers of extrachromosomal origin. Microbiology 151(Pt 8):2551–2561

Bouzroud S, Gasparini K, Hu G, Barbosa MAM, Rosa BL, Fahr M, Bendaou N, Bouzayen M, Zsögön A, Smouni A, Zouine M (2020) Down regulation and loss of Auxin response factor 4 function using CRISPR/Cas9 alters plant growth, stomatal function and improves tomato tolerance to salinity and osmotic stress. Genes 11(3)

Butler NM, Baltes NJ, Voytas DF, Douches DS (2016) Geminivirus-mediated genome editing in potato (*Solanum tuberosum* L.) using sequence-specific nucleases. Front Plant Sci 7

Cao W, Dong X, Ji J, Yang L, Fang Z, Zhuang M, Zhang Y, Lv H, Wang Y, Sun P, Liu Y, Li Z, Han F (2021) BoCER1 is essential for the synthesis of cuticular wax in cabbage (*Brassica oleracea* L. var. *capitata*). Sci Hortic 277:109801

Chandrasekaran J, Brumin M, Wolf D, Leibman D, Klap C, Pearlsman M, Sherman A, Arazi T, Gal-On A (2016) Development of broad virus resistance in non-transgenic cucumber using CRISPR/Cas9 technology. Mol Plant Pathol 17(7):1140–1153

Chen W, Qian Y, Wu X, Sun Y, Wu X, Cheng X (2014) Inhibiting replication of begomoviruses using artificial zinc finger nucleases that target viral-conserved nucleotide motif. Virus Genes 48(3):494–501

Choi SH, Ahn WS, Jie EY, Cho HS, Kim SW (2022) Development of late-bolting plants by CRISPR/Cas9-mediated genome editing from mesophyll protoplasts of lettuce. Plant Cell Rep 41(7):1627–1630

Cong L, Ran FA, Cox D, Lin S, Barretto R, Habib N, Hsu PD, Wu X, Jiang W, Marraffini LA, Zhang F (2013) Multiplex genome engineering using CRISPR/Cas systems. Science 339(6121):819–823

Danilo B, Montes É, Archambeau H, Lodé M, Rousseau-Gueutin M, Chèvre AM, Mazier M (2022) I-SceI and customized meganucleases-mediated genome editing in tomato and oilseed rape. Transgenic Res 31(1):87–105

Ding H, He J, Wu Y, Wu X, Ge C, Wang Y, Zhong S, Peiter E, Liang J, Xu W (2018) The tomato mitogen-activated protein kinase SlMPK1 is as a negative regulator of the high-temperature stress response. Plant Physiol 177(2):633–651

Entine J, Felipe MSS, Groenewald JH, Kershen DL, Lema M, McHughen A, Nepomuceno AL, Ohsawa R, Ordonio RL, Parrott WA, Quemada H, Ramage C, Slamet-Loedin I, Smyth SJ, Wray-Cahen D (2021) Regulatory approaches for genome edited agricultural plants in select countries and jurisdictions around the world. Transgenic Res 30(4):551–584

Gago C, Drosou V, Paschalidis K, Guerreiro A, Miguel G, Antunes D, Hilioti Z (2017) Targeted gene disruption coupled with metabolic screen approach to uncover the LEAFY COTYLEDON1-LIKE4 (L1L4) function in tomato fruit metabolism. Plant Cell Rep 36(7):1065–1082

Ganapathy S, Nirmalakumari A, Natesan S, Jegadeesan S, Raveendran T (2008) Isolation of macromutations and mutagenic effectiveness and efficiency in little millet varieties. World J Agri Sci 4:483–486

Gao Y, Zhu N, Zhu X, Wu M, Cai-Zhong J, Grierson D, Luo Y, Shen W, Zhong S, Fu DQ, Qu G (2019) Diversity and redundancy of the ripening regulatory networks revealed by the fruit ENCODE and the new CRISPR/Cas9 CNR and NOR mutants. Hortic Res 6(1)

Gao Y, Wei W, Fan Z, Zhao X, Zhang Y, Jing Y, Zhu B, Zhu H, Shan W, Chen J, Grierson D, Luo Y, Jemrić T, Jiang CZ, Fu DQ (2020) Re-evaluation of the nor mutation and the role of the NAC-NOR transcription factor in tomato fruit ripening. J Exp Bot 71(12):3560–3574

Goell JH, Hilton IB (2021) CRISPR/Cas-based epigenome editing: advances, applications, and clinical utility. Trends Biotechnol 39(7):678–691

Gomez MA, Lin ZD, Moll T, Chauhan RD, Hayden L, Renninger K, Beyene G, Taylor NJ, Carrington JC, Staskawicz BJ, Bart RS (2019) Simultaneous CRISPR/Cas9-mediated editing of cassava eIF4E isoforms nCBP-1 and nCBP-2 reduces cassava brown streak disease symptom severity and incidence. Plant Biotechnol J 17(2):421–434

Hong JK, Suh EJ, Park SR, Park J, Lee YH (2021) Multiplex CRISPR/Cas9 mutagenesis of BrVRN1 delays flowering time in Chinese cabbage (*Brassica rapa* L. ssp. *pekinensis*). Agriculture 11(12):1286

Hu B, Li D, Liu X, Qi J, Gao D, Zhao S, Huang S, Sun J, Yang L (2017) Engineering non-transgenic Gynoecious cucumber using an improved transformation protocol and optimized CRISPR/Cas9 system. Mol Plant 10(12):1575–1578

Hundleby PAC, Harwood WA (2019) Impacts of the EU GMO regulatory framework for plant genome editing. Food Energy Secur 8(2):e00161

Ishikawa M, Yoshida T, Matsuyama M, Kouzai Y, Kano A, Ishibashi K (2022) Tomato brown rugose fruit virus resistance generated by quadruple knockout of homologs of TOBAMOVIRUS MULTIPLICATION1 in tomato. Plant Physiol 189(2):679–686

Ishino Y, Shinagawa H, Makino K, Amemura M, Nakatura A (1987) Nucleotide sequence of the iap gene, responsible for alkaline phosphatase isozyme conversion in *Escherichia coli*, and identification of the gene product. J Bacteriol 169(12):5429–5433

Ito Y, Nishizawa-Yokoi A, Endo M, Mikami M, Toki S (2015) CRISPR/Cas9-mediated mutagenesis of the RIN locus that regulates tomato fruit ripening. Biochem Biophys Res Commun 467(1):76–82

Jansen R, van Embden JDA, Gaastra W, Schouls LM (2002) Identification of genes that are associated with DNA repeats in prokaryotes. Mol Microbiol 43(6):1565–1575

Jeong SY, Ahn H, Ryu J, Oh Y, Sivanandhan G, Won KH, Park YD, Kim JS, Kim H, Lim YP, Kim SG (2019) Generation of early-flowering Chinese cabbage (*Brassica rapa* spp. *pekinensis*) through CRISPR/Cas9-mediated genome editing. Plant Biotechnol Rep 13(5):491–499

Jinek M, Chylinski K, Fonfara I, Hauer M, Doudna JA, Charpentier E (2012) A programmable dual-RNA-guided DNA endonuclease in adaptive bacterial immunity. Science (New York, N.Y.) 337(6096):816–821

Jogam P, Sandhya D, Alok A, Peddaboina V, Allini VR, Zhang B (2022) A review on CRISPR/Cas-based epigenetic regulation in plants. Int J Biol Macromol 219:1261–1271

Jung YJ, Lee GJ, Bae S, Kang KK (2018) Reduced ethylene production in tomato fruits upon CRSPR/Cas9-mediated LeMADS-RIN mutagenesis. Hortic Sci Technol 36(3):396–405

Kampmann M (2020) CRISPR-based functional genomics for neurological disease. Nat Rev Neurol 16(9):465–480

Kathiria P, Eudes F (2014) Nucleases for genome editing in crops. Biocatal Agric Biotechnol 3(1):14–19

Kim YG, Cha J, Chandrasegaran S (1996) Hybrid restriction enzymes: zinc finger fusions to Fok I cleavage domain. Proc Natl Acad Sci USA 93(3):1156–1160

Kim Y-C, Kang Y, Yang E-Y, Cho M-C, Schafleitner R, Lee JH, Jang S (2021) Applications and major achievements of genome editing in vegetable crops: a review. Front Plant Sci 12:688980

Klimek-Chodacka M, Oleszkiewicz T, Lowder LG, Qi Y, Baranski R (2018) Efficient CRISPR/Cas9-based genome editing in carrot cells. Plant Cell Rep 37:575–586

Lawrenson T, Shorinola O, Stacey N, Li C, Østergaard L, Patron N, Uauy C, Harwood W (2015) Induction of targeted, heritable mutations in barley and *Brassica oleracea* using RNA-guided Cas9 nuclease. Genome Biol 16(1):1–13

Lee JE, Neumann Id M, Duro DI, Schmid Id M (2019) CRISPR-based tools for targeted transcriptional and epigenetic regulation in plants

Lee MH, Lee J, Choi SA, Kim YS, Koo O, Choi SH, Ahn WS, Jie EY, Kim SW (2020) Efficient genome editing using CRISPR–Cas9 RNP delivery into cabbage protoplasts via electro-transfection. Plant Biotechnol Rep 14(6):695–702

Li T, Yang X, Yu Y, Si X, Zhai X, Zhang H, Dong W, Gao C, Xu C (2018) Domestication of wild tomato is accelerated by genome editing. Nat Biotechnol 36(12):1160–1163

Li R, Liu C, Zhao R, Wang L, Chen L, Yu W, Zhang S, Sheng J, Shen L (2019) CRISPR/Cas9-mediated SlNPR1 mutagenesis reduces tomato plant drought tolerance. BMC Plant Biol 19(1):38

Lor VS, Starker CG, Voytas DF, Weiss D, Olszewski NE (2014) Targeted mutagenesis of the tomato PROCERA gene using transcription activator-like effector nucleases. Plant Physiol 166(3):1288–1291

Ma C, Liu M, Li Q, Si J, Ren X, Song H (2019a) Efficient BoPDS gene editing in cabbage by the CRISPR/Cas9 system. Hortic Plant J 5(4):164–169

Ma C, Zhu C, Zheng M, Liu M, Zhang D, Liu B, Li Q, Si J, Ren X, Song H (2019b) CRISPR/Cas9-mediated multiple gene editing in Brassica oleracea var. capitata using the endogenous tRNA-processing system. Hortic Res 6(1)

Maioli A, Gianoglio S, Moglia A, Acquadro A, Valentino D, Milani AM, Prohens J, Orzaez D, Granell A, Lanteri S, Comino C (2020) Simultaneous CRISPR/Cas9 editing of three PPO genes reduces fruit flesh robwning in *Solanum melongena* L. Front Plant Sci 11:1883

Meyer CM, Goldman IL, Grzebelus E, Krysan PJ (2022) Efficient production of transgene-free, gene-edited carrot plants via protoplast transformation. Plant Cell Rep 41(4):947–960

Murovec J, Guček K, Bohanec B, Avbelj M, Jerala R (2018) DNA-free genome editing of B*rassica oleracea* and *B. rapa* protoplasts using CRISPR-cas9 ribonucleoprotein complexes. Frontiers Plant Sci 871:1594

Nakamura M, Gao Y, Dominguez AA, Qi LS (2021) CRISPR technologies for precise epigenome editing. Nat Cell Biol 23(1):11–22

Nekrasov V, Wang C, Win J, Lanz C, Weigel D, Kamoun S (2017) Rapid generation of a transgene-free powdery mildew resistant tomato by genome deletion. Sci Rep 7(1)

Nonaka S, Arai C, Takayama M, Matsukura C, Ezura H (2017) Efficient increase of ɣ-aminobutyric acid (GABA) content in tomato fruits by targeted mutagenesis. Sci Rep 7(1):1–14

Nonaka S, Ito M, Ezura H (2023) Targeted modification of CmACO1 by CRISPR/Cas9 extends the shelf-life of *Cucumis melo* var. Reticulatus melon. Front Genome Editing 5

Odipio J, Alicai T, Ingelbrecht I, Nusinow DA, Bart R, Taylor NJ (2017) Efficient CRISPR/cas9 genome editing of phytoene desaturase in cassava. Front Plant Sci 8:1780

Ortigosa A, Gimenez-Ibanez S, Leonhardt N, Solano R (2019) Design of a bacterial speck resistant tomato by CRISPR/Cas9-mediated editing of SlJAZ2. Plant Biotechnol J 17(3):665–673

Pan C, Ye L, Qin L, Liu X, He Y, Wang J, Chen L, Lu G (2016) CRISPR/Cas9-mediated efficient and heritable targeted mutagenesis in tomato plants in the first and later generations. Sci Rep 6(1):1–9

Perroud P-F, Guyon-Debast A, Veillet F, Kermarrec M-P, Chauvin L, Chauvin J-E, Gallois J-L, Nogué F (2022) *Prime editing* in the model plant *Physcomitrium patens* and its potential in the tetraploid potato. Plant Sci 316:111162

Pickar-Oliver A, Gersbach CA (2019) The next generation of CRISPR-Cas technologies and applications. Nat Rev Mol Cell Biol 20(8):490–507

Pixley KV, Falck-Zepeda JB, Paarlberg RL, Phillips PWB, Slamet-Loedin IH, Dhugga KS, Campos H, Gutterson N (2022) Genome-edited crops for improved food security of smallholder farmers. Nat Genet 54(4):364–367

Robinson RW (2008) Rationale and methods for producing hybrid cucurbit seed. J New Seeds 1(3–4):1–47

Rodríguez-Leal D, Lemmon ZH, Man J, Bartlett ME, Lippman ZB (2017) Engineering quantitative trait variation for crop improvement by genome editing. Cell 171(2):470–480.e8

Santillán Martínez MI, Bracuto V, Koseoglou E, Appiano M, Jacobsen E, Visser RGF, Wolters AMA, Bai Y (2020) CRISPR/Cas9-targeted mutagenesis of the tomato susceptibility gene PMR4 for resistance against powdery mildew. BMC Plant Biol 20(1):1–13

Shimatani Z, Kashojiya S, Takayama M, Terada R, Arazoe T, Ishii H, Teramura H, Yamamoto T, Komatsu H, Miura K, Ezura H, Nishida K, Ariizumi T, Kondo A (2017) Targeted base editing in rice and tomato using a CRISPR-Cas9 cytidine deaminase fusion. Nat Biotechnol 35(5):441–443

Soyk S, Müller NA, Park SJ, Schmalenbach I, Jiang K, Hayama R, Zhang L, van Eck J, Jiménez-Gómez JM, Lippman ZB (2016) Variation in the flowering gene SELF PRUNING 5G promotes day-neutrality and early yield in tomato. Nat Genetics 49(1):162–168

Sun W, Wang Y (2022) SuperFi-Cas9: high Fidelity meets high activity. CRISPR J 5(2):171–173

Sun Z, Li N, Huang G, Xu J, Pan Y, Wang Z, Tang Q, Song M, Wang X (2013) Site-specific gene targeting using transcription activator-like effector (TALE)-based nuclease in *Brassica oleracea*. J Integr Plant Biol 55(11):1092–1103

Sun B, Jiang M, Zheng H, Jian Y, Huang WL, Yuan Q, Zheng AH, Chen Q, Zhang YT, Lin YX, Wang Y, Wang XR, Wang QM, Zhang F, Tang HR (2020) Color-related chlorophyll and carotenoid concentrations of Chinese kale can be altered through CRISPR/Cas9 targeted editing of the carotenoid isomerase gene BoaCRTISO. Hortic Res 7(1):161

Takeuchi A, Ohnuma M, Teramura H, Asano K, Noda T, Kusano H, Tamura K, Shimada H (2021) Creation of a potato mutant lacking the starch branching enzyme gene StSBE3 that was generated by genome editing using the CRISPR/dMac3-Cas9 system. Plant Biotechnol 38(3):345–353

Toinga-Villafuerte S, Vales MI, Awika JM, Rathore KS (2022) CRISPR/Cas9-mediated mutagenesis of the granule-bound starch synthase gene in the potato variety Yukon gold to obtain amylose-free starch in tubers. Int J Mol Sci 23(9):Article 9

Tuncel A, Corbin KR, Ahn-Jarvis J, Harris S, Hawkins E, Smedley MA, Harwood W, Warren FJ, Patron NJ, Smith AM (2019) Cas9-mediated mutagenesis of potato starch-branching enzymes generates a range of tuber starch phenotypes. Plant Biotechnol J 17(12):2259–2271

Türkoğlu A, Tosun M, Haliloglu K (2022) Evaluation of ethyl methanesulfonate-induced in vitro mutagenesis, polymorphism and genomic instability in wheat (*Triticum aestivum* L.). J Crop Sci Biotechnol 26

Ueta R, Abe C, Watanabe T, Sugano SS, Ishihara R, Ezura H, Osakabe Y, Osakabe K (2017) Rapid breeding of parthenocarpic tomato plants using CRISPR/Cas9. Sci Rep 7(1):1–8

van de Vossenberg BTLH, Dawood T, Woźny M, Botermans M (2021) First expansion of the public tomato Brown rugose fruit virus (ToBRFV) Nextstrain build; inclusion of new genomic and epidemiological data. Phyto Front™ 1(4):359–363

Vu TV, Nguyen NT, Kim J, Song YJ, Nguyen TH, Kim J-Y (2024) Optimized dicot prime editing enables heritable desired edits in tomato and Arabidopsis. Nat Plants 10(10):1502–1513

Waltz E (2022) GABA-enriched tomato is first CRISPR-edited food to enter market. Nat Biotechnol 40(1):9–11

Wang L, Chen L, Li R, Zhao R, Yang M, Sheng J, Shen L (2017) Reduced drought tolerance by CRISPR/Cas9-mediated SlMAPK3 mutagenesis in tomato plants. J Agric Food Chem 65(39):8674–8682

Wang R, Tavano EC d R, Lammers M, Martinelli AP, Angenent GC, de Maagd RA (2019) Re-evaluation of transcription factor function in tomato fruit development and ripening with CRISPR/Cas9-mutagenesis. Sci Rep 9(1):1–10

Wang R, Angenent GC, Seymour G, de Maagd RA (2020) Revisiting the role of master regulators in tomato ripening. Trends Plant Sci 25(3):291–301

Woo JW, Kim J, il Kwon S, Corvalán C, Cho SW, Kim H, Kim SG, Kim ST, Choe S, Kim JS (2015) DNA-free genome editing in plants with preassembled CRISPR-Cas9 ribonucleoproteins. Nat Biotechnol 33(11):1162–1164

Wood AJ, Lo T-W, Zeitler B, Pickle CS, Ralston EJ, Lee AH, Amora R, Miller JC, Leung E, Meng X, Zhang L, Rebar EJ, Gregory PD, Urnov FD, Meyer BJ (2011) Targeted genome editing across species using ZFNs and TALENs. Science 333(6040):307–307

Xu C, Liberatore KL, Macalister CA, Huang Z, Chu YH, Jiang K, Brooks C, Ogawa-Ohnishi M, Xiong G, Pauly M, van Eck J, Matsubayashi Y, van der Knaap E, Lippman ZB (2015) A cascade of arabinosyltransferases controls shoot meristem size in tomato. Nat Genetics 47(7):784–792

Xu C, Park SJ, van Eck J, Lippman ZB (2016) Control of inflorescence architecture in tomato by BTB/POZ transcriptional regulators. Genes Dev 30(18):2048–2061

Xu ZS, Feng K, Xiong AS (2019) CRISPR/Cas9-mediated multiply targeted mutagenesis in orange and purple carrot plants. Mol Biotechnol 61(3):191–199

Yang Y, Zhu G, Li R, Yan S, Fu D, Zhu B, Tian H, Luo Y, Zhu H (2017) The RNA editing factor SlORRM4 is required for normal fruit ripening in tomato. Plant Physiol 175(4):1690–1702

Yang T, Ali M, Lin L, Li P, He H, Zhu Q, Sun C, Wu N, Zhang X, Huang T, Li C-B, Li C, Deng L (2023) Recoloring tomato fruit by CRISPR/Cas9-mediated multiplex gene editing. Hortic Res 10(1):uhac214

Yu QH, Wang B, Li N, Tang Y, Yang S, Yang T, Xu J, Guo C, Yan P, Wang Q, Asmutola P (2017) CRISPR/Cas9-induced targeted mutagenesis and gene replacement to generate long-shelf life tomato lines. Sci Rep 7(1):1–9

Yu X, Yu J, Lu Y, Li W, Huo G, Zhang J, Li Y, Zhao J, Li J (2024) An efficient and universal protoplast-based transient gene expression system for genome editing in *Brassica* crops. Hortic Plant J 10(4):983–994

Zafirov D, Giovinazzo N, Bastet A, Gallois JL (2021) When a knockout is an Achilles' heel: resistance to one potyvirus species triggers hypersusceptibility to another one in *Arabidopsis thaliana*. Mol Plant Pathol 22(3):334–347

Zhan X, Zhang F, Zhong Z, Chen R, Wang Y, Chang L, Bock R, Nie B, Zhang J (2019) Generation of virus-resistant potato plants by RNA genome targeting. Plant Biotechnol J 17(9):1814–1822

Zhang H, Si X, Ji X, Fan R, Liu J, Chen K, Wang D, Gao C (2018) Genome editing of upstream open reading frames enables translational control in plants. Nat Biotechnol 36(9):894–898

Zhang H, Qin C, An C, Zheng X, Wen S, Chen W, Liu X, Lv Z, Yang P, Xu W, Gao W, Wu Y (2021) Application of the CRISPR/Cas9-based gene editing technique in basic research, diagnosis, and therapy of cancer. Mol Cancer 20(1):1–22

Zhu H, Li C, Gao C (2020) Applications of CRISPR–Cas in agriculture and plant biotechnology. Nat Rev Mol Cell Biol 21(11):661–677

Zsögön A, Cermak T, Voytas D, Peres LEP (2017) Genome editing as a tool to achieve the crop ideotype and de novo domestication of wild relatives: case study in tomato. Plant Sci Int J Exp Plant Biol 256:120–130

Zsögön A, Čermák T, Naves ER, Notini MM, Edel KH, Weinl S, Freschi L, Voytas DF, Kudla J, Peres LEP (2018) De novo domestication of wild tomato using genome editing. Nat Biotechnol 36(12):1211–1216

QTL Mapping for Drought Tolerance in Vegetables

Muhammad Farhan Yousaf, Muhammad Naeem, and Muhammad Abu Bakar Zia

Abstract Vegetables, which contain minerals and vitamins, are a major source of nutrition for humans, and are negatively impacted by a variety of abiotic stresses, particularly drought. This is a constraint that must be overcome to avoid yield losses while maintaining nutritional and global food security. Generation of drought tolerant vegetable varieties is an important and time-consuming challenge for crop breeders. Though conventional crop breeding methods have been used to their full potential for the development of drought tolerant varieties, their applications are limited in crop improvement due to time lag, environmental dependency, and low efficiency. With the introduction of low-cost advanced next-generation sequencing technology, plant breeders can now better understand the drought stress-tolerance mechanisms and associated genomic regions, thanks to QTL identification and mapping. This chapter will emphasize the importance and current status of QTL mapping in producing drought stress-tolerance crops in the near future.

Keywords Vegetables · Drought · QTL mapping · NGS

M. Farhan Yousaf (✉)
Department of Agricultural Genetic Engineering, Niğde Omer Halisdemir University, Niğde, Türkiye

Center for Quantitative Genetics and Genomics, Aarhus University, Aarhus, Denmark

Research and Development, EUROPLANT Innovation, GmbH & Co. KG, Lüneburg, Germany

M. Naeem
Agriculture Extension and Adaptive Research Wing, Punjab Agriculture Department, Lahore, Pakistan

M. A. Bakar Zia
Department of Plant Breeding and Genetics, Faculty of Agricultural Science and Technology, University of Layyah, Layyah, Pakistan

U. K. Chaudhry et al. (eds.), *Drought Stress*,
https://doi.org/10.1007/978-3-031-80610-0_14

1 Introduction

1.1 Importance of Vegetables

Vegetables are essential for healthy living and are an important part of a meal that can potentially increase energy levels. Carrot, sweet pepper, cabbage, cauliflower, lettuce, spinach, potato, tomato, onion, bottle gourd, and reddish provides essential minerals, vitamins, dietary fiber, and phytochemicals, while also lower the risk of dangerous diseases and other medical conditions. The interesting thing in vegetables are, their stems, leaves, fruit, flowers, roots, and seed are all important components and can be consumed in raw or cooked conditions. As cofactors of metabolic reactions, gene regulators, coenzymes and radical scavenging molecules, vegetables also help to maintain proper metabolic functions in cells and tissues (Bouchenak and Lamri-Senhadji 2013; Septembre-Malaterre et al. 2018). To alleviate poverty and malnutrition, it is necessary to increase vegetable consumption and introduce new varieties that are affordable to low-income people groups. Vegetables are grown around the world, and according to a global vegetable survey, 392 vegetable crops are grown globally. Asia produces the most vegetables, including, eggplants, spinach, lettuce, cucumbers, tomatoes, and peppers, while Europe produces the most chicory. In terms of per capita production, Asia is the leading producer of peppers, chilli, spinach, cucumbers, and eggplants, North America is the leading producer of lettuce, tomatoes, and sweet peppers, and Europe is the leading producer of chicory (FAOSTAT 2021).

1.2 Effect of Drought on Vegetable Crops

Most of the vegetables have small root systems and succulent nature, making them vulnerable to abiotic stress such as drought, which ultimately limits the adaptability and productivity (Table 1), susceptible to insects, weeds and diseases which significantly increases the losses (Sekhon et al. 2019).

Drought is caused by inadequate rainfall and/or changing rainfall patterns, resulting in increased evapotranspiration that affects photosynthetic kinetics, antioxidant production, protein levels, hormone composition, osmotic adjustment, root depth and extension, photosynthesis inhibition, stomata closing and opening, chlorophyll content decrease, transpiration decrease, and growth inhibition (Lawlor and Cornic 2002; Szegletes et al. 2000; Yordanov et al. 2000; Zhu 2002) (Fig. 1).

Drought can also cause pollen sterility, and abscisic acid accumulation. Additionally, rapid dehydration causes oxidative stress which impairs energy dissipation capacity and metabolic protection against reactive oxygen species (ROS) all of which are essential for survival in drought conditions and maintains crop yields or productivity (Ort 2001; Chaves and Oliveira 2004; Nahar and Ullah 2011).

Table 1 Drought stress at critical stages and its impact on vegetable crops

Vegetables	Critical stages	Consequences of a water shortage (Kemble and Sanders 2000)
Potato	Induction and enlargement of Potato	Splitting, poor tuber growth and yield
Tomato	Flowering, Fruit enlargement	Shedding of flower, fertilization problem, fruit size reduced and splitting
Chilli	Flowering, Fruit set	Shedding of flower, production of dry matter reduced, less nutrient uptake
Eggplant	Flowering, Fruit growth and development	Poor colour of fruit, yield reduces
Onion	Formation of bulb and enlargement	Doubling and splitting of bulb, poor storage life
Cauliflower and cabbage	Formation and enlargement/ development of head	Cauliflower browning and buttoning; cabbage tip burning and head splitting
Cucumber	Flowering and fruit setting	Pollen grains deformed and non-viable, bitterness and fruit deformity
Carrot, turnip, and radish	Root enlargement	Distorted, rough, and grow slowly or poorly, in carrots strong and pungent odour, build-up of harmful nitrates in the roots
Okra	Flowering, pod growth and development	Significant yield loss, fibre development, and mite infestation
Pea	Flowering and pod filling	reduced plant growth and root nodulation, along with inadequate grain fill
Melons	Flowering, fruit development that is consistent throughout	A decrease in total soluble sugars, a reduction in sugar and ascorbic acid, and an increase in nitrate content in watermelon fruit all contribute to the poor quality of muskmelons
Lettuce	Throughout the development process	Leaf toughness, poor plant growth, and tip burning
Leafy vegetables	Entire life of plant's growth and development	Leaf hardiness, poor vegetation/foliage growth, and accumulation of nitrate

Drastic dehydration tolerance is uncommon in most of the vegetable crops, but it is found in species that are native to dry environments (Ingram and Bartels 1996). When plants are exposed to an abiotic stress, such as drought, a variety of physiological, metabolic, and defence systems are activated to allow plants to survive and maintain growth and productivity (Fig. 2). Plants may avoid drought stress by shortening their life cycle when water is scarce (Chaves and Oliveira 2004; Nahar and Ullah 2011). Vegetables response to drought stress, affect by different factors includes, genotype, duration and stress severity, stage of growth, physiological process of growth, the different activity of respiration patterns and photosynthesis activity, different genes expression patterns, and environmental factors (Anjum et al. 2017; Eftekhari et al. 2017; El-Moneim et al. 2020; Itam et al. 2020).

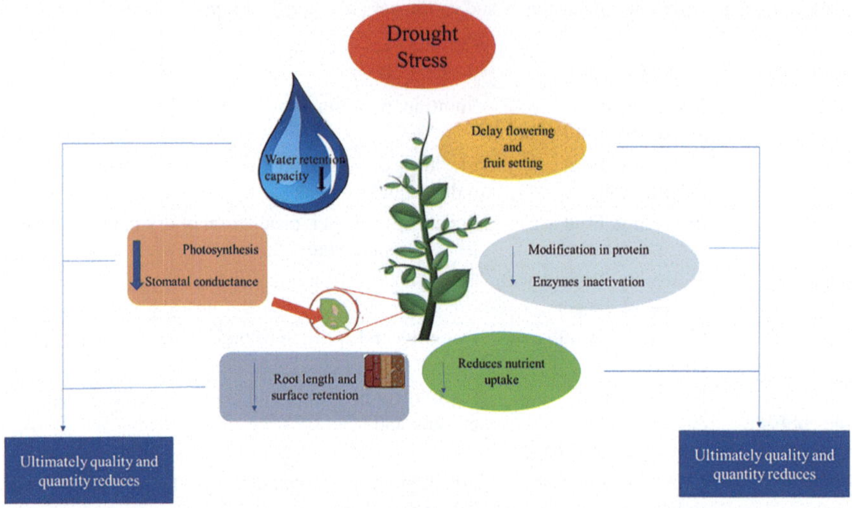

Fig. 1 Drought stress has a number of negative effects on vegetable crops, including a decrease in photosynthetic rate, nutrient uptake, and protein modification. Drought stress shortens root length and causes enzyme inactivation

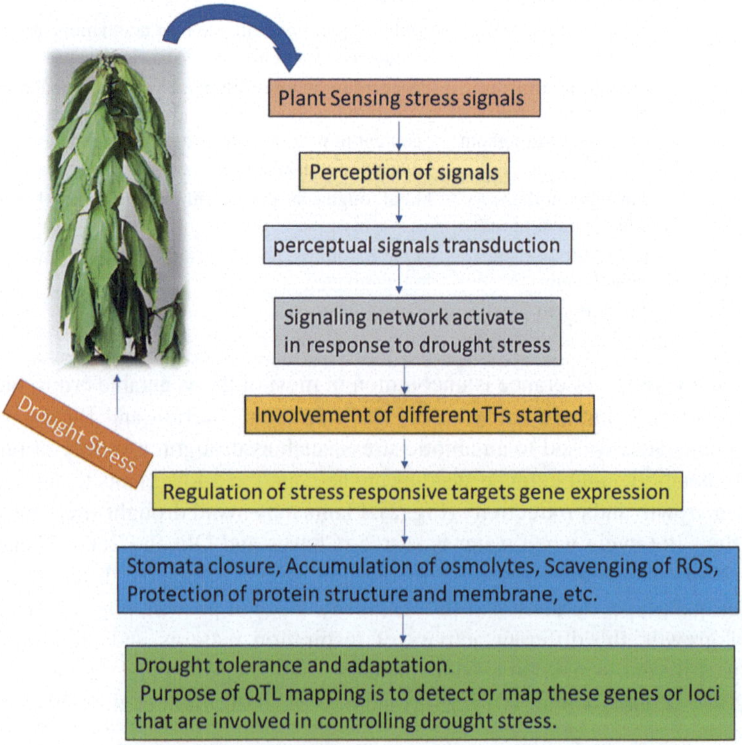

Fig. 2 The figure depicts drought stress tolerance, beginning with the perception of drought stress and progressing through transcription factors to the activation of genes involved in adaptation

1.3 Breeding of Drought Tolerant Varieties

The lack of high-yielding genotypes suitable for cultivation under drought conditions and the severity of droughts increasing globally are the main factors limiting vegetable production. Drought is a key challenge for researchers and breeders due to its complex genetics, and the associated traits are quantitative (controlled by minor genes/polygenic), which hinders in development of drought-tolerant cultivars. Conventional breeding techniques such as bulk, pedigree, single seed descent, mutation breeding, and backcrossing relied primarily on selection, using natural processes of sexual and asexual reproduction and contribute to the improvement of vegetable traits such as quality, nutrition, and yield while putting less emphasis on abiotic stresses. With traditional breeding methods, it is also possible to skip or delay the time of interested trait in cultivar development due to extensive field trials, obstacles to distant hybridization, time-consuming/lengthy screening processes, and environmental dependence (Fig. 3). Due to these shortcomings, old-fashioned breeding is insufficient for plant genome enhancement or genetic improvement which is necessary for the development of new varieties that can withstand drought (Ahmar et al. 2020; Sekhon et al. 2019).

With recent advances in genetics, successful production of new varieties has been accelerated, and pendulum has been shifted from traditional to modern methods, resulting in a new chapter of vegetable research and the discovery of new avenues (Aditika and Priyanka 2017; Dhaliwal et al. 2020). For many crop species,

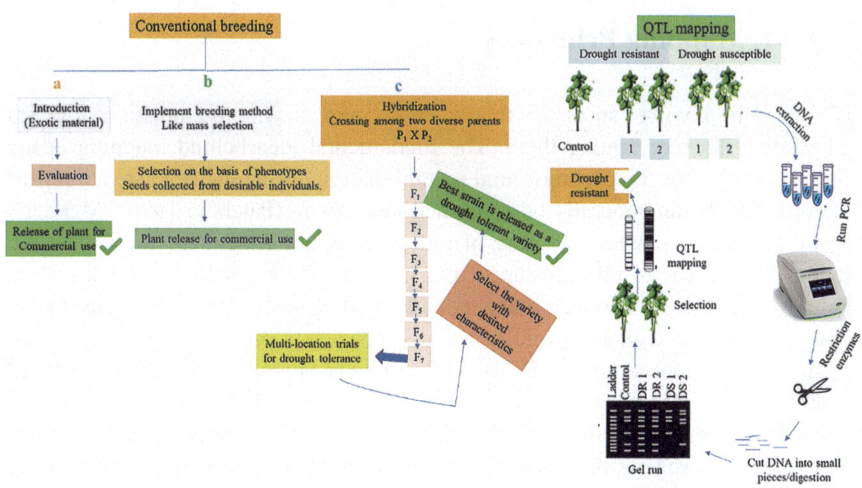

Fig. 3 Several conventional breeding techniques are depicted in the diagram, leading to the development of drought-tolerant genotypes in vegetables. Plant breeding methods such as introduction, hybridization, and mass selection are widely used. Modifying these methods has the potential to improve vegetables growth and yield under drought conditions (Left panel) Flow chart showing the mechanism of progeny screening for drought stress resistant using DNA based markers (Right panel)

including several vegetable species, marker assays, molecular markers, genetic linkage maps, and whole-genome sequence have been developed and published. (Singh 2007; Khan et al. 2015; Diouf et al. 2018; Thant et al. 2021). Molecular technologies create molecular markers that might be helpful in locating quantitative trait loci (QTLs) (Fig. 3). The ability to identify genetic variation in the germplasm pool for nearly any crop species has been greatly enhanced by the development of molecular markers, which are based on polymorphisms found in any given DNA sample. In contrast to traditional breeding, it provides advantages like as, the availability of easy assay, reproducibility, convenience in usage, stability regardless of environmental or external factors, and representation across the entire genomes. Furthermore, these molecular markers are used to create a QTL map. Quantitative trait loci analysis enables the discovery and evaluation of the genetic factors influencing any trait, including drought tolerance, through the investigation of marker genotype segregation and phenotypic values of individuals or lines. Depending on the resources at hand and the degree of complexity in the genetics and breeding of the species, QTL mapping can be used for precision breeding for each vegetable crop. Additionally, there are more QTLs, candidate genes, and genes linked to drought resistance, which could speed up or accelerate the genetic improvement of vegetable crops (Jacob et al. 2016).

This chapter describes the current state of QTL mapping technique in relation to drought stress in vegetables and how these helpful tools might be applied to enhance tolerance against significant constraints affecting vegetable crops.

2 QTL Mapping Principles

QTL mapping involves strategies that creates a map showing the position of genes and relative distance among them. The fundamental idea behind mapping is the segregation of genes by chromosomal recombination during meiosis (sexual reproduction), which can generally be examined in progeny (Paterson 1996). Mendel's second law, also known as the law of independent assortment, is implied by the random assortment of chromosomes into gametes during meiosis, which causes alleles of one gene to segregate independently of alleles of another gene. This law is definite within the scenario while genes are exclusively present on different chromosomes and might segregate freely and no longer be applicable when genes are present on identical chromosome as those are connected collectively and vice versa. Genes that are close together or tightly linked will be transmitted more frequently from parent to progeny than genes that are far apart. Homologous chromosomes may move over and alternate segments forming chiasma in early meiosis. This complete manner is known as recombination as it includes rearrangement of genetic information by using interplay of DNA molecules. Linked genes have a recombination frequency of less than 50%, whereas genes on different chromosomes assort independently (unlinked) and have a 50% recombination frequency (Hartl 1988). The distance between two genes affects the likelihood that a crossover will lead to gene recombination. The closer two markers are to one another on a chromosome, the less frequently they combine (In contrast, the farther apart two markers are located on a chromosome, the higher the frequency of recombination between them).

2.1 *Pre-requisite for QTL Mapping*

2.1.1 Mapping Population

Population is taken via controlled crosses and that creates gene map. The first step in mapping population is to select divergent parental traces with polymorphic developments. Screening genotypes panel for phenotype and to become áware of the extremes of the phenotypic distribution earlier than selecting the dad and mom of a mapping populace is really helpful. During linkage evaluation, it is important to include high heritability while avoiding motive sterility or high segregation distortion. The duplicate mode of plant is the second feature considered at some point in the mapping population process. Mapping populations in self-pollinating species arise from parents, each of them is quite homozygous (inbred). Yield reduction and suppression of wide cross can be interrupted via chromosomal pairing and recombination price (Singh and Singh 2015). The situation is more complicated in cross pollinating (outcrossing) species because most of these species no longer tolerate inbreeding. Self-incompatible plants exhibit high levels of genetic heterozygosity. Because of inbreeding depression, it is frequently no longer possible to produce natural traces; typically, only self-compatible plant life allows the generation of traces with the highest degree of homozygosity. Finally, the availability of plant material influences the selection of a mapping populace. Other factors include the amount of time available for population development and the mapping resolution required.

QTL mapping is performed by using 2nd filial generation (F_2), Recombinant Inbred Lines (RILs), Near Isogenic Lines (NILs), Backcross (BC) and Double Haploids (DHs) (Burr et al. 1988; Doerge 2002). Backcrossing populations are created by crossing F1 back into one of the parents, whereas F2 populations are created by selfing F1 hybrids created by mating the two diverse parents. Single-seed selections from individual plants within an F_2 population are used to produce RILs; these selections last for six to eight generations. More than 99% of the genome from BC_6 and above will be derived from recurrent parent if backcross selection is repeated for at least six generations (Babu et al. 2004). Selected BC_7F_1 individuals can be selfed to create BC_7F_2 lines that are homozygous for the target gene and are supposedly nearly isogenic with the recipient parent (NILs). DHs are commonly generated through duplicating the chromosomes via the utility of colchicine in anther subculture or pollen culture. All plant's seeds are harvested separately and maintained as DH traces containing identical chromosomal sets in every cell. DH lines are much appropriate for mapping population as these represent permanent resource for mapping, perfect for crossing pattern and no residual heterozygosity. Examples of their use in vegetables are found in Pink et al. (2008), Badu et al. (2017), and Singh et al. (2021). Because they are homozygous or true breeding lines that can be produced and multiplied without causing genetic change, DHs, RILs, and NILs are regarded as permanent populations. Although DH populations can only be produced in species with established haploid tissue culture protocols, they can be produced more quickly than RILs and NILs. Developing RILs and NILs populations takes a long time, which is a major drawback for mapping studies. In practice, initial genetic mapping studies employed populations of 50–250 individuals (Mohan et al. 1997), but high-resolution fine mapping necessitates a larger population size. Finally, population selection consists of excessive significance in creation of linkage map.

2.1.2 Selection of Molecular Markers for Mapping

Genetic diversity in conventional plant breeding was identified using observational selection. However, with the development of molecular biology, this work is now decided at the molecular level based on DNA changes and their effects on the phenotype. After the plant's DNA has been extracted, the samples are examined for changes using PCR or hybridization, and then different molecules are identified based on their size, chemical composition, or charges using agarose or polyacrylamide gel electrophoresis (Bernardo 2008; Jena and Mackill 2008). Genetic markers are organic substances that can be used as experimental probes or labels to follow an individual, cell, nucleus, tissue, gene, or chromosome through allele variations. In classical genetics, genetic polymorphism serves as a proxy for allelic diversity, whereas in modern genetics, genetic polymorphism refers to the relative variation among genetic loci in the genome. To better understand heredity and variation, genetic markers can be used (Avise 2012) PCR-based molecular markers and hybridization-based molecular markers are the two main categories of DNA markers. DNA markers are composed of multiple sets of markers. The restriction fragment length polymorphism (RFLP) is a molecular marker based on hybridization, whereas the other molecular markers, such as Simple Sequence Repeats (SSRs) or microsatellites, Cleaved Amplified Polymorphic Sequence (CAPS), Expressed Sequence Tags (ESTs), Amplified Fragment Length Polymorphisms (AFLPs), Random Amplified Polymorphic DNA (RAPD), that paved the way to construct mapping in plants. RFLP markers, the most well-known genetic markers at the time, were primarily used in the first large-scale attempts to produce genetic maps (Helentjaris et al. 1986). Codominant inheritance, good inter-laboratory transferability, locus-specificity that enables studies of synteny (conserved gene order between related organisms), and high reproducibility are among the major strengths of RFLP markers. The low level of polymorphism, lack of automation, time-consuming nature, and frequent requirement for radioactively labelled probes are some of the disadvantages of RFLP analysis. Other disadvantages include the high quantity and quality of DNA requirements, reliance on the creation of species-specific probe libraries, and dependence on human labor. The strategy for linkage mapping significantly changed when PCR-based markers were developed, though microsatellite markers are still used to create linkage maps today. SSRs have numerous advantages, including co-dominant inheritance, high information content, repeatability, high transferability, locus specificity, and ease/simplicity of automation for high throughput screening (Powell et al. 1996; Gupta et al. 1999). However, their use has been restricted to a small number of agriculturally significant crops due to the high development cost and effort required to generate functional SSR primers for a specific species (Squirrell et al. 2003). Single nucleotide polymorphism (SNP) markers are regarded as excellent for genetic mapping and diversity evaluation in crop plants due to their great abundance and relatively even distribution across the genome (Chagné et al. 2007). Additionally, there are numerous tools for analysing SNP loci, many of which can be easily automated for allele calling and data collection. SNP detection has evolved into a preferred marker technique due to the abundance of source polymorphisms and the ease with which allele calls can be automated.

2.1.3 Phenotyping of Mapping Population

The quantitative characteristics that are to be assessed must be as precise as possible. Dates shouldn't be missing, but small amounts of missing information are acceptable. The population's missing data lowers the effective sample size and weakens the power of QTL mapping. The data is pooled across location and replication to produce a single quantitative value for the line. It is also necessary to measure the target traits in experiments carried out in various locations in order to better understand the QTL x Environment (QTLxE) interaction.

2.2 QTL Identification

QTL mapping's main objective is to find QTL while reducing the likelihood of false positives (Type I error), or associations between a marker and a QTL that isn't real. The following methods are frequently used to perform QTL or trait association tests: (1) Single marker approach, (2) Simple interval mapping, (3) Composite interval mapping, and (4) Multiple interval mapping. Because the QTL has a significant impact and can be introgressed using marker-assisted back crossing (MABB), plant breeders may not need to be aware of its precise location (Stevens et al. 2007). The next section discusses how discovered QTLs for various traits in different vegetable crops have been used to improve crops, particularly to boost yield in elite lines that are drought tolerant.

3 Utility and Status of QTL Mapping Against Drought Stress in Various Vegetable Families

3.1 Solanaceae *Family*

3.1.1 Tomato (*Solanum lycopersicum*)

In cultivated tomato (*Solanum lycopersicum*), QTLs pertaining to chemical and physical properties of fruit quality have been mapped previously (Causse et al. 2001; Pascual et al. 2016). But little attention was paid to their interaction with the abiotic factors especially drought stress. First QTL mapping study involving the interaction of watering regime and fruit quality characteristics were examined by Albert et al. (2016a). 119 RILs were generated from a cross between two tomato accessions Cervil and Levovil. These were genotyped by 679 polymorphic SNP markers resulting in the identification of 56 QTLs corresponding to 19 plant and fruit traits under drought stress, colocalized on chromosomes 2, 3, 4, 6 and 11. Five major QTLs related to fruit weight (FW) were *fw2.1, fw2.3, fw2.2, fw3.2, fw11.2*, and *fw11.3*. In addition to this an antagonistic QTL (QTL having opposite allelic effects under control and drought treatments) were also identified, one such associated with pH was mapped on 38 Mbp on chromosome 6. This QTL lies in close proximity to the two putative malate transporters (*Solyc06g072910* and *Solyc06g072920*) (Sauvage et al. 2014). QTLs related to fruit weight holds primary significance under drought conditions. Large fruit weight suffers great losses under limited water situation but momentarily possesses more soluble solids and dry matter content. So, in the scenario of slight water deficit environment, to compensate yield losses it is advisable to grow large fruits for improved tomato flavor (Dorais et al. 2008). The antagonistic QTLs associated with soluble solids, dry matter content (*Y03_64701243*), pH, stem diameter (*Y04_63370382*), and fruit number were positioned on chromosomes 3, 4 and 6, respectively. These QTLs were best candidates for marker assisted selection (MAS) to build genotypes specifically for improved growth and quality traits under drought conditions (Albert et al. 2016a).

Three constitutive QTLs (identified in both control and drought watering regimes) positioned on chromosomes 6, 7 and 10, namely *MalicDM.Avi_6.3;MalicFM.Avi_6*.3, *GlucoseDM.Avi_7.1;MalicDM.Avi_7.2; MalicFM.Avi_7.2,* and *FructoseDM.Avi_10.2* were identified by Albert et al. (2016b) and Pascual et al. (2016). These constitutive QTLs were linked to tomato fruit quality traits like vitamin C, sugars, and acid contents with a putative functional role in carbon metabolism and malate compartmentation, sugar metabolism, and citric and malic acid accumulation (Guillet et al. 2002; Li et al. 2012; Sauvage et al. 2014). Apart from constitutive QTLs, 15 interactive (through interaction tests) and special QTLs (QTLs specific to either control or drought regime) were also detected. It includes *CitricDM.Avi_1.1, VitCDM.Avi_1.1, SSC_Avi_1.2, FructoseDM.Avi_1.1, SSC.Avi_2.2, pH.Avi_2.3, FructoseDM.Avi_4.1, MalicDM.Avi_4.1, GlucoseFM.Avi_6.1, FructoseFM.Avi_6.1, VitCFM.Avi_7.1, SSC.Avi_9.1, VitCFM.Avi_10.1,*

FructoseDM.Avi_10.1, and *FructoseFM.Avi_11.3*. These QTLs carry candidate genes related to solutes and water transport (aquaporins and other transporters), conservation of protein (chaperone; heat/cold shock proteins), hormonal signaling (auxin, ethylene, and gibberellin), and sugar metabolism (invertases and sucrose phosphate synthase), and play a significant role in water deficit response (Wang et al. 2003; Shinozaki and Yamaguchi-Shinozaki 2007). *Lin5* and *Lin 7* encodes for cell wall invertase located on chromosome 9 (*SSC.Avi_9.1*) involved in the transport of glucose and fructose into the cell after sucrose cleavage. Drought modifies interconnection between source and sink while invertases are important in integration, amplification, and regulation of source-sink relationships (Albert et al. 2016b). MAS utilizing these QTLs/genes will help to improve drought tolerance in tomato genotypes.

In near past, MAGIC (Multi parent Advanced Generation Intercross) population (having balanced allelic frequency) comprising of 250 tomato lines derived from 8 parental lines was utilized by Diouf et al. (2018) to detect QTLs associated with seven phenological and fruit characteristics under water deficit conditions. A genetic map was constructed using 1345 polymorphic SNPs covering 84% genome and spanned over 2156 cM (Pascual et al. 2016). Similar to Albert et al. (2016b), *Fw2.2* positioned on chromosome 2 was also reported by Diouf et al. (2018), indicating the utility of this genomic region for tomato breeding programs. Among the 54 QTLs identified by Diouf et al. (2018), 35 were treatment specific. QTLs pertaining to fruit quality traits like fruit weight and soluble solids are polygenic in nature, thus localized on chromosomes 6 and 7, carrying ten or more QTLs, respectively. QTLs responsible for the observed phenotypic variation depends upon environment i.e., QTL × E interaction. Out of 35 QTLs, only two QTLs were stable across environments along with the identification of 1/3 QTLs in at least two environments. It suggests that breeding strategy should be based on targeted environment to develop stable QTLs for achieving better results in that particular environment. Moreover, wild plant relatives contain huge gene pool of resistant germplasm against abiotic stress, as observed in *Solanum pennellii* drought resistant introgression lines. These lines have adapted to enhance antioxidants such as lycopene under drought stress situations to scavenge the free radicals. QTL mapping investigation involving a cross between wild tomato *S. pennellii* (LA716) and processing tomato variety (M82) resulted in the development of introgression lines (IL) which allows for the identification of QTLs associated with drought resistant. An insight into one such study reported IL 2-5, IL 3-5 and IL 7-1 to be resistant under deficit irrigation, by depicting enhanced levels of lycopene (Dariva et al. 2021).

3.1.2 Potato (*Solanum tuberosum*)

Potato, being a shallow rooted crop (85% root concentrated in upper 0.3 m soil) is more prone to scarce water conditions. Monneveux et al. (2013) reviewed about drought tolerance in potato and listed genes involved in alleviating drought stress in transgenic potato plants. Early studies on drought tolerance related QTLs were carried out by Anithakumari et al. (2011, 2012). Illumina GoldenGate genotyping array was used in these studies. This array is enriched in markers present within the genes involved in abiotic tolerance. 23 and 47 QTLs were identified pertaining to drought tolerance in potato carried out under greenhouse and PEG induced *in-vitro* environment, respectively (Anithakumari et al. 2011, 2012). Both studies utilized diploid potatoes including wild progenitors for QTL mapping. *Solanum phureja* is notorious diploid potato landrace that offers tolerance against abiotic stresses. A backcross biparental diploid population generated from an interspecific cross between *Solanum phureja* and *Solanum tuberosum* Group *Andigenum* along with genetic map (Sharma et al. 2013) was used for QTL analysis. This robust map identified potential QTLs associated with drought tolerance in potato. Drought specific QTLs linked to plant height, SPAD, tuber weight and biomass were located on chromosomes 3 (69.3 cM), 5 (24.5 cM), 8 (30.58 cM), and 12 (57.98 cM) (Khan et al. 2015). Chromosome 5 and 8 are considered hotspots for maturity, vigor, biomass and tuber traits under water deficit conditions in many studies (Visker et al. 2003; Anithakumari et al. 2011).

Apart from diploid potatoes, tetraploid germplasm was also used for QTL analysis. Tetraploid potato, Albatros (drought tolerant) and Ramses (drought sensitive) were crossed to obtain F_1 population, which was mapped with 73 SSRs and 505 AFLPs. Eight genomic regions conferring drought tolerance were identified and majority of them were linked with drought tolerance index DRYM i.e., deviation of relative starch yield from experimental median. The DRYM QTL contains candidate genes involved in abscisic acid, brassinosteroids, ethylene signaling pathways, jasmonic acid and enzymes relevant to cell wall remodeling. Enhanced cell membrane stability under drought stress conditions is one of the regulatory mechanisms in drought tolerant cultivars (Schumacher et al. 2021). These results suggests that potato breeding clones can be screened based on these metabolites (discussed earlier).

In order to accelerate the process of selection for drought tolerance characteristics, it is essential to utilize genome wide high density molecular markers to dissect genomic regions linked to drought tolerance. High confidence 69,011 SNPs were determined by Hamilton et al. (2011) that leads to the development of SolCAP arrays. These SNP arrays provide a great source to identify casual polymorphisms associated with water deficit conditions. 20 K SNP array developed by Vos et al. (2015) has been utilized for mapping of QTLs related to tuber size distribution, and marketable yield in drought prone conditions by using 14 K Infinium SNP array. A major SNP (PotVar0030768) controlling tuber size distribution was present on chromosome 3 at 55.65 Mbp having additive allelic effect (bearing A/G polymorphism) (Aliche et al. 2019). Under drought situations, it is recommended to grow genotypes yielding fewer tubers since less tubers will be more likely to be bulked thus resulting in enhancing the average size of tubers. It is the most effective when drought persists at later growth stages of growing season. Identification of genetic factors controlling mean tuber size affects marketable yield in potato was done by Harris (2012). Another important feature influencing drought is maturity type of potato cultivars, late maturity types will be more likely to bulk their tubers compared to early maturity group. Delayed senescence provides extended period of time for the translocation of photo-assimilates thereby higher yields and wider tuber size distribution under drought conditions. Aliche et al. (2019) were of the view that genomic areas on chromosome 3 control both the spread of tuber size and marketable yield in potato under water limiting conditions. Literature suggests that the casual SNP (PotVar0030768) was present within the candidate gene *PRR* (pentatricopeptide repeat-containing protein) (Ichinose and Sugita 2016; Schmitz-Linneweber and Small 2008). In Arabidopsis, this gene family is known to be involved in abiotic tolerance including drought (Jiang et al. 2015; Sharma and Pandey 2016). Another antioxidant metabolite which played crucial role in drought tolerance of tuber crops is proline. Besides serving as a drought marker, it acts as a key osmolyte maintaining osmotic balance, and minimizing the damage caused by ROS species by balancing $NADPH/NADP^+$ and GSH/GSSG under stress conditions (Furlan et al. 2020). Two significant QTLs (*solcap_snp_c1_13430* and *solcap_snp_c1_10750*) associated with proline concentration were positioned on chromosomes 1 (60.98 Mbp) and 4 (56.63 Mbp), respectively. These QTLs carry candidate genes involved in starch and glycogen hydrolysis along with biosynthesis of steroid and triterpenoid (Tagliotti et al. 2021). Marker assisted selection utilizing the significantly associated genomic regions will help to improve the drought tolerance in potato.

3.1.3 Eggplant (*Solanum melongena*)

Eggplant or brinjal is the third most widely grown vegetable (54 million Mt. production) in the *Solanaceae* family after potato and tomato (FAOSTAT 2022). Due to its sheer importance as food security crop, it is included in the Annex 1of the International Treaty on Plant Genetic Resources for Food and Agriculture (Fowler et al. 2003). First study pertaining to the genetic map of eggplant was published by Doganlar et al. (2002). The researchers designed RFLP based genetic map in F_2 population of *S. melongena* × *S. linneanum.* Improvements in genetic map was done by the inclusion of COSII (conserved ortholog set) markers (Wu et al. 2009). This genetic map was used to locate QTLs that control certain morphological characteristics in eggplant (Frary et al. 2014). First draft genome of eggplant was released in 2014, covering 833.1 Mbp (74% eggplant genome) (Hirakawa et al. 2014). Further improvements in the genome sequence were made by combining both optical mapping and Illumina sequencing approaches (https://solgenomics.net/organism/Solanum_melongena/genome). A most recent high quality genome assembly for eggplant was published by Wei et al. (2020) using inbred line HQ-1315, and several next generation sequencing platforms such as Nanopore, Illumina, Hi-C technology, 10X genomics sequencing technologies.

The availability of high-quality chromosome level genome sequence has fostered genetic mapping studies in eggplant. Several QTLs controlling male sterility (Khan and Isshiki 2016), resistance to *Ralstonia solanacearum* phylotype I (Salgon et al. 2017), and eggplant prickles (Qian et al. 2021) were identified. As drought severely impairs eggplant productivity leading to the reduction in plant height, dry matter and fruit yield, there is dire need of drought tolerant eggplant varieties. Wild eggplant relatives offer great source of drought tolerant germplasm. *S. incanaum* is a close wild relative of *S. melongena* having high amounts of bioactive phenolics and also offer drought tolerance. Mapping and transcriptomic analysis of *S. incanum* along with *S. aethiopicum* (cross compatible specie of *S. melongena)* revealed 36 single nucleotide variations (SNVs) and 976 validated SSRs. These SNVs and SSRs will be helpful for eggplant breeding program to attain drought tolerance (Gramazio et al. 2016). Backcrossing of *S. melongena* (recipient parent) and *S. incanum* (donor parent) resulted in 16 introgression lines advanced for QTL mapping analysis. Ten stable QTLs were detected on seven chromosomes (1, 2, 3, 4, 5, 8, and 12). A major QTL controlling fruit weight (*fw2*) was positioned chromosome 2. Since, introgression lines were used as a mapping population derived from *S. incanum*, these QTLs were equally important for drought tolerance in eggplant (Mangino et al. 2020). Characterization of eggplant genotypes based on drought tolerance utilizing SSR markers are available in the literature (Thant et al. 2021), but QTL analysis linked to drought tolerance is still at infant stage in eggplant.

3.2 Amaryllidaceae *Family*

3.2.1 Onion (*Allium cepa*)

Onion is widely grown vegetable in temperate, tropical, and boreal regions around the globe. It has culinary applications and nutritional significance. Despite its importance, genomic research on onion is limited due to high inbreeding depression, large genome size (16 Gbp), biennial life cycle, laborious breeding systems, and cross-pollinating nature (McCallum et al. 2006; Duangjit et al. 2013). Therefore, QTL mapping studies for drought tolerance in onion is limited despite of the fact that it seriously affects onion productivity and bulb weight. Onion bulbs store various polysaccharides (fructans and fructose) that acts as protectants against water deficit conditions. Genetic mapping of F2:3 population developed from a cross between W202A × Texas Grano 43′ was done using SSR and SSCP (Single-stranded conformation polymorphism) markers. A significant QTL was mapped on *Frc* locus positioned on chromosome 8 that depicts high fructan contents in the bulb (McCallum et al. 2006). This region may contribute to enhancing drought tolerance in onion. But due to lack of reference genome in onion, huge efforts are still undergoing to develop genome wide markers and genetic map that will offer fruitful opportunities in upgrading research, and breeding objectives. In near past, genotyping by sequencing (reference free) and SNP assay were done in F_2 segregating mapping population derived from a cross between two inbred lines (NW-001 and NW-002). *De novo* assembled contigs yielded 1,851,428 SNPs along with 10,091 high fidelity SNP markers. Eight linkage groups spanning over a genetic length of 1383 cM (average marker length 8.08 cM) were also developed by using GBS. 96

SNP markers were validated by using GBS-Fluidigm assay (Jo et al. 2017). A study conducted by Martin et al. (2005) validated 43 markers with a further addition of 93 and 930 markers developed by Baldwin et al. (2012) and Duangjit et al. (2013). These SNP markers and linkage map (Jo et al. 2017) will significantly contribute towards improvement of complex traits especially drought tolerance through QTL mapping, association mapping and MAS in onion breeding programs.

3.3 Cucurbitaceae *Family*

3.3.1 Cucumber (*Cucumis sativus*)

In cucumber, though draft genome sequence is available (Li et al. 2019; Osipowski et al. 2020) coupled with genome sequencing and high throughput genotyping technologies, QTL mapping for abiotic factors especially drought resistance is very limited. In cucumber, several QTLs contributing tolerance against low temperature stress; *qLTG1.1*, *qLTG1.2*, *qLTG4.1*, and *qLTG2.1*(Song et al. 2018; Yagcioglu et al. 2019), salinity; *qST6.1* (Liu et al. 2021) and waterlogged conditions; *ARN6.1, ARN5.1, and ARN3.1* (Xu et al. 2017), have been identified. Cucumber genome contains 79 *LEA* (late embryogenesis abundant) genes, known to be expressed in root and leaf tissues against drought stress conditions. The main *LEA* genes responsible for alleviating drought stress in cucumber includes *CsLEA-09*, *CsLEA-32*, *CsLEA-57*, and *CsLEA-54* (Celik Altunoglu et al. 2016). Studies also demonstrated the substantial roles of Lipoxyginases (LOXs) (Yang et al. 2012) and SAPs (stress-associated proteins, a types of zinc finger proteins) (Lai et al. 2020) against drought stress in cucumber cultivars. Future knowledge on drought tolerance related QTLs/genes should be generated based on genetic mapping studies in cucumber.

3.4 Brassicaceae *Family*

3.4.1 Cabbage (*Brassica oleracea*)

Cabbage is an important vegetable grown worldwide, and provide a source of minerals, fiber, and vitamins, particularly during cool season. It is regarded as a significant source of chemo-protectants phytochemicals in human diet (Sharma et al. 2018). However, cabbage breeding is still lagging behind due to use of conventional methods of breeding. Genome-wide SNP based genetic map was constructed by Huang et al. (2017) which revealed 711 bins displaying 3985 SNPs representing 16 agronomic traits mapped to 10 linkage groups in Chinese cabbage (*Brassica rapa* subsp. *pekinensis*). Likewise, a lot of QTL mapping research can be accessed online related to Chinese cabbage and cabbage (*B. oleracea*) linked to several agro-morphological and biotic resistance traits. Currently, no specific research entails

QTL mapping of cabbage in response to drought stress. It might be due to non-availability of high quality, comprehensively assembled cabbage (*B. oleracea* var. *capitata*) genome. Broekgaarden et al. (2018) used interval mapping (MapQTL 6) for construction of genetic linkage map that covered all 9 chromosomes (average marker interval 5.6 cM). Chromosomes 2 and 9 were linked with whitefly resistance in cabbage. Moreover, these regions expressed differentially expressed genes (DEGs) that are orthologous to abscisic acid (ABA) signaling in Arabidopsis such as *AAO1*, *MYB96* and *RD22* (Vishwakarma et al. 2017). *MYB96* provides drought tolerance in *B. oleracea* by mediating ABA signals (Seo et al. 2009).

A new database focused on the genomics and genetics of *Brassica* plants known as BRAD (Database of *Brassiceae* genomic data) has been developed and can be accessed through http://brassicadb.org (Cheng et al. 2011). Genomic work on *B. oleracea* based on NGS showed two genomes; 02–12 (*B. oleracea* var. *capitata*) assembly and TO1000 (kale-like; *B. oleracea* var. *alboglabra*) assembly (Liu et al. 2014; Parkin et al. 2014). Nevertheless, these genomes faced several gaps and errors, which make them difficult to use for many studies (Lee et al. 2016; Liu et al. 2017). *De novo* assembly of DH line D134 combined with SMRT (single molecule real-time) sequencing, 10× Genomics and Hi-C technologies for scaffolding yielded high quality draft genome of cabbage. High quality chromosome level D134 assembly showed a size of 529.92 Mbp which is 135 Mb longer as compared to 02–12 reference genome, containing annotation of 44,701 high quality protein coding genes. This high-quality draft genome together with availability of genome wide markers will result in significant improvement of cabbage crop against abiotic stress factors especially drought (Lv et al. 2020). Reference genomes can be used to identify high similarity flanking sequences by blasting against the assembled contigs, like *B. oleracea* genome sequence (Cheng et al. 2011; Lv et al. 2020). Resulting markers for instance SNPs will depict even distribution across genome enhancing the efficiency of QTL mapping for traits of interests.

3.5 Apiaceae *Family*

3.5.1 Carrot (*Daucus carota*)

Carrot is diploid (2n = 2x = 18) outcrossing, cool season vegetable. The discovery of CMS (cytoplasmic male sterility) in 1940s and 1950s shifted large scale major production area to hybrids (Simon 2000). Numerous molecular markers like AFLP and RAPD were used in genetic diversity and molecular varietal characterization studies of carrot (Le Clerc et al. 2000; Baranski et al. 2000; Que et al. 2019). With regard to QTL mapping, QTLs for leaf blight (*Alternaria dauci*), and carotenoids like α-carotenes, β-carotenes, and carotene lycopenes have been studied (Que et al. 2019). Few reports of genes pertaining to drought stress have been reported. Thaumatin-like proteins (TLP) are known to perform vital role in abiotic stress response. In carrot, gene encoding TLP i.e., *dcTLP*, was upregulated conferring tolerance against drought stress and is not regulated by salicylic acid, ABA, or jasmonic acid (Jung et al. 2005). Ahn and Song (2012) were of the view that heat shock protein *DcHSP17.7* also shown to accumulate in carrots under osmotic stress. Expression levels of *DcNCED1* and *DcNCED2* increased significantly, which suggests the role of *NCED* (9-cis-epoxycarotenoid dioxygenase) genes in drought tolerance. *NCED* genes catalyze the decomposition of apocarotenoids and carotenoids through degradation of carotenes and lutein. Thus, selection of carrot cultivars with relatively high carotenoids content will provide strong drought tolerance (Zhang et al. 2021). Though transcriptomic approach involving gene expression analysis have been done in carrots yet serious efforts are required to dissect drought stress related QTLs/genes in carrots.

4 Conclusion

Vegetables are an important source of minerals, vitamins and carbohydrates, as well as bioactive compounds that significantly contribute to promoting health. The consumption of either fresh or processed vegetables is expanding owing to the awareness of well-balanced diet among the consumers. Most vegetables have small root system and are succulent in nature, requiring adequate amounts of water. Water shortage scenario affects adaptability and productivity of vegetables, making them susceptible to insects, weeds, and illnesses, which dramatically elevates crop losses (Sekhon et al. 2019). It is deemed necessary to develop genotypes capable of drought tolerance in order to maintain crop quality and quantity. During the recent decades, a great work has been done in drought resistance in vegetables by implementing conventional breeding programs, but current knowledge about drought tolerance is still quite limited, particularly with regard to the subsequent questions: (1) How domestication resulted in an evolution of drought tolerance in crop species? (2) How to find out the genes offering drought tolerance and evaluate their breeding potential?, and (3) How to shorten the gap between theoretical research and practical work? Secondly, traditional breeding techniques for genetic improvement in vegetables are less effective or insufficient for the development of new plant varieties with drought resistance due to several factors, for instance, distant hybridization barriers, long-term field trials, lengthy procedures of screening, and dependence on the environment (Ahmar et al. 2020; Sekhon et al. 2019). With the introduction and advancement in next generation sequencing (NGS) techniques, and marker assisted selection (MAS), a new chapter has begun in vegetables research, and new avenues have been discovered which address the above scientific question on scientific grounds (Aditika and Priyanka 2017; Dhaliwal et al. 2020). Drought stress tolerance is a complex trait and controlled by minor genes or polygenes. Identifying the genetic elements involved in plant responses to drought stress will provide a good platform for improving drought tolerance. Genome mapping or QTL mapping is one of the ways to identify several genomic regions that are associated with drought tolerance as discussed earlier in the chapter. Screening of ideal mapping populations is critical for the identification of putative QTLs and dissecting the gene of interest that may be transferred in the development of tolerant varieties. Most QTL mapping studies in vegetables focused on biotic, agro-morphological traits with less attention on drought-tolerant studies. As a result, there exists a gap for the development of drought tolerant vegetable cultivars. While QTL mapping has its advantages, it also has some drawbacks which can be improved by using association mapping. Association mapping analyze genetic diversity among natural populations to link phenotype with genotype rather than relying on bi-parental population. Future improvements in trait and germplasm conservation along with allele mining benefit greatly from the ability to map QTLs in diverse collections of wild and weedy relatives, landraces, breeding lines, elite lines or natural populations. Techniques such as introgression library (IL) and advanced backcross QTL (AB-QTL) possess potential for mining of alleles from foreign germplasm to enhance crop productivity, quality, adaptability, and nutritional value, when exploring natural diversity (Tanksley and McCouch 1997; Zamir 2001). Recent genomic studies showed that detection of novel QTLs influencing drought tolerance enables the use of QTL pyramiding via MAS for the development of drought-tolerant cultivars. MAS is expected to develop as a main tool in vegetable crops in the near future, as it is a primary choice of breeding for trait enhancement in most breeding programmes, due to the ease of marker development and application, as well as the low cost of marker genotyping. Additionally, the precise evaluation of drought tolerance under field conditions for a large number of genotypes is a common difficulty for researchers studying this trait, combined with lack of suitable genotyping and mapping platforms. It is therefore important to include high-throughput phenotyping and genotyping technologies for the development of drought tolerant vegetable cultivars, which has recently gain attention from both consumers and researchers, around the globe.

References

Aditika KH, Priyanka SA (2017) Vegetable improvement in India; recent past, present and future: a review. Int J Curr Microbiol Appl Sci 6:3246–3255

Ahmar S, Gill RA, Jung KH, Faheem A, Qasim MU, Mubeen M, Zhou W (2020) Conventional and molecular techniques from simple breeding to speed breeding in crop plants: recent advances and future outlook. Int J Mol Sci 21(7):2590

Ahn YJ, Song NH (2012) A cytosolic heat shock protein expressed in carrot (*Daucus carota* L.) enhances cell viability under oxidative and osmotic stress conditions. HortScience 47(1):143–148

Albert E, Gricourt J, Bertin N, Bonnefoi J, Pateyron S, Tamby JP, Causse M (2016a) Genotype by watering regime interaction in cultivated tomato: lessons from linkage mapping and gene expression. Theor Appl Genet 129(2):395–418

Albert E, Segura V, Gricourt J, Bonnefoi J, Derivot L, Causse M (2016b) Association mapping reveals the genetic architecture of tomato response to water deficit: focus on major fruit quality traits. J Exp Bot 67(22):6413–6430

Aliche EB, Oortwijn M, Theeuwen TP, Bachem CW, van Eck HJ, Visser RG, van der Linden CG (2019) Genetic mapping of tuber size distribution and marketable tuber yield under drought stress in potatoes. Euphytica 215(11):1–19

Anithakumari AM, Dolstra O, Vosman B, Visser RG, van der Linden CG (2011) *In vitro* screening and QTL analysis for drought tolerance in diploid potato. Euphytica 181(3):357–369

Anithakumari AM, Nataraja KN, Visser RG, van der Linden CG (2012) Genetic dissection of drought tolerance and recovery potential by quantitative trait locus mapping of a diploid potato population. Mol Breed 30(3):1413–1429

Anjum SA, Ashraf U, Tanveer M, Khan I, Hussain S, Shahzad B, Wang LC (2017) Drought induced changes in growth, osmolyte accumulation and antioxidant metabolism of three maize hybrids. Front Plant Sci 8:69

Avise JC (2012) Molecular markers, natural history and evolution. Springer Science & Business Media

Babu R, Nair SK, Prasanna BM, Gupta HS (2004) Integrating marker-assisted selection in crop breeding–prospects and challenges. Curr Sci 87:607–619

Badu M, Tripathy B, Sahu GS, Jena AK (2017) Role of doubled haploids in vegetable crop improvement. J Pharmacogn Phytochem 6(6):384–389

Baldwin S, Pither-Joyce M, Wright K, Chen L, McCallum J (2012) Development of robust genomic simple sequence repeat markers for estimation of genetic diversity within and among bulb onion (*Allium cepa* L.) populations. Mol Breed 30(3):1401–1411

Baranski R, Jagosz B, Michalik B, Simon PW, Grzebelus D (2000) Comparison of RAPD and AFLP techniques used for the evaluation of genetic diversity of carrot breeding materials. In: International symposium on molecular markers for characterizing genotypes and identifying cultivars in horticulture, vol 546, pp 413–416

Bernardo R (2008) Molecular markers and selection for complex traits in plants: learning from the last 20 years. Crop Sci 48(5):1649–1664

Bouchenak M, Lamri-Senhadji M (2013) Nutritional quality of legumes, and their role in cardiometabolic risk prevention: a review. J Med Food 16(3):185–198

Broekgaarden C, Pelgrom KT, Bucher J, van Dam NM, Grosser K, Pieterse CM, van Wees SC (2018) Combining QTL mapping with transcriptome and metabolome profiling reveals a possible role for ABA signaling in resistance against the cabbage whitefly in cabbage. PLoS One 13(11):e0206103

Burr B, Burr FA, Thompson KH, Albertson MC, Stuber CW (1988) Gene mapping with recombinant inbreds in maize. Genetics 118(3):519–526

Causse M, Saliba-Colombani V, Lesschaeve I, Buret M (2001) Genetic analysis of organoleptic quality in fresh market tomato. 2. Mapping QTLs for sensory attributes. Theor Appl Genet 102(2):273–283

Celik Altunoglu Y, Baloglu P, Yer EN, Pekol S, Baloglu MC (2016) Identification and expression analysis of LEA gene family members in cucumber genome. Plant Growth Regul 80(2):225–241

Chagné D, Batley J, Edwards D, Forster JW (2007) Single nucleotide polymorphism genotyping in plants. In: Association mapping in plants. Springer, New York, NY, pp 77–94

Chaves MM, Oliveira MM (2004) Mechanisms underlying plant resilience to water deficits: prospects for water-saving agriculture. J Exp Bot 55(407):2365–2384

Cheng F, Liu S, Wu J, Fang L, Sun S, Liu B, Wang X (2011) BRAD, the genetics and genomics database for *Brassica* plants. BMC Plant Biol 11(1):1–6

Dariva FD, Pessoa HP, Copati MGF, de Almeida GQ, de Castro Filho MN, de Toledo Picoli EA, Nick C (2021) Yield and fruit quality attributes of selected tomato introgression lines subjected to long-term deficit irrigation. Sci Hortic 289:110426

Dhaliwal SK, Talukdar A, Gautam A, Sharma P, Sharma V, Kaushik P (2020) Developments and prospects in imperative underexploited vegetable legumes breeding: a review. Int J Mol Sci 21(24):9615

Diouf IA, Derivot L, Bitton F, Pascual L, Causse M (2018) Water deficit and salinity stress reveal many specific QTL for plant growth and fruit quality traits in tomato. Front Plant Sci 9:279

Doerge RW (2002) Mapping and analysis of quantitative trait loci in experimental populations. Nat Rev Genet 3(1):43–52

Doganlar S, Frary A, Daunay MC, Lester RN, Tanksley SD (2002) A comparative genetic linkage map of eggplant (*Solanum melongena*) and its implications for genome evolution in the *Solanaceae*. Genetics 161(4):1697–1711

Dorais M, Ehret DL, Papadopoulos AP (2008) Tomato (*Solanum lycopersicum*) health components: from the seed to the consumer. Phytochem Rev 7(2):231–250

Duangjit J, Bohanec B, Chan AP, Town CD, Havey MJ (2013) Transcriptome sequencing to produce SNP-based genetic maps of onion. Theor Appl Genet 126(8):2093–2101

Eftekhari A, Baghizadeh A, Yaghoobi MM, Abdolshahi R (2017) Differences in the drought stress response of DREB2 and CAT1 genes and evaluation of related physiological parameters in some bread wheat cultivars. Biotechnol Biotechnol Equip 31(4):709–716

El-Moneim DA, Alqahtani MM, Abdein MA, Germoush MO (2020) Drought and salinity stress response in wheat: physiological and TaNAC gene expression analysis in contrasting Egyptian wheat genotypes. J Plant Biotechnol 47(1):1–14

FAOSTAT (2021) Available online: https://www.fao.org/faostat/en/#data. Accessed on 12 Aug 2022

FAOSTAT (2022) Available online: http://faostat3.fao.org/home. Accessed on 2 Sept 2022

Fowler C, Moore G, Hawtin G (2003) The international treaty on plant genetic resources for food and agriculture: a primer for the future harvest center of the CGIAR. International Plant Genetic Resources Institute, Rome

Frary A, Frary A, Daunay MC, Huvenaars K, Mank R, Doğanlar S (2014) QTL hotspots in eggplant (*Solanum melongena*) detected with a high resolution map and CIM analysis. Euphytica 197(2):211–228

Furlan AL, Bianucci E, Giordano W, Castro S, Becker DF (2020) Proline metabolic dynamics and implications in drought tolerance of peanut plants. Plant Physiol Biochem 151:566–578

Gramazio P, Blanca J, Ziarsolo P, Herraiz FJ, Plazas M, Prohens J, Vilanova S (2016) Transcriptome analysis and molecular marker discovery in *Solanum incanum* and *S. aethiopicum*, two close relatives of the common eggplant (*Solanum melongena*) with interest for breeding. BMC Genomics 17(1):1–17

Guillet C, Just D, Bénard N, Destrac-Irvine A, Baldet P, Hernould M, Rothan C (2002) A fruit-specific phosphoenolpyruvate carboxylase is related to rapid growth of tomato fruit. Planta 214(5):717–726

Gupta PK, Varshney RK, Sharma PC, Ramesh B (1999) Molecular markers and their applications in wheat breeding. Plant Breed 118(5):369–390

Hamilton JP, Hansey CN, Whitty BR, Stoffel K, Massa AN, Van Deynze A, Buell CR (2011) Single nucleotide polymorphism discovery in elite North American potato germplasm. BMC Genomics 12(1):1–12

Harris PM (ed) (2012) The potato crop: the scientific basis for improvement. Springer Science & Business Media

Hartl DL (1988) A primer of population genetics, 2nd edn. Sinauer Associates, Inc, Sunderland

Helentjaris T, Slocum M, Wright S, Schaefer A, Nienhuis J (1986) Construction of genetic linkage maps in maize and tomato using restriction fragment length polymorphisms. Theor Appl Genet 72(6):761–769

Hirakawa H, Shirasawa K, Miyatake KOJI, Nunome T, Negoro S, Ohyama AKIO, Fukuoka H (2014) Draft genome sequence of eggplant (*Solanum melongena* L.): the representative *Solanum* species indigenous to the old world. DNA Res 21(6):649–660

Huang L, Yang Y, Zhang F, Cao J (2017) A genome-wide SNP-based genetic map and QTL mapping for agronomic traits in Chinese cabbage. Sci Rep 7(1):1–14

Ichinose M, Sugita M (2016) RNA editing and its molecular mechanism in plant organelles. Genes 8(1):5

Ingram J, Bartels D (1996) The molecular basis of dehydration tolerance in plants. Annu Rev Plant Biol 47(1):377–403

Itam M, Mega R, Tadano S, Abdelrahman M, Matsunaga S, Yamasaki Y, Tsujimoto H (2020) Metabolic and physiological responses to progressive drought stress in bread wheat. Sci Rep 10(1):1–14

Jacob C, Carrasco B, Schwember AR (2016) Advances in breeding and biotechnology of legume crops. Plant Cell Tissue Organ Cult 127(3):561–584

Jena KK, Mackill DJ (2008) Molecular markers and their use in marker-assisted selection in rice. Crop Sci 48(4):1266–1276

Jiang SC, Mei C, Liang S, Yu YT, Lu K, Wu Z, Zhang DP (2015) Crucial roles of the pentatricopeptide repeat protein SOAR1 in Arabidopsis response to drought, salt and cold stresses. Plant Mol Biol 88(4):369–385

Jo J, Purushotham PM, Han K, Lee HR, Nah G, Kang BC (2017) Development of a genetic map for onion (*Allium cepa* L.) using reference-free genotyping-by-sequencing and SNP assays. Frontiers. Plant Sci 8:1606

Jung YC, Lee HJ, Yum SS, Soh WY, Cho DY, Auh CK, Lee SC (2005) Drought-inducible-but ABA-independent-thaumatin-like protein from carrot (*Daucus carota* L.). Plant Cell Rep 24(6):366–373

Kemble JK, Sanders DC 2000. Basics of vegetable crop irrigation Alabama Co-operative Extension System ANR1169

Khan MMR, Isshiki S (2016) Cytoplasmic male sterility in eggplant. Hortic J 85(1):1–7

Khan MA, Saravia D, Munive S, Lozano F, Farfan E, Eyzaguirre R, Bonierbale M (2015) Multiple QTLs linked to agro-morphological and physiological traits related to drought tolerance in potato. Plant Mol Biol Report 33(5):1286–1298

Lai W, Zhou Y, Pan R, Liao L, He J, Liu H, Liu S (2020) Identification and expression analysis of stress-associated proteins (SAPs) containing A20/AN1 zinc finger in cucumber. Plan Theory 9(3):400

Lawlor DW, Cornic G (2002) Photosynthetic carbon assimilation and associated metabolism in relation to water deficits in higher plants. Plant Cell Environ 25(2):275–294

Le Clerc V, Mausset AE, Veret A, Briard M (2000) A comparative study on the use of ISSR, microsatellites and RAPD markers for varietal identification of carrot genotypes. In International symposium on molecular markers for characterizing genotypes and identifying cultivars in horticulture, vol 546, pp 377–385

Lee J, Izzah NK, Choi BS, Joh HJ, Lee SC, Perumal S, Yang TJ (2016) Genotyping-by-sequencing map permits identification of clubroot resistance QTLs and revision of the reference genome assembly in cabbage (*Brassica oleracea* L.). DNA Res 23(1):29–41

Li Z, Palmer WM, Martin AP, Wang R, Rainsford F, Jin Y, Ruan YL (2012) High invertase activity in tomato reproductive organs correlates with enhanced sucrose import into, and heat tolerance of, young fruit. J Exp Bot 63(3):1155–1166

Li Q, Li H, Huang WU, Xu Y, Zhou Q, Wang S, Zhang Z (2019) A chromosome-scale genome assembly of cucumber (*Cucumis sativus* L.). GigaScience 8(6):giz072

Liu S, Liu Y, Yang X, Tong C, Edwards D, Parkin IA, Paterson AH (2014) The *Brassica oleracea* genome reveals the asymmetrical evolution of polyploid genomes. Nat Commun 5(1):1–11

Liu XP, Gao BZ, Han FQ, Fang ZY, Yang LM, Zhuang M, Zhang YY (2017) Genetics and fine mapping of a purple leaf gene, BoPr, in ornamental kale (*Brassica oleracea* L. var. *acephala*). BMC Genomics 18(1):1–9

Liu D, Dong S, Bo K, Miao H, Li C, Zhang Y, Gu X (2021) Identification of QTLs controlling salt tolerance in cucumber (*Cucumis sativus* L.) seedlings. Plan Theory 10(1):85

Lv H, Wang Y, Han F, Ji J, Fang Z, Zhuang M, Yang L (2020) A high-quality reference genome for cabbage obtained with SMRT reveals novel genomic features and evolutionary characteristics. Sci Rep 10(1):1–9

Mangino G, Plazas M, Vilanova S, Prohens J, Gramazio P (2020) Performance of a set of eggplant (*Solanum melongena*) lines with introgressions from its wild relative *S. incanum* under open field and screenhouse conditions and detection of QTLs. Agronomy 10(4):467

Martin WJ, McCallum J, Shigyo M, Jakse J, Kuhl JC, Yamane N, Havey MJ (2005) Genetic mapping of expressed sequences in onion and in silico comparisons with rice show scant colinearity. Mol Gen Genomics 274(3):197–204

McCallum J, Clarke A, Pither-Joyce M, Shaw M, Butler R, Brash D, Havey MJ (2006) Genetic mapping of a major gene affecting onion bulb fructan content. Theor Appl Genet 112(5):958–967

Mohan M, Nair S, Bhagwat A, Krishna TG, Yano M, Bhatia CR, Sasaki T (1997) Genome mapping, molecular markers and marker-assisted selection in crop plants. Mol Breed 3(2):87–103

Monneveux P, Ramírez DA, Pino MT (2013) Drought tolerance in potato (*S. tuberosum* L.): Can we learn from drought tolerance research in cereals? Plant Sci 205:76–86

Nahar K, Ullah SM (2011) Effect of water stress on moisture content distribution in soil and morphological characters of two tomato (*Lycopersicon esculentum* Mill) cultivars. J Sci Res 3(3):677–682

Ort DR (2001) When there is too much light. Plant Physiol 125(1):29–32

Osipowski P, Pawełkowicz M, Wojcieszek M, Skarzyńska A, Przybecki Z, Pląder W (2020) A high-quality cucumber genome assembly enhances computational comparative genomics. Mol Gen Genomics 295(1):177–193

Parkin IA, Koh C, Tang H, Robinson SJ, Kagale S, Clarke WE, Sharpe AG (2014) Transcriptome and methylome profiling reveals relics of genome dominance in the mesopolyploid *Brassica oleracea*. Genome Biol 15(6):1–18

Pascual L, Albert E, Sauvage C, Duangjit J, Bouchet JP, Bitton F, Causse M (2016) Dissecting quantitative trait variation in the resequencing era: complementarity of bi-parental, multi-parental and association panels. Plant Sci 242:120–130

Paterson AH (1996) Making genetic maps. In: Paterson AH (ed) Genome mapping in plants. Academic, London, pp 23–39

Pink D, Bailey L, McClement S, Hand P, Mathas E, Buchanan-Wollaston V, Teakle G (2008) Double haploids, markers and QTL analysis in vegetable Brassicas. Euphytica 164(2):509–514

Powell W, Machray GC, Provan J (1996) Polymorphism revealed by simple sequence repeats. Trends Plant Sci 1(7):215–222

Qian Z, Zhang B, Chen H, Lu L, Duan M, Zhou J, Li D (2021) Identification of quantitative trait loci controlling the development of prickles in eggplant by genome re-sequencing analysis. Front Plant Sci 12:731079

Que F, Hou XL, Wang GL, Xu ZS, Tan GF, Li T, Xiong AS (2019) Advances in research on the carrot, an important root vegetable in the *Apiaceae* family. Hortic Res 6:69

Salgon S, Jourda C, Sauvage C, Daunay MC, Reynaud B, Wicker E, Dintinger J (2017) Eggplant resistance to the *Ralstonia solanacearum* species complex involves both broad-spectrum and strain-specific quantitative trait loci. Front Plant Sci 8:828

Sauvage C, Segura V, Bauchet G, Stevens R, Do PT, Nikoloski Z, Causse M (2014) Genome-wide association in tomato reveals 44 candidate loci for fruit metabolic traits. Plant Physiol 165(3):1120–1132

Schmitz-Linneweber C, Small I (2008) Pentatricopeptide repeat proteins: a socket set for organelle gene expression. Trends Plant Sci 13(12):663–670

Schumacher C, Thümecke S, Schilling F, Köhl K, Kopka J, Sprenger H, Horn R (2021) Genome-wide approach to identify quantitative trait Loci for drought tolerance in tetraploid potato (*Solanum tuberosum* L.). Int J Mol Sci 22(11):6123

Sekhon BS, Sharma A, Chahota RK (2019) Marker-assisted breeding for disease resistance in legume vegetable crops. In: Handbook of plant and crop stress, 4th edn. CRC Press, pp 575–592

Seo PJ, Xiang F, Qiao M, Park JY, Lee YN, Kim SG, Park CM (2009) The MYB96 transcription factor mediates abscisic acid signaling during drought stress response in Arabidopsis. Plant Physiol 151(1):275–289

Septembre-Malaterre A, Remize F, Poucheret P (2018) Fruits and vegetables, as a source of nutritional compounds and phytochemicals: changes in bioactive compounds during lactic fermentation. Food Res Int 104:86–99
Sharma M, Pandey GK (2016) Expansion and function of repeat domain proteins during stress and development in plants. Front Plant Sci 6:1218
Sharma SK, Bolser D, de Boer J, Sønderkær M, Amoros W, Carboni MF, Bryan GJ (2013) Construction of reference chromosome-scale pseudomolecules for potato: integrating the potato genome with genetic and physical maps. G3: Genes Genomes Genetics 3(11):2031–2047
Sharma A, Rathore JP, Ali A, Qadri I, Hussain SM, Angmo T (2018) Major diseases and pathogen ecology of cabbage. Pharma Innov J 7(7):574–578
Shinozaki K, Yamaguchi-Shinozaki K (2007) Gene networks involved in drought stress response and tolerance. J Exp Bot 58(2):221–227
Simon PW (2000) Domestication, historical development, and modern breeding of 394 Carrot. Plant Breed Rev 7:157–190
Singh RJ (2007) Genetic resources, chromosome engineering, and crop improvement: vegetable crops. CRC Press, Boca Raton
Singh BD, Singh AK (2015) Mapping populations. In: Marker-assisted plant breeding: principles and practices. Springer, New Delhi, pp 125–150
Singh S, Singh SP, Singh A, Singh KP (2021) Application of double haploid in vegetable improvement. Agriblossom 1:1–7
Song Z, Wang W, Shi L, Zhang S, Xie Q, Wei S, Gu X (2018) Identification of QTL s controlling low-temperature tolerance during the germination stage in cucumber (*Cucumis sativus* L.). Plant Breed 137(4):629–637
Squirrell J, Hollingsworth PM, Woodhead M, Russell J, Lowe AJ, Gibby M, Powell W (2003) How much effort is required to isolate nuclear microsatellites from plants? Mol Ecol 12(6):1339–1348
Stevens B, Allen NJ, Vazquez LE, Howell GR, Christopherson KS, Nouri N, Barres BA (2007) The classical complement cascade mediates CNS synapse elimination. Cell 131(6):1164–1178
Szegletes ZS, Erdei L, Tari I, Cseuz L (2000) Accumulation of osmoprotectants in wheat cultivars of different drought tolerance. Cereal Res Commun 28(4):403–410
Tagliotti ME, Deperi SI, Bedogni MC, Huarte MA (2021) Genome-wide association analysis of agronomical and physiological traits linked to drought tolerance in a diverse potatoes (*Solanum tuberosum*) panel. Plant Breed 140(4):654–664
Tanksley SD, McCouch SR (1997) Seed banks and molecular maps: unlocking genetic potential from the wild. Science 277(5329):1063–1066
Thant HM, Thwe AA, Thu MK, Htwe NM, Myint KT (2021) Drought tolerance eggplant (*Solanum* spp.) selection by molecular markers (SSRs). In: Proceedings of the National Workshop on Green Ecology in Myanmar: Yezin Agricultural University, 29 August 2019. Food & Agriculture Organization, p 102
Vishwakarma K, Upadhyay N, Kumar N, Yadav G, Singh J, Mishra RK, Sharma S (2017) Abscisic acid signaling and abiotic stress tolerance in plants: a review on current knowledge and future prospects. Front Plant Sci 8:161
Visker MHPW, Keizer LCP, Van Eck H, Jacobsen E, Colon L, Struik P (2003) Can the QTL for late blight resistance on potato chromosome 5 be attributed to foliage maturity type? Theor Appl Genet 106(2):317–325
Vos PG, Uitdewilligen JG, Voorrips RE, Visser RG, van Eck HJ (2015) Development and analysis of a 20K SNP array for potato (*Solanum tuberosum*): an insight into the breeding history. Theor Appl Genet 128(12):2387–2401
Wang W, Vinocur B, Altman A (2003) Plant responses to drought, salinity and extreme temperatures: towards genetic engineering for stress tolerance. Planta 218(1):1–14
Wei Q, Wang W, Hu T, Hu H, Wang J, Bao C (2020) Construction of a SNP-based genetic map using SLAF-Seq and QTL analysis of morphological traits in eggplant. Front Genet 11:178

Wu F, Eannetta NT, Xu Y, Tanksley SD (2009) A detailed synteny map of the eggplant genome based on conserved ortholog set II (COSII) markers. Theor Appl Genet 118(5):927–935

Xu X, Ji J, Xu Q, Qi X, Chen X (2017) Inheritance and quantitative trail loci mapping of adventitious root numbers in cucumber seedlings under waterlogging conditions. Mol Gen Genomics 292(2):353–364

Yagcioglu M, Jiang B, Wang P, Wang Y, Ellialtioglu SS, Weng Y (2019) QTL mapping of low temperature germination ability in cucumber. Euphytica 215(4):1–11

Yang XY, Jiang WJ, Yu HJ (2012) The expression profiling of the lipoxygenase (LOX) family genes during fruit development, abiotic stress and hormonal treatments in cucumber (*Cucumis sativus* L.). Int J Mol Sci 13(2):2481–2500

Yordanov I, Velikova V, Tsonev T (2000) Plant responses to drought, acclimation, and stress tolerance. Photosynthetica 38(2):171–186

Zamir D (2001) Improving plant breeding with exotic genetic libraries. Nat Rev Genet 2(12):983–989

Zhang RR, Wang YH, Li T, Tan GF, Tao JP, Su XJ, Xiong AS (2021) Effects of simulated drought stress on carotenoid contents and expression of related genes in carrot taproots. Protoplasma 258(2):379–390

Zhu JK (2002) Salt and drought stress signal transduction in plants. Annu Rev Plant Biol 53:247–273

Radiation-Induced Mutation for Drought Tolerance In Vegetables

K. Yaprak Kantoğlu, Hayrettin Peşkircioğlu, Ş. Şebnem Ellialtiğlu, and Şule Sarıçam Kökpınar

Abstract Changing climatic conditions bring many problems in terms of agricultural production in terms of the loss of yield increment. For this reason, many studies are carried out to develop new plant varieties that can tolerate drought and have high adaptability to cultivation in arid conditions by using different plant breeding methods. The mutation breeding method emerges among the others as a method used in the creation of alternative variations in the limited gene pool, both in vitro and in vivo. When the IAEA mutant variety database is examined, it is seen that mutation breeding practices, which is one of the conventional breeding techniques in breeding studies for drought tolerance, come up with successful results in many plant species, including vegetable crops (especially tomato), as well. It is obvious that the studies to be carried out using mutation breeding in vegetable crops in order to create genetic variability with new biotechnological techniques by keeping up with changing world conditions are going to yield much faster and more target-based results. This chapter will discuss methodologies for induced mutation and biotechnological approaches in the breeding of vegetable crops for drought tolerance.

Keywords Drought tolerance · Induced mutation · Vegetable crops

K. Y. Kantoğlu (✉) · H. Peşkircioğlu
Turkish Energy Nuclear and Mineral Research Agency, Nuclear Energy Research Institute, Ankara, Türkiye

Ş. Ş. Ellialtiğlu
Doqutech Academy Llc. Co., Ankara University Technopolis, Ankara, Türkiye

Ş. S. Kökpınar
Geçitkuşağı Agricultural Research Institute, Eskişehir, Türkiye

U. K. Chaudhry et al. (eds.), *Drought Stress*,
https://doi.org/10.1007/978-3-031-80610-0_15

1 Introduction

Changes in the world's climate and weather patterns that occur over long periods of time are defined as climate change. Although it was not believed in the past that the climate could change as a result of human-induced activities, a number of studies have been conducted by Eunice Newton Foote to determine how heat can be absorbed in the earth's atmosphere and the effects of carbon dioxide on its role (Anonymous 2022a). Irish scientist John Tyndall with his tests in the 1860s determined that CO_2, methane, and volatile hydrocarbons were particularly effective at absorbing energy from the sun. As a result, it was revealed that CO_2 alone acts like a sponge and can absorb more than one wavelength of sunlight (Anonymous 2022b). Following this, in the 1930s, British engineer Guy Stewart Callendar noted that the North Atlantic region warmed significantly just after the industrial revolution. With his calculations, Callendar suggested that doubling the CO_2 in the earth's atmosphere could warm the earth by 2 °C (3.6 °F) (Callendar 1938). Until the 1960s, discussions continued about the ongoing greenhouse warming of the planet. By the 1980s, a sharp increase in global temperatures began to be accepted. Many experts reported that turning point events brought global warming to the fore, and 1988 was a critical turning point in this sense (Kogan 1997; Mo et al. 1997). As a result of these developments, it has been revealed as an inevitable fact that the increase in greenhouse gases in the world, the increase in global temperature, and, as a result, a water shortage will occur.

Although three quarters of the world is covered with water, the amount of fresh water required for human needs is very limited. The total amount of fresh water in the world is approximately 35 million km^3 and constitutes only 2.5% of the total water existence. With the increasing population, the inefficient use of water, and the evaporation that comes with the increase in temperature, the need for water is gradually increasing. In the climate change that the world has entered in the last 30 years, drought is the most severe abiotic stress in terms of agricultural production and is one of the biggest problems in today's climate scenario (Bechtold 2018). The challenging task of solving this problem and developing drought-resistant plant varieties is also in high demand. Plant breeders, on the other hand, use these strategies to develop new plant varieties by understanding and learning the physiological, morphological, and biological strategies that have developed during millions of years of evolution of plants that are resistant in nature and survive in all environmental conditions. However, difficulties arising from genetic barriers may cause studies to progress more slowly. This limitation on water, which is the most basic input in agricultural production, forces plants to a more sustainable life in terms of water. Plants have developed various tolerance mechanisms against thirst with the operation of various natural selection mechanisms during millions of years of evolution (Blum 2005).

The species in the vegetable crops, which have important contributions to human health and nutrition, are more sensitive to this stress factor than other plant groups. In today's world, where changing climatic conditions and the limitation of irrigation resources are in question, the importance of the effects of this stress factor is increasing in terms of agricultural production. For this reason, the development of vegetable varieties that can better evaluate environmental conditions in terms of drought ecology is a focus issue all over the world (Shamim et al. 2014). Due to the high temperature factor accompanying drought, it is important to develop genetic material that can tolerate drought and high temperatures in breeding studies. Thus, it will be easier to overcome the problems encountered in productivity. The research revealed that the plant tries to protect itself against arid and high temperatures in the molecular, cellular, and whole plant levels with the internal mechanisms. However, there are many points that cannot be explained due to the opposition with a very complex mechanism. Carrying out breeding studies in a way to activate these mechanisms is one of the important issues that attract the attention of breeders. These genetic mechanisms can be observed in plants as drought escape, avoidance, and tolerance. Especially in the avoidance mechanism, the formation of cuticular wax in the physical structure of the plant, the closure of stomata, and the development of the root structure towards deeper levels in the soil by fringing are the most striking features (Kumar et al. 2012).

In the drought tolerance mechanism, it has been revealed in the studies that the plant takes turgor under control by balancing the osmotic pressure, increases the elasticity of the cells and decreases in size, and develops features such as protoplasmic tolerance. Before starting studies to provide drought resistance or tolerance, it is also important to know the factors triggered by drought stress in the plant, apart from the above-mentioned avoidance mechanisms. These effects are briefly summarized below.

2 Morphological Modifications

(a) Short stature, stunted growth (dwarf plant): Studies that reveal the decrease and variation in plant habitus due to drought stress (Shamim et al. 2014; Saima et al. 2018; Sallam et al. 2018; Khanzada et al. 2020).
(b) Early maturity: It is important for the plant to complete its life cycle without entering the dry period or being exposed to too much, and it also provides an escape from the drought, since the generative period of the plants in general coincides with the potentially drier period.
(c) Leaf-related characteristics: reduced leaf area, limited leaf extension, leaf rolling, stoma position, waxiness, hairiness (leaf hair), and leaf color.
(d) High root to shoot ratio.
(e) Reduction in shoot ratio.
(f) Reduction in plant height.

The formations that will serve this function are morphological modifications in drought-resistant plants or plants that are substantially more drought-resistant, such as minimizing the perspiration area and feathering as a result of the contracting and thinned surfaces in touch with the atmosphere. Under the ground, on the contrary, morphological changes occur, such as capillary elongation and increase, and deeper penetration of the root.

It is the primary stress manifested by loss of turgor in the plant when significant water loss occurs from plant cells (Shamim et al. 2014). The structure of the plasma membrane is a result of the aqueous environment in the cell; this structure is formed by the repulsion of the hydrophobic phospholipid tails in the membrane by the water (liquid-solid phase). With the loss of water from the cell, the membrane structure changes; the hydrophilic heads of the phospholipids converge, and the membranes take on a compact appearance (gel phase). In this new structure, membrane lipids have lateral and rotational movement with less kinetic energy than in the liquid-solid phase. Due to the loss of water, the volume of the cell also decreases, and the plasma membrane separates from the cell wall and maintains its relationship only through plasmodesms (plasmolysis). Collapse of the plasma membrane and tonoplast under tension may cause ruptures (Kumar et al. 2012), which may result in the release of hydrolytic enzymes located on the membranes, and thus autolysis of the cytoplasm (Solankey et al. 2015). This damage often permanently disrupts normal cellular metabolism.

3 Physiological Changes

(a) Closure of stomata
(b) Stopping of photosynthesis
(c) Increase in oxidative stress
(d) Changes in cell wall
(e) Decrease in leaf water potential
(f) Decrease in stomatal conductivity
(g) Decrease in intrinsic CO_2 ratio
(h) Stopping of plant growth
(i) Decrease in transpiration rate
(j) Increase in efficiency of plant water using capacity
(k) Increase in proline accumulation
(l) Stopping of translocation

Most of the physiological changes we listed above in the face of drought are the result of events that trigger each other with the closure of stomata. (Tambussi et al. 2000). The two main factors that cause plants to close their stomata during drought are the leaf water potential; hydraulic signals coming from changes in cell turgor, and chemical signals sent by abscisic acids. Abscisic acid (ABA), which is synthesized in the roots and transported to the guard cells by transpiration current, binds to the hypothetical ABA receptor in the guard cells and ensures the closure of stomata under drought stress conditions (Solankey et al. 2015; Salava et al. 2021). It has been observed that stomatal closure depends on the water potential of the soil rather than the water potential in the leaf. Non-stomatal limitation of photosynthesis; it may be associated with oxidative damage to chloroplast lipids, pigments, or proteins (Varshney et al. 2011). Photosynthetic capacity in plants is also affected by the change in light intensity and proportional water content of the environment.

4 Biochemical Changes

(a) Stopping Rubisco productivity
(b) Decreased photochemical efficiency
(c) ROS production
(d) Oxidative damage
(e) Antioxidant production
(f) Decreased chlorophyll content
(g) High proline accumulation
(h) Polyamine production
(i) Increase in antioxidant enzymes
(j) Carbohydrate production
(k) ABA accumulation

The biochemical changes mentioned above are part of a chain that started as a result of drought and ultimately affect photosynthesis activity. The most important of these changes is the cessation of Rubisco productivity. Rubisco is the abbreviated name of the ribulose bisphosphate carboxylase/oxygenase enzyme, and it is the enzyme that shows the most activity in the plant due to its function in photosynthesis. Ribulose (ribulose 1,5 disphosphate) combines with carbon dioxide in the air to form 3 phosphoglycerates. At this stage, the rubisco, or ribulose 1,5 biphosphate carboxylase-oxygenase enzyme, starts the Calvin cycle by binding carbon dioxide. If the oxygen in the environment is more than carbon dioxide, it binds the oxygen and causes photorespiration, not photosynthesis. In summary, it is the enzyme that starts the Calvin cycle by holding carbon dioxide in the air. The amount of Rubisco in the leaves is controlled by the rate of synthesis and degradation. Even under drought stress, the Rubisco holoenzyme is relatively stable with a half-life of several days (Smirnof 1993; Webber et al. 1994). However, in tomato (Bradford 1976), Arabidopsis (Williams et al. 1994), and rice (Vu et al. 1998), it led to a rapid reduction in the abundance of the Rubisco small subunit (rbc S). Rubisco's short-term responses to drought stress are unclear as different studies have produced conflicting results. One study found that while studying sunflower and tobacco, respectively, drought had little effect on Rubisco (Gimenez et al. 1992). In another study, loss of Rubisco activity was thought to be a rapid and very early response to drought stress in soybeans. However, increasing the severity and duration of drought stress decreases Rubisco activity and protein content in sunflower (Tezara and Lawlor 1995) and wheat (Kicheva et al. 1994), respectively.

When genetically, the inheritance of drought resistance is examined; it appears that this trait is managed in a very complex way and that there are interactions among many related gene groups. It is also reported by researchers that some morphological characters play a key role in the differentiation of drought-tolerant plants. These morphological features are wax content, narrowing of the leaf area, strong root system development, and earliness. In addition to morphological markers, biochemical activations in the plant are also important reagents. The amount of proline, trehalose accumulation, changes in nitrate activity, and the amount of stored carbohydrates are also considered as distinguishing criteria in selection. However, genetic information regarding biochemical changes has not been fully resolved in plants.

In the studies carried out, the genes controlling drought stress have been identified for many plant species. According to these studies, it is known that AVP1 stimulates root system development in arid conditions and that wilty and wilty dwarf genes in tomatoes are activated under arid conditions and are effective in decreasing the amount of leaf fall and evaporation (Gaxiola et al. 2001). In pepper, it was determined that the gene controlling the ER5 protein increased the production of cytoplasmic protein by promoting hydrophobic plant behavior at a higher rate (Kim et al. 2005), while the AREB cupida dehydrin gene increased abscisic acid production in tomato with the bzip transcription factor (Hsieh et al. 2010). It is important for breeding studies to know the periods when drought or water restriction is effective, especially for vegetable species (Kumar et al. 2012). The periods when different vegetable species are affected by drought stress are presented in Table 1.

It is evident from Table 1, especially the vegetable species in the *Solanacea* and *Cucurbitace* families are highly affected by drought damage. Therefore, it is important to make the plant more resistant during these periods in breeding studies. The tomato, which is among the summer vegetable species and is widely grown in arid and semi-arid areas, is a species on which drought studies are concentrated. In particular, one of the wild species, *Solanum panelli*, is tolerant with high adaptation to arid conditions. For this reason, it is considered a model species for molecular identification of the strength mechanism.

Based on all this information, today, breeding studies for the development of new varieties with drought resistance or tolerance in vegetables are carried out very intensively. For this purpose, conventional breeding methods, biotechnological methods, transgene technology, and omic and proteomic technologies have been put into practice in different species (Babiye 2019; Salava et al. 2021). However, the fact that the characters transferred from wild species used in crossbreeding studies cause the transmission of undesirable traits due to linkage (Daskalov 1986; Kharkwal et al. 2004; Solankey et al. 2015), incompatibility, and the effects of some preventive internal mechanisms (Zijlstra et al. 1991), they appear to be very limiting factors in the creation of new variations within the population. For these reasons, researchers use mutation breeding as an alternative method to create new variations through mutation by using physical and chemical mutagens in many plant species and to select new individuals with desired characteristics (Ahloowalia and Maluszynski 2001; Salava et al. 2021). With the determination of the role of spontaneous mutations in creating new alleles by scientists in the last period of the

Table 1 Periods when drought is effective on different types of vegetables and their potency. (Kumar et al. 2012; Potopa et al. 2012; Solankey et al. 2015; Muñoz et al. 2021)

Vegetable crops	Period	Resulting impact
Green bean and tepary bean	Flowering	Growth retardation, reduced root nodulation
Lettuce	Plant growth	Coarse leaf formation, tip blight and poor seed formation, poor head development
Asparagus	Spear production	Decrease in yield and quality
Leafy vegetables	Plant growth	Nitrate accumulation in leaves, poor plant growth
Cabbage	Plant growth and flowering	Decreased pollen viability, fruit deformity
Melon	Flowering and fruit set	Poor fruit set, decrease in yield and the ratio of ascorbic acid and sugars
Watermelon	Flowering and fruit set	Nitrate accumulation, reduced seed viability
Cucumber	Flowerind and fruit growth	Decreased pollen viability, fruit shaped deformation, reduced seed viability
Onion	Head development	Multiple and small head growth, reduced shelf and storage life
Tuberous vegetables	Root development	Formation of rough, weak, amorphous tubers, harmful nitrate build-up and hardening of internal tissues
Cruciferous	Head development	Head blight in head cabbage, formation of brown spots on cauliflower
Okra	Flowering and pod formation (set)	Serious loss in yield, formation of pods with coarse and fibrous structure, insufficient seed development
Pepper	Flowering and fruit set	Early flower and fruit fall, poor plant growth due to insufficient nutrient uptake from the soil, decrease in dry matter
Eggplant	Flowering and fruit set	Low yield, discoloration of fruit, poor seed development
Tomato	Early flowering, fruit set and development	Flower drop, blossom end rot due to Mg/Ca imbalance, fruit cracking, small fruit formation and poor seed development

eighteenth century and their effects on increasing genetic diversity (van Harten 1998), it emerges as a technique used in practice. When the mutation breeding technique and the biotechnological methods that have been predominantly carried out in the last 50 years are compared in terms of cost, the mutation breeding technique is a cheap and environmentally friendly method. Today, both *in vivo* and *in vitro* conditions, physical (X-ray, gamma-ray, ultraviolet radiation, beta radiation, fast and thermal neutrons) (Kharkwal et al. 2004; Oladosu et al. 2016; Kumawat et al. 2019; Salava et al. 2021) and chemical mutagens [basic analogues and related components (hydroxylamine) -NH_2OH-, acridine dye-acridine orange, ethidium bromide-EB, ethylene diamine tetra-acetic acid-EDTA-, antibiotics (streptomycin, mitomycin), alkylated substances {ethyl methanesulphonate-EMS-, dimethyl sulphonate-DMS-, triethylene melamine -TEM-, and azides (sodium azide)}] are used in mutation breeding studies (Kantoğlu and Kunter 2021). Mutation breeding studies carried out with chemical and physical mutagen applications appear as an important breeding method, especially in expanding the limited gene pool. When a comparison is made between the frequency of occurrence of induced mutations and mutations that occur spontaneously in nature, it is seen that the mutation frequency obtained with induced mutations is higher. A total of 3448 (MVD 2024) mutant registered cultivars have been developed in field crops, horticultural crops, and ornamental crops by the mutation breeding method till today. This is the number of cultivars registered with the IAEA Mutant Variety Database on a voluntary basis, but it should be noted that this number is higher when combined with the ones not registered. According to IAEA MVD (2024) data, there are a total of 10 registered varieties developed within the vegetable species that are tolerant to drought and high temperatures by the mutation breeding method. It is seen that 6 of these registered varieties are tomatoes, 1 of them is eggplant, 1 of them is Chinese garlic, 2 of them lettuce and 1 of them is a pea variety. It has been recorded that these developed varieties have high yield capacities apart from drought tolerance.

Mutants developed by mutation breeding can be easily selected within the population for monogenic or oligogenic traits such as plant architecture, productivity, and disease resistance. However, there are mutant cultivars that have also been developed for improving polygenic traits, including yield and tolerance to abiotic stresses (Ahloowalia et al. 2004; Salava et al. 2021). Mutation breeding has been and is still successfully used in a wide variety of plant species, including vegetatively propagated species such as ornamental plants and fruit species, and seed propagated species such as tomatoes, peppers, lettuce, beans, barley, chickpeas, wheat, and cotton. Mutation breeding is used not only to develop new varieties with new traits but also to study the genetic basis of plant traits. New methods developed with advancing technology have opened new doors for screening "drought-insensitive mutants" that are more productive under drought stress conditions (Khalil et al. 2018). In recent years, targeted local lesions in the genome have been screened with various molecular screening techniques such as Targeted Induced Locel Lesions in Genoms (TILLING) and Next Generation Sequence (NGS), allowing applications to identify and transfer gene regions of interest in breeding studies (Salava et al. 2021). This advantage of the mutation breeding technique has gained popularity among molecular geneticists and has led to an increase in mutation breeding applications today. In addition to these, the identification of many new and desired characters brought to plants by mutation breeding, using omic technology, has gained speed and value in the last 20 years (Egea et al. 2018). Protein separation and definition are also important in breeding studies, especially in terms of drought, which is one of the stress factors. Identification of selected proteins by mass spectrometry (MS) and processing them into the data bank provides valuable key information for researchers working on the same species but in different varieties (Çelik et al. 2021).

In today's modern world, it is important to adapt biotechnological methods and testing methods to breeding programs in order to achieve faster results in plant breeding studies carried out to obtain new varieties with the desired quality. It has been observed that genetically modified plant studies have recently increased in breeding studies carried out to obtain plants that are tolerant or resistant to drought stress factors that adversely affect the production potential. However, one of the most important limiting factors in these studies is the low percentage of transformation in gene expression that occurs after transformation. In addition, there are additional limiting factors, such as the difficulties experienced in plant regeneration and the cost burden. Therefore, with the addition of tissue culture and molecular techniques in current breeding studies, classical breeding methods still maintain their effectiveness and attractiveness (Rai et al. 2011). Although the application efficiency differs according to the plant species, the use of *in vitro* techniques for biotic and abiotic stress studies in vegetable species is important in terms of shortening the testing period and obtaining effective results. As a result of revealing the totipotency properties of cells with cell and callus culture methods, which are plant tissue culture techniques that have been studied since the 1920s, a number of biological events in plants that could not be resolved before were clarified by obtaining new individuals from each cell with the characteristics of the mother plant (Sağel et al. 2002). Today, by using callus and cell suspension cultures in resistance breeding studies, resistant/tolerant plants are obtained from these cells by selecting cells that are resistant to some environmental stress factors (such as drought). In addition, by creating mutations with chemical mutagens (EMS, sodium azide, etc.) added to the nutrient medium in *in vitro* conditions and with physical mutagens applied to *in vitro* donor plants (such as UV, X-ray, ^{60}Co or ^{137}Cs irradiation facilities, electron and proton accelerators) that will form the beginning of culture. It is also possible to expand the existing variation (Rai et al. 2011; Çelik et al. 2021; Kantoğlu and Kunter 2021), and to regenerate plants with desired characteristics by selection within this variation *in vitro* conditions. Although these methods do not seem very popular when compared to the studies on obtaining genetically modified plants, which are very popular today, it emerges as a method that requires a lower cost, does not create any disadvantage in terms of human health, and makes it possible to achieve a definite result in a shorter time compared to the production of transgenic plants, which are controversial all over the world (Kantoğlu et al. 2014).

In vitro mutation breeding can be applied successfully in many plant species in order to increase genetic diversity and test a wider variation. For this purpose, different substance applications are carried out to determine the resistance or tolerance limits in the testing studies carried out on the mutant variation obtained after *in vitro* mutagen application (Kantoğlu et al. 2010). It is also possible to test the promising lines obtained from mutation breeding studies under field conditions as a classical breeding study, or with the vegetative mutant material excluded from the trial by negative selection *in vitro* conditions (Seth and Kendurkar 2015). Thus, an investigative idea can be obtained as a result of *in vitro* testing studies on whether an ignored breeding material has a rootstock value (Kumar et al. 2017). In studies on drought stress, polyethylene glycol (PEG) or mannitol can be used. In these test studies, two different applications are carried out *in vitro*. The first is the incremental testing method that spans a long testing process. In this application, in plants that maintain their ability to survive by giving low-dose stressor to the nutrient medium, it is determined to what extent they can maintain this ability by increasing the application dose. The second application is to compare plants with a very high dose of selective agent applied directly to *in vitro* nutrient media and to identify plants that retain their viability. During *in vitro* selection studies, non-tolerant cells may show an adaptation as a result of epigenetic effects, especially in cell and callus culture studies. This condition can sometimes become inherited by mitosis. With the use of selective agents, cells showing tolerance or resistance can be captured *in vitro*, and plants regenerated from these cells can be defined as true tolerant or resistant. DNA methylation is also an important factor in the emergence of modifications because of epigenetic reasons. For this reason, many researchers today agree that the step-by-step method of stress pressure is a more effective selection method to minimize the problem of the situation arising due to epigenetic effects *in vitro* conditions (Rai et al. 2011; Kantoğlu et al. 2021). The persistence of many commercial plant species is correlated with the ability to assess adverse conditions such as drought and high temperatures, which are two of the abiotic stress factors. For this reason, as a good breeding strategy, it is important to know the mechanism of the feature that is aimed to be developed before using biotechnological methods in studies and designing studies accordingly. Drought resistance/tolerance, which is one of the abiotic stress factors, especially *in vitro* conditions, is a factor that can be easily studied in vegetable, fruit, cereal, and ornamental plants. Therefore, creating mutant populations to be created after both *in vivo* and *in vitro* mutagen applications is a technique that allows for the development of drought-tolerant plants. In this context, findings and methods regarding the application of the technique will be presented in this section.

5 Analytical Framework

Many plant species are sensitive to drought, and this environmental problem causes significant crop losses worldwide. Although efforts have been made to obtain tolerant genotypes in many plant species, especially for this problem, the gene pool that can be used for breeding studies is quite limited due the control of drought tolerance by complex and multiple genes in the plant genotype. Accelerating the improvement works is unavoidable, especially given the increase rate in arid and semi-arid production areas due to changing climatic conditions. In breeding studies, it is important to determine whether the key gene regions for drought tolerance are present in the plant. Environmental stresses including intense light, UV, heat stress, hot/cold treatments, drought, and herbicides have all been shown to increase the production of reactive oxygen production (ROS) (Miller et al. 2010; Suzuki et al. 2012). Ironically, the oxygen required for aerobic life—in its reduced forms—is one of the most toxic substances all organisms have to deal with. In this sense, both molecular and physiological analyses are made and the result is reached. Throughout evolution, plants gain a high degree of control over ROS toxicity and use ROS as a signaling molecule to activate the defense mechanism to maintain balance within the cell. Water stress conditions increase the amount of active oxygen in the tissue, triggering the damage of enzymes, chloroplast pigments, membrane lipids and proteins with active sulfhydric groups, thereby disrupting their structural integrity (Anderson et al. 1990; Navari-Izzo et al. 1994). Thus, the process that initiates drought damage in the plant begins. In the event of a water shortage, the plant closes its stomata in order not to lose water, which limits the CO2 uptake required for photosynthesis, and in this case, the quantum efficiency of light decreases. There is an increased rate of O2- formation under drought stress in many species; it causes lipid peroxidation, fatty acid saturation, and ultimately complete damage to membranes (Smirnof 1993; Foyer et al. 1997; Alexieva et al. 2003). In tests with H_2O_2, one of the ROS that are produced under various abiotic stresses and cause eventual death in plants, it was determined that low-level (10 mM) foliar applications increased pyrroline-5-carboxylate synthase (P5CS) and sucrose-phosphate synthesis in barley and rice species and increased resistance to stress conditions by triggering the systems that regulate osmotic pressure (Maggio et al. 2018).

So much research is carried out in order to solve the mechanism of resistance against abiotic stress, especially drought in plants, and to give direction to breeding studies in this sense. In particular, the tomato species, which is among the vegetable species, is the model plant considered in this respect. Studies continue by using many different molecular techniques for the application of the information obtained in the field of genetic engineering through genome mapping studies (Tanksley and McCouch 1997; Frary et al. 2005; Mir et al. 2012; Solankey et al. 2015; Çelik et al. 2021). In this context, in the study of the researcher group on tomatoes, comparative transcriptomic analyses were performed with the GeneChip Tomato Genome Array (Affymetrix) on plant leaves exposed to water stress. Samples were taken at 0 (control) and 4 days before dehydration before symptoms of dehydration appeared. For Solanum pennellii (Sp) and Solanum lycopersicum (Sl), the effects on Sp in drought tolerant under stress conditions were tried to be defined by examining wild and cultivated species. For transcriptomic information, it was determined by qRT-PCR analysis that up-and down-regulated diffentially expressed genes (DEGs) showed differential expression in the Sp and Sl comparison (Egea et al. 2018).

Due to the rapid change in climatic conditions and the limitation of irrigable agricultural areas day by day, important problems arise, especially in vegetable agriculture due to product losses (Kumar et al. 2012). For this reason, it is of great importance to develop new drought-tolerant varieties for vegetable species affected by drought damage, especially tomatoes (Rai et al. 2011; Solankey et al. 2015; Sánchez-Romera et al. 2018). For this purpose, breeding studies are carried out using both classical breeding methods and biotechnological methods (Mishra and Singh 2014). Although it is known that the drought tolerance of wild species is higher than that of cultivated species, studies have mainly shifted towards biotechnological applications due to the low success rate in hybridizations with wild species. However, there are some limiting factors here too. Therefore, the mutation breeding technique, which is included in the classical breeding methods, is a technique that helps researchers enrich genetic diversity and expand the limited gene pool (Kantoğlu et al. 2014). In this context, studies on the selection of mutants with the desired characteristics in the gene pool developed by induced mutation both in vitro and in vivo conditions and to obtain new varieties from these mutants are still up to date (Kantoğlu et al. 2014; Çelik et al. 2021). In addition, the mutants obtained have the chance to be used as an alternative gene source for hybridization programs (Solankey et al. 2015; Jankuloski et al. 2018). Apart from vegetable species, there are 226 varieties developed and registered as high temperature and drought-tolerant as a result of breeding studies carried out by mutation induced mutation in species such as bread wheat, corn, mulberry, Bermuda grass, Sudan grass, groundnut, grasspea, mungbean, cotton, chickpea, lentil, rice, iris, oat, millet, alfalfa, barley, and soybean (Jankuloski et al. 2018; MVD 2024). Biotechnology, genetic engineering approaches, and conventional breeding studies are important in terms of improving, revealing, and bringing functionally usable gene regions to the plant. However, the costs, environmental effects, the need for high-consistency equipment and time, especially of new generation biotechnological methods, are a significant disadvantage and can limit the breeder. On the other hand, mutation breeding is still a popular method for many species because it is an environmentally friendly breeding method (Solankey et al. 2015; Çelik et al. 2021; Kantoğlu and Kunter 2021).

With the introduction of in vitro techniques in the twentieth century, important developments have been made in the field of plant breeding. In vitro techniques, especially vegetative propagation of plant material under controlled conditions, emerge as a technique that allows controlled selection of the propagated material. For this purpose, embryos, anthers, cells, shoot tips, tubers, bulbs, rhizomes, and cuttings can be easily used for reproduction of mutant plants in in vitro conditions. In this way, it can be very easy to obtain the material that is reproduced fastly. Apart from this, in vitro approaches in plant propagation also serve the purpose of obtaining disease-free plants and preserving genetic material. Obtaining haploid and cybrid plants, in crossbreeding breeding studies, in vitro techniques are also used for purposes such as resque immature embryos and culturing them in artificial nutrient media. Again, in terms of genetic transformation studies, it is possible to obtain transgenic plants by regeneration from plant organs or parts of which gene transfer is made by using in vitro applications (van Harten 1998). However, there are advantages to in vitro techniques in terms of testing abiotic stress factors. Today, in in vitro mutation breeding studies, in vitro techniques open new doors to researchers working on the subject, allowing faster and more effective results (Kantoğlu et al. 2021). Water stress has negative effects on the plant's water and nutrient uptake in the soil by changing the osmotic pressure balance in the root zone (Table 1). This adversely affects plant growth and development. In breeding studies, especially in research on drought stress, in vitro techniques are used to shorten plant testing time and to quickly test a large number of plant materials against drought. For this purpose, polyethylene glycol (PEG) and manitol are intensively used in in vitro nutrient media, and a selection process is carried out according to the development of plants in vitro and their responses to agents (as PEG and manitol) (Kumar et al. 2012). PEG has been used successfully in in vivo studies as well as in drought tests in in vitro studies. Masheva et al. (2022) conducted drought stress tests on 19 mutant M3 green bean lines and two cultivars (Mastillen and Evros) with pot experiments.

Irrigation of the plants was continued by adding 20% PEG to the Murashige and Skoog (MS) nutrient solution, which was applied for irrigation every 2 days after 14 days following the seed sowing and prepared at half strength. Drought tolerance indices of mutant lines were determined by determining leaf, root development, and proline amounts in plants. As a result of the study, it was determined that mutant lines 9, 16, and 20 were the lines that showed the best viability in PEG applications. Kulkarni and Deshpande (2007) conducted an experiment on MS nutrient medium containing 0, 20, 40, and 60 g/l PEG to determine the drought tolerance of mutant and hybrid genotypes in tomato. As a result of the study, it was determined that the Hy-3 hybrid genotype and the MTG-1-4 mutant line performed better than the control and 16 other genotypes. Similarly, Çelik et al. (2021) cultured leaf explants from 7-day old mutant plants in vitro in MS nutrient medium containing 5% PEG. Researchers examined the performance of calluses from mutant explants under dry conditions. It was determined in the study that M34–1 and M34–11 mutant lines from the cultured lines formed calli with higher viability compared to the control. In another study conducted on peppers, shoot tips in vitro were irradiated with gamma rays at doses of 0, 10, and 20 Gy and cultured in MS medium containing PEG. It was determined that some of the mutant plants obtained from both dose applications within the in vitro variation obtained by irradiation maintained their viability due to increased PEG amounts (Salim et al. 2019).

6 Methodological Approach

6.1 Material Determination

Before starting the mutation breeding study, it is important to start the study by knowing the answer of the species planned to be studied for the mutation breeding study with a good literature review. Knowing whether there is a genetic predisposition is a seemingly unimportant factor for researchers to achieve more effective results. Knowing this factor is one of the key points in the success of the study (Anonymous 1977; Sarsu et al. 2018). After determining the genetic predisposition, it is absolutely necessary to start the study with a standard variety. It is important that the breeder knows all the properties of the starting material well, because the differences within the developed mutant population can be observed quickly by the breeder in this way.

6.2 Mutagen Treatment and Breeding Steps

After the material selection is completed, it is necessary to decide what type of mutagen should be applied in the mutation breeding study. Generally, physical mutagen application is the preferred method. Apart from this, chemical mutagen application and sometimes both physical and chemical mutagen applications can be done together in order to change the genetic characteristics that cannot be broken (van Harten 1998; Sarsu et al. 2018; Kantoğlu and Kunter 2021). After the mutagen to be applied is determined, the effective mutagen dose must be determined for the standard variety to be studied. A dose determination study should be performed for this. If seed irradiation is to be performed in the dose determination study, irradiation should be carried out at doses ranging from 0 to 600 Gy (0, 50, 100, 200, 300, 400, 500, and 600 Gray (Gy)) and the trials should be repeated at least three replications and at least 10 seeds in each replication. It should be established to have 10 to 30 seeds. Similar trials are also carried out in chemical mutagen applications. However, in chemical mutagen applications, since factors such as mutagen solution pH, application time, and ambient temperature are effective on mutagen activity, it is important to determine the specified parameters correctly according to the mutagen and variety.

Root lenght, plant height, fresh seedling and root weight, and dry seedling and root weights should be measured in seedlings to determine the effective mutagen dose (EMD50) 30 days after seed sowing following mutagen application. It is also important to monitor seed germination to understand the pre-effects of mutagen following planting. It is a value that is mainly used in the calculation of the effective mutagen dose for the plant height in the seedling measurements made 30 days after the mutagen application. EMD50 is calculated by performing linear regression analysis with the average seedling height values obtained. Here, in the calculation, the dose that gives 50% seedling height of the control plants is determined as a result of linear regression analysis as the effective mutagen dose. After the EMD50 is determined according to the variety to be studied, at least 1000 seeds should be treated with the determined dose of mutagen in order to establish the main mutant population (Anonymous 1977). If the physical conditions are suitable, it is preferred that the number of seeds to be irradiated be greater than 1000 to increase the mutation frequency. Because the mutation rate, which is 1/10,000 in nature, becomes 1/1000 with the application of induced mutation. Therefore, the more material mutagen is applied, the larger the population will be formed (Anonymous 1977; Kantoğlu and Kunter 2021; Kantoğlu 2022). Although the EMD50 dose is known as the basic dose in practice, 10% lower and upper values of this dose can also be used in practice for a large population (van Harten 1998). Seeds should be planted quickly following mutagen application. Seeds can be stored in refrigerator conditions for a maximum of 1 week after mutagen application. Mutagen-induced radicals in seeds stored for a long time due to irradiation. Prolongation of this period is not a desired situation, as there will be substance formation. Following the irradiation, the main population is established with the material called M1 and the seeds are taken from the inbred plants. It is important to store the seed from each plant separately so that the material does not mix from the start. At the M2 stage, taking into account the physical conditions, the seedlings obtained from at least 10 seeds obtained from each plant should be identified and transferred to the field, and observations should be started. One of the most important mistakes we encounter in mutation breeding studies is that the groups doing the study start selection, molecular and chemical analysis at the M1 stage. This is a completely wrong practice. Because, even though the plants are very sensitive due to mutagen application at the M1 stage, many physical damage caused by mutagens occurs in plants. Therefore, the main mutagen-induced effects were observed starting from the M2 stage; the differences in plants (chimeric formations, number of blind plants, differences in plant growth, abnormal formations in flower structure, sterility, etc.) should be recorded and the mutation frequency should be calculated (Fig. 1). As can be seen in Fig. 1, mutant lines are created at the M3 stage with seeds taken from single plants selected at the M2 stage, and the relevant analysis can be started while the necessary observations continue.

In vitro mutation breeding is one of the methods preferred by researchers, depending on the plant species, especially in recent years. It is an important advantage that in vitro conditions allow the selection of more materials by testing plants under laboratory conditions against abiotic stress factors. Figure 2 presents in vitro mutation breeding and the steps to be followed. If it is aimed to develop a character, especially for abiotic stress conditions in mutation breeding studies carried out in both in vitro and in vivo conditions, it is important that field tests and laboratory test results give the same result. The tests should be repeated under field conditions for at least 2 years, taking into account the ecological conditions, and the statistical results should be stable. Otherwise, the results will be misleading and may mislead the researcher.

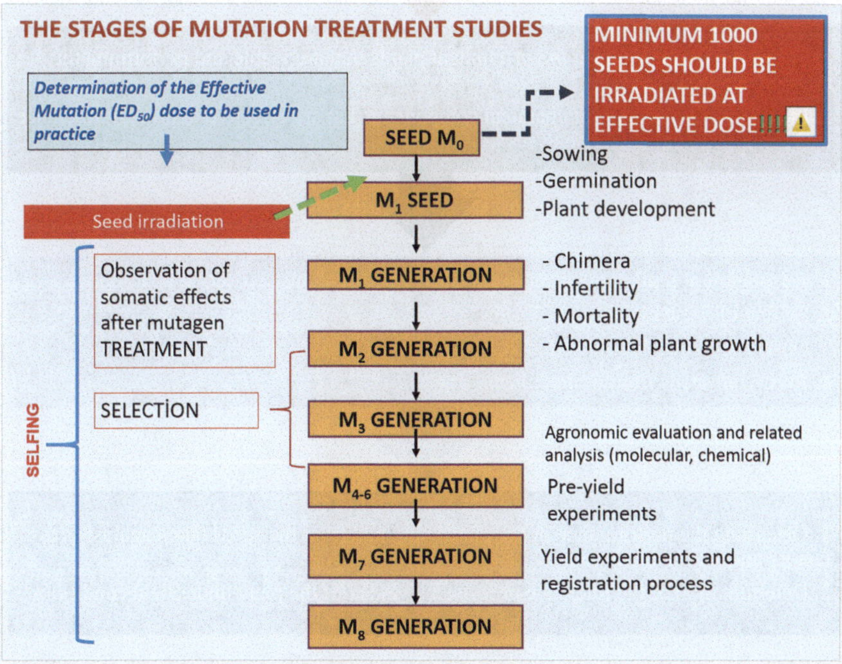

Fig. 1 Stages of mutation breeding in vegetable crops

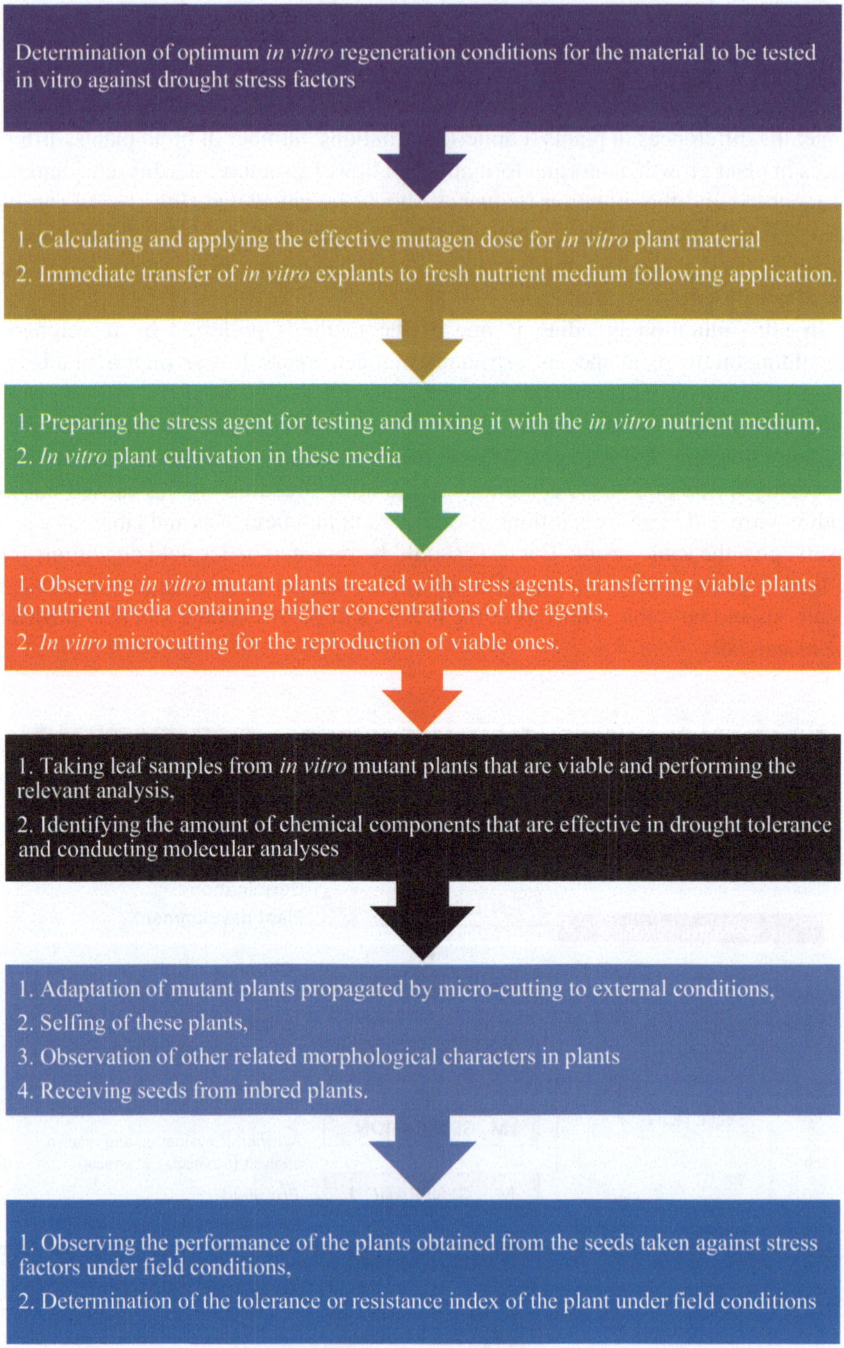

Fig. 2 *In vitro* mutation breeding steps

7 Outcomes of Earlier Studies

Effects that occur mainly due to climatic conditions such as drought, high temperature, salinity, and low temperature damage, which are defined as abiotic stress, are very effective in reducing productivity in terms of agricultural production. Especially with global warming, the increasing temperatures and the decrease in the amount of irrigation water during the growing period cause serious problems in terms of vegetable production. For these reasons, breeding studies for vegetables are shifting to the development of new varieties that are tolerant to abiotic stress conditions and can evaluate these conditions (Rai et al. 2011). It is obvious that considering the shrinking cultivation areas and yield losses, the current production may reach its limits and will not be enough to feed the world population in the next 50 years (Solankey et al. 2015). Tomato (the mostly studied), pepper, and eggplant members of the Solanaceae family, which are in the group of summer vegetables, appear as the species most affected by drought damage. Apart from these species, green beans, peas, and leafy vegetable species are also adversely affected by drought (Kumar et al. 2012; Masheva et al. 2022). Due to the problems that cannot be overcome in studies carried out with classical breeding methods, the use of biotechnological methods such as tissue culture in hybridization breeding, transgene technology, and genome editing methods have also been integrated into breeding studies in order to get faster and more effective results (Salava et al. 2021). Although wild species such as L. pimpinelli L. is known to be drought tolerant, drought resistance has been tried to be transferred to cultivars, no significant improvement has been achieved yet, especially in tomato (Salava et al. 2021). For these and similar reasons, benefiting from mutation breeding techniques contributes to developing an alternative gene pool for breeders. Breeding studies are carried out for the determination of drought tolerance mutants and their breeding as varieties by using in vitro and in vivo testing and testing methods within the variation created by the chemical and physical mutagens used for this purpose. In addition, studies are carried out on mutants to determine the genomic factors controlling drought by using different molecular screening techniques (Fig. 3).

Mutant lines are developed in vivo by mutation breeding. Many studies have been conducted and are still being carried out to identify tolerant mutants after PEG applications at different rates in in vitro conditions (Table 2). Thus, testing of more mutant genotypes and selection of genetic material with the desired tolerance can be achieved. The mutant population obtained with mutagens applied in vitro conditions has been successfully tested in a similar way, and hopeful material selection is achieved (Table 2). In in vitro cultures, callus cultures are mainly used, as well as techniques such as direct seed sowing and shoot tip culture are used in practice. The most important point is to test the selected genetic material under field conditions by evaluating the results of the tolerant material after chemical analysis (like proline, ABA). In this way, it is ensured that mutants that show successful performance in both in vitro and in vivo tests are brought into production as varieties. Mutant cultivars registered by successfully passing all these tests are presented in Table 3.

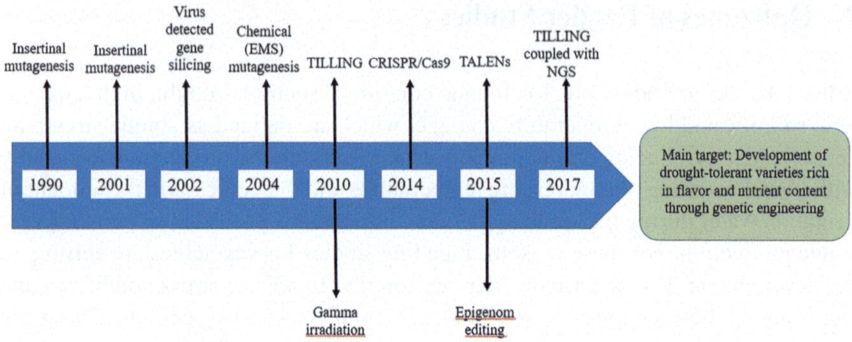

Fig. 3 The progress of studies carried out with mutation breeding, especially tomatoes, after 1990. (Modified from Salava et al. (2021)

Table 2 Examples of *in vitro* drought tolerance tests on vegetable species. (Modified from Kantoğlu et al. 2021)

Species	Stress factor and agent	*In vitro* culture type	Referances
Capsicum annuum L.	Drougt tolerance (PEG and mannitol treatment)	Callus and suspension culture	Santos-Díaz and Ochoa-Alejo (1994)
Dacus carota L.	Drougt tolerance (PEG)	Suspension culture	Fallon and Phillips (1989)
Lycopersicum esculentum L.	Drougt tolerance (PEG)	Callus culture	Çelik et al. (2021)
Lycopersicum esculentum L.	Drougt tolerance (PEG)	Seed culture	Roy et al. (2017)
Lycopersicum esculentum L.	Drougt tolerance (PEG)	Callus culture	Aazami et al. (2010)
Lycopersicum esculentum L.	Drougt tolerance (PEG)	Seed culture	Kulkarni and Deshpande (2007)
Lycopersicum esculentum L.	Drougt tolerance (PEG)	Callus culture	Abdel-Raheem et al. (2007)
Lycopersicum esculentum L.	Drougt tolerance (PEG)	Seed culture	Manoj and Uday (2007)
Lycopersicum esculentum L.	Drougt tolerance (PEG)	Callus culture	Al Hattab et al. (2015)
Solanum melongena L.	Drougt tolerance (PEG)	Callus culture	Siaga et al. (2016)
Solanum melongena L.	Drougt tolerance (PEG)	*In vitro* plantlet	Zayova et al. (2017)
Capsicum annum L.	Drougt tolerance (PEG)	Shoot tip culture	Salim et al. (2019)

Although the number of cultivars developed seems to be low, studies on mutation breeding and drought stress continue within the scope of many national and international projects. As can be seen in Table 3, tomato is dominant among the mutant cultivars developed, which indicates the genetic susceptibility of tomato to mutation. Because a species' genetic predisposition is an important factor in the success

Table 3 Drought tolerant mutant vegetable cultivars registered in the IAEA database (MVD 2024)

Species	Variety name	Country	Mutagen type	Characteristics	Registration date
Lycopersicum esculentum L.	Giron 50	Cuba	Gamma Ray (300 Gy)	High temperature tolerance	2021
Lycopersicum esculentum L.	Domi	Cuba	Gamma Ray (500 Gy)	Drought tolerance, High temperature tolerance, High yield, Disease resistence	2015
Lycopersicum esculentum L.	Maybel	Cuba	Gamma Ray (500 Gy)	Drought tolerance,	2007
Lycopersicum esculentum L.	Magine	Cuba	Gamma Ray (300 Gy)	Drought tolerance, Good quality	2007
Lycopersicum esculentum L.	Summer Star	Mauritius	Gamma Ray (390 Gy)	High temperature tolerance	2019
Solanum melongena L.	PKM	India	Gamma Ray (250 Gy)	Drought tolerance, High yield, Long storage period	1985
Pisum sativum L.	Talovets 60	Russian Federation	Chemical mutagen	Adaptation to different climates (high temperature), Resistance to loding	1993
Allium macrostemon L.	Ningsuan 1	China	Gamma Ray (15 Gy)	Drought tolerance, High yield, Good quality, Disease resistance, Good storage and transportation	1990
Lactuca sativa L.	Evergreen	Japan	^{32}P and gamma rays (100 μC + 420 Gy)	Tolerance to high temperature	1966
Lactuca sativa L.	Giantgreen	Japan	^{32}P	Tolerance to high temperature	1966

of mutation breeding studies. Therefore, mutation breeding studies are especially valuable for craeting mutant gene pools for breeding studies (Rai et al. 2011; Kumar et al. 2012; Salava et al. 2021; Muñoz et al. 2021; Kantoğlu 2022; Masheva et al. 2022). Therefore, it is important not to ignore the mutation breeding technique for breeding studies carried out under abiotic stress conditions and to integrate it more effectively into studies.

8 Conclusion

The increase in the world population and the parallel increase in food demand is an inevitable result. It is seen that agricultural production areas have decreased all over the world due to both climate change and the negative consequences of industrialization, and especially good quality agricultural lands have been lost. The prevalence of abiotic stress conditions such as drought, high temperature, and salinity, which is in question due to climate change, also causes an increase in losses in agricultural production. Therefore, it is important for vegetable species to develop new varieties that can tolerate these negative factors in production. In the history of humanity, during the transition period from a hunter-gatherer life to an agricultural society and afterwards, human beings have instinctively selected plants with higher appeal and produced better yields. However, due to the negative developments in the change of climatic conditions, the direction of improvement in the problem of water scarcity and the accompanying drought began to differ. Different kinds of studies have been done and continue to be done in order to understand the reactions of the plant and these reaction mechanisms, especially under stress conditions. By using conventional breeding methods, biotechnological approaches, transgene technology, CRISPR/Cas9, epigenome editing, genome editing techniques, studies are carried out to understand the working mechanisms of genes and gene groups that control drought, to develop more adaptable and tolerant varieties. The mutation breeding technique, which is included in the conventional breeding methods, especially for the enrichment of the shrinking gene pool, still maintains its popularity and contributes to the researchers. In addition to enriching the gene pool, TILLING, NGS, omic, and proteomic screenings for determining the functions of genes or gene groups changed by mutation breeding provide valuable information about the explanation of the drought tolerance mechanism and its transmission. Today, both *in vivo* and *in vitro* breeding studies and biotechnological applications are important in terms of accelerating the breeding process and obtaining more effective results. As a result of both *in vivo* and *in vitro* mutation breeding studies, it was seen that the number of breeding studies carried out with the mutation breeding method has increased with the increase in the number of superior vegetable varieties brought to agricultural production and the technique still maintains its popularity. In particular, the International Atomic Energy Agency, which supports the peaceful use of nuclear techniques in different scientific disciplines and operates under the umbrella of the United Nations, provides great support for the use of nuclear techniques in the field of agriculture and the development of new varieties. In recent years, it has increased its support programs for the development of new tolerant varieties in order to produce solutions for climate change and the problems that come with it. It is known that countries active in space studies, such as China, the United States of America, and the Russian Federation, are working to develop new varieties adaptable to space conditions by exposing the seeds of different species to cosmic radiation. For this reason, the use of fast neutron and electron accelerator sources, which act at different penetration levels in the plant material to be irradiated, apart from UV, X-ray, and gamma sources, which are used as physical mutagens today, is on an increasing trend in countries such as Korea, China, and Israel. Therefore, it will be meaningful to plan breeding studies with the use of different mutagen sources in order to create richer variation in the studies for the development of new vegetable varieties with tolerance to abiotic stress conditions for the future.

References

Aazami MA, Torabi M, Jalili E (2010) In vitro response of promising tomato genotypes for tolerance to osmotic stress. Afr J Biotechnol 9(26):4014–4017

Abdel-Raheem AT, Ragab AR, Kasem ZA, Omar FD, Samera AM (2007) In vitro selection for tomato plants for drought tolerance via callus culture under polyethylene glycol (PEG) and mannitol treatments. In: African crop science conference proceedings, vol. 8., No. 8, pp 2027–2032

Ahloowalia BS, Maluszynski M (2001) Induced mutations-a new paradigm in plant breeding. Euphytica 118(2):167–173

Ahloowalia BS, Maluszynski M, Nichterlein K (2004) Global impact of mutation-derived varieties. Euphytica 135:187–204

Al Hattab ZA, Qaudhy WK, Razaq AA, Al Kaaby EA, Al Ani JA (2015) In vitro regeneration of tomato (*Lycopersicon esculentum*) plants under drought stress. Int J of Multidiscip Curr Res 3:1194–1198

Alexieva V, Ivanov S, Sergiev I, Karanov E (2003) Interaction between stresses. Bulg J Plant Physiol 29(3–4):1–17

Anderson JV, Hess JL, Chevone BI (1990) Purification, characterization, and immunological properties for two isoforms of glutathione reductase from eastern white pine needles. Plant Physiol 94(3):1402–1409

Anonymous (1977) Manual on mutation breeding, International atomic energy agency technical reports series no: 119. IAEA, Vienna., ISBN: 92-0-115077-6, p 288

Anonymous (2022a) https://www.mpg.de/18190093/eunice-newton-foote Excessed on July 2022

Anonymous (2022b) https://theconversation.com/john-tyndall-the-forgotten-co-founder-of-climate-science-143499. Excessed on July 2022

Babiye B (2019) Application of biotechnology as a breeding tool in crop improvement for drought tolerances. Int J Res Stud Microbiol Biotechnol (IJRSMB) 5(1):32–39

Bechtold U (2018) Plant life in extreme environments: how do you improve drought tolerance? Front Plant Sci 9:543

Blum A (2005) Drought resistance, water-use efficiency, and yield potential-are they compatible, dissonant, or mutually exclusive? Aust J Agric Res 56(11):1159–1168

Bradford MM (1976) A rapid and sensitive method for the quantitation of microgram quantities of protein utilizing the principle of protein-dye binding. Anal Biochem 72(1–2):248–254

Callendar GS (1938) The artificial production of carbon dioxide and its influence on temperature. Q.J.R Meteorol Soc 64:223–240

Çelik Ö, Ayan A, Meriç S, Atak Ç (2021) Comparison of tolerance related proteomic profiles of two drought tolerant tomato mutants improved by gamma radiation. J Biotechnol 330:35–44

Daskalov S (1986) Mutation breeding in pepper. Mutat Breed Rev 4:1–26

Egea I, Albaladejo I, Meco V, Morales B, Sevilla A, Bolarin MC, Flores FB (2018) The drought-tolerant *Solanum pennellii* regulates leaf water loss and induces genes involved in amino acid and ethylene/jasmonate metabolism under dehydration. Sci Rep 8:2791

Fallon KM, Phillips R (1989) Responses to water stress in adapted and unadapted carrot cell suspension cultures. J Exp Bot 40(6):681–687

Foyer CH, Lopez-Delgado H, Dat JF, Scott IM (1997) Hydrogen peroxide-and glutathione-associated mechanisms of acclimatory stress tolerance and signalling. Physiol Plant 100(2):241–254

Frary A, Xu Y, Liu J, Mitchell S, Tedeschi E, Tanksley S (2005) Development of a set of PCR-based anchor markers encompassing the tomato genome and evaluation of their usefulness for genetics and breeding experiments. Theor Appl Genet 111:291–312

Gaxiola RA, Li J, Undurraga S, Dang LM, Allen GJ, Alper SL, Fink GR (2001) Drought- and salt-tolerant plants result from overexpression of the AVP1 H+-pump. Proc Natl Acad Sci U S A (PNAS) 98(20):11444–11449

Gimenez C, Mitchell VJ, Lawlor DW (1992) Regulation of photosynthetic rate of two sunflower hybrids under water stress. Plant Physiol 98(2):516–524

Hsieh T-H, Li C-W, Su R-C, Cheng C-P, Tsai SY-C, Chan M-T (2010) A tomato bZIP transcription factor, SlAREB, is involved in water deficit and salt stress response. Planta 231:1459–1473

Jankuloski L, Foster BP, Nakagawa H (2018) Major traits to be improved by mutation breeding: Manual on mutation breeding, 3rd edn, pp 177–204

Kantoğlu KY (2022) Effects of gamma irradiation on seed germination in watermelon (*Citrullus lanatus* (Thunb.) Matsum & Nakai) and determination of effective mutation dose. Alatarim 21(1):10–17

Kantoğlu KY, Kunter B (2021) Mutasyon ıslahı: Süs bitkileri ıslahı (klasik ve biyoteknolojik yöntemler), 145–202 p

Kantoğlu Y, Seçer E, Erzurum K, Tutluer İ, Kunter B, Peşkircioğlu H Sağel Z (2010) Mass screening techniques for selecting crops resistant to disease. Improving tolerance to F*usarium oxysporum* f sp *melonis* in melon using tissue culture and mutation techniques: Mass screening techniques for selecting crops resistant to disease, 235–243

Kantoğlu KY, Tepe A, Kunter B, Firat AF, Peşkircioğlu H (2014) Vegetable crops breeding by induced mutation and a practical case study of *Capsicum annuum* L. In: Tomlekova NB, Kozgar MI, Wani MR (eds) Mutagenesis: exploring genetic diversity in crops. Wageningen Publishers, pp 41–56

Kantoğlu KY, Kuşvuran Ş, Ellialtıoğlu ŞŞ (2021) Doku kültürü yöntemlerinin stres çalışmalarında kullanım: Sebzelerde stres toleransı ve ıslah stratejileri, pp 708–739

Khalil F, Naiyan X, Tayyab M, Pinghua C (2018) Screening of EMS-induced drought-tolerant sugarcane mutants employing physiological, molecular and enzymatic approaches. Agronomy 8(10):226

Khanzada H, Wassan GM, He H, Mason AS, Keerio AA, Khanzada S, Rasheed A (2020) Differentially evolved drought stress indices determine the genetic variation of *Brassica napus* at seedling traits by genome-wide association mapping. J Adv Res 24:447–461

Kharkwal MC, Pandey RN, Pawar SE (2004) Mutation breeding for crop improvement. In: Jain HK, Kharwal MC (eds) Plant breeding. Springer, Dordrecht, pp 601–645

Kicheva MI, Tsonev TD, Popova IP (1994) Stomatal and nonstomatal limitations to photosynthesis in two wheat cultivars subjected to water stress. Photosynthetica (Czech Republic) 30:107–116

Kim H-S, Lee JH, Kim JJ, Kim C-H, Jun J-S, Hong Y-N (2005) Molecular and functional characterization of CaLEA6, the gene for a hydrophobic LEA protein from *Capsicum annuum*. Gene 344:115–123

Kogan FN (1997) Global drought watch from space. Bull Am Meteorol Soc 78(4):621–636

Kulkarni M, Deshpande U (2007) In vitro screening of tomato genotypes for drought resistance using polyethylene glycol. Afr J Biotechnol 6(6):691–696

Kumar R, Solankey SS, Singh M (2012) Breeding for drought tolerance in vegetables. Veg Sci 39(1):1–15

Kumar P, Rouphael Y, Cardarelli M, Colla G (2017) Vegetable grafting as a tool to improve drought resistance and water use efficiency. Front Plant Sci 8:1130

Kumawat S, Rana N, Bansal R, Vishwakarma G, Mehetre ST, Das BK, Deshmukh R (2019) Expanding avenue of fast neutron mediated mutagenesis for crop improvement. Plan Theory 8(6):164

Maggio A, Bressan RA, Zhao Y, Park J, Yun DJ (2018) It's hard to avoid avoidance: Uncoupling the evolutionary connection between plant growth, productivity and stress tolerance. Int J Mol Sci 19(11):3671

Manoj K, Uday D (2007) Gradient in vitro testing of tomato (*Solanum Lycoersicon*) genotypes by inducing water deficit: a new approach to screen germplasm for drought tolerance. Asian J Plant Sci 6(6):934–940

Masheva V, Spasova-Apostolva V, Aziz S, Tomlekova N (2022) Variations in proline accumulation and relative water content under water stress characterize bean mutant lines (*P. vulgaris* L.). Bulgarian J Agr Sci 28(3):430–436

Miller G, Suzuki N, Ciftci-Yilmaz S, Mittler R (2010) Reactive oxygen species homeostasis and signalling during drought and salinity stresses. Plant Cell Environ 33(4):453–467

Mir RR, Zaman-Allah M, Sreenivasulu N, Trethowan R, Varshney RK (2012) Integrated genomics, physiology and breeding approaches for improving drought tolerance in crops. Theor Appl Genet 125:625–645

Mishra D, Singh B (2014) Studies on effectiveness and efficiency of gamma rays in green gram (*Vigna radiata* (L.) Wilczek). SABRAO J Breed Genet 46(1):34–43

Mo KC, Paegle JN, Higgins RW (1997) Atmospheric processes associated with summer floods and droughts in the central United States. J Clim 10(12):3028–3046

Muñoz LC, Debouck DG, Rivera M, Muñoz JE, Alpala D, Sarsu F, Rao IM (2021) Mutation breeding for heat and drought tolerance in tepary bean ('*Phaseolus acutifolius*' A. Gray). Aust J Crop Sci 15(9):59–67

MVD (2024) IAEA Mutant Variety Data Base. https://nucleus.iaea.org/sites/mvd/SitePages/Search.aspx. Accessed on 18 Apr 2024

Navari-Izzo F, Pinzino C, Quartacci MF, Sgherri CLM, Izzo R (1994) Intracellular membranes: kinetics of superoxide production and changes in thylakoids of resurrection plants upon dehydration and rehydration. Proc R Soc Edinburgh, Sect B: Biol Sci 102:187–191

Oladosu Y, Rafii MY, Abdullah N, Hussin G, Ramli A, Rahim HA, Usman M (2016) Principle and application of plant mutagenesis in crop improvement: a review. Biotechnol Biotechnol Equip 30(1):1–16

Potopa V, Mozn M, Soukup J (2012) Drought evolution at various time scales in the lowland regions and their impact on vegetable crops in The Czech Republic. Agric For Meteorol 156:21–133

Rai MK, Kalia RK, Singh R, Gangola MP, Dhawan AK (2011) Developing stress tolerant plants through in vitro selection-an overview of the recent progress. Environ Exp Bot 71(1):89–98

Roy MR, Rashed MRU, Mitu AS (2017) Screening and diversity analysis of drought tolerant genotypes in vitro in tomato. Agric Res Technol 4(2):1–6

Sağel Z, Peşkircioğlu H, Tutluer İ, Uslu N, Şenay A, Taner KY, Yalçın S (2002) Bitki Islahında Mutasyon ve Doku kültürü Teknikleri. III Ulusal Mutasyon Kursu Kurs Notları:1–60

Saima S, Li G, Wu G (2018) Effects of drought stress on hybrids of *Vigna radiata* at germination stage. Acta Biol Hung 69(4):481–492

Salava H, Thula S, Mohan V, Kumar R, Maghuly F (2021) Application of genome editing in tomato breeding: Mechanisms, advances, and prospects. Int J Mol Sci 22(2):682

Salim HG, El Kaaby EA, Al-Ajeel SA (2019) Response of irradiated chilli pepper shoots (*Capsicum annuum*. L.) propagated in vitro to drought stress tolerance. Int J Multidisc Curr Res 7:258–260

Sallam A, Mourad AM, Hussain W, Stephen Baenziger P (2018) Genetic variation in drought tolerance at seedling stage and grain yield in low rainfall environments in wheat (*Triticum aestivum* L.). Euphytica 214(9):1–18

Sánchez-Romera B, Calvo-Polanco M, Ruiz-Lozano JM, Zamarreño ÁM, Arbona V, García-Mina JM, Aroca R (2018) Involvement of the def-1 mutation in the response of tomato plants to arbuscular mycorrhizal symbiosis under well-watered and drought conditions. Plant Cell Physiol 59(2):248–261

Santos-Díaz MDS, Ochoa-Alejo N (1994) PEG-tolerant cell clones of chili pepper: growth, osmotic potentials and solute accumulation. Plant Cell Tissue Organ Cult 37(1):1–8

Sarsu F, Ghanim AMA, Das P, Bahuguna RN, Kusolwa PM, Ashraf M, Singla-Pareek SL, Pareek A, Forster BP, Ingelbrecht I (2018) Pre-field screening protocols for heat-tolerant mutants in rice. Springer, Cham, pp 1–39

Seth R, Kendurkar SV (2015) In vitro screening: An effective method for evaluation of commercial cultivars of tomato towards salinity stress. Int J Curr Microbiol App Sci 4(1):725–730

Shamim F, Farooq K, Waheed A (2014) Effect of different water regimes on biometric traits of some tolerant and sensitive tomato genotypes. J Anim Plant Sci 24(4):1178–1182

Siaga E, Maharijaya A, Rahayu MS (2016) Plant growth of eggplant (*Solanum melongena* L.) in vitro in drought stress polyethylene glycol (PEG). Biovalentia: Biol Res J 2(1):10–17

Smirnof N (1993) The role of active oxygen in the response of plants to water deficit and desiccation. New Phytol 125:27–58

Solankey SS, Singh RK, Baranwal DK, Singh DK (2015) Genetic expression of tomato for heat and drought stress tolerance: an overview. Int J Veg Sci 21(5):496–515

Suzuki N, Koussevitzky SHAI, Mittler RON, Miller GAD (2012) ROS and redox signalling in the response of plants to abiotic stress. Plant Cell Environ 35(2):259–270

Tambussi EA, Bartoli CG, Beltrano J, Guiamet JJ, Araus JL (2000) Oxidative damage to thylakoid proteins in water-stressed leaves of wheat (*Triticum aestivum*). Physiol Plant 108(4):398–404

Tanksley SD, McCouch SR (1997) Seed banks and molecular maps: unlocking genetic potential from the wild. Science 277(5329):1063–1066

Tezara W, Lawlor DW (1995) Effects of water stress on the biochemistry and physiology of photosynthesis in sunflower. In: Photosynthesis: From Light to Biosphere, vol 4. Kluwer Academic Publishers, pp 625–628

van Harten AM (1998) Mutation breeding: theory and practical applications. Cambridge University Press, Cambridge, UK, p 353

Varshney RK, Pazhamala L, Kashiwagi J, Gaur PM, Krishnamurthy L, Hoisington D (2011) Genomics and physiological approaches for root trait breeding to improve drought tolerance in chickpea (Cicer arietinum L.). In: Root Genomics. Springer, Berlin/Heidelberg, pp 233–250

Vu JC, Baker JT, Pennanen AH, Allen LH Jr, Bowes G, Boote KJ (1998) Elevated CO_2 and water deficit effects on photosynthesis, ribulose bisphosphate carboxylase-oxygenase, and carbohydrate metabolism in rice. Physiol Plant 103(3):327–339

Webber AN, Nie GY, Long SP (1994) Acclimation of photosynthetic proteins to rising atmospheric CO_2. Photosynth Res 39(3):413–425

Williams J, Bulman MP, Neill SJ (1994) Wilt-induced ABA biosynthesis, gene expression and down-regulation of rbcS mRNA levels in *Arabidopsis thaliana*. Physiol Plant 91(2):177–182

Zayova E, Philipov P, Nedev T, Stoeva D (2017) Response of in vitro cultivated eggplant (*Solanum melongena* L.) to salt and drought stress. AgroLife Sci J 6(1):276–282

Zijlstra S, Purimahua C, Lindhout P (1991) Pollen tube growth in interspecific crosses between *Capsicum* species. HortScience 26(5):585–586

An Overview of Doubled Haploid Plant Production to Improve Drought Tolerance

Yıldız Aydın, Ezgi Çabuk Şahin, Yunus Emre Aktaş, and Ahu Altınkut Uncuoğlu

Abstract Drought stress adversely affects plant growth and causes extensive losses in agricultural production. The inheritance of traits is important for developing effective breeding program and the most cost-effective and environmentally friendly approach to improve drought tolerance is deployment of effective resistant genes in cultivars. F1 cultivars allow plant breeders to combine the best traits and multiple host plant resistance. If the parents are homozygous, the hybrids will be uniform, an increasingly important trait in production. However, producing homozygous pure parents takes a long time, a lot of effort and cost. The haploid form is providing a good basis for producing a pure line of plants (DH-doubled haploids) allowing new varieties to be regenerated. The aim of this review is to provide an overview of DH production and different potential methods for haploid induction and their potential effect in the development of drought tolerant germplasm.

Keywords Haploid · Double haploid · Drought tolerance

1 Introduction

Factors such as the rapid increase in the world population and climate change threaten the stable supply of food production on a global scale. The gradual decrease of arable land on earth due to unplanned urbanization, desertification, increased salinity and global warming makes new regulations to expand agricultural areas and obtain more products from unit area inevitable. Quantitative increase in plants,

Y. Aydın (✉)
Science Faculty, Department of Biology, Marmara University, Istanbul, Türkiye
e-mail: ayildiz@marmara.edu.tr

E. Ç. Şahin · Y. E. Aktaş
Institute of Pure and Applied Sciences, Marmara University, Istanbul, Türkiye

A. A. Uncuoğlu
Engineering Faculty, Department of Bioengineering, Marmara University, Istanbul, Türkiye

U. K. Chaudhry et al. (eds.), *Drought Stress*,
https://doi.org/10.1007/978-3-031-80610-0_16

which is the main source of nutrition for the world population, is only possible if more products are obtained per unit area. Sustainable increase in agricultural production is necessary to feed the world population and the amount of product obtained from a unit area in plant production depends on both the genotypic potential of the plant and environmental factors. Due to their sessile nature, plants encounter many stress factors that will negatively affect their growth and development throughout their life cycle. These stress factors, which may be of biotic or abiotic origin, may adversely affect the quantity and quality of the product by creating physiological and biochemical damages in plants.

Abiotic stress conditions include environmental factors such as temperature (extreme heat and cold), water stress (drought and flooding), soil pH (salinity or acidic), nutrients deficiency and toxicity (minor and major elements), mechanical stress, and excess light (Singh et al. 2021). These factors act as primary stress factors for plants and trigger secondary stress factors such as osmotic, oxidative, and ionic stresses. Drought is the most limiting of all abiotic stressors. It impacts more than one-third of soils globally and lowers agricultural productivity more than any other environmental stress (Lambers et al. 2008).

The polygenic complexity of plant breeding and a lack of understanding of the underlying processes of drought tolerance are causing poor progress in enhancing agricultural production in drought-prone environments (Tuberosa and Salvi 2006; Cattivelli et al. 2008).

For sustainable agricultural development and economic success of breeding activities, it is necessary to develop varieties with desirable characteristics. Several breeding approaches are presented, with a focus on developing appropriate selection criteria, parental selection, hybridization and selection, mutation breeding, tissue culture, and doubled haploids to enhance plant drought tolerance. The production of homozygous plants is important for both genetic studies and hybrid seed production, especially in plants where foreign pollination is high. Dihaploid technology is an important application in obtaining homozygous lines used to accelerate plant breeding programs.

2 Enhancement of Plant Drought Tolerance by Double Haploids

Agricultural production is expected to decrease further in many regions in the future due to the decrease in arable land and water resources that can be used in agriculture and the intensifying global climate change (Mickelbart et al. 2015). Considering the rapidly increasing world population with global climate change, the development of plants that can maintain their yield under stress conditions and influencing global agricultural systems and requiring the development of stress-tolerant crop varieties (Takeda and Matsuoka 2008; Newton et al. 2011).

To maintain high yields throughout irregular growing seasons, agricultural crops depend especially heavily on the availability of water. Thus, agricultural drought, or a lack of water, has a negative impact on plant and crop productivity by affecting leaf size, stem extension, and root proliferation, disrupting plant water and nutrient relationships, and reducing water-use efficiency (Chaves et al. 2003; Liang et al. 2016). Plant responses to abiotic stress are dynamic and complex (Kosová et al. 2011).

Plants can stretch their growth and development mechanisms in a way that will be least damaged by these changes that may occur in environmental conditions, and they can even adapt in a way that will be least affected by environmental factors when grown in the same climatic conditions for long periods of time.

Plants need water, light, mineral nutrients, and carbon for growth. Environmental conditions that lower development and yield below optimal levels are described as abiotic stress. Environmental factors such as water, temperature, and salinity affect plants according to their habitat may differ from species to species (Ashraf and Foolad 2007).

The conventional and modern biotechnology tools support breeding programs by responding quickly to biotic and abiotic restrictions including the development drought of tolerance and water saving plant varieties (Tefera 2017). For breeding purpose, the rapid and low-cost generation of hybrid seeds, the assessment of variety in genetic pools, and the production of homozygous lines are critical (Maraschin et al. 2005).

Due to its various superior agronomic benefits (high yield and quality, resistance/tolerance to biotic and/or abiotic stress factors, and wide adaptability), hybrid cultivars are now used in the production of plants. F1 cultivars have great uniformity, high yields due to heterosis, and allow for quick production and appropriate genotype selection. Without homozygous pure parental lines with well-defined characteristics and strong combining ability, hybrids cannot be produced. However, it takes a long time, a lot of effort, and a huge amount of cost to produce homozygous pure parents. Additionally, traditional breeding techniques need 8–10 years to produce an F1 hybrid variety, especially in a highly open-pollinated species like *Cucurbita* spp. (Kurtar et al. 2020).

As a result of the integration of biotechnological approaches into breeding programs, crucial characteristics like as increased yield and quality, resistance to diseases and pests, and adaptability to unfavorable environmental conditions can be imparted to new varieties more rapidly and effectively (Sarıkamış 2017). This integration is also significant in cultivar breeding in terms of identifying dominant and recessive genes (Ferrie and Möllers 2011).

Using homozygous lines, it is possible to improve selection efficiency or additive genetic variation for quantitative traits like drought tolerance. Since haploids contain only one set of alleles at each locus, they allow to development of pure lines resulting from breeding studies in a shorter time. The biotechnological techniques for doubled haploidy have several significant benefits for both basic and applied research (Dunwell 2010).

Plant breeders have traditionally achieved homozygosity through self-fertilization, a time-consuming method, although plants are rarely entirely homozygous (Sopory and Munshi 1996). In breeding studies, it is aimed to transfer the variations in the genetic structure and natural distribution of plants to the offspring, usually through heredity. Here, the transfer of the desired traits is carried out by interspecies hybridization and the development of a new variety is a long and laborious process. In this process, a large number of individuals obtained in each generation are observed and tested using morphological markers based on phenotypic characteristics, and breeding work is continued by selecting suitable genotypes for the next generation.

However, in traditional selections based on phenotypic characteristics, problems such as the scarcity of markers, presence of pleiotropic and epistatic effects, and morphological characters being affected by variable environmental conditions are encountered (Melchinger et al. 1990). Intermediate expressions due to heterozygosity are not observed in double haploid lines. Therefore, identification of phenotypic markers can be much more effective. The purification process, which usually takes many years, can be completed in a shorter time with the dihaploidization method, in which 100% homozygous pure dihaploid (DH) lines can be obtained by doubling the chromosome numbers of haploid plants and these haploid plants (Fig. 1).

The function of DH in the breeding process is primarily determined by whether the plant is self-pollinated or cross-pollinated. There are several techniques for producing haploids, and subsequently DHs (after a process of chromosome doubling), using both female and male gametophytes based on *in vitro* and *in vivo* haploid induction processes, or a combination of the two to produce haploid embryos *in vivo* and then rescue them *in vitro*.

Haploid plants have as many chromosomes in their somatic cells as there are in the gamete cells of the plant species to which they belong.

In higher plants, natural sporophytic haploids were first observed in pipe flower (*Datura stramonium* L.) (Blakeslee et al. 1922), then in tobacco (*Nicotiana tabacum*), wheat (*Triticum aestivum*) and various other plants (Forster et al. 2007). It is not possible to plan and carry out breeding programs based on natural haploid formation due to the insufficient and irregular emergence of naturally occurring haploids. It is absolutely necessary to obtain haploids regularly and in numbers that can be used in breeding programs. For this purpose, researchers according to the species in obtaining haploid plants have developed different methods. Haploids can be formed by themselves, by *in vivo* (interspecific or intraspecific hybridization, centromere-mediated haploidization), or *in vitro* (immature female or male gametophyte culture) methods.

The starting materials used to obtain haploid plants, which contain only one series of alleles at each locus and have an important place in breeding studies with these characteristics, are gametes (Dunwell 2010), and can be obtained *in vitro* by stimulating male or female organs against haploidy. (Ellialtıoğlu et al. 2001; Germana 2011a). However, in order to use haploidization technique in a breeding program and to be successful, first, the predisposition of genotypes to haploidy is important. A haploid plant can be obtained by in situ induction, hybridization

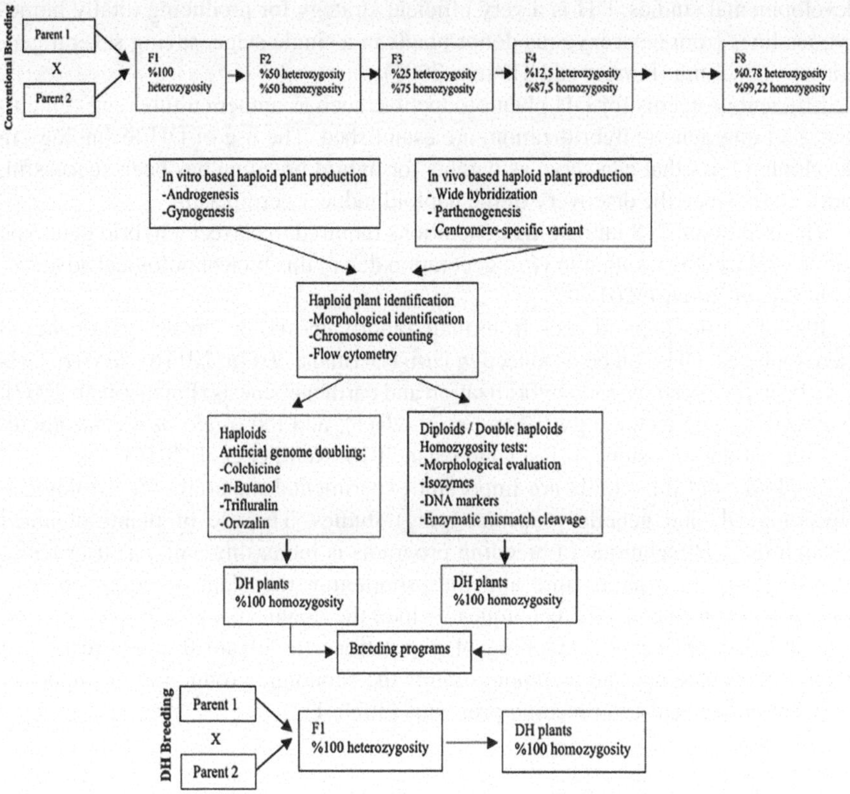

Fig. 1 The schem for the production of doubled haploidy (DH) both *in vitro* and *in vivo* and its applications in plant breeding

between distant relatives, delaying pollination, pollination with abortive or irradiated pollen, temperature shocks, different chemicals, hormone applications, X- or UV-rays. On the other hand, haploid plants can be obtained by in vitro stimulation, androgenesis (anther or microspore culture) or gynogenesis (ovule culture, ovarian culture, or embryo recovery) methods (Arı 2006).

A DH cell is created through induced or spontaneous chromosomal doubling, which can then be developed into a doubled haploid plant (Yali and Mitiku 2022). Haploid plants can occur spontaneously in some plant species in various natural ways (Kasha 1974). The frequency of spontaneous haploid formation, which is formed by one of the ways of androgenesis, gynogenesis, semigamy, polyembryony or chromosome elimination in nature, remains at a low level of 0.001–0.01%, while natural haploid formation is not encountered in many species (Pochard and Dumas de Vaulx 1971).

Haploid plants must be transformed into diploid plants to be used in plant breeding programs. Haploids and completely homozygous doubled haploid populations are valuable sources to analyze the important agronomic traits in agricultural and developmental studies. DH is a very efficient strategy for producing totally homozygous lines from heterozygous donor plants in a single stage, saving several generations of selfing (Forster and Thomas 2005).

Efficient protocols for DH plant production, such as anther culture, egg cell culture, and intergeneric hybridization, are established. The use of DH technology in developing lines that can serve as parents for hybrid breeding has been successful, particularly since the discovery of the haploid inducer germplasm.

The traditional 7–8 inbreeding generations required to correct a hybrid genotype can be replaced by a single *in vitro* generation due to this biotechnological advancement (Seguí-Simarro 2015).

By cultivating haploid cells from male (androgenesis) or female (gynogenesis) gametophytes, DHs can be produced *in vitro* (Germana 2011a, 2011b). *In vivo*, DHs have been produced by wide hybridization and parthenogenesis (Forster et al. 2007), synthesizing DH technology (Zhang et al. 2011), and the usage of a centromere-specific variant of histone 3 (Ravi and Chan 2010; Kelliher et al. 2016).

Haploids and dihaploids are important experimental materials for cytological, physiological, and genetic (gene mapping) studies. The use of plants obtained through the DH technique in breeding programs is increasing daily, as it provides homozygosity in a short time and thus shortening the plant breeding process. Recent applications of DH techniques include the diagnosis of recessive mutants, genetic selection, genetic mapping of populations for obtained stress resistance plants and double haploid techniques allow the economic production of important crops, mainly cereals, in breeding programs (Table 1).

3 Case Studies

3.1 Brassica Species

Brassica vegetables have caught the interest of breeders and seed firms in recent years, owing to enhanced nutritional attributes such as anti-cancer, antioxidants, and a variety of other traits that enhance their nutritional profile. Modern cultivars of most of the Brassica vegetables are F1 hybrids. Development of F1 hybrids differs from inbred ones, in having higher yields and better qualities. Due to their high cross-pollination rates and the presence of saprophytic self-incompatibility, Brassica crops can develop inbred lines by recurrent selfing, a process that requires time (6–7 generations of selfing) and is labor-intensive. The production of DH lines by microspore culture is another method for obtaining pure inbred lines in one generation (Sharma et al. 2020).

Drought is one of the stress factors that limits the growth of canola. This environmental factor has a considerable impact on seed fatty acid composition, notably oleic and erucic acid levels (Xiao-yan et al. 2015) and yield at the flowering stage (Xie et al. 2013). The use of polyethylene glycol (PEG) is one of the ideal ways to

Table 1 Studies on doubled haploid plant production to improve drought tolerance together with main haploid induction system used on different plant species

Species	Methods used to produce haploids	Outcomes	References
Oryza sativa subsp. *japonica*	Anther culture	The association between various root traits and rice drought resistance under various water stress levels was examined, and QTLs that influence root features at various developmental stages were found. The drought resistance index, which is defined as the yield in typical lowland environments to yield under the stress of high field conditions, was highly connected with basal root thickness.	Li et al. (2005)
Triticum aestivum	Anther culture	The doubled haploid transgenics outperformed the natural type in terms of abiotic stress tolerance, as measured by nitrate reductase enzyme activity and plant output under post-anthesis water constraint.	Chauhan and Khurana (2011)
Hordeum vulgare L.	Anther culture	During drought stress, antioxidative enzymes (SOD and CAT) are moderately active, digesting reactive oxygen species and preventing oxidative damage to plant cells. The rate of protein synthesis and its activity level determine barley's great resistance to drought stress, which is a constitutively controlled characteristic.	Gołębiowska-Pikania et al. (2017)
Oryza sativa	Anther culture	The results showed that four lines had better yields than Ciherang (3.6 t/ha) and Inpari 18 (3.3 t/ha) under ideal field conditions and were drought resistant. These lines had significant agronomic traits, including high total tiller number, medium day to harvest, and high yield.	Akbar et al. (2018)
Brassica napus X *Brassica rapa*	Microspore culture	The proportion of interspecies hybrid seeds that germinated in an osmotic solution at different concentrations demonstrated their superiority to parental *B. napus* and *B. rapa* cultivars. The produced doubled haploid plants were resistant to dry growth conditions, and the seed quality complied with canola technology standards.	Daurova et al. (2020)
Camelina sativa L.	Interspecific cross	Different doubled haploid lines of camelina were studied to evaluate the water deficit tolerance for in the greenhouse and laboratory conditions. Due to its relative water content and faster callus development rate, the Soheil cultivar was characterized as a high tolerant cultivar among other genotypes.	Ghorbani et al. (2020)

(continued)

Table 1 (continued)

Species	Methods used to produce haploids	Outcomes	References
Triticum aestivum L.	Interspecific cross	The findings of QTL analysis revealed that four and seven QTL for root characteristics were discovered in the DrExp 15 and DrExp 16 trials, respectively. The QTL analysis results indicated that the contributions of the three genomes to drought tolerance were not equal.	Salarpour et al. (2020)
Hordeum vulgare L.	Anther culture	Young seedlings and heading plants were treated to drought stress during the most crucial periods of barley plant growth. The produced population of DH lines showed an increased level of variation in drought tolerance level and provided valuable genetic resource for breeding programs.	Malaga et al. (2020)
Triticum aestivum	Anther culture	In two water regimes over the course of three years, 54 QTL were found that were associated to 10 agronomic characteristics relevant to drought resistance.	Nagy et al. (2021)
Zea mays	Interspecific cross	The significant accumulation of the advantageous alleles in four locations was seen in a group of 64 drought-tolerant BC3F6 lines, indicating a link between the four regions and maize drought tolerance.	Hu et al. (2021)
Triticum aestivum	Anther culture	Drought stress tolerance were demonstrated in doubled haploid wheat plants overexpressing the HVA1 gene. The discovery that the overexpression lines are ABA-sensitive suggests that HsfA6 functions in the ABA-mediated pathway that provides heat tolerance.	Samtani et al. (2022)
Oryza sativa	Anther culture	Androgenic doubled haploid lines (DHLs) were produced using anther culture from a cross between drought-sensitive (*Khandagiri*) and drought-tolerant (*Dular*) rice genotypes.	Tripathy and Maharana (2024)
Zea mays	Colchicine treatment	A total of 129 DHLs were recovered, with 55 showing significant drought tolerance and yield traits. Among these, DHL 41 stood out for its short growth duration, semi-dwarf stature, high grain weight, fertility rate, and yield potential, making it ideal for upland conditions and suitable for mapping drought tolerance-related QTLs/genes.	Chen et al. (2024)

create drought conditions in order to screen some drought tolerance measures such as germination, plant height, root length, and relative water content (RWC) (Basha et al. 2015). Daurova et al. (2020) used microspore culture to create doubled haploids of hybrids between B. napus and B. rapa and tested their drought resistance using PEG-6000.

3.2 *Allium Species*

The most effective alternative to traditional breeding for creating homozygous inbred lines for crop breeding, including onion, is DH technology. Onion cultivation is typically restricted to places with sufficient irrigation systems due to its sensitivity to drought, which affects the production price (Brouwer et al. 1989).

In order to explore the transcriptome and related metabolome variability and find factors that might aid in abiotic stress protection and the generation of stress-tolerant varieties, Abdelrahman et al. (2015) generated double haploid parents of onion, shallot, and their F1 hybrid.

3.3 *Cereal Species*

Cereals, according to their resistance to drought; rye, triticale, barley, wheat, and oats. Seed germination and seedling establishment are faster in DH transgenic plants. They perform better in terms of drought tolerance than nontransgenic DH plants, as determined by percent germination, seedling growth, and biomass accumulation. Due to its wide adaptation and place in human nutrition, the decrease in the yield of wheat, one of the most important crops in the world, due to drought, threatens the sustainability of agricultural production and studies on drought resistance are very intense on wheat.

Chauhan and Khurana (2011) produced transgenic DH plants in wheat through anther culture and Agrobacterium-mediated transformation, in commercial Indian bread wheat varieties. They demonstrated that the HVA1 transgenic plants have higher tolerance to simulated water stress. The development of DH transgenic wheat using anther culture method in a commercial cultivar for a desired characteristic is the subject of the first study in this field. This technique would also be beneficial in the functional genomics of wheat and other agronomically important allopolyploids.

4 Conclusion

Stress-tolerant crop types will be required as a result of the continuing global climate change scenario's expected additional rise in drought severity and frequency, which can have a damaging effect on plant development and yield. One of the most important research studies (irrigation and other cultural processes) to minimize the effects of drought is the breeding of drought-resistant varieties. However, it is difficult to develop studies on this subject unless the mechanism of drought resistance and the parameters that indicate it are well understood.

The effects of abiotic stress conditions are generally examined individually in the laboratory. However, these stresses are rarely observed alone in natural conditions. Plant responses to abiotic stress are dynamic and complex. For example, under arid conditions, high temperature and high light stresses can also occur. Another similarity between different abiotic stresses is that they can have the same effects on plants. For example, drought, salt stress, and freezing temperatures all negatively affect plant metabolism and growth, among other effects, by preventing plants from taking up water. Therefore, it is very important to develop double haploid lines that can be a source of resistance for more than one abiotic stress factor, rather than for a single abiotic stress factor, in order to increase the stress resistance in plants.

References

Abdelrahman M, Sawada Y, Nakabayashi R, Sato S, Hirakawa H, El-Sayed M, Shigyo M (2015) Integrating transcriptome and target metabolome variability in doubled haploids of *Allium cepa* for abiotic stress protection. Mol Breed 35(10):1–11

Akbar MR, Purwoko BS, Dewi IS, Suwarno WB (2018) Agronomic and drought tolerance evaluation of doubled haploid rice breeding lines derived from anther culture. SABRAO J Breed Genet 50(2)

Arı E (2006) Anther culture studies on *Anemone coronaria* var. *coccinea* Native to Turkey. Doctoral dissertation, PhD Thesis. Cukurova University

Ashraf MFMR, Foolad MR (2007) Roles of glycine betaine and proline in improving plant abiotic stress resistance. Environ Exp Bot 59(2):206–216

Basha PO, Sudarsanam G, Reddy MMS, Sankar S (2015) Effect of PEG induced water stress on germination and seedling development of tomato germplasm. Int J Recent Sci Res 6(5):4044–4049

Blakeslee AF, Belling J, Farnham ME, Bergner AD (1922) A haploid mutant in the jimson weed, "*Datura stramonium*". Science 55(1433):646–647

Brouwer C, Prins K, Heibloem M (1989) Irrigation water management: irrigation scheduling, FAO. Water Resources, Development and Management Service Land and Water Development Division

Cattivelli L, Rizza F, Badeck FW, Mazzucotelli E, Mastrangelo AM, Francia E et al (2008) Drought tolerance improvement in crop plants: an integrated view from breeding to genomics. Field Crop Res 105(1–2):1–14

Chauhan H, Khurana P (2011) Use of doubled haploid technology for development of stable drought tolerant bread wheat (*Triticum aestivum* L.) transgenics. Plant Biotechnol J 9(3):408–417

Chaves MM, Maroco JP, Pereira JS (2003) Understanding plant responses to drought-from genes to the whole plant. Funct Plant Biol 30(3):239–264

Chen YR, Lübberstedt T, Frei UK (2024) Development of doubled haploid inducer lines facilitates selection of superior haploid inducers in maize. Front Plant Sci 14:1320660

Daurova A, Daurov D, Volkov D, Zhapar K, Raimbek D, Shamekova M, Zhambakin K (2020) Doubled haploids of interspecific hybrids between *Brassica napus* and *Brassica rapa* for canola production with valuable breeding traits. OCL 27(1):45

Dunwell JM (2010) Haploids in flowering plants: origins and exploitation. Plant Biotechnol J 8(4):377–424

Ellialtıoğlu Ş, Kaplan F, Abak K (2001, April) The effect of carrot extract and activated charcoal on the androgenesis of pepper. In: Proceedings of XIth EUCARPIA meeting on genetics and breeding of Capsicum & Eggplant, vol 9, p 13

Ferrie AM, Möllers C (2011) Haploids and doubled haploids in brassica spp. for genetic and genomic research. Plant Cell Tissue Organ Cult 104(3):375–386

Forster BP, Thomas WT (2005) Doubled haploids in genetics and plant breeding. Plant Breed Rev 25:57–88

Forster BP, Heberle-Bors E, Kasha KJ, Touraev A (2007) The resurgence of haploids in higher plants. Trends Plant Sci 12(8):368–375

Germana MA (2011a) Anther culture for haploid and doubled haploid production. Plant Cell Tissue Organ Cult 104(3):283–300

Germana MA (2011b) Gametic embryogenesis and haploid technology as valuable support to plant breeding. Plant Cell Rep 30(5):839–857

Ghorbani M, Kahrizi D, Chaghakaboodi Z (2020) Evaluation of Camelina sativa doubled haploid lines for the response to water-deficit stress. J Med Plants By-Prod 9(2):193–199

Gołębiowska-Pikania G, Kopeć P, Surówka E, Janowiak F, Krzewska M, Dubas E et al (2017) Changes in protein abundance and activity induced by drought during generative development of winter barley (*Hordeum vulgare* L.). J Proteome 169:73–86

Hu X, Wang G, Du X, Zhang H, Xu Z, Wang J et al (2021) QTL analysis across multiple environments reveals promising chromosome regions associated with yield-related traits in maize under drought conditions. Crop J 9(4):759–766

Kasha KJ (1974) Haploids in higher plants: advances and potential. In: Proceedings of the 1st international symposium. University of Guelph, Guelph

Kelliher T, Starr D, Wang W, McCuiston J, Zhong H, Nuccio ML, Martin B (2016) Maternal haploids are preferentially induced by CENH3-tailswap transgenic complementation in maize. Front Plant Sci 7:414

Kosová K, Vítámvás P, Prášil IT, Renaut J (2011) Plant proteome changes under abiotic stress-contribution of proteomics studies to understanding plant stress response. J Proteome 74(8):1301–1322

Kurtar ES, Seymen M, Ünal KAL (2020) An overview of doubled haploid plant production in Cucurbita species. Yüzüncü Yıl Üniversitesi Tarım Bilimleri Dergisi 30(3):510–520

Lambers H, Raven JA, Shaver GR, Smith SE (2008) Plant nutrient-acquisition strategies change with soil age. Trends Ecol Evol 23(2):95–103

Li Z, Mu P, Li C, Zhang H, Li Z, Gao Y, Wang X (2005) QTL mapping of root traits in a doubled haploid population from a cross between upland and lowland japonica rice in three environments. Theor Appl Genet 110(7):1244–1252

Liang C, Meng Z, Meng Z, Malik W, Yan R, Lwin KM et al (2016) GhABF2, a bZIP transcription factor, confers drought and salinity tolerance in cotton (*Gossypium hirsutum* L.). Sci Rep 6(1):1–14

Malaga S, Janeczko A, Janowiak F, Waligórski P, Oklestkova J, Dubas E, Żur I (2020) Involvement of homocastasterone, salicylic and abscisic acids in the regulation of drought and freezing tolerance in doubled haploid lines of winter barley. Plant Growth Regul 90(1):173–188

Maraschin SDF, De Priester W, Spaink HP, Wang M (2005) Androgenic switch: an example of plant embryogenesis from the male gametophyte perspective. J Exp Bot 56(417):1711–1726

Melchinger AE, Lee M, Lamkey KR, Woodman WL (1990) Genetic diversity for restriction fragment length polymorphisms: relation to estimated genetic effects in maize inbreds. Crop Sci 30(5):1033–1040

Mickelbart MV, Hasegawa PM, Bailey-Serres J (2015) Genetic mechanisms of abiotic stress tolerance that translate to crop yield stability. Nat Rev Genet 16(4):237–251

Nagy É, Szabó-Hevér Á, Lehoczki-Krsjak S, Lantos C, Kiss E, Pauk J (2021) Detection of drought tolerance-related QTL in the plainsman V./Cappelle Desprez doubled haploid wheat population. Cereal Research Communications, pp 1–10

Newton AC, Johnson SN, Gregory PJ (2011) Implications of climate change for diseases, crop yields and food security. Euphytica 179(1):3–18

Pochard E, Dumas de Vaulx R (1971) Monoploidy in bell-pepper (*Capsicum annum* L.); practical results obtained in a shortened breeding cycle involving the use of monoploid plants in the third generation. Z Pflanzenzucht 65:23–46

Ravi M, Chan SW (2010) Haploid plants produced by centromere-mediated genome elimination. Nature 464(7288):615–618

Salarpour M, Pakniyat H, Abdolshahi R, Heidari B, Razi H, Afzali R (2020) Mapping QTL for agronomic and root traits in the kukri/RAC875 wheat (*Triticum aestivum* L.) population under drought stress conditions. Euphytica 216(7):1–19

Samtani H, Sharma A, Khurana P (2022) Overexpression of HVA1 enhances drought and heat stress tolerance in *Triticum aestivum* doubled haploid plants. Cells 11(5):912

Sarıkamış G (2017) Biotechnology in the development of vegetable varieties [Sebze çeşitlerinin geliştirilmesinde biyoteknoloji]. TÜRKTOB Dergisi 24:15–17

Seguí-Simarro JM (2015) Doubled haploidy in model and recalcitrant species. Front Plant Sci 6:1175

Sharma VK, Srivastava A, Mangal M (2020) Recent Trends in Sweet Pepper Breeding. In: Accelerated plant breeding, volume 2. Springer, Cham, pp 417–444

Singh DP, Singh AK, Singh A (2021) Breeding for resistance to abiotic stresses. In: Singh DP, Singh AK, Singh A (eds) Plant breeding and cultivar development. Academic, pp 399–424

Sopory SK, Munshi M (1996) Anther culture. In: Jain SM, Sopory SK, Veilleux RE (eds) In vitro haploid production in higher plants. Kluwer Academic Publishers, Dordrecht, pp 145–176

Takeda S, Matsuoka M (2008) Genetic approaches to crop improvement: responding to environmental and population changes. Nat Rev Genet 9(6):444–457

Tefera A (2017) Review on concept and impact of double haploid techniques in crop improvement. J Nat Sci Res 7:23

Tripathy SK, Maharana M (2024) Doubled haploid breeding-a potential strategy for development of high yielding drought tolerant genotypes in Upland rice. J Crop Sci Biotechnol: 1–9

Tuberosa R, Salvi S (2006) Genomics-based approaches to improve drought tolerance of crops. Trends Plant Sci 11(8):405–412

Xiao-yan Y, Ming-lian F, Gen-ze L, Jing-qiao W, Sheng-guang W, De-rong C (2015) Evaluation of drought resistance of DH line from hybrid of *B. napus* and *B. juncea*. Chinese J Oil Crop Sci 37(1):62

Xie X, Zhang X, He Q (2013) Identification of drought resistance of rapeseed (*Brassica napus* L.) during germination stage under PEG stress. J Food Agric Environ 11(2 Part 2):751–756

Yali W, Mitiku T (2022) Mutation breeding and its importance in modern plant breeding. J Plant Sci 10(2):64–70

Zhang L, Zhang L, Luo J, Chen W, Hao M, Liu B et al (2011) Synthesizing double haploid hexaploid wheat populations based on a spontaneous alloploidization process. J Genet Genomics 38(2):89–94

A Hydroponic Approach to Overcome the Drought Stress Problem for Vegetable Production

Muhammad Abbas Khan, Tayyaba Zainab, Tajwar Alam, Zohaib Younas, Sajid Hussain, Chudary Saddam Hussain, Maria Fayyaz, and Khajista Tahira

Abstract During the past two decades, hydroponics plant production has increased tremendously all over the world. Climate change and heavy industrialization is affecting cultivable soil day by day. Moreover, use of chemicals and heavy fertilizers have depleted the soil resources and has negative effects on human health. Recent studies showed that many vegetable crops including potato, tomato, peppers, and leafy vegetables showed good potential to be grown under hydroponics system. This system also provides sustainable vegetable growth with minimum input. The use of controlled environment growth also provides plants safety against pests and weeds. It can also play important role against abiotic stresses especially drought. As studies have shown that it requires 70% less moisture compared to soil moisture needs. Currently, world needs a plant growth system which requires least labor and resources to produce quality vegetables. In this book chapter, we explained about the use of hydroponics techniques and their types that can be used for sustainable vegetable production.

M. A. Khan (✉)
National Centre of Industrial Biotechnology, PMAS Arid Agriculture University, Rawalpindi, Pakistan

Institute of Biochemistry and Biotechnology, PMAS Arid Agriculture University, Rawalpindi, Pakistan
e-mail: abbasiha@uaar.edu.pk

T. Zainab · K. Tahira
Institute of Biochemistry and Biotechnology, PMAS Arid Agriculture University, Rawalpindi, Pakistan

T. Alam
Institute of Hydroponics Agriculture, PMAS Arid Agriculture University, Rawalpindi, Pakistan

Z. Younas · S. Hussain · C. S. Hussain
Department of Botany, PMAS Arid Agriculture University, Rawalpindi, Pakistan

M. Fayyaz
National Centre of Industrial Biotechnology, PMAS Arid Agriculture University, Rawalpindi, Pakistan

U. K. Chaudhry et al. (eds.), *Drought Stress*,
https://doi.org/10.1007/978-3-031-80610-0_17

Keywords Hydroponics · Vegetable · Growth · Drought stress

1 Introduction

Pakistan was a water surplus country, but currently due to excessive demand, it is facing water scarcity problems that is also predicted to increase in the future. Population in Pakistan is increasing at a rapid rate compared to developed countries that is resulting in more water demands; contrarily water resources are becoming limited (Liaquat et al. 2020). Pakistan intensely relies upon yearly rainfall precipitation, which statistically checks up to 70% of harvest water need (Abd El-Raheem et al. 2015). There are numerous reasons for water deficiency, such as absence of capacity structures, risky trench framework, and old water system framework. Soilless cultivation is a good alternative for vegetable production, particularly in areas with severe soil problems and limited water supply (Ammar et al. 2019). Currently, soilless horticulture methods are based on, hydroponic arrangement (such as floating systems, nutrient film technique (NFT)) and aeroponics. Soilless cultures are used in about 3.5% of the total area developed under passages and nurseries for vegetable growth (Liaquat et al. 2020). Despite the more efficient use of nutritional components like water, this considerable diffusion at the field scale certainly shows the presence of various benefits of this production technique. In Pakistan, where per capita land accessibility has dynamically dropped to 0.15 ha and is expected to contract further to 0.06 ha by 2050, the situation is quite concerning. Furthermore, per capita water availability has decreased from 5600 cu meters to 1200 cu meters, indicating that the water supply may be decreased further (Ammar et al. 2019). Along these lines, the potential and the capacity to fulfil developing needs for food as innovation versatility should be objectively set out upon with a solid responsibility and devotion. The best test, which lies ahead, is to continue creating food and energy while maintaining the health of the biological system and restoring the climate (Treftz and Omaye 2016). The issues being emerging out of this situation are to be defeated in such a way, so that everybody can get enough and nutritiously take care of without exaggerating the normal assets. Hydroponic is one such significant method for developing new vegetables and natural products incorporating blossoms not just in nations having contracting area and water assets, but also for those having enormous population. Hydroponic can essentially be characterized as the growth of plants in a water and manure combination supplying essential nutrients for plant growth. It is sometimes referred to as soilless horticulture or growing plants in a soilless substrate (Ammar et al. 2019).

2 Drought Stress Effect on Quality of Vegetables

Drought enormously affects plants universally and is bound to improve with climatic changes. Plant improvement is truly affected by abiotic stresses like drought, pungency, and temperature (Ahmad et al. 2021). Drought stress is one of the major abiotic stresses limiting growth yield and deteriorating the quality of vegetables. Water shortage is a serious restricting element during dry seasons, especially on the weight of vegetables. Drought cause colossal problems in efficiency of vegetable yields for the most part contingent on the length and seriousness of the dry spell pressure (Mishra and Singh 2010). Drought stress during vegetative or early growth stages generally reduces yield by diminishing seed germination, and seedling establishment (Ahmad et al. 2014).

Drought stress is lethal to yield in field production of vegetables. Stresses, such as biotic and abiotic factors, in agriculture causes considerable productivity deprivation and reflecting more epidemic because of climate change (Stefanelli et al. 2010). Within the abiotic factors, drought conditions are dominating as it effects development of plant (Ferrara et al. 2011), restraining ACP (Agricultural Crop Production). Climate change catastrophes associated to water shortage and depletion of irrigation water attainability is prominent. Studies concerning plant-water connection have been principally oriented respecting the comprehension of the roles played by water in crop development and productivity (Ferrara et al. 2011). Plant's reaction to drought is initiated with the triggering of several signaling cascades (Jahangir et al. 2009). Phytohormone abscisic acid (ABA) is one of the regulators of these cascades to change plant metabolome and gene expression in responding to osmotic stress. The crop reaction to drought condition, and use of different irrigation strategies significant attributes to quality, particularly the metabolites with nutritional values (proteins, mineral elements, lipids, and sugars) and secondary metabolites for their health benefits (Zhao et al. 2009; Wang et al. 2015).

Water stress throughout growth depress the level of carbohydrate gathered in leaves and stems deciding a mass restriction in seed, since the biomass composed in such organs from the vegetative to reproductive phases is requiring to development of adequate resources for translocation (Marcaida III et al. 2014). Water constraint during the reproductive stages directly limit seed development, resulting an untimely grain desiccation (Fábián et al. 2011). Water stress has been demonstrated to cause an assemblage of total soluble solids (TSS) in mostly crop plants as a significant adjustment mechanism (Babita et al. 2010). Soluble solids like sugars, proline, glycine, or glycerol are also affective in turgor maintenance (i.e., accumulation of osmotically active solutes) as elemental physiological manner for lowering the negative results of drought (Aldesuquy et al. 2013). In tomato fruits a lower quantity of irrigation water generated greater TSS contents as °Brix (average amount of TSS) of 5.78 for low water, and 4.30 for the normal water use. Contrary, in onion, water stress was shown to not to imply to alter the quantity of TSS (Kumar et al. 2007). In many situations higher volume of fruit soluble solids can be demonstrated with lesser water accumulation in fruits, despite an enlarged active biosynthesis is frequently well defined.

Medicinal and aromatic plants have a growing interest in many areas, involving agroalimentary, perfumes, pharmaceutical industries, and natural cosmetics. Their value comes from the biosynthesis of secondary metabolites to be used in industry which is, however, steadily influenced by environmental stresses (Kim et al. 2012; Steinbauer et al. 2015). Drought substantially impacts flavonoid content in various aromatic plants generating adjustments in necessary oil yield and formulation (Petropoulos et al. 2008; Kleinwächter and Selmar 2014).

Vegetables are viewed as exceptionally sensitive to dry seasons and dry spells supplemented with increased evapotranspiration in high temperature increments which affects photosynthesis of the plants and influence the harvests yield (Pervez et al. 2009). More than 1.2 billion people are directly affected of the land degradation in dry territories, which prompts declining natural and financial helpfulness. The dry season pressure predominantly decreases the water potential, free water, bound water, and growth of plants by diminishing the stomatal conductance, photosynthetic rate and intercellular CO2 which eventually lessens the yield. In dry season conditions, plants for the most part produce reactive oxygen species (ROS) which triggers oxidative stress in plants. This oxidative burst damages lipids, nucleic acids, and proteins (Mishra and Singh 2010).

3 Potential of Hydroponic System in the Era of Climate Change

Nutritious and good eating routine is a significant means to keep up with great well-being. The nourishing security, which is regularly less esteemed than food security by specialists, can be accomplished with the utilization of vegetables. Low vegetable consumption is frequently connected with hunger in kids and other health related problems. Vegetables are a significant supplement of vitamins A, C, K, thiamine, pyridoxine, folate, carotenes, minerals, and other minor components. They have fundamental bioactive mixtures like phenolic (flavonoids) which assumes a huge part in avoidance of harmful infections. World Health Organization (WHO) suggests the intake of 200–250 g of vegetables in a day, however the normal utilization of vegetables on our planet is a long way behind this proposal. There are a few studies demonstrating that low vegetable consumption resulting in weight gain and it is clear that healthy nutrition requires the utilization of fiber rich and low glycemic load (Nour et al. 2018).

The need of hydroponic system increases with the passage of time because of several reasons such as lack of land due to increasing population and soil destruction due to mankind lifestyle. India is one of the examples where population is increasing very fast (Dazzi et al. 2019). The utilization of hydroponic system is 90% more effective in United Arab Emirates (UAE) than traditional farming, because this system helps in controlling pH and use nutrients from the water where remaining water can be recycled and reused. This help in preventing the wastage of water. Hydroponics is a supportable system and safer as well as suitable for any crop and climate which makes it preferred in the dry and desert areas (Sharma et al. 2018). United Arab Emirates' land is consisting of 80% of sand desert and 3% of agriculture land. To increase the agriculture production in such deserted area can be possible with hydroponic system that helps in enhancing the national productivity and limits the food imports. Nutrient Film technique (NFT) is mostly used in UAE because it can be enlarging and maintained easily. This system helps in crops growth in channels without soil along with the protection of plants from soil forming diseases (Dandona et al. 2017). NFT technique can be manageable to the climate and preferred for certain crops like herbs, tomatoes, and eggplants. The current and future requirements of Kingdom of Saudi Arabia are to increase the national production by hydroponic technique without using the desert agriculture due to land and water deficiency (Abdallah et al. 2020). For the agriculture, the available resources are key requirements that are very limited in Kingdom of Saudi Arabia. To reduce the import rate and to enhance the national productivity, hydroponics and greenhouse systems are mostly preferred and used these days and in near future. In these systems, there is a demand of nutrient rich water instead of soil for the plants (Bridgewood 2003). This system helps in producing the high standard and bulk quantity of fruits and vegetables which is helpful in decreasing the burden on cultivated land and irrigation water. The average yield by hydroponic system is up to 200 tons for each acre obtained. Vegetable production, such as tomato and cucumber, and hay yield is in high quantity by utilizing the less water in hydroponic systems (Fiaz et al. 2018).

This technique is also utilized in the Qatar due to its beneficial and profitable applications, where the current hydroponic system constructed in the middle of desert give rise the cucumber, tomato, green beans, cherry, vegetables, and many fruits. United Arab Emirates are also taking interest in the hydroponic agriculture and invest so much on utilization of this technology (Fiaz et al. 2018). The production of lettuce, strawberries, cucumber, and capsicums are also done on the large scale by the hydroponic system. Organic mushrooms are grown by the hydroponic system gives yield of about 8000 kg. Similarly, in Baniyas one of the largest hydroponic centers has been introduced by Abu Dhabi initiated the lettuce growth and extended the vegetable mass production (Dandona et al. 2017).

In Kuwait, the national food mass production is short because of severe weather and inaccessible fresh water. Hydroponic system with the applications of water and fertilizers for the good growing conditions and great potential in arid climate proven as a safe agriculture system in Kuwait. Farmers in Kuwait have realized that the production of the vegetables by the hydroponic system contain the food items with good taste, uniformity, more nutrient content, and good color, and texture (Al-hassawi et al. 1996).

Customary horticultural frameworks use enormous volumes of water, jeopardizing the water supply of people in the future. These frameworks' ecological and economical impracticality features the critical need to move towards creative agronomic techniques, like shut circle aquaculture (Touliatos et al. 2016).

4 Hydroponic Technology

Hydroponics is a technique of modern science. It was started to be used for business purposes in the past 40 years (Ammar et al. 2019), and in various cases it has been adapted ranging from outside field culture to indoor nursery culture to grow vegetables (Mishra and Singh 2010). It is a space-age science, but it may be used in under developing agricultural countries for maximum food production in a constrained area. Its limitations are wellsprings of new water and supplements. In regions, where even freshwater is in short supply, hydroponics can be used by desalinating seawater. Its production has been used in the non-arable lands, like deserts and dry coastal belts (Liaquat et al. 2020).

The term Hydroponics was derived from the Greek word "hydro" which means water and "pones" which means work (work for water). In hydroponic systems, the amount of water, nutrients, and photoperiod are controlled with automated systems according to the need of plant. Hydroponic system is a contemporary technique that usually applies to nutrient solution without the soil substrates. This technique was developed in 1940 by researchers at Purdue University. It was adopted in Arizona, Abu Dhabi, Belgium, California, Denmark, German, and Holland at commercial level in early 60 s (Bhattarai et al. 2008). It has a capacity to reuse water and nutrients, controlled atmosphere along with higher production yield, and successive prevention of soil-borne diseases and pests (Lommen 2007).

Innovation will undoubtedly achieve huge changes and advancements in various fields including hydroponics as new advancements arise at an always expanding pace, and the regions wherein this happens are both different, yet important, for pretty much every business. The aim of this hypothesis is to direct new farming firms on how reasonable the hydroponic innovation is for the future of agriculture (Treftz and Omaye 2016). Due to ongoing problems in conventional cultivation method, it is important to advise farmers to invent on Hydroponic Agriculture of Pakistan (Ammar et al. 2019).

A hydroponic system is a watery layout of growing plants in a tube for regulating a set degree of sustenance arrangement. Those supplements were later distributed down the chain of the box, as directed by the developer. Every outlet has a related suction tool whereby air is entrained by the arrangement giving from the power source consequently giving generously uniform air circulation of supplement arrangement (Liaquat et al. 2020).

As the world population is increasing day by day, there is expectation to rise in the global population around 10 billion and urbanization may reach to 68%. New methods of agriculture are required, hydroponic is one of them which helps to sustain our economy throughout the year. "Sustainability means fulfilling the needs of the present without compromising the ability of future generations to meet their own needs" (Borowy 2013).

Food scarcity is to happen to be a global problem, production of food in the world needs to increase by 70% till 2060s to meet with population demand which is going to be 9.1 billion in 2050 and that need can be fulfilled by using hydroponic techniques in all over the world. When we irrigate the crops in the open field water losses ratio is very high, that is about 80%. While in the hydroponic system water losses ratio reduced about to 1.5% (Mehmood et al. 2019). Thus, by using hydroponic system we would be able to use water more efficiently. Acceptability is an important element in the modern society as the world population is increasing exponentially. The major issue comes with the requirement of food. Because of many problems in the land-based cultivation hydroponics is becoming more popular in the world due to the controlled environment. Using hydroponic techniques, we can overcome food scarcity and increase water conservation. Thus, our future generations can meet up the demand of food and water for longer term, and this will ultimately lead us towards sustainability. Hydroponic system is providing the benefits in the cultivation of the different crops like tomato and cucumber. Humans need food for their survival. For that purpose, human is dependent for cultivation and other processes, with minimum consumption of natural resources and minimum cost it fulfils the demand. Soils Culture offers efficient nutrients, high density plants, with better quality of products, bringing an increase in yield. It is very effective for areas of the world where agriculture and fertile land are insufficient for agriculture (Lommen 2007). There is opportunity for home gardeners as well as commercial producers to produce at minimum cost by hydroponics and the yield is 20–25% more relative to conventional farming (AlShrouf 2017).

Pakistan is currently importing the tomatoes of 64 million US$ in the off-season and exporting the 3.63 million US$. If we start producing tomatoes hydroponically; we can reduce our imports in off-season as well. Production system mention in Fig. 1 used for tomato production. Production of cucumber is more in hydroponics as compared to open-field system and its efficiency varies among various cultivation systems (AlShrouf 2017). Commercialized hydroponic was started in the world during 1960s and 1970. With the passage of time due to rapid urbanization it become the important part in the cultivation techniques. Now hydroponics is becoming an alternative way to produce healthy food plants to cover up the gap of production (Kumar et al. 2018).

4.1 Types of Hydroponic Systems

The different techniques used in hydroponics are Wicking, Deep Water Culture (DWC), Nutrient Film Technique (NFT), Deep Flow Technique (DFT), Ebb and Flow, Drip Systems, Aeroponics. As indicated by; there are fundamentally two kinds of hydroponic systems: static arrangement culture and constant flow arrangement culture. In static arrangement culture plants are filled in holders of supplement

Fig. 1 Hydroponics facility operated by PMAS Arid Agriculture University, Rawalpindi, Pakistan

solution. The supplement arrangement is either changed on a timetable, for example, once each week, or when the focus dips under a specific level. At whatever point the arrangement is drained under a specific level, either water or new supplement arrangement is added to make the important change. In consistent flow arrangement culture, the supplement arrangement continually streams past the roots (Kumar et al. 2018; AlShrouf 2017; Chow et al. 2018).

4.2 *Ebb and Flow*

Ebb and Flow, which are called as Flood and Drain, it is one of the admitted hydroponics systems. The versatile system is comparatively budget friendly to erect and needs an intermediary level of skill to nurture. That methodology authorizes to effortlessly adjust garden, doing addition and removal of plants as required without influencing the surrounding crops. Just like other methodologies the core concept is very straightforward; plants are settled in a rack, which is recurrently filled up with rich nutritive value water forced out of a reservoir beneath. The system utilizes attraction to return the water to the reservoir for recycling purpose. This system is complex for the beginners as it involves multiple components, but once it is assembled, it produces plant effectively. Following are the components of this system

- Plant Tray
- Reservoir
- Submersible Pump
- Timer

4.3 Drip Irrigation System

Drip Irrigation is one of the effective waters and nutrient distribution system for growing crops. It gives farmers full privilege to produce high yields and at the same time saving water in addition to fertilizers, energy. Water and nutrients are distributed covering the field in pipes called 'driplines' promoting minor units known as 'drippers. Every dripper discharge drops accommodating water and fertilizer, that results in the constant implementation of water and nutrients straight to root zone of every plant covering a whole field. With drip irrigation system, farmers are assured that there will be:

- High saving of water
- Greater compatible quality yield
- Effective usage of fertilizers
- Low dependency on weather
- Minimum risks

4.4 Nutrient Film Technology (NFT)

The NFT system is mainly popular for the home hydroponic growers. Largely due to its straightforward pattern. Though NFT systems are better appropriate for most frequently utilization for growth of the small rapidly grown plants that may include multiple types of lettuce. Alongside with growing lettuce, several commercial growers grow divergent types of herbs and baby greens with the usage of NFT systems. In this system, pump is used to distribute fertilized water to grow tray and recycling of unutilized solution of nutrients. In NFT, there is a constant flow of solution of nutrients over the roots. This is completed with usage of gravity. The grow tray is put down at an angle that permits the water to flow towards the drainpipe, and a new solution is continuously pumped into the higher end of the tube.

4.5 E-Marketing in Hydroponics

In modern enterprises, marketing lags behind sales and operations. Today, without marketing, the brand message is unheard. In the digital age, building relationships with customers and building trust with them is more likely to be done online than ever before. E-Marketing is the latest marketing method that can be accessed by any consumer over the phone. E-Marketing refers to the strategies used to sell products and services online and through other digital means. Simply put, a user-friendly and technically sound website helps to attract more potential customers. For customers of hydroponics equipment suppliers, this made a big difference. Good marketing, on the other hand, is a real opportunity for hydroponics companies that want to create strong value propositions and a visible market presence.

During Covid-19, mostly small businesses were struggling, so the government was struggling. Encouraged all activities to support SMEs in all Indonesian states. In addition to this, specific hydroponics communities have been established around the city of Bandung to expand the activities of planting and harvesting hydroponics media. The agricultural hydroponic cultivation business had reached the stage where the production was very high but can not be supplied to customers on a regular basis (Hermanus et al. 2021).

Through hydroponic we grow the crops of irrespective of seasons. We can grow the crops for the whole year. We can provide the necessary environmental factors artificially through a controlled way production. Thus, there would be no impact of external environment on the production of our crops so the production level can reach up to 100%. To further increase yields, in a closed greenhouse, inject carbon dioxide around to support growth (CO2 enrichment) or add lighting to extend the day and control vegetative growth (Sardare and Admane 2013).

Generally 26% production of vegetable is irrigated with wastewater (untreated). When the untreated water is given to crops, the health issues become questionable and beyond the control. So, by using hydroponics techniques we can also overcome this problem (Lee et al. 2017).

4.6 Overview of Current Literature

This review provides up-to-date information from evidence-based secondary data sources on hydroponics, analyses hydroponics using various studies from 2010 to 2021 shown in the Table 1, and is satisfied with hydroponics techniques. It shows that it creates sustainability in a way that works. Prove how hydroponics and sustainability will help the world economy in the future. This work attempted to apply trophoblast technology to grow selected food crop models under controlled conditions.

4.7 Hydroponics and Sustainability

The yields of greenhouse hydroponic crops can be developed all year round, independent of the seasons, making crops more available in nearby business sectors and lowering expenses. As a result of being better in appearance as well as quality, as

Table 1 Modern research reviews and outcomes

Referances	Outcomes
Saha (2010)	Awareness of adequacy of this procedure is needed amid Bangladeshi farmers.
Premanandh (2011)	Many modern technologies like hydroponics are now used to address food insecurity.
Schnitzler (2012)	There would be reduction in use of water for farming through hydroponics in urban areas.
Sardare and Admane (2013)	World population increasing rapidly. To fulfil the demand of food soil-less cultivation is becoming more popular as it is adoptable all over the world.
Fitch et al. (2014)	Hydroponics is the way to attain self-sufficiency and increase in the productivity.
Lages Barbosa et al. (2015)	Cost benefit ratio of hydroponic is high as compare to conventional farming.
Treftz and Omaye (2016)	We become able to have nutritious food production in non-arable areas.
Alshrouf (2017)	The main advantage of using hydroponics is the conservation of water, increasing productivity per unit area and less or no use of agrichemicals.
Khan et al. (2018)	Hydroponic cultivation is far better than traditional cultivation.
Sharma et al. (2018)	The prime objective of hydroponics is to highlight the problems associated with soil cultivation.
Qiu et al. 2020	Indoor agriculture offers great potential benefits to food and resource sustainability.
Ciriello et al. (2021)	Hydroponic growing systems facilitate the bioaccumulation of these nutritional compounds.

well as being devoid of content (Raheem et al. 2015). With hydroponic system innovation and a regulated nursery climate, one may grow premium-grade produce with the least amount of area, water, and manure. Although hydroponic crops became in excess of 40,000 sections of land everywhere on the world with U.S. representing in excess of 1000 sections of land alone, yet innovation needs a superior force for its saddle and reception. Hydroponic farming is an escalated type of horticulture that may satisfy the purchaser's interest for premium deliver and furnish the producer with a beneficial business (Liaquat et al. 2020).

Numerous analysts have been worried about this innovation and have attempted to analyse the hydroponic framework to the conventional soil-based method for crop development with respect to sustainability. In hydroponics, ranchers who use mechanized frames can handle the exact nutrition of the plant, as opposed to soil development, allowing farmers to produce better products. As per facts, the nature of hydroponic yields can be half better contrasted with soil-based development which can incredibly expand the benefit of ranchers. Since the plant's development is controlled in the right conditions with hydroponic farming, the organic product develops quicker than in soil-based agribusiness (Al-Karaki and Al-Hashimi 2012).

Stated soil-based developed items expect acaricides to ensure plants from disease attacks and insects which is harmful both to the environment, and hurtful to the climate, the purchaser and the farmer, while with hydroponic development, there is decreased requirement for pesticides since crops are grown in a controlled climate. Likewise, soil-based cultivating requires more compost contrasted with the hydroponic farming framework to support the dirt with the suitable supplements needed by plants for development, while, hydroponic farming frameworks don't need a lot of compost since the supplements needed for plants development is provided through the supplement arrangement. According to, in soil-based crop production, forests are cleared to make room for crops. Cultivation itself loosens the soil, thereby increasing the risk of erosion. Since trees are part of the earth's "lungs," these actions are harmful to the environment (Kato et al. 2008).

4.8 Materials Used in Hydroponic

Vermiculite saw dust, sand, and peat moss are just a few of the hydroponic growing mediums that have been used. Recently, hydroponic materials such as rock wool, perlite and swollen clay pebbles have become popular. Perlite is a volcanic rock roasted at a very high temperature. It crumbles like popcorn and releases a porous white material suitable for hydroponics. In addition to hydroponics, Perlite is used in many non-horticultural industries such as commercial compost, construction and packaging materials. Many commercial potting soil combinations contain perlite.as well as in non-horticultural industries such as building and as a packaging material, in addition to hydroponics (Chow et al. 2018). Basalt rock is used to make Rockwool. It, too, is heated to high degrees before being spun into insulation-like fibbers. Plant propagation is frequent with the cubes, and perlite grow bags are common with the slabs. Because of a neutral pH and great capillary action, many hobby hydroponic gardeners utilize expanded clay pebbles as their growing medium. Expanded clay pebbles are often thought to be too expensive for commercial endeavours.

People's rising awareness of the environment and ecosystem makes hydroponics technology the best option for preventing natural resource deterioration. This approach yields the greatest food while requiring no soil, less water for cultivation, and less area. It is also disease and pest-free. It is crucial in developing countries with water scarcity, limited arable land, pollution and groundwater quality concerns, and food security concerns. It can be utilized in backyard farming and house-top agriculture activities as a beneficial tool. Hydroponics could be the best solution to earn a living by growing vegetables in an urban environment with restricted soil accessibility and even flowers on a small scale in a controlled environment (Balqiah et al. 2020).

Produce of superior quality is produced in a regulated setting. Consumers choose greenhouse hydroponic crops because they are free of herbicides and pesticides. In hydroponic produce, there are no soil-borne diseases or weeds (Ciriello et al. 2021). Only uses 1/10th to 1/30th of the water that would normally be utilized to grow equal amounts of field produce. Hydroponic vegetables provide superior taste, quality, appearance, consistency, and shelf life. Hydroponics is a method of growing plants with a mineral fertilizer solution dissolved in water. As a result, we urge that user-friendly systems be developed, as well as destiny research evaluating the monetary viability of traditional and natural hydroponic systems (Khan et al. 2018).

4.9 Scope of Hydroponic Around the Globe

To overcome the food shortages of the next era, Pakistan can use hydroponics to overcome the food shortage crisis. Ensuring an appropriate and sustainable supply of vegetables will be an innovative technology for Pakistan. It reduces water and area by 70–90% less than irrigated soil and land farming horticulture. Without water loss or absorbed by weeds or loss in evaporation. Pakistan is facing challenges of food security and water scarcity at the national level. Thus, there we face a problem of agricultural machinery for developing hydroponic system. This might need to develop machinery at low cost at local level which also results in the increase in productivity and help our farming community.

There is opportunity for farmer as businessman are investing in hydroponics. Because for farmer there is opportunity to get higher yields as compere to land agriculture. While for the businessman there is opportunity to invest for the making of equipment's/tools used in the hydroponic system (Munir et al. 2021).

The evolution of Hydroponic has been more prominent with the development of vertical agricultural structures that has attracted supporting attentions from different global regions, South Korea, Japan, and China. Singapore, Italy, Holland, United Kingdom, Jordan, Saudi Arabia, Canada, United Arab Emirates, USA, India, South Africa, Australia, and Pakistan. Annual productions form hydroponics of some countries around the globe are UK 54, USA 47.5, Australia 40, India 38, and Japan 27.3 million pounds (De Clercq et al. 2018). Practice of vertical farming is increasing day by day in USA; there is public private partnership to boost this industry. Government also provides funds to encourage this type of farming (Esfe et al. 2015), However in developing countries least focus has been given to this farming system.

4.9.1 Market Size

The market value of vertical farming in Europe during 2021 is estimated at $ 850 million. However, according to market data forecast reports, the European vertical farming market will grow at 22% annually to reach US $ 2.31 billion by 2026 (Anonymous 2021).

United Kingdom

In Europe market share of vertical farming was $850 million in 2021. There is the growth rate of 22%. Through these techniques yield increases up to 11% and intakes 82 times greater energy than traditional lettuce (De Clercq et al. 2018).

Australia

Like other countries they are adopting hydroponic system, Australia has also adopted hydroponic system to fulfil their population's food needs. In Australia, strawberries have better quality and getting benefits from it (De Clercq et al. 2018).

India

In India, vertical agriculture is mainly polygonal agriculture. Polyhouse agriculture is a protected road that gives vegetables and fruit productivity. The range of hydroponic cultivation in India is resolved by the hydroponic agricultural law, and the field of Indian outdoor problems is resolved in the future. Further types of staple cultures can be grown, soil and water consumption are reduced or not required. Highly profitable hydroponic harvest growing in India such as tomato, salad, cucumber, kale, spinach, pepper, capsica, ginger and peppermint. You can also grow herbs such as chives and basil. Other highest-yielding plants that can be grown on hydroponic farms are dill, mint, cannabis, strawberries, and cabbage. The most profitable hydroponic crop is lettuce.

The last expense for setting up commercial hydroponic farming in India was Rs. 110 lakhs to Rs. 150 lakhs, barring the cost of land. This expense fluctuates as indicated by innovation and the technology utilized (Anonymous 2021).

According to research, India has more than 40 commercially functioning hydroponic farms, with some new farms in the works. Due to low yields, most farms cater to adjacent metro cities, and businesses locate farms near demand centres for logistical convenience (Anonymous 2021).

Japan

In Japan there are more than 150 hydroponic plant factories. They have the very old history of hydroponics, which is about 2500 years ago in which they used to gardening in the hanging bags (Zhang et al. 2018). Government providing subsidies and relaxation in tariffs and imports controls. Shimamura is the world largest indoor farm in Japan. Its production is 10,000 heads of lettuce each day in a 25,000 square-foot factory (Anonymous 2021).

4.10 Future Prospective of Hydroponics

Hydroponic farming is experiencing a comeback. As food poverty increases, global calorie consumption rises, and newer, more valuable crops expand in volume, this trend appears to be set to continue through 2020. Drifting habits toward more plant-based foods will also aid the development of hydroponic systems in densely populated areas. Therefore, hydroponic farms are seen to meet the need for a more plant-based diet in a more insecure food supply chain (Balqiah et al. 2020). Meanwhile, as demand for more plant-based proteins grows, food prices continue to rise. In addition, suitable food production space is being converted to cultivate recreational crops, thus reducing the amount of food available. The most important benefit is food safety. Most hydroponic farms are housed in climate-controlled facilities (Ciriello et al. 2021). Products are grown under rigorous quality control conditions, with preventative actions in place. This also eliminates the need for insecticides for hydroponic producers. There are consistently fewer recalls for hydroponically farmed vegetables than for field-grown produce.

Companies can also use hold release techniques for farms that are close to their consumer base, in which the product is held until it is proven to be pathogen-free and then released to the public (Rodríguez-Ortega et al. 2017).

4.10.1 Challenges

Many food production systems face issues like those faced by the hydroponic farming business, such as a shrinking labour pool, increased pressure to keep food prices low, and rising production costs. One of the most pressing issues is labour. The agriculture sector is experiencing a growing manpower shortage, with more vacancies to fill than in past years. Cultivating global demand for North American-grown agricultural food is allowing more agricultural enterprises to open in North America, with many hydroponic farms expanding a wide variety of vegetables to support the growing plant-based diet trends (Treftz and Omaye 2016; Ammar et al. 2019).

4.10.2 Potential Approach to Overcome Future Drought Condition

Severe sociological, ecological, and economic problems may occur from droughts. Various scenario initiating mechanisms can cause droughts to occur, such as low precipitation or significant evapotranspiration losses. However, in order to establish management strategies designed to eliminate adverse societal and economic impacts, adaption techniques to accommodate, Predictions of droughts and quick drought-flood transitions are necessary at different time scales (Woodhouse et al. 2016).

Agricultural trends cause droughts when the water sources on existence are insufficient to meet the needs of crops or livestock at a specific moment. This kind of drought might be caused by temperature droughts, inadequate access to water resources, or even just unfortunate timing. Diurnal and seasonal forecasts are essential for early warning, estimating the frequency of events, and making long-term estimates for establishing adaption plans for the future conditions. Forecasts, frequency estimates, and projections are the three different forms of predictions, and they can all be examined using related methods and face comparable difficulties

(Mirza 2010). Combining global and regional climate models is a common practice for future change assessments and modifications in water management policies.

Predictions of economic and societal effects could be simpler for the general public to understand than those of hazards (Bachmair et al. 2016). However, it is challenging to establish a connection between risk and impact since benefit data is sometimes absent or limited (Kratzert et al. 2019; Silvestro et al. 2019), farmer could use hydroponics to implement drought remedy. Instead of diminishing existing water sources like rivers, farmers could save the water they obtain from rain and use it when they need it in dry situations by using rainwater harvesting. The other alternative is to switch to environmentally friendly energy sources, such as renewable power, which have negligible to no impact on the environment and won't cause droughts. This involves training the next generation about the necessity of preserving the environment, recycling it, and planting more trees, among other things. Drought prevention will become a personal responsibility when environmental stewardship is highlighted in the academic program, the media, and businesses (Hanel et al. 2018).

Observation, process understanding, analysis and forecasting, and human–water interactions are four interrelated areas that can be used to identify and classify common challenges to drought prediction. Managing the data availability challenge, which is a significant issue brought on by the effects of changing climate, land use, and water management, involves enhancing data sharing, utilizing new data sources like satellite data or crowd—sourcing data, and expanding sample sizes for future conditions.

5 Conclusion

Around the globe, sufficient production of green vegetables and grain to feed general populous and livestock has emerge as a big challenge. Reason is due to lack of water either natural or artificial calamity. Furthermore, allocation of skills, fertilizer and land requirements for growing crops. To tame this obstacle, hydroponics vegetables production technology is a rising as alternative to grow sufficient quality vegetables with supply of water only needed to grow plants without wasting. Hydroponic production mechanism of vegetable production, in which plant seed is germinated into a highly nutritious, highly quality and disease free product. It is also more pleasant to taste with high digestible index. Scarcity in the amount of water with influx in climate change needed for greatest vegetables production convicted to be benefited for zones where agriculture is difficult and due to densely populated land is scarce for growing different types of vegetables. Hydroponic vegetable production system has the capability of allowing farmers to get better and greater yield of vegetables that has the potential to supply huge ecological merits. Therefore, governmental and nongovernmental organization should give emphasis for effective utilization and advancement of hydroponics technology. Formal training should be given for the producers to increase widely used of this technology. Furthermore, mechanisms should be described to enhance the engagement of stakeholders in the activities of hydroponic vegetables production in any pats of the world without limited to agro-ecology and adverse effect of climate change to overcome the alarming situation of drought stress.

References

Abd El-Raheem RM, Kamel RM, Ali MF (2015) Reliability of using Kinovea program in measuring dominant wrist joint range of motion. Trends Appl Sci Res 10(4):224

Abdallah MMS, El Sebai TN, Ramadan AAEM, El-Bassiouny HMS (2020) Physiological and biochemical role of proline, trehalose, and compost on enhancing salinity tolerance of quinoa plant. Bull Natl Res Cent 44(1):1–13

Ahmad P, Jamsheed S, Hameed A, Rasool S, Sharma I, Azooz MM, Hasanuzzaman M (2014) Drought stress induced oxidative damage and antioxidants in plants. In: Oxidative damage to plants. Academic, pp 345–367

Ahmad A, Aslam Z, Naz M, Hussain S, Javed T, Aslam S et al (2021) Exogenous salicylic acid-induced drought stress tolerance in wheat (*Triticum aestivum* L.) grown under hydroponic culture. PLoS One 16(12):e0260556

Aldesuquy HS, Abbas MA, Abo-Hamed SA, Elhakem AH (2013) Does glycine betaine and salicylic acid ameliorate the negative effect of drought on wheat by regulating osmotic adjustment through solutes accumulation? J Stress Physiol Biochem 9(3):5–22

Al-hassawi WM, Abbasi NH, Mansour MM (1996, May) A neural-network-based approach for fault classification and faulted phase selection. In: Proceedings of 1996 Canadian conference on electrical and computer engineering, vol 1. IEEE, pp 384–387

Al-Karaki GN, Al-Hashimi M (2012) Green fodder production and water use efficiency of some forage crops under hydroponic conditions. Int Sch Res Notices

AlShrouf A (2017) Hydroponics, aeroponic and aquaponic as compared with conventional farming. Am Acad Sci Res J Eng Technol Sci 27(1):247–255

Ammar AM, Mohamed HS, Yousef MM, Abdel-Hafez GM, Hassanien AS, Khalil AS (2019) Dye-sensitized solar cells (DSSCs) based on extracted natural dyes. J Nanomater 2019:1–10

Anonymous (2021) Market analysis report: Hydroponics market size, share & trends analysis report by type (aggregate systems, liquid systems), by crops (tomatoes, lettuce, peppers, cucumbers, herbs), by region, and segment forecasts, 2021–2028. Report ID: GVR-4-68038-162-7 Available online: https://www.grandviewresearch.com/industry-analysis/hydroponics-market/methodology. Accessed on 10 Oct 2021

Babita M, Maheswari M, Rao LM, Shanker AK, Rao DG (2010) Osmotic adjustment, drought tolerance and yield in castor (*Ricinus communis* L.) hybrids. Environ Exp Bot 69(3):243–249

Bachmair S, Svensson C, Hannaford J, Barker LJ, Stahl K (2016) A quantitative analysis to objectively appraise drought indicators and model drought impacts. Hydrol Earth Syst Sci 20(7):2589–2609

Balqiah TE, Pardyanto A, Astuti RD, Mukhtar S (2020) Understanding how to increase hydroponic attractiveness: economic and ecological benefit. In: E3S web of conferences, vol 211. EDP Sciences, p 01015

Bhattarai KK, Xie QG, Mantelin S, Bishnoi U, Girke T, Navarre DA, Kaloshian I (2008) Tomato susceptibility to root-knot nematodes requires an intact jasmonic acid signaling pathway. Mol Plant-Microbe Interact 21(9):1205–1214

Borowy I (2013) Defining sustainable development for our common future: a history of the world commission on environment and development (Brundtland commission). Routledge

Bridgewood L (2003) Nutrient film technique. Hydroponics: Soilless gardening explained. Crowood Press, pp 17–24

Chow YN, Lee LK, Zakaria NA, Foo KY (2018) Phytotoxic effects of trivalent chromium-enriched water irrigation in *Vigna unguiculata* seedling. J Clean Prod 202:101–108

Ciriello M, Formisano L, Pannico A, El-Nakhel C, Fascella G, Duri LG et al (2021) Nutrient solution deprivation as a tool to improve hydroponics sustainability: yield, physiological, and qualitative response of lettuce. Agronomy 11(8):1469

Dandona L, Dandona R, Kumar GA, Shukla DK, Paul VK, Balakrishnan K et al (2017) Nations within a nation: variations in epidemiological transition across the states of India, 1990–2016 in the global burden of disease study. Lancet 390(10111):2437–2460

Dazzi C, Galati A, Crescimanno M, Papa GL (2019) Pedotechnique applications in large-scale farming: economic value, soil ecosystems services and soil security. Catena 181:104072

De Clercq M, Vats A, Biel A (2018) Agriculture 4.0: the future of farming technology. In: Proceedings of the world government summit, Dubai, pp 11–13

Esfe MH, Wongwises S, Naderi A, Asadi A, Safaei MR, Rostamian H et al (2015) Thermal conductivity of cu/TiO_2–water/EG hybrid nanofluid: experimental data and modeling using artificial neural network and correlation. Int Commun Heat Mass Transfer 66:100–104

Fábián A, Jäger K, Rakszegi M, Barnabás B (2011) Embryo and endosperm development in wheat (*Triticum aestivum* L.) kernels subjected to drought stress. Plant Cell Rep 30(4):551–563

Ferrara A, Lovelli S, Di Tommaso T, Perniola M (2011) Flowering, growth and fruit setting in greenhouse bell pepper under water stress. J Agron 10(1):12–19

Fiaz S, Noor MA, Aldosri FO (2018) Achieving food security in the Kingdom of Saudi Arabia through innovation: potential role of agricultural extension. J Saudi Soc Agric Sci 17(4):365–375

Fitch K, Jorgensen E, Riart N, Vandenberg H (2014) Increasing self-sufficiency in agriculture with hydroponics. Worcester Polytechnic Institute, Bangkok

Hanel M, Rakovec O, Markonis Y, Máca P, Samaniego L, Kyselý J, Kumar R (2018) Revisiting the recent European droughts from a long-term perspective. Sci Rep 8(1):1–11

Hermanus DR, Davincy Y, Ghilchrist B, Sudrajata J (2021) Design website and mobile app with E-commerce and digital marketing of micro small and medium enterprises (MSMEs) hydroponics in Bandung City, vol 3., No. 1. ICCD, pp 168–172

Jahangir M, Abdel-Farid IB, Kim HK, Choi YH, Verpoorte R (2009) Healthy and unhealthy plants: the effect of stress on the metabolism of Brassicaceae. Environ Exp Bot 67(1):23–33

Kato Y, Hirotsu S, Nemoto K, Yamagishi J (2008) Identification of QTLs controlling rice drought tolerance at seedling stage in hydroponic culture. Euphytica 160(3):423–430

Khan FA, Kurklu A, Ghafoor A, Ali Q, Umair M, Shahzaib (2018) A review on hydroponic greenhouse cultivation for sustainable agriculture. Int J Agric Environ Food Sci 2(2):59–66

Kim YC, Glick BR, Bashan Y, Ryu CM (2012) Enhancement of plant drought tolerance by microbes. In: Plant responses to drought stress. Springer, Berlin/Heidelberg, pp 383–413

Kleinwächter M, Selmar D (2014) Influencing the product quality by applying drought stress during the cultivation of medicinal plants. In: Physiological mechanisms and adaptation strategies in plants under changing environment. Springer, New York, NY, pp 57–73

Kratzert F, Klotz D, Herrnegger M, Sampson AK, Hochreiter S, Nearing GS (2019) Toward improved predictions in ungauged basins: exploiting the power of machine learning. Water Resour Res 55(12):11344–11354

Kumar S, Imtiyaz M, Kumar A, Singh R (2007) Response of onion (*Allium cepa* L.) to different levels of irrigation water. Agric Water Manag 89(1–2):161–166

Kumar A, Kumar K, Ghosh PK, Yadav KL (2018) MWCNT/TiO_2 hybrid nano filler toward high-performance epoxy composite. Ultrason Sonochem 41:37–46

Lages Barbosa G, Almeida Gadelha FD, Kublik N, Proctor A, Reichelm L, Weissinger E et al (2015) Comparison of land, water, and energy requirements of lettuce grown using hydroponic vs. conventional agricultural methods. Int J Environ Res Public Health 12(6):6879–6891

Lee LK, Fleegler EW, Farrell C, Avakame E, Srinivasan S, Hemenway D, Monuteaux MC (2017) Firearm laws and firearm homicides: a systematic review. JAMA Intern Med 177(1):106–119

Liaquat M, Malik AM, Ishaq M, Ashraf HMQ, Naseem I (2020) Evaluation of growth parameters and germination behaviour of different tomato (*Lycopersicon esculentum* Mill.) cultivars under hydroponic condition. Pak J Agric Res 33(3):544

Lommen WJ (2007) The canon of potato science: 27. Hydroponics Potato Res 50(3):315–318

Marcaida M III, Li T, Angeles O, Evangelista GK, Fontanilla MA, Xu J et al (2014) Biomass accumulation and partitioning of newly developed green super Rice (GSR) cultivars under drought stress during the reproductive stage. Field Crop Res 162:30–38

Mehmood A, Hussain A, Irshad M, Hamayun M, Iqbal A, Rahman H et al (2019) Cinnamic acid as an inhibitor of growth, flavonoids exudation and endophytic fungus colonization in maize root. Plant Physiol Biochem 135:61–68

Mirza MMQ (2010) A review on current status of flood and drought forecasting in South Asia. In: Mitra A, Sharma C (eds) Chapter 7 global environmental changes in South Asia. Springer, Dordrecht, pp 233–243

Mishra AK, Singh VP (2010) A review of drought concepts. J Hydrol 391(1–2):202–216

Munir MU, Ahmad A, Hopmans JW, Belgacem AO, Baig MB (2021) Water scarcity threats to National Food Security of Pakistan-issues, implications, and way forward. In: Emerging challenges to food production and security in Asia, Middle East, and Africa. Springer, Cham, pp 241–266

Nour M, Lutze SA, Grech A, Allman-Farinelli M (2018) The relationship between vegetable intake and weight outcomes: a systematic review of cohort studies. Nutrients 10(11):1626

Pervez MA, Ayub CM, Khan HA, Shahid MA, Ashraf I (2009) Effect of drought stress on growth, yield and seed quality of tomato (*Lycopersicon esculentum* L.). Pak J Agric Sci 46(3):174–178

Petropoulos SA, Daferera D, Polissiou MG, Passam HC (2008) The effect of water deficit stress on the growth, yield and composition of essential oils of parsley. Sci Hortic 115(4):393–397

Premanandh J (2011) Factors affecting food security and contribution of modern technologies in food sustainability. J Sci Food Agric 91(15):2707–2714

Qiu J, Bayabil HK, Li Y (2020) Indoor vertical farming systems for food security and resource sustainability. School of forest resources and conservation. EDIS, University of Florida, Gainesville

Raheem A, Azlina WW, Yap YT, Danquah MK, Harun R (2015) Thermochemical conversion of microalgal biomass for biofuel production. Renew Sust Energ Rev 49:990–999

Rodríguez-Ortega T, Bernués A, Olaizola AM, Brown MT (2017) Does intensification result in higher efficiency and sustainability? An emergy analysis of Mediterranean sheep-crop farming systems. J Clean Prod 144:171–179

Saha SK (2010) Soilless cultivation for landless people: an alternative livelihood practice through indigenous hydroponic agriculture in flood-prone Bangladesh. Ritsumeikan Asia Pacific University, Beppu

Sardare MD, Admane SV (2013) A review on plant without soil-hydroponics. Int J Res Eng Technol 2(3):299–304

Schnitzler WH (2012, May) Urban hydroponics for green and clean cities and for food security. In International symposium on soilless cultivation 1004 (pp 13–26)

Sharma N, Acharya S, Kumar K, Singh N, Chaurasia OP (2018) Hydroponics as an advanced technique for vegetable production: an overview. J Soil Water Conserv 17(4):364–371

Silvestro F, Rossi L, Campo L, Parodi A, Fiori E, Rudari R, Ferraris L (2019) Impact-based flash-flood forecasting system: sensitivity to high resolution numerical weather prediction systems and soil moisture. J Hydrol 572:388–402

Stefanelli D, Goodwin I, Jones R (2010) Minimal nitrogen and water use in horticulture: effects on quality and content of selected nutrients. Food Res Int 43(7):1833–1843

Steinbauer MJ, Wallis IR, Davies NW, Watson SJ (2015) Foliar quality of co-occurring mallee eucalypts: balance of primary and secondary metabolites reflects past growing conditions. Chemoecology 25(4):179–191

Touliatos D, Dodd IC, McAinsh M (2016) Vertical farming increases lettuce yield per unit area compared to conventional horizontal hydroponics. Food Energy Secur 5(3):184–191

Treftz C, Omaye ST (2016) Hydroponics: potential for augmenting sustainable food production in non-arable regions. Nutr Food Sci 46(5):672–684

Wang ZY, Gehring C, Zhu J, Li FM, Zhu JK, Xiong L (2015) The Arabidopsis vacuolar sorting receptor1 is required for osmotic stress-induced abscisic acid biosynthesis. Plant Physiol 167(1):137–152

Woodhouse CA, Pederson GT, Morino K, McAfee SA, McCabe GJ (2016) Increasing influence of air temperature on upper Colorado River streamflow. Geophys Res Lett 43(5):2174–2181

Zhang X, He D, Niu G, Yan Z, Song J (2018) Effects of environment lighting on the growth, photosynthesis, and quality of hydroponic lettuce in a plant factory. Int J Agric Biol Eng 11(2):33–40

Zhao CX, He MR, Wang ZL, Wang YF, Lin Q (2009) Effects of different water availability at post-anthesis stage on grain nutrition and quality in strong-gluten winter wheat. C R Biol 332(8):759–764

An Alternative Tool Against Drought Stress: Nanotechnology

İrem Aycan Şentürk and Ayten Kübra Yağiz

Abstract Plants have developed highly complex response mechanism to drought stress, which can have lethal effects under severe conditions. In order to increase the drought tolerance of plants different technologies have been used. Nanotechnology, the novel approach to increase drought tolerance, is simply defined as using tools and materials at nano size, which is employed in agriculture in different areas like nanofertilizers, nanopesticides, nanosensors and nanoparticles. The effect of nanoparticles starts with increasing the water uptake ability of seeds, which raises seed germination and development under drought stress and continuous with increasing seed vigor index. Besides, application of nanoparticles increases the photosynthesis rate, chlorophyll content, photosynthetic enzyme activities, nitrogen metabolism and leaf activities like MDA, proline index of plants. Moreover, nanoparticles have the ability of penetrating cell nucleus and changing the gene expression levels related with drought stress response. Recent studies present the possible usage of nanotechnology to increase the drought stress tolerance of plants and the results show that the nanotechnology is an alternative tool against drought stress.

Keywords Nanotechnology · Nanoparticles · Nanomaterilas · Droght stress · Agriculture

1 Introduction

The studies estimated that the world population will increase to approximately 10 billion people by 2050 (Leridon 2020), which will be another problem for agricultural production along with climate change, insufficient and polluted water sources and lands, intensive usage of agrochemicals. These problems, which agricultural

İ. A. Şentürk · A. K. Yağiz (✉)
Department of Agricultural Genetic Engineering, Niğde Ömer Halisdemir University, Niğde, Türkiye
e-mail: ayagiz@ohu.edu.tr

U. K. Chaudhry et al. (eds.), *Drought Stress*,
https://doi.org/10.1007/978-3-031-80610-0_18

production has to face with, create various stresses on crop plants, like drought stress. Drought stress causes the closure of stomata, reduces photosynthesis rate, plant development and loss of yield, even can have lethal effects under severe conditions. There are various strategies to decrease the effects of drought stress on crops. Development of high drought tolerant varieties can be solution of decreasing in global crop production. However, conventional breeding strategies has low success in this view due to the complexity of response mechanisms of plants to drought stress. Gene transfer or genome editing technologies are another alternative for mitigation the effects of drought stress. Yet, the molecular mechanism behind the stress response of the plants is not clear and current studies obtain limited information. For this purpose, nanotechnology is used as alternative tool against drought stress.

Nanotechnology is an emerging field with various using areas. It can be defined as controlling and restructuring of materials in nanoscale (<100 nm) to gain materials new properties and functions (NTSC, 2007). Rebuilding of materials in nanoscale revolutionized in physics, chemistry and all material related areas like agriculture (Leridon 2020).

2 Nanotechnology in Agriculture

Nanomaterials differs from their bulk form with unique properties. Nanomaterials have roles in agricultural production with less contamination, more efficiency, and having controlled releasing systems compared to other agrochemicals. There are different forms of nanomaterials, used in agriculture (Fig. 1). Nanofertilizers, nanopesticides, and nanosensors are used in crop production for different purposes.

2.1 Nanofertilizers

Obtaining high yielding crops and boosting in the world population will increase the fertilizer need. However, using high amounts of fertilizer is not a solution, as even more than half of the fertilizers are not uptaken by plants, and released to the soil and water sources (Bashir et al. 2022). Nanofertilizers are useful candidates with low nutrient loss, slow release ability and high uptaken capacity by targetted plants (Kah et al. 2018). Nanotechnology is used for nanofertilizer production in two ways; first one is the production of nano-sized micro or macronutrients like nanoiron; second one is the encapsulation of agrochemicals in nanocapsules (DeRosa et al. 2010; Kah et al. 2018).

Nanofertilizers have two important features; (i) increasing nutrient use efficiency, and (ii) reducing the cost (Kah et al. 2018). Nanofertilizer treatment increases the nutrient uptake ability of plants by developing nanoscale pores by molecular transporters and novel nanopores in cuticule layer (Rico et al. 2011). Moreover,

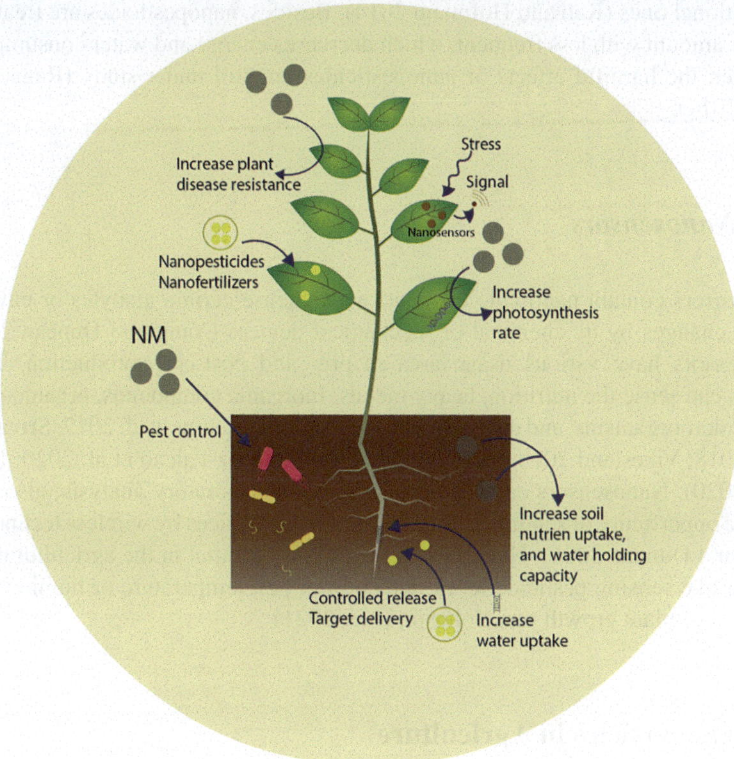

Fig. 1 Roles of nanotechnology in agriculture

nanotechnology gives the opportunity of slow or controlled release of fertilizers, which increases nutrient use efficiency and decreases the possible contamination of environment by intense fertilizer usage (Liu and Lal 2014). Slow release of the nanofertilizers prevents conversation of fertilizers into gas form or leaching in the soil, also fertilizers are released via signals, spread from plants (DeRosa et al. 2010). Another advantage of using nanotechnology is development of intelligent fertilizers, which communicates with plants and microbiota in the soil (Mastronardi et al. 2015).

2.2 *Nanopesticides*

Pesticide treatment is a standard procedure of crop production. Nanopesticides treatment increases the efficiency of pesticides by slow degradation and controlled release (Chhipa 2017). Apart from the physical differences between conventional pesticides and nanopesticides there are chemical differences like dissolving and colloidal states (Kumpiene et al. 2008). The solubility of pesticides increases at nanoscale, which decreases the pollution level of pesticide compared to conventional ones (Kah and Hofmann 2014). Besides, nanopesticides are treated in smaller amount with less frequent, which decreases energy and water consumption. However, the harmful effects of nanopesticides are still under study (Ragaei and Sabry 2014).

2.3 *Nanosensors*

Nanosensors contain nanoscale element, which sense certain analytes or environmental changes by its chemical or mechanical devices (Yang and Duncan 2021). Nanosensors have various usage area in pre- and post-crop production. These devices can sense the nutrition, heavy metals, inorganic compounds, organic molecules, microorganisms, and environmental conditions (Kearns et al. 2017; Srivastava et al. 2018; Vikesland 2018; Zhang et al. 2018; Jimenez-Falcao et al. 2020; Rippa et al. 2020). Nanosensors can be used for standard laboratory analysis; also, they give the opportunity of sending information to smart devices by wireless technology (Yang and Duncan 2021). Nanosensors have high potential in the agricultural production like sensing pesticide level, detection of pH, temperature or humidity and detection of plant growth rate (Johnson et al. 2021).

3 Nanoparticles in Agriculture

The concept of nanoparticles (NP) contains different products like polymer based, lipid based and metal particles at nanoscale (Fig. 2). Nanotechnology gives the opportunity of designing target focusing NPs with high solubility ability and preventing premature degradation of active ingredients. NPs can be designed by two strategies; (i) improving the effects of active ingredients or (ii) introducing novel gradients, which are under study (Kah et al. 2013).

3.1 *Polymer Based NPs*

Polymers are used for encapsulation of agrochemicals at nanoscale. Three different techniques are used for polymer encapsulation; sphere, capsule, and fibers (Hellmann et al. 2011; Perlatti et al. 2013; Nuruzzaman et al. 2016). Nanocapsules are used for encapsulation of agrochemicals in a liquid matrix by a polymeric membrane. This kind of capsules gives better results for target delivery systems (Nuruzzaman et al. 2016). Moreover, choosing right polymer material for encapsulation gains new characteristics to agrochemicals and improve their efficiency (Choudhary and Gupta 2011; Feng and Zhang 2011). Nanospheres are another encapsulation technique which contains agrochemicals homogeneously in the polymeric matrix (Haynes and Van Duyne 2001). Nanospheres are used for slow

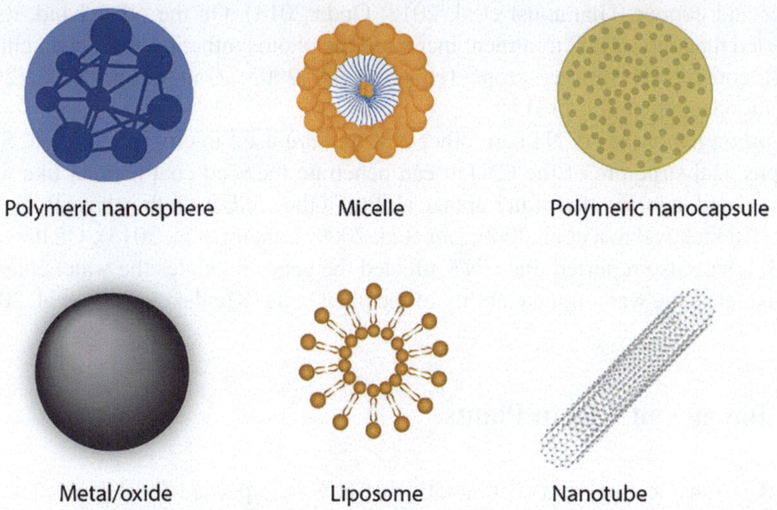

Fig. 2 Different nanoparticles, mostly used in agricultural applications

releasing of the ingredients, which decrease the releasing of free agrochemicals to environment (Wanyika 2013). Nanofibers consist of fibers with less than 100 nm diameters in web structure (Huang et al. 2003). Due to its structure, nanofibers have high stiffness and strength, which provides stability to carry agrochemicals (Huang et al. 2003; Hellmann et al. 2011).

3.2 *Nanoparticle-Aided Harvesting*

Hyrophobic and lipophilic ingredients are carried by lipid based NPs. In order to encapsulation of this kind of agrochemicals, liposomes are used. This encapsulation system is designed at nanoscale and named as nanoliposome (Nuruzzaman et al. 2016). Apart from nanoliposomes, micelles, which are single layered liposome capsules, are used for encapsulation of agrochemicals (Lucy and Glauert 1964). Micelles are preferred for encapsulation of water-insoluble ingredients (Nuruzzaman et al. 2016). They also can be used for controlled release of chemicals (Adak et al. 2012).

3.3 *Metal/Oxide-Based NPs*

Nanosized materials have different functions compare to their bulk structure (Roduner 2006). There are different using area of nanomaterials (NMs) in agricultural production. Since their unique properties, they can be used as pesticides or fertilizers. Various studies showed the antimicrobial activity of different NMs like silver and copper (Giannousi et al. 2013; Ouda 2014). On the other hand, it was reported that nano-TiO2 treatment increased the photosynthetic activity and chlorophyll content of various crops (Hong et al. 2005; Gao et al. 2006, 2008; Mahmoodzadeh et al. 2013).

Carbon nanotubes (CNT) are other NM that are used in crop production. Since the physical structure of the CNT it can penetrate the seed coat and act like water channel and increases the water uptake ability of the seeds even they have thick seed coat (Khodakovskaya et al. 2009; Liu et al. 2009; Lahiani et al. 2013). On the other hand, it was also reported that CNT affected the genes regulates the water channels and increase the water uptake ability of the plant cells (Khodakovskaya et al. 2012).

4 Journey of NPs in Plants

Uptake, translocation, and accumulation of NPs is important for determination of phytotoxicity and role in food security of NPs. However, the uptake and translocation of NPs is not systematic, and the uptake and translocation pathways of NPs is not clear yet. There are two treatment systems of NPs to plants: foliar and root exposure.

After foliar treatment, NPs are uptaken by leaves and the direction of translocation is from top to the bottom (top-down) (Lv et al. 2019). There are various limiting factors of uptaken of NP uptake like size and chemical structure of NPs (Eichert et al. 2008; Eichert and Goldbach 2008; Popp et al. 2005). During the foliar applicaiton, the first barrier will be cuticle. NPs smaller than 5 nm can penetrate leaves by cuticle layer (Lv et al. 2019). Another alternative for hydrophilic NPs is stomata. When stomata are in open form, the diameter is larger than 20 nm and let penetration of NPs up to 43 nm (Kim et al. 2015). It is reported that NPs are accumulated in leaf stomata and in some plant species, like Arabidopsis thaliana, Cucurbita pepo, Allium porrum, Lactuca sativa, and Citrullus lanatus, NPs are observed in inner tissues. On the other hand, NPs can penetrate plant cells by aquaporin, ion channels, and endocytosis, or by creating by new pores (especially CNT) (Rico et al. 2011). The translocation of NPs is performed by vascular tissue of plants. Various studies showed translocation of NPs to shoots or roots after stomatal penetration of the NPs, and it is estimated that translocation of NPs happens via phloem (Lv et al. 2015; Schymura et al. 2017).

Uptaken of NPs via root cells is not illuminated yet. Sabo-Attwood et al. (2012) reported that 3.5 nm Au NPs was uptaken via root; however, 18 nm Au NPs accumulated in the outer side of the roots of tobacco. In another study, 200 nm SiO2 nanoparticles were uptaken via Arabidopsis roots (Slomberg and Schoenfisch 2012). These show that apart from the size of the NPs, the type of NPs and plant have important role of uptaken of NPs by roots. Root surface is mostly negatively charged due to secreting of small molecules like organic acids. Thus, positively charged NPs are more capable of penetrating in root cells or accumulating around root (Zhou et al. 2011). The translocation of NPs can be in both apoplastic or symplastic pathways, which are not illuminated yet. On the other hand, the translocation of NPs from root to leaves are performed via xylem (Lv et al. 2019).

Apart from chemical component, stability and functionalization and type of NPs, crop type also affects the accumulation and translocation of NPs in plants. There are different ways of penetration of NPs in plant cells. Having more surface area of NPs increases the reactivity of NPs with their environment compared to their bulk structure. High reactivity of NPs creates complexes with membrane transportes and increases the transportation of NPs into the plant cells (Watanabe et al. 2008). The main transpotation way of metal-based NPs is ion transporters (Hall and Williams 2003). There are different pathways of uptaken of NPs by a cell and delivery to other cells like apoplastically, symplastically, or through plasmodesmata. However, current studies do not present the exact mechansims, and more studies are needed to conduct for exploring the mechanisms of translocation of nanoparticles in plants (Rico et al. 2011).

5 Plant Responses to Drought Stress

Plants are exposed to many abiotic stress factors throughout their lives because of the sessile living conditions. Abiotic or environmental stress is defined as the limiting conditions which influences the growth and development of plants by the environmental factors (Bakhsh and Hussain 2015). Drought has become even more important by increasing its impact especially with population growth and global warming in recent years (Holden et al. 2003). In agricultural terms, drought is defined as the insufficient water content in the soil for obtaining the water need of the plant. High temperature, decreasing in the amount of precipitation, delay in the rainy season, strong wind, and low humidity are the important factors can be listed as reasons of drought (Passioura 1996). Drought causes a slowdown in the development of plants and a decrease in the number of crops.

Plants have developed different adaptation mechanisms in order to survive in limited water conditions (Basu et al. 2016). There are two main mechanisms developed within the plant against drought stress which ensure the survival of the plant in drought conditions: stress avoidance and stress tolerance (Ors and Ekinci 2015).

Avoidance is a crucial adaptation mechanism for survival during drought stress. Some morphological changes occur on the surfaces of the plant in contact with the environment, and some changes happen in its chemical structure (Selda and Ekinci 2015). These changes are the development of the cuticle, the covering of the leaf surfaces with hairs, the change in the size, distribution and location of the stomata, the increase in the proportional water content, the changes in the vegetative and vascular tissues, the development of the proliferation and fertilization strategies, and the ligninization of the xylem canals. Plants are able to continue their growth and development in drought conditions under favour of these changes (Agarwal and Zhu 2008).

Table 1 Morphological, physiological and molecular response of plants to drought stress

Morphological responses of plants to drought stress	Physiological responses of plants to drought stress	Molecular responses of plants to drought stress
Acceleration of root development and deceleration of stem growth to reach water sources in the first periods of drought Decreasing the number of leaves and leaf area Abscission (shedding) of leaves Changing in color and wilting occurs in the plant	The turgor pressure decreases The stomata are closed in order to reduce water loss Photosynthesis decreases Photorespiration increases Abscisic acid (ABA) syhthesis increases Compatible osmolythes accumulation increases	Enzyme activities are reduced Drought responsive genes and photosynthetic genes are regulated

During drought stress, various defense mechanisms are regulated at physiological, biochemical, and molecular levels (Table 1; Tuteja 2007). Plants have developed these tolerance mechanisms in order to protect themselves from the negative effects of drought stress and to continue their lives. Reduction of leaf area, deep root development towards moist soil layers, and closure of stomata are the changes sighted in the plant as the first steps of defense against drought. Leaf growth is inhibited, and new leaf formation is limited against drought stress in plants (Mahajan and Tuteja 2005).

In higher plants, the water taken from the soil through the roots is transported to the top organs of the plant by osmotic rules. If the amount of water lost by transpiration is greater than the amount of water taken in, competition begins between plant organs for water due to the negative pressure. In other words, the balance between the water potentials of the different organs of the plant is disturbed. The water potential is the sum of the turgor potential and the osmotic potential in the plant. The turgor potential is formed as a result of the collision of water molecules, which are independently present in the cell. The turgor pressure must always be positive in the cell. When the turgor pressure decreases, wilting occurs. Therefore, turgor potential is the basic element of water potential in plants under drought stress. Due to the loss of turgor potential, stomatal movements become irregular, as a result photosynthesis and respiration are affected, and cell membranes rupture. Increases in drought stress result in lower chlorophyll accumulation, increased mesophyll resistance, increased stomatal diffusion resistance, decreased photosynthetic enzyme activities, and decreased photochemical and biochemical activities of chloroplasts (Chetal et al. 1982; Toraman 1999). Drought stress directly affects the photosynthetic performance and thus the productivity levels of plants (Levitt 1980; Acar 1999).

In order to reduce the water loss; first, plants close their stomata (Redondo-Gómez 2013; Verma et al. 2013). Closure of stomata is performed by abscisic acid (ABA), a plant hormone, and ABA is involved in the regulation of stomal behavior and metabolism during drought stress (Hartung et al. 2002). When plants sense the changes of water content in the soil under drought stress, they start to transmit water deficiency signal from roots to leaves through ABA accumulation for adaptation of drought stress conditions. Drought triggers the production of ABA in the roots, which is carried to the shoots, causing the stomata to close and eventually to growth restriction. During stress adaptation, ABA has important roles in stress response and signaling mechanim, helps regulation of genes involved in osmotic regulation and ion compartmentation. Under normal conditions, ABA level is low; however, ABA production increases and plays an important role in the activation of transcription factors when the drought stress is first detected in plants. The first detection of the stress factor takes place in the cytoplasm. The presence of the stressor is then transferred to the nucleus via signal transduction, and drought-sensitive genes are divided into two groups, ABA-dependent and ABA-independent, according to their dependence on ABA for induction. Drought stressor-specific transcription factors play an important role in stressor tolerance in plant cells.

Osmotic regulation, which is accepted as one of the most important responses of plants to drought stress and an important component of resistance mechanisms, occurs as a large accumulation of interchangeable different types of ions and molecules (osmolytes) (Ashraf and Harris 2004; Sanchez et al. 2004). Accumulated osmolytes under drought stress conditions include amino acids such as proline, glutamate, glycine-betaine, and sugars (mannitol, sorbitol and trehalose). Proline accumulation exposed to drought stress is higher than other amino acids in plants (Ashraf and Harris 2004). Proline may also play a role in attenuating cytoplasmic acidity (Srinivas and Balasubramanian 1995) and maintaining NADP+/NADPH ratios compatible with metabolism (Hare and Cress 1997). Proline accumulation in response to drought and salt stress in plants takes place in the cytosol, where proline mainly contributes to cytoplasmic osmotic regulation (Ketchum et al. 1991).

Specific stress-related genes have roles on molecular control mechanisms of drought stress. These genes are classified into thre main categories; (1) genes involving ion and water uptake transport (eg ion transporters and aquaporins); (2) genes directly involved in the maintenance of proteins and membranes (e.g. LEA proteins (late embryogenesis dependent), osmoprotectants, chaperones, and heat shock proteins); and (3) genes involved in transcriptional control and signaling cascades (e.g., transcription factors such as the HSF, DREB, ABF, and MYB families, phospholipases, SOS kinases, and MAP kinase) (Mahajan and Tuteja 2005).

Defence strategies help plants to develop their adaptation mechanisms in low water levels. Plant varieties, plant growth stages, genetic potential and severity of drought stress are factors that determine drought adaptation strategies. First, the plant's response to drought begins with the perception of stress and continues with signal transmission that activates stress-specific regulation mechanisms. Various secondary messengers such as Ca2+, ROS, ABA, phosphoglycerol, diacylglycerol, and transcriptional regulators play important roles in various signal transduction pathways. Ca2+ can activate transcription factors in response to drought stress by binding to some transcription factors, resulting in changes on the expression of genes related to drought stress.

In some cases, Ca2+ sensors bind to their respective interaction partners and then initiate a phosphorylation cascade. Drought stress-sensitive genes or transcription factors are targets of regulating drought stress-sensitive gene expression. Receptor kinases sense stress factors and have the ability to bind phosphorus. After the receptor detects the stress factor, it binds phosphorus to MAPKKK. Secondary messengers then bind to the active site of MAPKKKs, which activate MAPKK by phosphorylation, and then phosphorylate MAPK and the final phosphorylation transcription factors, causing changes in gene expression (Taj et al. 2010).

6 Effect oF NPs on Drought Stress

Abiotic stresses (drought, salinity, high temperature) effect the growth, productivity, and yield of crop plants. Scientists have searched for a solution to prevent adverse effects of abiotic stresses on crops. Drought is one of the most important factors limiting plant growth and development. It is predicted that it will become a bigger problem due to reasons such as global warming. Since the drought is one of the biggest problem in agriculture, agricultural nanotechnology focuses on the applications that will make plants tolerant to drought stress. Application of nanoparticles has shown promising potential for manipulating and improving crop production, and reducing related stresses to achieve sustainability. Some nanoparticles can improve plant growth and drought stress tolerance based on their unique physicochemical properties (Tarigholizadeh 2021).

The use of nanoparticles increases the accumulation of free proline and other amino acids, facilitates water and nutrient uptake, and contributes to the tolerance of drought conditions in plants by contributing to the activation of antioxidant enzymes such as catalase, superoxide dismutase, and peroxidase (Fig. 3). On the other hand, nanoparticles affect the expression of genes related to drought stress (Ali et al. 2021).

6.1 Effect of TiO2 Nanoparticle on Drought Stress

There are many studies about the application of titanium dioxide (TiO2) nanoparticle on different plants under drought condition. The studies showed that TiO2 nanoparticle can reduce the effects of stress in plants under drought stress. In one study, TiO2 nanoparticle was applied to spinach seeds, and reported that the TiO2 nanoparticle has a positive effect on the spinach plant. As a result of this study, it was thought that the effects of stress on plants under drought stress of TiO2 nanoparticles could be reduced (Zheng et al. 2005). It has been reported that the effects of drought stress are reduced when TiO2 nanoparticle is applied to Linum usitatissium plant (Aghdam et al. 2016). Mohammadi et al. (2016) studied on two different concentrations (10–40 ppm) of TiO2 nanoparticles. TiO2 nanoparticles were also

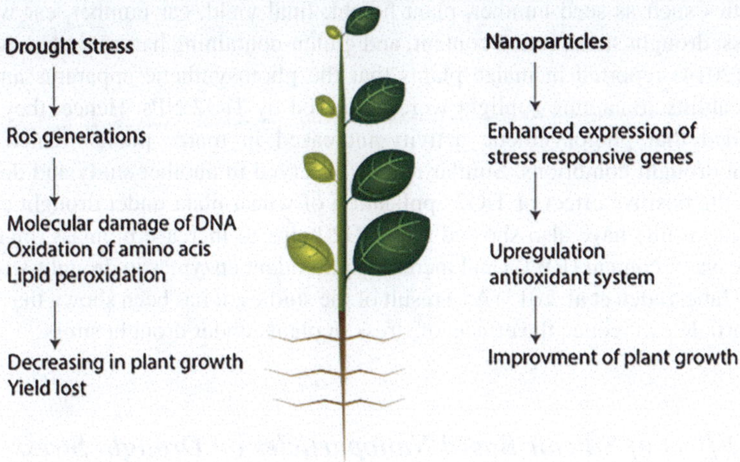

Fig. 3 The importance of using nanoparticles in drought stress

applied to the Dracocephalum moldavica L. plant. It was observed that the group treated with low concentration TiO2 nanoparticle (10 ppm) under drought stress was higher than the dry control group in terms of root dry weight, shoot dry weight, and essential oil content parameters. Aghdam et al. (2016) applied TiO2 nanoparticles at 3 different concentrations (10–100 and 500 mg/L) to Linum usitatissimum plant. The samples were examined in terms of parameters such as plant height, protein content in seeds, chlorophyll content, and seed weight. It was observed that TiO2 nanoparticle applied at low concentration showed a positive effect in the drought groups compared to the control group. However, it was determined that this positive effect decreased as the concentration of TiO2 nanoparticle increased. In another study, TiO2 nanoparticle was used in field trials in wheat plant. It was applied to both drought stressed groups and well irrigated groups. Economically important parameters such as starch content, seed number, biomass, and plant height were investigated, and it was observed that the TiO2 nanoparticle showed a positive effect in drought group with increased yield, which, however, decreased as the concentration of TiO2 nanoparticle increased (Jaberzadeh et al. 2013). When evaluated in general, it is seen that the application of TiO2 nanoparticles to plants has a positive effect under drought stress. However, as the applied concentration increases, the observed positive effect also decreases. This indicates that the effect of TiO2 nanoparticle is dose dependent.

Owolade et al. (2008) have reported that application of TiO2 in the leaves of Vigna unguiculata resulted in enhacement of photosynthesis rate and activation of antioxidant enzymes, such as catalase and peroxidase. Khan and Bano (2018) have resulted that the foliar application of TiO2 nanoparticle on wheat reduces the adverse effect of drought stress, and showed positive results for some agricultural properties such as starch and gluten content. The findings showed that the application of 0.02% TiO2 nanoparticles indicated improvement in different agronomic properties such as seed number, plant height, final yield, ear number, ear weight biomass, drought stress starch content, and gluten-containing harvest index. Akbari et al. (2014) reported in maize plants that the photosynthetic apparatus and the plant's ability to capture sunlight were improved by TiO2 NPs. Hence, they have concluded that photosynthetic activity increased in maize plants treated with TiO2 in drought conditions. Similar results observed in another study and demonstrated the positive effect of TiO2 application of wheat plant under drought condition. The results have also showed that TiO2 helps to increase biomass, maintain relative water content (RWC) and increase antioxidant enzymes in drought-stressed plants (Jaberzadeh et al. 2013) As a result of the studies, it has been shown that TiO2 nanoparticle can reduce the effects of stress in plants under drought stress.

6.2 *Effect of Silicon-Based Nanoparticles on Drought Stress*

Silicon-based materials are abundant in soil, so plants have them naturally in significant concentrations (1–10%). In plants, silicon (Si) is found in the form of amorphous silica (SiO2) in the cell wall (Bhatt and Sharma 2018). Silicium can be accumulated in high concentrations in plants. It has an increasing effect on plant growth. It is known that water losses decrease in plants that uptake sufficient silicium, and its accumulation increases resistance to fungal diseases and damage by insects in plants (Ma et al. 2001a, b).

As mentioned before, drought stress causes the stomata closure and then the photosynthetic rate is decreased. Silicon can reduce drought stress by reducing transpiration. Leaf transpiration occurs mainly in the stomata and partly in the cuticle. When Si accumulates beneath the cuticle layer, transpiration through the cuticle may be reduced by silicium deposition. In one study, it was shown that silicon can reduce the transpiration rate by 30% in rice with a thin cuticle. It has been observed that the effect of silicon in rice plants grown under drought stress is more significant than those grown under normal (non-stressed) condition (Ma et al. 2001a, b).

It has been proven that the Si content in the hull of the rice grain is 7% and in the barley grain it is 1.5%. It plays an important role in maintaining a high humidity condition by reducing the transpiration rate in the stem. Therefore, the accumulation of silicon is critical to gain resistance in drought stress of plants (Ma and Takahashi 2002). Drought stress can also damage plant cell membranes and the structure of cell walls, as well as inhibit photosynthesis and cell division (Riederer and Schreiber 2001). The closure of stomata in the cuticle layer of the plant during drought stress contributes to minimizing uncontrolled water loss (Ma 2004). Si helps to prevent water loss by accumulating in the leaf cuticle (Ma et al. 2001a, b). Therefore, it contributes to drought resistance by helping to prevent water loss of the plant.

Studies showed that SiO2 NPs regulate protein and phenol contents in maize plants as key features in crop growth and development research. Also, a higher Si level accumulated in the roots, which is considered a modulation for drought tolerance in maize. Si NPs deposited in plant tissues can increase the expression of key biochemical components, growth, and yield parameters in maize when applied at an optimal dosage (Suriyaprabha et al. 2012). In another study, it was observed that the application of different concentrations of SiO2 NP to hawthorn seedlings increased drought tolerance and preserved critical physiological and biochemical properties. This study revealed that pretreatment of hawthorn seedlings with SiO2 nanoparticle increases photosynthesis rate, stomatal conductivity, and water potential in xylem. A positive effect was observed on photosynthesis parameters, malondialdehyde (MDA), RWC, carotenoid, carbohydrate, and proline contents with SiO2 treatment (Ashkavand et al. 2015). Pei et al. (2010) stated that the application of SiO2 NPs at a concentration of 1.0 mM in wheat plant contributed to the stem growth, and water holding capacity was higher in drought-stressed plants. In some studies, TiO2 and SiO2 nanoparticles were applied together to the plants. For example, the combination of 50 mg/L TiO2 and 3200 mg/L SiO2 was applied to the leaves of the cotton plant and observed under drought stress. As a result of the study, it was concluded that the tolerance to drought stress increased in cotton (Shallan et al. 2016).

6.3 *Effect of Other Nanoparticles on Drought Stress*

Many studies have shown that silver (Ag) NPs affect drought stress tolerance. For example, one related study on drought stress in lentils, the application of AgNPs reduced drought stress in lentils (Hojjat and Ganjali 2016). Nanoparticles are known to eliminate reactive oxygen species (ROS). Mahdi Nezhad et al. (2018) showed that the need for antioxidant metabolism decreased with the application of AgNPs under drought stress.

As mentioned earlier, nanoparticles affect the expression of stress-related genes. For example, Gupta et al. (2018) determined that microarray analyzes in the expression of genes involved on stress tolerance increased after the application of AgNPs in Arabidopsis. It has been shown that metal and oxidative stress tolerance genes are also increased their expression. These genes are superoxide dismutase, peroxidase, and cytochrome P450-dependent oxidase.

In one study, the application of 50 mg/L AgNPs to tomato plant played an important role against drought stress. In the experiment, it was determined that there was a toxic effect in the application of AgNPs at a dose of 100 mg/L. Accordingly, it was once again emphasized that the effects of nanoparticles on drought response give different results at different concentrations (Akhoundnejad 2021; Akhoundnejad et al. 2022).

Zinc (Zn) is an important heavy metal which is an essential nutrient source for all organisms and transition metal with important functions in enzymatic activities (Ergen and Tunca 2018). It is required for chlorophyll production, pollen performance, fertility and germination, as well as for the regulation of lipid metabolites, nucleic acid, RNA metabolism, stability and DNA simulation and gene expression

(Hafizi and Nasr 2018). ZnO NPs is an important nanoparticle in drought stress tolerance. In one study, application of ZnO NPs to wheat plant under drought stress facilitated rapid utilization of seed reservoirs for seedling growth and reduced drought stress. It also contributed to tolerance by improving antioxidant enzyme activation (Sedghi et al. 2013).

In another study, the application of mixture of ZnO, B2O3 and CuO nanoparticles to the soybean plant reduced the negative effect of drought stress on the plant. The results have showed that stem growth increased by 33%, while grain yield increased by 36% with the effect of this under drought stress (Dimkpa et al. 2019).

Dimkpa et al. (2020) worked on drought stress in wheat plants. They were observed that application of ZnO nanoparticles developed yield grains and accumulation of Zn heavy metal. van Nguyen et al. (2022) studied that application of Cu NPs on maize plants developed grain productivity, biomasss level and enhanced anthocynain contents which helps to reduce oxidative stress under drought stress condition.

Application of Fe nanoparticles is also one of the strategies used to help plants affected by drought stress. In addition, it has been reported that it contributes to the increase in yield in plants (Davar-Zareii et al. 2014). Application of Fe NPs on Arabidopsis helped to promote changes of stomatas and pigment contents under drought stress condition (Kim et al. 2015).

CeO2 nanoparticles were applied to sorghum leaves under control condition (spraying water) and drought conditions. As a result of CeO2 application under drought stress, hydrogen peroxide (H2O2) decreased by 36%, while superoxide radical (O2•–) decreased by 41% as compared to control plants (Djanaguiraman et al. 2018).

In one study, CaNP (hydroxyapatite nanoparticle) applied to the Jute seeds resulting in drought tolerance by accumulation of proline (Das et al. 2016).

Behboudi et al. (2019) experienced that application of Chitosan NPs on wheat plants promoted productivity, improved photosynthesis rate and enhanced antioxidant ezymes activities such as SOD and CAT under drought stress conditions. Similarly, Behboudi et al. (2018) studied that application of the same molecule on barley plants showed enhanced grain yield and increased relative water content (RWC) under drought stress conditions.

7 Effect of NPs on Physiological Responses of Plants

Physiological responses against water deficiency in plants varies depending on the species, variety, the severity and duration of water loss, the developmental status of the plant, its age, organ and cell type, and cellular integrity (such as cell wall and cell membrane). These induced responses can occur in a matter of seconds or take minutes or hours to occur (Bray 1997). NPs are involved in plant growth development against abiotic stresses. Studies have shown that the application of Si NPs contributes to the development of drought tolerance in plants. For example, SiO2

NP was applied (0, 10, 50, and 100 mg L) to the Prunus mahaleb. Plant growth, physiological and biochemical parameters were investigated under drought stress. It has been observed that the application of high-dose SiO2 NPs have given more effective results than the application of low-dose in drougth tolerance (Tripathi et al. 2017). It has been reported that the growth in leaves is accelerated, the balance between Na + and K+ ions and the level of chlorophyll increase, and the photosynthesis process accelerates as a result of the application of Si NPs in Musa acuminata exposed to drought and salt stress (Mahmoud et al. 2020). Ashkavand et al. (2015) examined the hatworn seedlings under drought stress. Biomass content and photosynthetic pigment level increased, and stomatal conductivity and net photosynthesis rate improved with the application of SiO2 NP to the plant based on the results of scientific studies. In another study, researchers investigated the effects of synthesized SiO2 NPs on growth and physiological responses of maize to drought stress. The findings showed that SiO2 NPs regulate protein and phenol content which play an important role in plant growth and development. Moreover, excessive Si accumulation in the roots of the maize was found to be a contributing factor to drought tolerance. Researchers stated that the accumulation of Si in plant cells increased the expression of biochemical components and they found that optimum dosage of SiO2 application contributed to plant growth and increase yield (Suriyaprabha et al. 2012). The increase in enzymatic system activation, the accumulation of free proline and other amino acids of nutrients, and the improvement of water uptake by the plant indicate the contribution of NPs to the drought stress tolerance of the plant (Shallan et al. 2016: Wang et al. 2020).

Biochemical mechanisms of ZnO NPs affect photosynthesis in plants. In many studies, different concentrations of ZnO NPs were tested to understand the photosynthesis efficiency. According to the studies, the chlorophyll content increased with the application of this NP in cotton (Venkatachalam et al. 2017), in peanuts (Prasad et al. 2012), and in cilantro (Pullagurala et al. 2018). Contrary to these results, researchers have found a decrease in chlorophyll content with the application of ZnO in wheat (Tripathi et al. 2017) and in maize (Zhao et al. 2016). It is thought that the difference between the results depends on the application time, severity, and age of the plant.

Burman et al. (2013) and Pandey et al. (2010) compared the application of ZnO NPs and bulk ZnSO4 application in chickpea plant. They found that the application of ZnO NPs were more effective way in terms of biomass production, seed germination and plant development. Also, ZnO NPs improved auxin level and thus plant growth. In another study, it was observed that ZnS NP increased the growth of Brassica juncea. According to the results of this study, physiological parameters such as chlorophyll and water content and plant biomass increased significantly. It has been suggested that these results are related to the improvement of the antioxidant system in plants treated with ZnS NPs. In addition, another result of this study, application of ZnS in low doses was given more effective results compared to high concentrations (Nayan et al. 2016). In another experiment, ZnO NPs showed an increase in photosynthetic pigments and a decrease in lipid peroxidation in Coriandrum sativum. These pigments are considered as a part of defense mechanism against harmful singlet oxygen. Therefore, it has been said that photosynthetic pigments developed by the application of ZnO cope with the drought stress mechanism by stabilizing the production of ROS in plants (Bhatt et al. 2020). On the other hand, application of CuNP to the maize plants enhanced biomass grain yield and growth parameters under drought stress condition (van Nguyen et al. 2022).

In one experiment, it was stated that metal NPs could be effective in coping with abiotic stress in plants (Khan et al. 2017). Generally, it is known that NPs are taken into the plant via plasmodemata, and NP translocation by apoplast and symplast. The researchers emphasized that the NP application affects the biomass level, the amount of chlorophyll and thus the photosynthesis process. In addition, when NPs were taken into plant cells, it was observed that they regulate not only the N2 level and protein content, but also the expression of genes involved in stress (Rajput et al. 2021).

According to a study conducted in wheat under drought conditions, the dry matter ratio, RWC, chlorophyll and carotenoid contents increased as a result of the application of TiO2 NP to the plant, and this promoted the transpiration rate (Faraji and Sepehri 2020).

Yelderem and Demirci (2022) have resulted that the application of Ag NP to tomato is important in drought mechanisms. A wide leaf width is formed, stomata are closed, and a more effective water is used by plant especially at the dose of 50 mg/L of Ag.

8 Effect on NPs on Biochemical Responses of Plants

Water is the most important liquid for all known life forms where biochemical activities take place. Therefore, there are some biochemical responses caused by drought stress in plants. It is known that NP applications create some biochemical responses as well. They improve plant tolerance by increasing root hydraulic conductivity which accelerates water uptake to the plant, while also affecting ROS detoxification and stress signaling (Das and Das 2019).

NPs can reduce damage to plants exposed to abiotic stress. They are generally applied as foliar and/or soil. The antioxidant response developed by nanoparticles depends on the level of ROS damage. The emergence of antioxidant compounds indicates a positive response to abiotic stress conditions (Marslin et al. 2017; Fig. 4). It is thought that the underlying reason of plant damage under stress is about some biochemical mechanisms such as the detoxification pathway. These mechanisms play a role in stress damage due to enzymatic antioxidant activation (Teske and Detweiler 2015).

The application of Si NPs activates antioxidant activity and helps tolerance mechanism against drought stress in plants. For example, CAT, SOD, and GR activity increased in tomato (Shi et al. 2016), and in wheat (Gong et al. 2005) by the

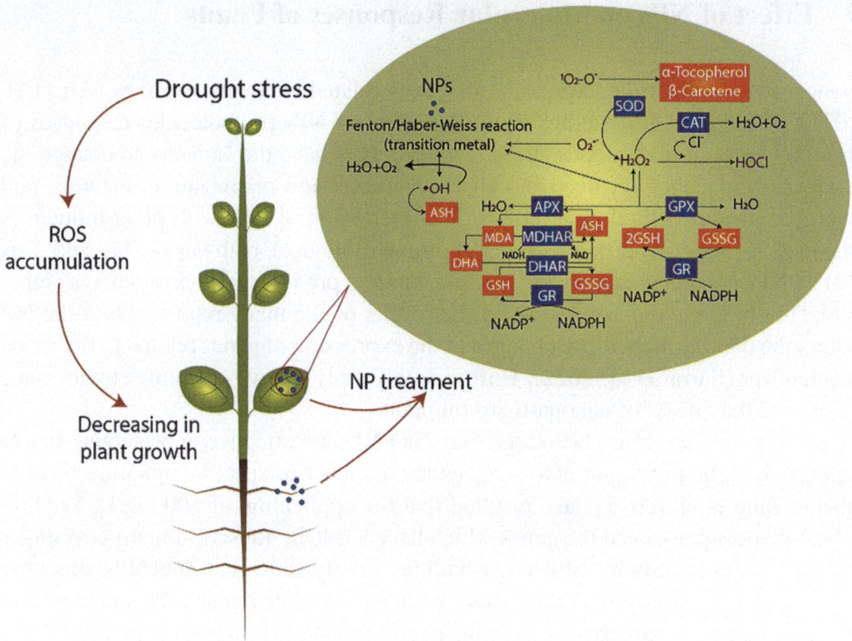

Fig. 4 Increase in antioxidant capacity via NPs application reduces abiotic stresses in plants such as drought and salinity

application of Si NPs under drought stress. On the other hand, ascorbate contents increased in wheat (Ma et al. 2006) while APX and MDA activities reduced in sunflower (Gunes et al. 2008). It is proved that SOD activity of plants increased by the application of different NPs, TiO2 NPs (Ghasemlou et al. 2019), ZnO NPs (Khan et al. 2019), and Cu NPs (Maswada et al. 2020).

In rice seedlings, nano-CuO treatment increased antioxidant enzymes and MDA concentration (Shaw and Hossain 2013). Similarly, nano-CuO treatment in barley regulated photosynthetic performance and antioxidant defense mechanism, however root and stem growth restricted (Shaw et al. 2014). In addition to that antioxidative enzymes were enchanced by reason of Zn and Cu application reduced MDA accumulation in wheat under drought stress (Taran et al. 2017). In a study conducted with stevia plant, low concentration of ZnO NP increased antioxidant activation while application of low concentrations resulted in oxidative stress (Javed et al. 2017). The stimulation of enzymatic antioxidants by the application of ZnO in plants is also of great importance in terms of tolerance to drought stress. In addition to all these, application of selenium nanoparticles (SeNPs) enhances ROS enzymes acitivity and reduces oxidative stress under drought stress condition in plants (Jung et al. 2019; Kaya et al. 2019).

9 Effect of NPs on Molecular Responses of Plants

Nanoparticles affect the expression of genes related to abiotic stresses (Ali et al. 2021). There are a few studies about the effect of NPs on molecular responses of drought stress in plants. As a stress response, NPs have the capacity to change the expression of genes involved in cell biosynthesis and organization, electron and energy transport (Mittal et al. 2020). The interaction of NPs with plant influences changes in gene expressions and important biological pathways (Das and Das 2019). NPs are recognized by the calcium-binding protein complex or act as a signal molecule in the cytoplasm based on the results of the many reports. Once the NP enters the plant cell, there are changes in the expression of genes related to the stress mechanism (Sarraf et al. 2022). Unfortunately, only a few proteomic studies have examined the effects of nanoparticles on plants.

In many studies, it has been seen that ZnO NPs are effective in protecting plants against drought stress and activating genes involved in specific signaling mechanisms. Yang et al. (2017) have resulted that the application of 500 mg/kg ZnO on wheat plants upregulated the genes which have a role in transcription process under drought stress conditions. Molecular findings clearly show that ZnO NPs affect the expression of transcription factors and many stress-related genes. This also affects plant immunity, productivity, metabolism, and growth (Mirakhorli et al. 2021).

When the effect of TiO2 NPs on gene expressions in Arabidopsis roots was examined, 80 up- and 74 down-regulated gene changes were observed. Beside this, 660 up- and 826 down-regulated genes were found when the effect of ZnO NPs was examined. While these genes are mainly involved in drought and oxidative stress tolerance, it has also been observed that some biotic stress factors stimulate these genes (Landa et al. 2012).

Kim et al. (2015) studied that Fe NPs contributed to drought tolerance by upregulating AHA2 gene in Arabidopsis plant. This is related to the mechanism of opening and closing of stomata in terms of tolerance to stress. Jin et al. (2017) experienced that Al2O3 NPs promoted the transcription of genes related to root development in Arabidopsis plant. When all these results are considered, it is seen that NP application in plants is a mechanism that trigger stress tolerance.

Application of Ag NPs in tomato plant triggered tolerance in salinity and drought stresses. The overexpression of AREB (encode ABA inducing tolerance to osmotic stress), MAPK2 (regulates antioxidant and hormonal plant defense system), P5CS (encodes proline synthesis) and CRK1 genes, which are involved in ABA regulation; and the suppression of TAS14 (reduces the osmotic potential), DDF2 (biosynthesis of gibberellic acid), and ZFHD (biosynthesis of ABA) genes (Almutairi 2016), contributes to stress tolerance. Syu et al. (2014) showed that Ag NPs were observed to suppress ACC 7 synthase (ACS7) and ACC oxidase 2 (ACO2) genes and overexpress indoleacetic acid 8 (IAA8), 9-cisepoxycarotenoid dioxygenase (NCED3), and RD22 proteins in Arabidopsis under drought stress.

The application of thiol gated mesoporous silica NPs to the Arabidopsis seedlings enhanced AtGALK2 which is ABA-inducible marker gene. Thus, tolerance mechanism improved against drought stress condition (Sun et al. 2018).

Linh et al. (2020) applied ZnO, Cu, Fe, and Co NPs on soybean leaves and roots under drought stress conditions. Many genes involved in drought tolerance mechanism are tested with all nanoparticules such as GmRD20A, GmERD1, GmNAC11, and GmMYB118. According to results, the expression of RD20A increased, especially Fe and Co applied plants. RD20 gene is known as responsible for desiccation (Sarraf et al. 2022). Application of Fe increased the expression of all genes in roots and expression of three drought responsible genes increased in leaves with all NPs applied plants. The expression of GmERD1 (early responsive to dehydration 1) gene increased in shoots and roots with all NPs application. ERD1 gene have a role in independent pathway of ABA (Stolf-Moreira et al. 2011). The expression of GmNAC11 enhanced in roots with all NP applications and DREB2 enhanced in roots of the ZnO, Cu, and Fe NPs treated plants. The expression of GmMYB118 enhanced in leaves with all NP applications and only the application of Fe NP increased its application in leaves and roots. They examined that application of NPs enhanced the drought-responsive genes and helps soybean plants to cope with drough stress.

10 Conclusion

Environmental conditions have important role on plant development and growth. Undesirable environmental factors can cause stress on plants, like drought. Plant have developed different response mechanims to drought stress; however, these mechanisms is not clear yet. Various methods have been developed to increase the drought tolerance of plants. Nanotechnology, using of current technology and materials in nanosize, is one of these methods. There are different application methods of nanotechnology; like nanofertilizers, nanopesticides, nanosensors and nanomaterils. Nanomaterials differs from their bulk structure in many properties. Being at very small size is the most important one. Thus, nanomaterials can penetrate into the cell and nucleus. Thus, they can create new water channels and causes changes in gene expression, which increases the water uptake capacity of the plants. On the other hand, nanomaterials are able to change molecular mechanisms in the plant cell and increase the antioxidant capacity and some of them can decrease the ROS element, which helps to decrease the possibile damage of drought response mechanisms to the plant. Nevertheless, this novel topic still needs to be studied and how the nanoparticles can affect plant cells at molecular and genetic level should be illuminated. In conclusion, nanomaterials have a great potential as an alternative tool against drought stress.

References

Acar O (1999) Kuraklığa dayanıklı bazı arpa (Hordeum spp.) çeşitlerinde süperoksit dismutaz (SOD) aktivitelerinin araştırılması [Investigation of superoxide dismutase (SOD) activities in some drought-resistant barley (Hordeum spp.) varieties]. Ege Üniversitesi. Fen Bilimleri Enstitüsü, Doktora Tezi, İzmir

Adak T, Kumar J, Shakil NA, Walia S (2012) Development of controlled release formulations of imidacloprid employing novel nano-ranged amphiphilic polymers. J Environ Sci Health B 47(3):217–225

Agarwal M, Zhu JK (2008) Integration of abiotic stress signaling pathways. In: Jenks MA, Hasegawa PM (eds) Plant abiotic stress. Wiley, Pondicherry, India pp 215–247

Aghdam MTB, Mohammadi H, Ghorbanpour M (2016) Effects of nanoparticulate anatase titanium dioxide on physiological and biochemical performance of *Linum usitatissimum* (Linaceae) under well-watered and drought stress conditions. Rev Bras Bot 39(1):139–146

Akbari GA, Morteza E, Moaveni P, Alahdadi I, Bihamta MR, Hasanloo T (2014) Pigments apparatus and anthocyanins reactions of borage to irrigation, methylalchol and titanium dioxide. Int J Biosci 4(7):192–208

Akhoundnejad Y (2021) Kuraklık stresinde gümüş (Ag) nanopartiküllerinin domateste kalitesi üzerine etkileri [Effects of silver (Ag) nanoparticles on tomato quality under drought stress]. Şırnak Üniversitesi Fen Bilimleri Dergisi 1(2):1–13

Akhoundnejad Y, Karakas O, Demirci O (2022) Response of lettuce to silver nanoparticles under drought conditions. Iran J Sci Technol Trans A: Sci 46(1):111–120

Ali S, Mehmood A, Khan N (2021) Uptake, translocation, and consequences of nanomaterials on plant growth and stress adaptation. J Nanomater 2021:1–17

Almutairi ZM (2016) Influence of silver nano-particles on the salt resistance of tomato (*Solanum lycopersicum*) during germination. Int J Agric Biol 18(2):449–457

Ashkavand P, Tabari M, Zarafshar M, Tomásková I, Struve D (2015) Effect of SiO2 nanoparticles on drought resistance in hawthorn seedlings. Leśne Prace Badawcze 76(4):350–359

Ashraf MPJC, Harris PJC (2004) Potential biochemical indicators of salinity tolerance in plants. Plant Sci 166(1):3–16

Bakhsh A, Hussain T (2015) Engineering crop plants against abiotic stress: current achievements and prospects. Emir J Food Agric 27:24–39

Bashir M, Bahar FA, Bhat MA, Singh L, Mahdi SS, Palmo T et al (2022) Review: nanotechnology in agriculture: prospects and problems. Biol Forum Int J 14(1):1–10

Basu S, Ramegowda V, Kumar A, Pereira A (2016) Plant adaptation to drought stress. F1000Res 5:1554

Behboudi F, Tahmasebi Sarvestani Z, Kassaee MZ, Modares Sanavi SAM, Sorooshzadeh A, Ahmadi SB (2018) Evaluation of chitosan nanoparticles effects on yield and yield components of barley (*Hordeum vulgare* L.) under late season drought stress. J Water and Environ Nanotechnol 3(1):22–39

Behboudi F, Tahmasebi-Sarvestani Z, Kassaee MZ, Modarres-Sanavy SAM, Sorooshzadeh A, Mokhtassi-Bidgoli A (2019) Evaluation of chitosan nanoparticles effects with two application methods on wheat under drought stress. J Plant Nutr 42(13):1439–1451

Bhatt D, Sharma G (2018) Role of silicon in counteracting abiotic and biotic plant stresses. Int J Chem Stud 6(2):1434–1442

Bhatt S, Nagappa AN, Patil CR (2020) Role of oxidative stress in depression. Drug Discov Today 25(7):1270–1276

Bray EA (1997) Plant responses to water deficit. Trends Plant Sci 2(2):48–54

Burman U, Saini M, Kumar P (2013) Effect of zinc oxide nanoparticles on growth and antioxidant system of chickpea seedlings. Toxicol Environ Chem 95(4):605–612

Chetal S, Wagle DS, Nainawatee HS (1982) Alterations in glycolipids of wheat and barley leaves under water stress. Phytochemistry 21(1):51–53

Chhipa H (2017) Nanofertilizers and nanopesticides for agriculture. Environ Chem Lett 15(1):15–22

Choudhary V, Gupta A (2011) Polymer/carbon nanotube nanocomposites. In: Carbon nanotubes-polymer nanocomposites. InTech, Shanghai, pp 65–90

Das A, Das B (2019) Nanotechnology a potential tool to mitigate abiotic stress in crop plants. In: Abiotic and biotic stress in plants. IntechOpen. https://doi.org/10.5772/intechopen.83562

Das A, Ray R, Mandal N, Chakrabarti K (2016) An analysis of transcripts and enzyme profiles in drought stressed jute (*Corchorus capsularis*) and rice (*Oryza sativa*) seedlings treated with $CaCl_2$, hydroxyapatite nano-particle and β-amino butyric acid. Plant Growth Regul 79(3):401–412

Davar-Zareii F, Roozbahani A, Hosnamidi A (2014) Evaluation the effect of water stress and foliar application of Fe nanoparticles on yield, yield components and oil percentage of safflower (*Carthamus tinctorious* L.). Int J Adv Biol Biomed Res 2(4):1150–1159

DeRosa MC, Monreal C, Schnitzer M, Walsh R, Sultan Y (2010) Nanotechnology in fertilizers. Nat Nanotechnol 5(2):91–91

Dimkpa CO, Singh U, Bindraban PS, Elmer WH, Gardea-Torresdey JL, White JC (2019) Zinc oxide nanoparticles alleviate drought-induced alterations in sorghum performance, nutrient acquisition, and grain fortification. Sci Total Environ 688:926–934

Dimkpa CO, Andrews J, Fugice J, Singh U, Bindraban PS, Elmer WH et al (2020) Facile coating of urea with low-dose ZnO nanoparticles promotes wheat performance and enhances Zn uptake under drought stress. Front Plant Sci 11:168

Djanaguiraman M, Nair R, Giraldo JP, Prasad PVV (2018) Cerium oxide nanoparticles decrease drought-induced oxidative damage in sorghum leading to higher photosynthesis and grain yield. ACS Omega 3(10):14406–14416

Eichert T, Goldbach HE (2008) Equivalent pore radii of hydrophilic foliar uptake routes in stomatous and astomatous leaf surfaces–further evidence for a stomatal pathway. Physiol Plant 132(4):491–502

Eichert T, Kurtz A, Steiner U, Goldbach HE (2008) Size exclusion limits and lateral heterogeneity of the stomatal foliar uptake pathway for aqueous solutes and water-suspended nanoparticles. Physiol Plant 134(1):151–160

Ergen ŞF, Tunca EÜ (2018) Nanotoxicity modelling and removal efficiencies of ZnONP. Int J Phytoremediation 20(1):16–26

Faraji J, Sepehri A (2020) Exogenous nitric oxide improves the protective effects of TiO2 nanoparticles on growth, antioxidant system, and photosynthetic performance of wheat seedlings under drought stress. J Soil Sci Plant Nutr 20(2):703–714

Feng BH, Zhang ZY (2011) Carboxymethy chitosan grafted ricinoleic acid group for nanopesticide carriers. Adv Mater Res 236:1783–1788

Gao F, Hong F, Liu C, Zheng L, Su M, Wu X et al (2006) Mechanism of nano-anatase TiO2 on promoting photosynthetic carbon reaction of spinach. Biol Trace Elem Res 111(1):239–253

Gao F, Liu C, Qu C, Zheng L, Yang F, Su M, Hong F (2008) Was improvement of spinach growth by nano-TiO2 treatment related to the changes of Rubisco activase? Biometals 21(2):211–217

Ghasemlou F, Amiri H, Karamian R, Mirzaie-asl A (2019) Alleviation of the effects of on drought stress *Verbascum nudicuale* by methyl jasmonate and titanium dioxide nanoparticles. Iran J Plant Physiol 9(4):2911–2920

Giannousi K, Avramidis I, Dendrinou-Samara C (2013) Synthesis, characterization and evaluation of copper based nanoparticles as agrochemicals against *Phytophthora infestans*. RSC Adv 3(44):21743–21752

Gong H, Zhu X, Chen K, Wang S, Zhang C (2005) Silicon alleviates oxidative damage of wheat plants in pots under drought. Plant Sci 169(2):313–321

Gunes A, Pilbeam DJ, Inal A, Coban S (2008) Influence of silicon on sunflower cultivars under drought stress, I: growth, antioxidant mechanisms, and lipid peroxidation. Commun Soil Sci Plant Anal 39(13–14):1885–1903

Gupta SD, Agarwal A, Pradhan S (2018) Phytostimulatory effect of silver nanoparticles (AgNPs) on rice seedling growth: an insight from antioxidative enzyme activities and gene expression patterns. Ecotoxicol Environ Saf 161:624–633

Hafizi Z, Nasr N (2018) The effect of zinc oxide nanoparticles on safflower plant growth and physiology. Eng Technol Appl Sci Res 8(1):2508–2513

Hall JÁ, Williams LE (2003) Transition metal transporters in plants. J Exp Bot 54(393):2601–2613

Hare PD, Cress WA (1997) Metabolic implications of stress-induced proline accumulation in plants. Plant Growth Regul 21(2):79–102

Hartung W, Sauter A, Hose E (2002) Abscisic acid in the xylem: where does it come from, where does it go to? J Exp Bot 53(366):27–32

Haynes CL, Van Duyne RP (2001) Nanosphere lithography: a versatile nanofabrication tool for studies of size-dependent nanoparticle optics. J Phys Chem B 105(24):5599–5611

Hellmann C, Greiner A, Wendorff JH (2011) Design of pheromone releasing nanofibers for plant protection. Polym Adv Technol 22(4):407–413

Hojjat SS, Ganjali A (2016) The effect of silver nanoparticle on lentil seed germination under drought stress. Int J Farm Allied Sci 5(3):208–212

Holden NM, Brereton AJ, Fealy R, Sweeney J (2003) Possible change in Irish climate and its impact on barley and potato yields. Agric For Meteorol 116(3–4):181–196

Hong F, Zhou J, Liu C, Yang F, Wu C, Zheng L, Yang P (2005) Effect of nano-TiO2 on photochemical reaction of chloroplasts of spinach. Biol Trace Elem Res 105(1):269–279

Huang ZM, Zhang YZ, Kotaki M, Ramakrishna S (2003) A review on polymer nanofibers by electrospinning and their applications in nanocomposites. Compos Sci Technol 63(15):2223–2253

Jaberzadeh A, Moaveni P, Moghadam HRT, Zahedi H (2013) Influence of bulk and nanoparticles titanium foliar application on some agronomic traits, seed gluten and starch contents of wheat subjected to water deficit stress. Notulae Botanicae Horti Agrobotanici Cluj-Napoca 41(1):201–207

Javed R, Usman M, Yücesan B, Zia M, Gürel E (2017) Effect of zinc oxide (ZnO) nanoparticles on physiology and steviol glycosides production in micropropagated shoots of *Stevia rebaudiana* Bertoni. Plant Physiol Biochem 110:94–99

Jimenez-Falcao S, Villalonga A, Arevalo-Villena M, Briones-Perez A, Martinez-Manez R, Martinez-Ruiz P, Villalonga R (2020) Enzyme-controlled mesoporous nanosensor for the detection of living *Saccharomyces cerevisiae*. Sensors Actuators B Chem 303:127197

Jin Y, Fan X, Li X, Zhang Z, Sun L, Fu Z et al (2017) Distinct physiological and molecular responses in Arabidopsis thaliana exposed to aluminum oxide nanoparticles and ionic aluminum. Environ Pollut 228:517–527

Johnson MS, Sajeev S, Nair RS (2021) Role of nanosensors in agriculture. In: In 2021 international conference on computational intelligence and knowledge economy (ICCIKE). IEEE, March, pp 58–63

Jung HI, Kong MS, Lee BR, Kim TH, Chae MJ, Lee EJ et al (2019) Exogenous glutathione increases arsenic translocation into shoots and alleviates arsenic-induced oxidative stress by sustaining ascorbate–glutathione homeostasis in rice seedlings. Front Plant Sci 10:1089

Kah M, Hofmann T (2014) Nanopesticide research: current trends and future priorities. Environ Int 63:224–235

Kah M, Beulke S, Tiede K, Hofmann T (2013) Nanopesticides: state of knowledge, environmental fate, and exposure modeling. Crit Rev Environ Sci Technol 43:1823–1867

Kah M, Kookana RS, Gogos A, Bucheli TD (2018) A critical evaluation of nanopesticides and nanofertilizers against their conventional analogues. Nat Nanotechnol 13(8):677–684

Kaya C, Okant M, Ugurlar F, Alyemeni MN, Ashraf M, Ahmad P (2019) Melatonin-mediated nitric oxide improves tolerance to cadmium toxicity by reducing oxidative stress in wheat plants. Chemosphere 225:627–638

Kearns H, Goodacre R, Jamieson LE, Graham D, Faulds K (2017) SERS detection of multiple antimicrobial-resistant pathogens using nanosensors. Anal Chem 89(23):12666–12673

Ketchum RE, Warren RS, Klima LJ, Lopez-Gutiérrez F, Nabors MW (1991) The mechanism and regulation of proline accumulation in suspension cell cultures of the halophytic grass *Distichlis spicata* L. J Plant Physiol 137(3):368–374

Khan N, Bano A (2018) Effects of exogenously applied salicylic acid and putrescine alone and in combination with rhizobacteria on the phytoremediation of heavy metals and chickpea growth in sandy soil. Int J Phytoremediation 20(5):405–414

Khan MN, Mobin M, Abbas ZK, AlMutairi KA, Siddiqui ZH (2017) Role of nanomaterials in plants under challenging environments. Plant Physiol Biochem 110:194–209

Khan ZS, Rizwan M, Hafeez M, Ali S, Javed MR, Adrees M (2019) The accumulation of cadmium in wheat (*Triticum aestivum*) as influenced by zinc oxide nanoparticles and soil moisture conditions. Environ Sci Pollut Res 26(19):19859–19870

Khodakovskaya M, Dervishi E, Mahmood M, Xu Y, Li Z, Watanabe F, Biris AS (2009) Carbon nanotubes are able to penetrate plant seed coat and dramatically affect seed germination and plant growth. ACS Nano 3(10):3221–3227

Khodakovskaya MV, De Silva K, Biris AS, Dervishi E, Villagarcia H (2012) Carbon nanotubes induce growth enhancement of tobacco cells. ACS Nano 6(3):2128–2135

Kim JH, Oh Y, Yoon H, Hwang I, Chang YS (2015) Iron nanoparticle-induced activation of plasma membrane H+-ATPase promotes stomatal opening in Arabidopsis thaliana. Environ Sci Technol 49(2):1113–1119

Kumpiene J, Lagerkvist A, Maurice C (2008) Stabilization of As, Cr, Cu, Pb and Zn in soil using amendments–a review. Waste Manag 28(1):215–225

Lahiani MH, Dervishi E, Chen J, Nima Z, Gaume A, Biris AS, Khodakovskaya MV (2013) Impact of carbon nanotube exposure to seeds of valuable crops. ACS Appl Mater Interfaces 5(16):7965–7973

Landa P, Vankova R, Andrlova J, Hodek J, Marsik P, Storchova H et al (2012) Nanoparticle-specific changes in *Arabidopsis thaliana* gene expression after exposure to ZnO, TiO2, and fullerene soot. J Hazard Mater 241:55–62

Leridon H (2020) World population outlook: explosion or implosion? Population Societies 573(1):1–4

Levitt J (1980) Responses of plants to environmental stress, volume 1: chilling, freezing, and high temperature stresses. Academic, Cambridge

Linh TM, Mai NC, Hoe PT, Lien LQ, Ban NK, Hien LTT et al (2020) Metal-based nanoparticles enhance drought tolerance in soybean. J Nanomater 2020:4056563

Liu R, Lal R (2014) Synthetic apatite nanoparticles as a phosphorus fertilizer for soybean (Glycine max). Sci Rep 4(1):1–6

Liu Q, Chen B, Wang Q, Shi X, Xiao Z, Lin J, Fang X (2009) Carbon nanotubes as molecular transporters for walled plant cells. Nano Lett 9(3):1007–1010

Lucy JA, Glauert AM (1964) Structure and assembly of macromolecular lipid complexes composed of globular micelles. J Mol Biol 8(5):727–748

Lv J, Zhang S, Luo L, Zhang J, Yang K, Christie P (2015) Accumulation, speciation and uptake pathway of ZnO nanoparticles in maize. Environ Sci Nano 2(1):68–77

Lv J, Christie P, Zhang S (2019) Uptake, translocation, and transformation of metal-based nanoparticles in plants: recent advances and methodological challenges. Environ Sci Nano 6(1):41–59

Ma JF (2004) Role of silicon in enhancing the resistance of plants to biotic and abiotic stresses. Soil Sci Plant Nutr 50(1):11–18

Ma JF, Takahashi E (2002) Soil, fertilizer, and plant silicon research in Japan. Elsevier, Amsterdam

Ma JF, Goto S, Tamai K, Ichii M (2001a) Role of root hairs and lateral roots in silicon uptake by rice. Plant Physiol 127(4):1773–1780

Ma JF, Miyake Y, Takahashi E (2001b) Silicon as a beneficial element for crop plants. Stud Plant Sci 8:17–39

Ma JF, Tamai K, Yamaji N, Mitani N, Konishi S, Katsuhara M et al (2006) A silicon transporter in rice. Nature 440(7084):688–691

Mahajan S, Tuteja N (2005) Cold, salinity and drought stresses: an overview. Arch Biochem Biophys 444(2):139–158

Mahmoodzadeh H, Aghili R, Nabavi M (2013) Physiological effects of TiO2 nanoparticles on wheat (*Triticum aestivum*). Tech J Eng Appl Sci 3(14):1365–1370

Mahmoud LM, Dutt M, Shalan AM, El-Kady ME, El-Boray MS, Shabana YM, Grosser JW (2020) Silicon nanoparticles mitigate oxidative stress of in vitro-derived banana (*Musa acuminata* 'Grand Nain') under simulated water deficit or salinity stress. S Afr J Bot 132:155–163

Marslin G, Sheeba CJ, Franklin G (2017) Nanoparticles alter secondary metabolism in plants via ROS burst. Front Plant Sci 8:832

Mastronardi E, Tsae P, Zhang X, Monreal C, DeRosa MC (2015) Strategic role of nanotechnology in fertilizers: potential and limitations. In: Nanotechnologies in food and agriculture. Springer, Cambridge, pp 25–67

Maswada HF, Mazrou YS, Elzaawely AA, Eldein SMA (2020) Nanomaterials. Effective tools for field and horticultural crops to cope with drought stress: a review. Span J Agric Res 18(2):15

Mirakhorli T, Ardebili ZO, Ladan-Moghadam A, Danaee E (2021) Bulk and nanoparticles of zinc oxide exerted their beneficial effects by conferring modifications in transcription factors, histone deacetylase, carbon and nitrogen assimilation, antioxidant biomarkers, and secondary metabolism in soybean. PLoS One 16(9):e0256905

Mittal D, Kaur G, Singh P, Yadav K, Ali SA (2020) Nanoparticle-based sustainable agriculture and food science: recent advances and future outlook. Front Nanotechnol 2:10

Mohammadi H, Esmailpour M, Gheranpaye A (2016) Effects of TiO2 nanoparticles and water-deficit stress on morpho-physiological characteristics of dragonhead (*Dracocephalum moldavica* L.) plants. Acta Agriculturae Slovenica 107(2):385–396

Nayan R, Rawat M, Negi B, Pande A, Arora S (2016) Zinc sulfide nanoparticle mediated alterations in growth and anti-oxidant status of *Brassica juncea*. Biologia 71(8):896–902

Nezhad NM, Mousavi H, Fakheri B, Heidari F (2018) The effect of some nanoparticles on the activity of antioxidant enzymes and parthenolide yield of feverfew plant (*Tanacetum parthenium* L.) under water deficit stress. Env Stresses Crop Sci 11(4):917–929

NTSC (2007) The National Nanotechnology Initiative strategic plan. The National Science and Technology Council, p 46

Nuruzzaman MD, Rahman MM, Liu Y, Naidu R (2016) Nanoencapsulation, nano-guard for pesticides: a new window for safe application. J Agric Food Chem 64(7):1447–1483

Ors S, Ekinci M (2015) Kuraklık stresi ve bitki fizyolojisi [Drought stress and plant physiology]. Derim, p 237

Ouda SM (2014) Antifungal activity of silver and copper nanoparticles on two plant pathogens, *Alternaria alternata* and *Botrytis cinerea*. Res J Microbiol 9(1):34

Owolade OF, Ogunleti DO, Adenekan MO (2008) Titanium dioxide affects disease development and yield of edible cowpea. J Agric Food Chem 7(50):2942–2947

Pandey AC, Sanjay SS, Yadav RS (2010) Application of ZnO nanoparticles in influencing the growth rate of *Cicer arietinum*. J Exp Nanosci 5(6):488–497

Passioura JB (1996) Drought and drought tolerance. Plant Growth Regul 20(2):79–83

Pei ZF, Ming DF, Liu D, Wan GL, Geng XX, Gong HJ, Zhou WJ (2010) Silicon improves the tolerance to water-deficit stress induced by polyethylene glycol in wheat (*Triticum aestivum* L.) seedlings. J Plant Growth Regul 29(1):106–115

Perlatti B, de Souza Bergo PL, Fernandes JB, Forim MR (2013) Polymeric nanoparticle-based insecticides: a controlled release purpose for agrochemicals. In: Insecticides-development of safer and more effective technologies. IntechOpen, Rijeka, Croatia, pp 23–35

Popp C, Burghardt M, Friedmann A, Riederer M (2005) Characterization of hydrophilic and lipophilic pathways of *Hedera helix* L. cuticular membranes: permeation of water and uncharged organic compounds. J Exp Bot 56(421):2797–2806

Prasad TNVKV, Sudhakar P, Sreenivasulu Y, Latha P, Munaswamy V, Reddy KR et al (2012) Effect of nanoscale zinc oxide particles on the germination, growth and yield of peanut. J Plant Nutr 35(6):905–927

Pullagurala VLR, Adisa IO, Rawat S, Kalagara S, Hernandez-Viezcas JA, Peralta-Videa JR, Gardea-Torresdey JL (2018) ZnO nanoparticles increase photosynthetic pigments and decrease lipid peroxidation in soil grown cilantro (*Coriandrum sativum*). Plant Physiol Biochem 132:120–127

Ragaei M, Sabry AKH (2014) Nanotechnology for insect pest control. Int J Sci Environ Technol 3(2):528–545

Rajput VD, Minkina T, Kumari A, Singh VK, Verma KK, Mandzhieva S et al (2021) Coping with the challenges of abiotic stress in plants: new dimensions in the field application of nanoparticles. Plan Theory 10(6):1221

Redondo-Gómez S (2013) Abiotic and biotic stress tolerance in plants. In: Molecular stress physiology of plants. Springer, India, pp 1–20

Rico CM, Majumdar S, Duarte-Gardea M, Peralta-Videa JR, Gardea-Torresdey JL (2011) Interaction of nanoparticles with edible plants and their possible implications in the food chain. J Agric Food Chem 59(8):3485–3498

Riederer M, Schreiber L (2001) Protecting against water loss: analysis of the barrier properties of plant cuticles. J Exp Bot 52(363):2023–2032

Rippa M, Castagna R, Brandi S, Fusco G, Monini M, Chen D et al (2020) Octupolar plasmonic nanosensor based on ordered arrays of triangular Au nanopillars for selective rotavirus detection. ACS Appl Nano Mater 3(5):4837–4844

Roduner E (2006) Size matters: why nanomaterials are different. Chem Soc Rev 35(7):583–592

Sabo-Attwood T, Unrine JM, Stone JW, Murphy CJ, Ghoshroy S, Blom D et al (2012) Uptake, distribution and toxicity of gold nanoparticles in tobacco (*Nicotiana xanthi*) seedlings. Nanotoxicology 6(4):353–360

Sanchez FJ, De Andres EF, Tenorio JL, Ayerbe L (2004) Growth of epicotyls, turgor maintenance and osmotic adjustment in pea plants (*Pisum sativum* L.) subjected to water stress. Field Crop Res 86(1):81–90

Sarraf M, Vishwakarma K, Kumar V, Arif N, Das S, Johnson R et al (2022) Metal/metalloid-based nanomaterials for plant abiotic stress tolerance: an overview of the mechanisms. Plan Theory 11(3):316

Schymura S, Fricke T, Hildebrand H, Franke K (2017) Elucidating the role of dissolution in CeO2 nanoparticle plant uptake by smart radiolabeling. Angew Chem Int Ed 56(26):7411–7414

Sedghi M, Hadi M, Toluie SG (2013) Effect of nano zinc oxide on the germination parameters of soybean seeds under drought stress. Ann West Univ Timisoara Ser Biol 16(2):73

Selda ÖRS, Ekinci M (2015) Kuraklık stresi ve bitki fizyolojisi [Drought stress and plant physiology]. Derim 32(2):237–250

Shallan MA, Hassan HM, Namich AA, Ibrahim AA (2016) Biochemical and physiological effects of TiO2 and SiO2 nanoparticles on cotton plant under drought stress. Res J Pharm Biol Chem Sci 7(4):1540–1551

Shaw AK, Hossain Z (2013) Impact of nano-CuO stress on rice (*Oryza sativa* L.) seedlings. Chemosphere 93(6):906–915

Shaw AK, Ghosh S, Kalaji HM, Bosa K, Brestic M, Zivcak M, Hossain Z (2014) Nano-CuO stress induced modulation of antioxidative defense and photosynthetic performance of Syrian barley (*Hordeum vulgare* L.). Environ Exp Bot 102:37–47

Shi Y, Zhang Y, Han W, Feng R, Hu Y, Guo J, Gong H (2016) Silicon enhances water stress tolerance by improving root hydraulic conductance in *Solanum lycopersicum* L. Front Plant Sci 7:196

Slomberg DL, Schoenfisch MH (2012) Silica nanoparticle phytotoxicity to *Arabidopsis thaliana*. Environ Sci Technol 46(18):10247–10254

Srinivas V, Balasubramanian D (1995) Proline is a protein-compatible hydrotrope. Langmuir 11(7):2830–2833

Srivastava AK, Dev A, Karmakar S (2018) Nanosensors and nanobiosensors in food and agriculture. Environ Chem Lett 16(1):161–182

Stolf-Moreira R, Lemos EGDM, Abdelnoor RV, Beneventi MA, Rolla AAP, Pereira SDS et al (2011) Identification of reference genes for expression analysis by real-time quantitative PCR in drought-stressed soybean. Pesquisa Agropecuaria Brasileira 46:58–65

Sun D, Hussain HI, Yi Z, Rookes JE, Kong L, Cahill DM (2018) Delivery of abscisic acid to plants using glutathione responsive mesoporous silica nanoparticles. J Nanosci Nanotechnol 18(3):1615–1625

Suriyaprabha R, Karunakaran G, Yuvakkumar R, Prabu P, Rajendran V, Kannan N (2012) Growth and physiological responses of maize (*Zea mays* L.) to porous silica nanoparticles in soil. J Nanopart Res 14(12):1–14

Syu YY, Hung JH, Chen JC, Chuang HW (2014) Impacts of size and shape of silver nanoparticles on Arabidopsis plant growth and gene expression. Plant Physiol Biochem 83:57–64

Taj G, Agarwal P, Grant M, Kumar A (2010) MAPK machinery in plants: recognition and response to different stresses through multiple signal transduction pathways. Plant Signal Behav 5(11):1370–1378

Taran N, Storozhenko V, Svietlova N, Batsmanova L, Shvartau V, Kovalenko M (2017) Effect of zinc and copper nanoparticles on drought resistance of wheat seedlings. Nanoscale Res Lett 12(1):1–6

Tarigholizadeh S (2021) Role of nanoparticles in various abiotic stresses for sustainable crop productivity. In: Rajput VD, Singh A, Singh AK, Minkina TM (eds) Emerging tools for sustainable agriculture and food security. 31-6831-68, ISBN 978-81-932436-6-4

Teske SS, Detweiler CS (2015) The biomechanisms of metal and metal-oxide nanoparticles' interactions with cells. Int J Environ Res Public Health 12(2):1112–1134

Toraman ES (1999) Characterization of membrane phospholipids and proteins of some wheat (T*riticum aestivum* L.) varieties under salt and drought stress conditions (Master's thesis). Middle East Technical University

Tripathi DK, Singh S, Singh S, Pandey R, Singh VP, Sharma NC et al (2017) An overview on manufactured nanoparticles in plants: uptake, translocation, accumulation and phytotoxicity. Plant Physiol Biochem 110:2–12

Tuteja N (2007) Abscisic acid and abiotic stress signaling. Plant Signal Behav 2(3):135–138

van Nguyen D, Nguyen HM, Le NT, Nguyen KH, Nguyen HT, Le HM et al (2022) Copper nanoparticle application enhances plant growth and grain yield in maize under drought stress conditions. J Plant Growth Regul 41(1):364–375

Venkatachalam P, Priyanka N, Manikandan K, Ganeshbabu I, Indiraarulselvi P, Geetha N et al (2017) Enhanced plant growth promoting role of phycomolecules coated zinc oxide nanoparticles with P supplementation in cotton (*Gossypium hirsutum* L.). Plant Physiol Biochem 110:118–127

Verma S, Nizam S, Verma PK (2013) Biotic and abiotic stress signaling in plants. In: Stress signaling in plants: genomics and proteomics perspective, vol 1. Springer, New York, pp 25–49

Vikesland PJ (2018) Nanosensors for water quality monitoring. Nat Nanotechnol 13(8):651–660

Wang D, Wang Z, Wang X, Zhuang X, Tian C, Luan F, Fu X (2020) Functionalized copper nanoclusters-based fluorescent probe with aggregation-induced emission property for selective detection of sulfide ions in food additives. J Agric Food Chem 68(40):11301–11308

Wanyika H (2013) Sustained release of fungicide metalaxyl by mesoporous silica nanospheres. In: Nanotechnology for sustainable development. Springer, Cham, pp 321–329

Watanabe T, Misawa S, Hiradate S, Osaki M (2008) Root mucilage enhances aluminum accumulation in *Melastoma malabathricum*, an aluminum accumulator. Plant Signal Behav 3(8):603–605

Yang T, Duncan TV (2021) Challenges and potential solutions for nanosensors intended for use with foods. Nat Nanotechnol 16(3):251–265

Yang J, Cao W, Rui Y (2017) Interactions between nanoparticles and plants: phytotoxicity and defense mechanisms. J Plant Interact 12(1):158–169

Yelderem AK, Demirci Ö (2022) Response of lettuce to silver nanoparticles under drought conditions. Iran J Sci Technol 46:111–120

Zhang L, Peng D, Liang RP, Qiu JD (2018) Graphene-based optical nanosensors for detection of heavy metal ions. Trends Anal Chem 102:280–289

Zhao L, Huang Y, Hu J, Zhou H, Adeleye AS, Keller AA (2016) 1H NMR and GC-MS based metabolomics reveal defense and detoxification mechanism of cucumber plant under nano-Cu stress. Environ Sci Technol 50(4):2000–2010

Zheng L, Hong F, Lu S, Liu C (2005) Effect of nano-TiO2 on strength of naturally aged seeds and growth of spinach. Biol Trace Elem Res 104(1):83–91

Zhou D, Jin S, Li L, Wang Y, Weng N (2011) Quantifying the adsorption and uptake of CuO nanoparticles by wheat root based on chemical extractions. J Environ Sci 23(11):1852–1857

A Novel Potential Resource for Drought Tolerance: Antarctic Plants

Mehtap Vural Aytekin, Ali Fuat Gökçe, Sedat Serçe, and Zahide Neslihan Öztürk

Abstract Antarctica is an extreme environment with limited availability of favorable conditions for plant growth and reproduction. Thereby, species richness is highly reduced to some mosses, lichen, liverworts and only two vascular plants. These vascular plant species, *Deschampsia antarctica* É. Desv. named as Antarctic hair grass (*Poaceae*) and *Colobanthus quitensis* (Kunth) Bartl. known as Antarctic pearlwort (*Caryophyllaceae*) are found on the Northern and Western parts of the Antarctic Peninsula. They can survive under extreme climatic conditions defined by extremely low temperatures, drought, low nutrition availability, wind, high salinity, and high UV. Adaptation to such harsh environmental conditions is possible by regulation of metabolic mechanisms and gene expression at different levels. Understanding the adaptation mechanisms of these species to their unfavorable habitat is quite important to generate new resources to cope with adverse effects of abiotic stresses, especially drought, which is recently becoming the main yield-limiting factor in agriculture due to global warming. In this chapter, we aim to introduce our collection of these plants to the scientific community while discussing the potential of Antarctic plants to be used as novel resources for development of abiotic stress tolerant cultivars.

Keywords Antarctica · *D. antarctica* É. Desv · *C. quitensis* (Kunth) Bartl. · Abiotic stress · Drought · Salt

---

M. V. Aytekin (✉) · A. F. Gökçe · S. Serçe · Z. N. Öztürk
Department of Agricultural Genetic Engineering, Ayhan Şahenk Faculty of Agricultural Sciences and Technologies, Niğde Ömer Halisdemir University, Niğde, Türkiye

U. K. Chaudhry et al. (eds.), *Drought Stress*,
https://doi.org/10.1007/978-3-031-80610-0_19

1 Introduction

Globally available water for irrigation purposes has declined and it is predicted to further decline in consequence of climate change. Most crops need sufficient water for growth and development and require additional irrigation to flourish well and produce desired yields (Pequeno et al. 2021). Drought tolerance is the rate at which a plant is able to adapt to drought conditions. Plants in dry ecosystems are constantly exposed to random drought situation, and it is quite impossible for them not to experience the negative effects of these conditions. Hence, plants in these ecosystems have certain biochemical and/or morphological adaptive features that enable them to endure water stress and prevent cell injury. To tolerate drought conditions, plants must maintain turgor pressure by using osmotic adjustment, enhance cell elasticity and reduce cell volume by protoplasmic resistance (Azhar and Rehman 2018).

Many studies have been conducted on physiological and morphological mechanisms that various crops adopt to tolerate the adverse influence of drought (Kapoor et al. 2020; Rijal et al. 2021). Plants respond to and deal with drought in two major ways, either by stress avoidance or by stress tolerance. In stress escape, water absorbance is promoted through the accumulation of solutes to decrease tissue water potential as well as by root size increase, while loss of water is reduced by stomatal closure, restricted shoot development and modifications to leaf senescence. However, in stress tolerance, the strategies are focused on protecting cells from getting damaged as stress conditions get worse and avoidance strategies can't be further applied. Tolerance involves the detoxification of reactive oxygen species and the concentration of solutes like proline, which plays a double role as either an osmolyte or an osmoprotectant (Claeys and Inze 2013). Proline is accumulated in plant species in response to several abiotic stresses acting as an important factor of redox balance and a signaling molecule. It is also regarded as one of the primary osmolytes with the potential of scavenging, controlling membrane stability and protein protection from degradation during adverse osmotic stress (Szabados and Savouré 2010; Shabala et al. 2012). Different stress factors have similar osmotic responses, such as accumulation of osmolytes and ions (Na, K^+) that permits plants to keep the required cellular turgor pressure for cell enlargement in response stress conditions (McCubbin et al. 2014).

Antarctica has extreme environmental conditions to suppress plant growth and development such as low temperature, high soil salinity, fluctuating in annual precipitation and a very short-day length. Nevertheless, two plant species, namely *D. antarctica* and *C. quitensis*, has an ability to survive aforementioned environmental pressures (Cui et al. 2020). In this chapter we aim to introduce these species to the scientific community while discussing the potential of Antarctic plants to be used as novel resources for development of abiotic stress tolerant cultivars.

2 Climate of Antarctica

The Antarctic region of the earth is witnessing harsh climatic conditions ranging from freezing cold temperatures to the driest environment (Cavieres et al. 2016). Location and height are the main factors that results in extreme climatic conditions of this region. The reason for low temperatures in Antarctica is the decline of latitude and altitude resulted in very low angle of incidences of the solar radiation on the Earth's surface (Convey et al. 2009). Accordingly, East Antarctica is the coldest region as compared to West Antarctica. Its northernmost side ends into larger peninsula known as Antarctic Peninsula, the coastal region around Antarctica. Antarctic Peninsula is warmer and contains some ice and snow-free areas that favors the plant growth (Cavieres et al. 2016). This part is called as maritime Antarctic (Convey 2013), with a climate comparatively different than east Antarctica (Bargagli 2005). Generally, average temperature of Maritime Antarctic is approximately 7 °C higher than east cost of Antarctica where it can get hotter up to four months as compared to the other months of the year (Convey 2013). In previous years, air temperature has been increased in all over the world due to global warming, as well as in Antarctica (Cannone et al. 2022). It is reported that increase of air temperature in Antarctic Peninsula is faster than other parts of the Antarctica over the past ten years (Turner et al. 2014). Globally the average temperature increase is 0.6–1.0 °C (IPCC 2007), whereas in Antartica Peninsula this increase was observed to be by 3.7 °C (Vaughan et al. 2003).

3 The Flowering Plants of Antarctica

Antarctica is an extreme environment with limited availability of favorable conditions for plant growth and reproduction. Generally, the plants growing in Maritime Antarctic are continuously dealing with lower temperature conditions that suggests their natural tolerance to cold climate (Cavieres et al. 2016). Additionally, cold temperature has significantly influenced plant colonization in this particular region. Among vascular plant species, two plants showed natural potential of colonization in maritime Antarctic region. It includes *D. antarctica* É. Desv. (Antarctic hair grass) plant that belongs to *Poaceae* family. The other one is *Colobanthus quitensis* (Kunth) Bartl. (pearlwort) which is a perennial herb belonging to *Caryophyllaceae* family (Smith 2003). They are vascular plants (Cavieres et al. 2016), which means they have certain complex vascular structures that help them absorb nutrients through their leaves, stems, and roots. This feature, among others, accounts for their consistent survival on this continent. It is difficult to think that plants can survive in the barren ecosystem of Antarctica because of its cold weather condition, less sunlight and moisture, poor soil characteristic, and a short growing period of time.

These unfavorable factors have prevented most flora species from growing there. However, some plants have managed to adapt over the years and survived in these conditions. They could survive under extreme climatic conditions defined by extremely low temperatures, drought, low nutrition availability, wind, high salinity and high UV (Bishop 2022). Parnikoza et al. (2007) hypothesized that the origin of vascular plants species of Antarctica (*D. antarctica* and *C. quitensis*) by using available climate data and genetic diversity with a migration to Antarctica region during Oligocene-Pliocene when it was partially separated from other continents and birds helped dispersal of these two species in Antarctica. *D. antarctica* and *C. quitensis* are mainly presented in Maritime Antarctic, however those plant do not colonize into the continental part of Antarctica (Greene and Holtom 1971).

3.1 Deschampsia antarctica *É. Desv. (Antarctic Hairgrass)*

D. antarctica is able to tolerate environmental changes and survive in different habitat conditions including mineral to organic soils and from arid to areas that surrounded by water (Cavieres et al. 2016). Furthermore, it has not only adapted these extreme conditions, but also can survive most of the year (6–7 months) under the snow. It is reported that *D. antarctica* can naturally grow in two different regions. The first one is the Antarctic tundra, which is referred as having low nutrition and a dry area due to wind. Accordingly, plants are subjected to intensive UV-B radiation and drought in this region. Other one is coastal area that has a high salinity. Therefore, it is known that there are morphological and physiological differences in *D. antarctica* due to different environmental stress (Giełwanowska et al. 2005; Cavieres et al. 2016; Clemente-Moreno et al. 2020).

Habitat of the Antarctic hairgrass plants are the Antarctic Peninsula in dense and small tufts throughout the rocky places of the Peninsula and commonly found in penguin colonies (Cavieres et al. 2016; Bishop 2022). They can tolerate high levels of disturbance without withering away, which is a survival mechanism in such an ecosystem. During the rather short summer where plant growth is maximized, they also need to cope with a bit of abuse from manure gathered from breeding colonies of penguins, elephant seals and high wind speeds,. The Antarctic hairgrass is slightly shorter than a foot in height with complex and deep root systems that enables them maintain life within their environment, as well as absorbing minerals and water easily from their habitat. During winter, the hairgrass loses its long, slender, green leaves, to be able to survive the freezing temperatures without dying (Bishop 2022).

D. antarctica is regarded a self-pollinating plant with closed flowers that enhance self-pollination making way for cleistogamy (Edwards 1972). However, recent data seems to suggest the existence of a possibility of cross-pollination because in South America both cleistogamic and chasmogamic flowers exist for Antarctic hair grass species. Chasmogamy is presumed to occur in the mild seasons in Antarctica. The development of *D. antarctica* initiates in November at the beginning of germination of seeds and recovery of the previous year's tufts (Parnikoza et al. 2011). The species also possesses the potential to be vegetatively propagated through the outgrowth of tuft and splitting off from the tuft parts (Parnikoza et al. 2008). *D. antarctica* plants mostly grow as a single dense and contiguous tuft covering an area ranging from one to several hundred square meters and an uprooted plant has the potential of growing again once it is transplanted in another conducive environment. This potential makes it most desirable for molecular studies (Parnikoza et al. 2008, 2011). *D. antarctica* is member of *Poaceae* family that has more than 700 genus and 11.000 species. The phylogenetic analysis was performed with agronomically important plants based on the plastid DNA sequence of those plants. The results showed that *D. antarctica* is a close relative of some crop cultivars such as wheat, barley and rice that is agricultural important plants (Lee et al. 2014).

3.2 **Colobanthus quitensis** *(Kunth) Bartl. (Antarctic Pearlwort)*

C. quitensis look like less tolerance to adverse environmental conditions as compared to *D. antarctica* due to favorable growth condition that is sparsely vegetated, sheltered, and moist well-drained mineral soils (Smith 2003). *D. antarctica* population abundantly exist and extensively spread in Antarctica (Smith 2003), whereas geographical distribution of *C. quitensis* is wider than *D. antarctica* and its colonization extends from Mexico down to Antarctica, in the same environmental conditions with cold and well-watered areas (Cavieres et al. 2016).

C. quitensis is a perennial flowering plant developing as a dense low rounded hemispheric tuft with a life span of 35 to 40 years. It has a tap root, with little potential of vegetative propagation and has a similar age structure as *D. antarctica* (Edwards 1975; Kozeretska et al. 2010). The species is largely a self-pollinator with stamens arranged in front of the pistil providing higher chances of cleistogamy, leading to inbreeding. As high as 43 seeds per capsule in a single plant is the seed yield from one self-pollination event (Moore 1983). The seeds of this species can stay viable over a long period under favorable conditions. The germination temperature of *C. quitensis* is between 9 °C to 2 °C. As other perennials, *C. quitensis* grows slowly in extreme environments and the emergence of new plants coming from seeds happens only in some years. Antarctic and highland plants have low reproduction rates because seed reproduction and seedling survival are dependent on climatic factors, but Antarctica possess unfavorable climatic conditions that severely inhibit reproduction (Vera et al. 2013; Convey 1996). Research revealed that the success of reproduction in *C. quitensis* on some islands depend on the climatic conditions experienced in a specific year. This confirms the irregular blossom of this species on the Argentine Islands and in other places of the maritime Antarctic with recording high seedling mortality in some years. Expecting an excellent seed reproduction in this species is rare or almost nonexistent (Parnikoza et al. 2011).

4 Abiotic Stress Studies Related to *D. antarctica* and *C. quitensis*

Plants expose to many environmental stress factors during their growth and developmental stage. Cellular dehydration is one of the important components to overcome against various stress factors for plants life cycle. Water deficiency symptoms in plants occur under certain conditions which include insufficient soil moisture, low relative humidity, and low or increased temperatures (Zhang et al. 2021). Therefore, in response to environmental stress, plants have evolved different adaptation mechanisms. A member of late embryogenesis abundant protein Group II (LEA) is dehydrins which is expressed in response to environmental adversities (Habib et al. 2022). There are very limited studies conducted with wild species from dry to cold region of Antarctica as compared to agronomically important crops. Olave-Concha et al. (2004) determined the accumulation and synthesis of dehydrins in *D. antactica*. It revealed that *D. antarctica* plants synthesized different numbers and different sizes of dehydrin proteins under low temperature, salt stress, osmotic stress, and ABA treatments.

Abiotic stresses such as salinity, drought and fluctuating of temperature induce oxidative stress which is negatively affecting plant productivity. Oxidative stress accelerates reactive oxygen species (ROS) accumulation in plants in response to abiotic stresses. Superoxide anion (O_2), hydrogen peroxide (H_2O_2) and hydroxyl ion (OH^-) molecules are ROS species which disturb cell membrane integrity. To suppress ROS and tolerate abiotic stress, plants have different defense systems as enzymatic and non-enzymatic antioxidants. Thus, it is possible that Antarctica plants that is exposed to harsh climatic conditions in Antarctica may have higher antioxidant capacity to prevent detrimental effects of environmental stresses during growth and development (Das and Roychoudhury 2014). Zamora et al. (2010) reported activity of antioxidants in *D. antarctica* and revealed that there were no differences between control plants and treated plants in terms of malondialdehyde (MDA), chlorophyll and H_2O_2 content. On the other hand, there were no differences observed regarding quantity of MDA, chlorophyll and H_2O_2whereas glutathione reductase, peroxide and catalase enzymes were higher in *D. antactica* plants collected from Antarctic tundra region. They concluded the adaptation of *D. antarctica* to the dry condition in Antarctic region is due to the reduction of negative effects of free radicals based on antioxidant enzyme activity. Moreover, Sánchez-Venegas et al. (2009b) studied the response of Cu/ZnSOD gene from *D. antarctica* to evaluate thermostability. It was revealed that plants exhibited medium stability above 40 °C, however stable at −20 °C and 20 °C which suggested that Cu/ZnSOD gene was responsible for tolerance nature. They stated that thermostability of enzymes in *D. antarctica* is extremely high. Similarly, another study also showed that expression of Cu/Zn-SOD gene determined by using RT-PCR was increased in response to cold and UV-B radiation (Sánchez-Venegas et al. 2009a). Additionally, one of the recent studies revealed that Antarctica plants exposed to heat shock stress (23 °C and 35 °C) under *in vitro* conditions showed that these plants had a potential to survive even under higher temperature stress with least damage to photosynthetic membrane probably due to higher enzymatic and non-enzymatic antioxidants that assisted in proper cellular functioning and controlled oxidative stress (Cortés-Antiquera et al. 2021). Results mentioned above indicated that *D. antarctica* has higher antioxidant capacity to survive in Antarctica climate.

The evolution of higher plants occurs with plants possessing complex physiological and biochemical processes with which they survive unfavorable environmental conditions that inhibit growth. *D. antarctica* or Antarctic hair grass, possesses both morphological and physiological adaptive characteristics that aid survival in the adverse Antarctic ecosystem (Cortés-Antiquera et al. 2021). Raffinose Family Oligosaccharides (RFOs) are the non-structural carbohydrates of plants helping to deal with abiotic stresses (Keller et al. 2021). Many researchers indicated that high storage of carbohydrates occur in *D. antarctica* during the growing and seed maturation seasons, however, there exists no connection between the biological functions of carbohydrates (RFOs) and abiotic stress tolerance. Subjecting *D. antarctica* species to glacial and dry conditions both in the Antarctic ecosystem and under laboratory conditions showed a high build-up of galactinol (a precursor of RFOs) and raffinose, as well as the upregulation of galactinol synthase (GolS). Type II galactinol synthases are greatly enlarged in cold adapted monocot plants. This study revealed better expressions of Type II *D. antarctica* GolS2 (DaGolS2) as well as rice GolS2 (OsGolS2) in transgenic rice plants, exhibiting tolerance to cold and drought stress when set side by side their wild relatives. There was higher amounts of RFO and GolS enzymatic action observed in the DaGolS2- and OsGolS2-overexpressing in offsprings than in their wild relatives under both stress and normal conditions. After cold and drought treatments, DaGolS2 and OsGolS2 overexpressing plants had limited reactive oxygen species (ROS) in comparison with their wild relatives. Results revealed that *Poaceae* type II GolS2s performs conservative roles in *D. antarctica* in reaction to drought and cold conditions through the action of RFO accumulation and reduced amounts of ROS. Hence, in adverse ecological circumstances, making *DaGolS* genes produce effective RFOs to balance cellular proteins and membranes as well as preventing ROS accumulation stands as an efficient support mechanism for *D. antarctica* in the Antarctic ecosystem (Cui et al. 2020).

Abiotic stress results in disruption of photosynthetic apparatus and ultimately reduction in physiological functioning of plants but *D. antarctica* has evolved adaptive mechanism to alleviate the negative effects in response to stress conditions. As revealed by Sáez et al. (2019) that this specie has an adapting photosynthetic apparatus against low temperature conditions. For instance, when the overall temperature is increased above 10 °C due to global warming and caused a decrease in water availability, *D. antarctica* can reduce its capacity to effectively manage light energy via photochemical process and movement of heat. Furthermore *D. antarctica* is continuously facing freezing events that also triggers physiological water stress in plants. Based on this information, the response of this plant to water deficiency was explained by acclimatization that exhibited stable photochemical activity and photosynthetic apparatus by managing the absorption of light via PSII (Horton et al. 1996; Maxwell and Johnson 2000).

Climate change has affected the amount of water for crop irrigation globally and from the current weather variation models, dry and somewhat dry areas are projected to experience drastically low levels of precipitation (Chaudhry and Sidhu 2021). In view of this, fungi linked to plant roots seem to a unique plan to be explored when considering improvement of eco-physiological activity and quantity of crops under abiotic stress conditions. Hence, fungal endophytes from habitats already experiencing constant and consistent drought is highly encouraged to be used. In a study by Molina-Montenegro et al. (2016), Antarctic root-endophytes were collected from *C. quitensis* species and assessed on how inoculating isolated fungal endophytes can affect an increase in the overall photosynthesis, water use efficiency and production of fress biomass in a cultivar of lettuce cultivated under totally varied irrigation conditions. Lettuce (*Lactuca sativa*) was used because it is one of the most sensitive crops to drought. Also, assessment was based on whether biochemical mechanisms and environmentally tolerant gene expressions are positively impacted by the fungal endophytes. The results of this experiment showed that plants inoculated with root-endophytes from the Antarctica plant had greater net photosynthesis with greater efficiency in water use efficiency and osmoprotectant strategies under drought conditions and this associated with the significant fresh and dry biomass produced just as the significant growth of the root system. Additionally, the introduction of fungal endophytes resulted in an increase in proline concentrations, lowered lipid peroxidation and up-/down-regulation of ion homeostasis. With these results, it can be concluded that the presence of fungal endophytes could reduce the adverse impacts of drought on species and thereby enhancing drought tolerance by biochemical processes and enhancing the nutritional status of species. This projects root-endophytes as an excellent biotechnological instrument for upholding consistently greater degrees of eco-physiological activity and output in drought dominated areas. Knowing the effects of root-endophytes on the physiological tolerance and productivity in crops could be an excellent way of maximizing water use efficiency in aid of optimizing yields in drought zones. Root-endophytes can improve plant performance through diverse means, take into consideration the proposed rise in abiotic and biotic tolerance for *Geomyces* and *Lecanicillium* which are also fungal root-endophytes. (Molina-Montenegro et al. 2016).

Plants in extreme environments use functional symbiosis (an outstanding mechanism) to improve their tolerance potential under different stress conditions. Endophyticism is one of the most popular kinds of symbiosis which plays an important role in *C. quitensis* species adaptation to the severe temperatures and aridity of Antarctica (Prekrasna et al. 2021). Forecasts of the climate situation in Antarctica suggests that due to temperature increases and soil water availability, abiotic stress conditions will increase. The advantageous role of fungal endophytes play role in the performance of *C. quitensis* plants would decrease as stress conditions decrease (Morales-Quintana et al. 2021). Investigation of the function of endophytic fungi in osmoprotective molecules (producing sugar, proline, and oxidative stress), the appearance of some genes as well as physiological features (opening of the stomata, overall photosynthesis, and stomatal conductance) in *C. quitensis* species were performed as isolated species of *C. quitensis* were treated with endophytic fungi and others left as control in one instance while in the other instance some plants were exposed to much water and others to drought as control based on the current Antarctica situation. From the results of this experiment, the endophyte-mediated samples were found to experience reduced oxidative stress, increased plants sugars and proline production as well as varied gene expressions in drought stress response, which happened more under drought conditions than in when water was abundant. These increases were also observed in the physiological mechanisms as greater overall photosynthesis, opening of the stomata and conductance where endophytes are present and with the suitable water condition projected for Antarctica. These outcomes mean that where fungal endophytes exist, *C. quitensis* species exhibit more tolerance to drought. However, the endophytic advantage will be lost in the absence of drought. This implies that the role of endophytes could be less desired in the near future when climatic conditions in Antarctica rise resulting in higher water content in soils (Hereme et al. 2020).

5 A Unique Collection of Antarctica Plants

The importance of understanding the adaptation mechanisms of plants to extreme conditions is always appealing to improve the abiotic stress tolerance of agriculturally important plants. Both seeds and alive plants of Antarctic hair grass and Antarctic pearlwort were collected by Prof. Dr. Sedat Serçe from Niğde Ömer Halisdemir University, Department of Agricultural Genetic Engineering (Niğde, Türkiye) during the third National Antarctic Scientific Expedition of Türkiye in the 2019. The collection represents samples from six islands (King George, Ardley, Livingston, Galindez, Ridge, Lagotellerie) exemplifying two different climatic conditions observed in Antarctica (Antarctic tundra and Antarctic seaside). Dr. Serçe collected sample of two plant species, *D. antarctica* and *C. quitensis*, from six islands in Antarctica, King George, Livingston, Ardley, Ridge, Lagotellerie, and Galindez (Fig. 1). The collection of *D. antarctica* and *C. quitensis* maintained in Niğde Ömer Halisdemir University, Türkiye are unique for being a representative of different islands and ecosystems of Antarctica and a valuable source to be able to perform more studies on these plants adapted to one of the world's most extreme environments. As can be seen from the previous chapters, a limited number of studies were performed on these species' response to environmental changes and on understanding the adaptation mechanisms of them to adverse conditions. Here, we would like to present our collection to the scientific community.

The reason of collection of plant species from different locations was to be able to observe effect of ecosystems on those locations. Aforementioned locations represent two different regions in Antarctica; Antarctic tundra that is defined as having scarce of nutrition and a very dry area due to strong winds, and Antarctic coast

	Island	Sampling Height	Ecosystem
1	King George	15 m	Antarctic seaside
2	Ardley	5 m	Antarctic seaside
3	Livingston	5 m	Antarctic seaside
4	Galindez	10 m	Antarctic seaside
5	Ridge	100 m	Antarctic tundra
6	Lagotellerie	5 m	Antarctic seaside
7	Lagotellerie	200 m	Antarctic tundra

Fig. 1 Sampling locations of *D. antarctica* plants collected from Antarctica during the third National Antarctic Science Expedition

Fig. 2 *D.antarctica* plants collected from Antarctica during the third National Antarctic Science Expedition

having high salinity. Plant samples was taken with their soils and preserved under cold conditions for the transfer. All plant samples were safely brought to Türkiye. Then, plants were transferred into pots containing a mix of peat and perlite (3:1) in Laboratory of Department of Agricultural Genetic Engineering, Faculty of Ayhan Şahenk Agricultural Sciences and Technologies, Nigde Ömer Halisdemir Universtiy (Figs. 2 and 3). Seeds of *D. antarctica* was collected during during the third National Antarctic Science Expedition by Prof. Dr. Sedat Serçe (Fig. 4).

These species have been multiplied based on their tillering properties for the next studies. Their phenotypic characteristic was observed based on their flowering and

Fig. 3 *C. colobanthus* plants collected from Antarctica during the third National Antarctic Science Expedition

Fig. 4 Dry flowers (**a**) and mature seeds of *D. antarctica* (**b**) collected from Antarctica during the third National Antarctic Science Expedition

when the seeds of *C. quitensis* became mature, they were collected further storage (Fig. 5).

Objectives of plant sampling during the third National Antarctic Science Expedition was to determine genome and ploidy differences of two vascular plants (*D. antarctica* and *C. quitensis*) collected from different locations in Antarctica and compare with closely related species found in Türkiye in a comparative karyotyping analysis. The genome size evolution of these organisms that are found in the Antarctica should be directly correlated with the environment that they occupied. Therefore, studying the genome size and repeat sequence of the Antarctica plants and vegetation will obviously reveal the secret of genome dynamics in different environmental conditions. Flow cytometry was used for the first time to estimate

Fig. 5 Flowers of *C. quitensis* (**a**) and mature seeds (**b**)

genome size *D. antarctica* and *C. quitensis* in this study (Pascual-Díaz et al. 2020). The genome size of *D. antarctica* was found to be 2C = 10.63 pg, and that of *C. quintesis* as 2C = 2.01 pg. Although the expected genome size of *D. antarctica* was higher than 10.63 pg, they explained that might be due to the existence of B-chromosome. No significant difference for genome size was found among the tested populations of both species. The manuscript further suggested the colonization history of the Antarctic continent (Pascual-Díaz et al. 2020).

Plants were also maintained by using tissue culture techniques. Explants were transferred in MS0 media (half-strength) containing 1.1 g/L MS salt (Murashige and Skoog 1962), 5 g/L sucrose and 7 g/L phytoagar in glass jars. The pH was set between 5.7 or 5.8. Experiment was done in the Laboratories of the Faculty of Agricultural Sciences and Technologies, Niğde Ömer Halisdemir University. Plantlets were grown in a growth chamber and the temperatures were kept at 10 °C under 18/6 day and night condition (Fig. 6).

6 Potential of Antarctica Plants to Be a Novel Source for miRNAs

MicroRNAs (miRNA) are a class of short non-coding RNAs and were firstly identified from nematodes in 1993. In many eukaryotes, miRNAs are responsible for regulation of gene expression (Lee et al. 1993). They are single stranded small RNAs of 20–24 nucleotides long. They bind to their corresponding target mRNAs to suppress or enhance gene expression. MiRNAs lead to mRNA degradation by chopping phosphodiester bonds in mRNA or interrupting protein synthesis. Thus, expression of target mRNAs is post-transcriptionally regulated via miRNAs. The

Fig. 6 Plantlets of *C. quitensis* (**a**) and *D. antarctica* (**b**) under *in vitro* condition

miRNAs are reported to target more than one gene that regulate the functioning of plants in drought (Selbach et al. 2008).

Plant miRNAs were firstly found in *A. thaliana* (Reinhart et al. 2002). The regulation of miRNA has exhibited to participate in several important developmental processes of plants, including morphogenesis of leaf, change in vegetative phase, and time of flowering (Millar 2020). The miRNAs have important function on plants such as root formation and development, leaf formation, vascular bundle formation, alteration from vegetative to reproductive phase and floral transition (Chen 2004; Jones-Rhoades et al. 2006; Chuck et al. 2009). Besides, it is known that miRNAs have important function in hormonal signaling (Liu and Chen 2010) and in answer to abiotic and biotic stresses (Sunkar et al. 2006).

Most of the important crop plants miRNA studies have been conducted to reveal its beneficial role against abiotic stresses. For example, Gökçe et al. (2021) investigated the function of miRNAs in answer to single and combined stresses (drought, heat, heat and drought) in two contrasting potato cultivars. It revealed that stu-miR160a-5p was conserved which is related to the regulation of auxin response factor that is known to be responsible for drought tolerance. The study also revealed the role of Novel_105 that targets zinc finger family protein as well as MAD-box transcription factor. The tolerant potato cultivar showed the downregulation of Novel_105 that resulted in upregulation of its target gene and conferred drought tolerance.

Multifarious studies from *Poaceae* family have also revealed the function of miRNAs in answer to drought stress. For instance, in wheat transgenerational drought stress exhibited differential regulation miRNAs and their target genes; the bdi-miR394 was responsible for cytidylate kinase, ata-miR528-5p was responsible for F-box protein, and bdi-miR394 was responsible for glutathione reductase. The lower miRNA expression levels showed higher gene expression of the target genes to confer drought tolerance (Liu et al. 2021). Likewise, other study also showed the function of miRNAs in drought stress tolerance in wheat. The TaMIR1119 was reported to be drought responsive that played crucial role by transcriptional regulation that moderated accumulation of osmolyte, improved photosynthetic functioning and improved homeostasis of ROS (Shi et al. 2018).

Vascular plants of Antarctica (*D. antarctica* and *C. quitensis*) are well adapted to harsh climatic conditions and, thereby, they may reveal novel information that will be beneficial for the scientific community. However, there is no literature that has been reported in these Antarctic plants to reveal the miRNA profiling, although it may be beneficial for enhancing knowledge for *Poaceae* and *Caryophyllaceae* family plants regarding drought tolerance.

7 Importance of Plants of Antarctica as Novel Resources for Abiotic Stress Tolerance

Several experimental approaches have been adapted to render stress resilience in plants that included the transfer of stress tolerant gene from tolerant to resilience crop plants (Manna et al. 2021). However, studies have resulted in real outcome of the problems to cope with stress related yield losses, but these are also not stable due to gene flow. This might be due to suboptimal designing of expression strategies in heterologous functioning that evolved physiological functioning of the host plants (Doerner 2020). This chapter elaborated the adaptation mechanisms of Antarctica plants in response to harsh climatic conditions. Several studies have observed the different physio-biochemical functioning of these plants and compared their response to agronomically important crops (Lee et al. 2013; Park and Lee 2019; Hasanuzzaman et al. 2020). Antarctic plants do not have any economic importance, but they can be considered as model plant such as *D. antarctica* is a grass that belongs to Poaceae family. The agronomic crops such as corn, wheat, sorghum, oat, barley, and rice also fall in this family. It can be inferred from the available literature that *D. antarctica* plants can be beneficially used to generate novel information regarding stress tolerance in plants (Cui et al. 2020). The temperature of Antarctica is also increasing due to climatic changes, so these plants also have stress memory for their adaptation to the changing climatic conditions. Therefore, they have great potential for future abiotic stress studies. We believe recognition of this important resource by the scientific community will support the new approaches to prevent yield loss in agriculture due to global climate change.

Acknowledgements The study on investigation of genome size constancy in Antarctic populations of *Colobanthus quitensis* and *Deschampsia antarctica* supported by Turkish government during Third Turkish Antarctic expedition İTÜ/PolReC. The study on miRNAs is funded by TÜBİTAK-1001 Program with a project number of 220Z005.

References

Azhar MT, Rehman A (2018) Overview on effects of water stress on cotton plants and productivity. In: Biochemical, physiological and molecular avenues for combating abiotic stress tolerance in plants, pp 297–316

Bargagli R (2005) Antarctic ecosystems: environmental contamination, climate change, and human impact. Springer, Berlin

Bishop C (2022) The plants of Antarctica. Oceanwide Expeditions. https://oceanwideexpeditions.com/blog/the-plants-of-antarctica

Cannone N, Malfasi F, Favero-Longo SE, Convey P, Guglielmin M (2022) Acceleration of climate warming and plant dynamics in Antarctica. Curr Biol 32:1599–1606

Cavieres LA, Sáez P, Sanhueza C, Sierra-Almeida A, Rabert C, Corcuera LJ, Bravo LA (2016) Ecophysiological traits of Antarctic vascular plants: their importance in the responses to climate change. Plant Ecol 217:343–358

Chaudhry S, Sidhu GPS (2021) Climate change regulated abiotic stress mechanisms in plants: a comprehensive review. Plant Cell Rep 41:1–31

Chen X (2004) A microRNA as a translational repressor of APETALA2 in Arabidopsis flower development. Science 303:2022–2025

Chuck G, Candela H, Hake S (2009) Big impacts by small RNAs in plant development. Curr Opin Plant Biol 12:81–86

Claeys H, Inze D (2013) The agony of choice: how plants balance growth and survival under water-limiting conditions. Plant Physiol 162:1768–1779

Clemente-Moreno MJ, Omranian N, Sáez PL, Figueroa CM, Del-Saz N et al (2020) Low-temperature tolerance of the Antarctic species *Deschampsia antarctica*: a complex metabolic response associated with nutrient remobilization. Plant Cell Environ 43:1376–1393

Convey P (1996) Reproduction of Antarctic flowering plants. Antarct Sci 8:127–134

Convey P (2013) Antarctic ecosystems. In: Levin SA (ed) Encyclopedia of biodiversity, 2nd edn. Elsevier, San Diego, pp 179–188

Convey P, Bindschadler R, Di Prisco G, Fahrbach E, Gutt J, Hodgson DA et al (2009) Antarctic climate change and the environment. Antarct Sci 1:541–563

Cortés-Antiquera R, Pizarro M, Contreras RA, Köhler H, Zúñiga GE (2021) Heat shock tolerance in *Deschampsia antarctica* Desv. cultivated in vitro is mediated by enzymatic and non-enzymatic antioxidants. Front Plant Sci 12:635491

Cui LH, Byun MY, Oh HG, Kim SJ, Lee J, Park H et al (2020) Poaceae Type II galactinol synthase 2 from Antarctic flowering plant *Deschampsia antarctica* and rice improves cold and drought tolerance by accumulation of raffinose family oligosaccharides in transgenic rice plants. Plant Cell Physiol 61:88–104

Das K, Roychoudhury A (2014) Reactive oxygen species (ROS) and response of antioxidants as ROS-scavengers during environmental stress in plants. Front Environ Sci 2:53

Doerner P (2020) Extreme environments: crucibles of potent abiotic stress tolerance. J Exp Bot 71:3761–3764

Edwards JA (1972) Studies in *Colobanthus quitensis* (Kunth) Bartl. and *Deschampsia antarctica* Desv.: V. Distribution, ecology and vegetative performance on Signy Island. Br Antarct Surv Bull 28:11–28

Edwards JA (1975) Studies in *Colobanthus quitensis* (Kunth) Bartl. and *Deschampsia antarctica* Desv.: VII. Cyclic changes related to age in *Colobanthus quitensis*. Br Antarct Surv Bull 40:1–6

Giełwanowska I, Szczuka E, Bednara J, Gorecki R (2005) Anatomical features and ultrastructure of *Deschampsia antarctica* (*Poaceae*) leaves from different growing habitats. Ann Bot 96:1109–1119

Gökçe ZNÖ, Aksoy E, Bakhsh A, Demirel U, Çalışkan S, Çalışkan ME (2021) Combined drought and heat stresses trigger different sets of miRNAs in contrasting potato cultivars. Funct Integr Genomics 21:489–502

Greene DM, Holtom A (1971) Studies in *Colobanthus quitensis* (Kunth) Bartl. and *Deschampsia antarctica* Desv. III. Distribution, habitats and performance in the Antarctic botanical zone. Br Antarct Surv Bull 26:1–29

Habib I, Shahzad K, Rauf M, Ahmad M, Alsamadany H, Fahad S, Saeed NA (2022) Dehydrin responsive HVA1 driven inducible gene expression enhanced salt and drought tolerance in wheat. Plant Physiol Biochem 180:124–133

Hasanuzzaman M, Bhuyan MHM, Zulfiqar F, Raza A, Mohsin SM, Mahmud J et al (2020) Reactive oxygen species and antioxidant defense in plants under abiotic stress: revisiting the crucial role of a universal defense regulator. Antioxidants 9:681

Hereme R, Morales-Navarro S, Ballesteros G, Barrera A, Ramos P, Gundel PE, Molina-Montenegro MA (2020) Fungal endophytes exert positive effects on *Colobanthus quitensis* under water stress but neutral under a projected climate change scenario in Antarctica. Front Microbiol 11:264

Horton P, Ruban AV, Walters RG (1996) Regulation of light harvesting in green plants. Annu Rev Plant Biol 47:655–684

IPCC (2007) Climate change 2007: the physical science basis. In: Fourth IPCC assessment report working group 1. Cambridge University Press, Cambridge, UK

Jones-Rhoades MW, Bartel DP, Bartel B (2006) MicroRNAs and their regulatory roles in plants. Annu Rev Plant Biol 57:19–53

Kapoor D, Bhardwaj S, Landi M, Sharma A, Ramakrishnan M, Sharma A (2020) The impact of drought in plant metabolism: how to exploit tolerance mechanisms to increase crop production. Appl Sci 10:5692

Keller I, Rodrigues CM, Neuhaus HE, Pommerrenig B (2021) Improved resource allocation and stabilization of yield under abiotic stress. J Plant Physiol 257:153336

Kozeretska IA, Parnikoza IY, Mustafa O, Tyschenko OV, Korsun SG, Convey P (2010) Development of Antarctic herb tundra vegetation near Arctowski station, King George Island. Pol Sci 3:254–261

Lee RC, Feinbaum RL, Ambros V (1993) The C. elegans heterochronic gene lin-4 encodes small RNAs with antisense complementarity to lin-14. Cell 75:843–854

Lee J, Noh EK, Choi HS, Shin SC, Park H, Lee H (2013) Transcriptome sequencing of the Antarctic vascular plant *Deschampsia antarctica* Desv. under abiotic stress. Planta 237:823–836

Lee J, Kang Y, Shin SC, Park H, Lee H (2014) Combined analysis of the chloroplast genome and transcriptome of the Antarctic vascular plant *Deschampsia antarctica* Desv. PLoS One 9:92501

Liu Q, Chen YQ (2010) A new mechanism in plant engineering: the potential roles of microRNAs in molecular breeding for crop improvement. Biotechnol Adv 28:301–307

Liu H, Able AJ, Able JA (2021) Small RNAs and their targets are associated with the transgenerational effects of water-deficit stress in durum wheat. Sci Rep 11:1–17

Manna M, Thakur T, Chirom O, Mandlik R, Deshmukh R, Salvi P (2021) Transcription factors as key molecular target to strengthen the drought stress tolerance in plants. Physiol Plant 172:847–868

Maxwell K, Johnson GN (2000) Chlorophyll fluorescence-a practical guide. J Exp Bot 51:659–668

McCubbin T, Bassil E, Zhang S, Blumwald E (2014) Vacuolar Na+/H+ NHX-type antiporters are required for cellular K+ homeostasis, microtubule organization and directional root growth. Plan Theory 3:409–426

Millar AA (2020) The function of miRNAs in plants. Plan Theory 9:198

Molina-Montenegro MA, Oses R, Torres-Díaz C, Atala C, Zurita-Silva A, Ruiz-Lara S (2016) Root-endophytes improve the ecophysiological performance and production of an agricultural species under drought condition. AoB Plants 8:plw062

Moore DM (1983) Flora of Tierra del Fuego. Anthony Nelson Oswestry, UK

Morales-Quintana L, Barrera A, Hereme R, Jara K, Rivera-Mora C, Valenzuela-Riffo F et al (2021) Molecular and structural characterization of expansins modulated by fungal endophytes in the Antarctic *Colobanthus quitensis* (Kunth) Bartl. Exposed to drought stress. Plant Physiol Biochem 168:465–476

Murashige T, Skoog F (1962) A revised medium for rapid growth and bioassays with tobacco tissue culture. Physiol Plant 15:473–497

Olave-Concha N, Ruiz-Lara S, Munoz X, Bravo LA, Corcuera LJ (2004) Accumulation of dehydrin transcripts and proteins in response to abiotic stresses in *Deschampsia antarctica*. Antarct Sci 16:175–184

Park JS, Lee EJ (2019) Waterlogging induced oxidative stress and the mortality of the Antarctic plant, Deschampsia Antarctica. J Ecol Environ 43:1–8

Parnikoza IY, Maidanuk DN, Kozeretska IA (2007) Are *Deschampsia antarctica* Desv. and *Colobanthus quitensis* (Kunth) Bartl migratory relicts? Tsitol Genet 41:36–40

Parnikoza I, Kozeretska O, Kozeretska I (2008) Is a translocation of indigenous plant material successful in the Maritime Antarctic? Polarforschung 78:25–27

Parnikoza I, Kozeretska I, Kunakh V (2011) Vascular plants of the Maritime Antarctic: origin and adaptation. Am J Plant Sci 2:381

Pascual-Díaz JP, Serçe S, Hradecká I, Vanek M, Özdemir BS, Sultana N et al (2020) Genome size constancy in Antarctic populations of *Colobanthus quitensis* and *Deschampsia antarctica*. Polar Biol 43:1407–1413

Pequeno DN, Hernandez-Ochoa IM, Reynolds M, Sonder K, MoleroMilan A, Robertson RD et al (2021) Climate impact and adaptation to heat and drought stress of regional and global wheat production. Environ Res Lett 16:054070

Prekrasna IE, Dzhulai A, Parnikoza I (2021) Preliminary estimates of the number and diversity of the culturable endophytic bacteria from *Deschampsia antarctic*a and *Colobanthus quitensis*. Visnik Ukrains' Kogo Tovaristva Genetikiv i Selekcioneriv 19:21–30

Reinhart BJ, Weinstein EG, Rhoades MW, Bartel B, Bartel DP (2002) MicroRNAs in plants. Genes Dev 16:1616–1626

Rijal B, Baduwal P, Chaudhary M, Chapagain S, Khanal S, Khanal S, Poudel PB (2021) Drought stress impacts on wheat and its resistance mechanisms. Malays J Sustain Agric 5:67–76

Sáez PL, Rivera BK, Ramírez CF, Vallejos V, Cavieres LA, Corcuera LJ, Bravo LA (2019) Effects of temperature and water availability on light energy utilization in photosynthetic processes of *Deschampsia antarctica*. Physiol Plant 165:511–523

Sánchez-Venegas JR, Dinamarca J, Moraga AG, Gidekel M (2009a) Molecular characterization of a cDNA encoding Cu/Zn superoxide dismutase from *Deschampsia antarctica* and its expression regulated by cold and UV stresses. BMC Res Notes 2:1–7

Sánchez-Venegas JR, Navarrete A, Dinamarca J, Ramírez LAB, Moraga AG, Gidekel M (2009b) Cloning and constitutive expression of *Deschampsia antarctica* Cu/Zn superoxide dismutase in *Pichia pastoris*. BMC Res Notes 2:1–9

Selbach M, Schwanhausser B, Thierfelder N, Fang Z, Khanin R, Rajewsky N (2008) Widespread changes in protein synthesis induced by microRNAs. Nature 455:58–63

Shabala L, Mackay A, Tian Y, Jacobsen SE, Zhou D, Shabala S (2012) Oxidative stress protection and stomatal patterning as components of salinity tolerance mechanism in quinoa (*Chenopodium quinoa* Willd). Physiol Plant 146:26–38

Shi GQ, Fu JY, Rong LJ, Zhang PY, Guo CJ, Kai XIAO (2018) TaMIR1119, a miRNA family member of wheat (*Triticum aestivum*), is essential in the regulation of plant drought tolerance. J Integr Agric 17:2369–2378

Smith, R. I. L., 2003. The enigma of *Colobanthus quitensis* and *Deschampsia antarctica* in Antarctica. In: Huiskes AHL, Gieskes, W. W. C., Rozema, J., Schorno, R. M. L., van der Vies, S. M., Wolff, W. J. (eds) Antarctic biology in a global context. Backhuys Publishers, Leiden 234–239

Sunkar R, Kapoor A, Zhu JK (2006) Posttranscriptional induction of two Cu/Zn superoxide dismutase genes in Arabidopsis is mediated by downregulation of miR398 and important for oxidative stress tolerance. Plant Cell 18:2051–2065

Szabados L, Savouré A (2010) Proline: a multifunctional amino acid. Trends Plant Sci 15:89–97

Turner J, Barrand N, Bracegirdle T, Convey P, Hodgson D, Jarvis M, Klepikov A (2014) Antarctic climate change and the environment: an update. Polar Record 50:237–259

Vaughan D, Marshall G, Connelley W, Parkinson C, Mulvaney R, Hodgson D et al (2003) Recent rapid regional climate warming on the Antarctic Peninsula. Climate Change 60:243–274
Vera ML, Fernández-Teruel T, Quesada A (2013) Distribution and reproductive capacity of *Deschampsia antarctica* and *Colobanthus quitensis* on Byers Peninsula, Livingston Island, South Shetland Islands, Antarctica. Antarct Sci 25:292–302
Zamora P, Rasmussen S, Pardo A, Prieto H, Zúñiga GE (2010) Antioxidant responses of in vitro shoots of *Deschampsia antarctica* to polyethylene glycol treatment. Antarct Sci 22:163–169
Zhang H, Zhu J, Gong Z, Zhu JK (2021) Abiotic stress responses in plants. Nat Rev Genet 23:1–16

The manufacturer's authorised representative in the EU is Springer Nature Customer Service Centre GmbH, Europaplatz 3, 69115 Heidelberg, Germany. If you have any concerns regarding our products, please contact ProductSafety@springernature.com

Printed and bound by CPI Group (UK) Ltd, Croydon, CR0 4YY
30/09/2026
02225569-0017